LESSONS ON RINGS,
MODULES AND MULTIPLICITIES

Lessons on rings, modules and multiplicities

D. G. NORTHCOTT, F.R.S.

*Town Trust Professor of Pure Mathematics,
University of Sheffield*

CAMBRIDGE
AT THE UNIVERSITY PRESS
1968

CAMBRIDGE UNIVERSITY PRESS
Cambridge, New York, Melbourne, Madrid, Cape Town, Singapore, São Paulo, Delhi

Cambridge University Press
The Edinburgh Building, Cambridge CB2 8RU, UK

Published in the United States of America by Cambridge University Press, New York

www.cambridge.org
Information on this title: www.cambridge.org/9780521071512

First published 1968
This digitally printed version 2008

A catalogue record for this publication is available from the British Library

Library of Congress Catalogue Card Number: 68-21397

ISBN 978-0-521-07151-2 hardback
ISBN 978-0-521-09807-6 paperback

TO ROSE

who helped me most

CONTENTS

3. Rings and modules of fractions

4. Noetherian rings and modules

CONTENTS

PREFACE

This book has grown out of lectures and seminars held at the University of Sheffield in recent years. Its purpose is to give a virtually self-contained introduction to certain parts of Modern Algebra and to provide a bridge between undergraduate and postgraduate study.

The title, *Lessons on rings, modules and multiplicities*, was chosen partly because a certain emphasis has been placed on instruction. I have long been interested in problems involving the introduction of young mathematicians to relatively advanced topics and, in this book, I have endeavoured to present the chosen material in a manner which will not only make it interesting but also easy to assimilate.

One fact of general interest has emerged which I did not foresee when I started. It was my intention to write about Commutative Algebra, but the contents of the first chapter are of such generality that it seemed wrong to exclude non-commutative rings at that particular stage. From then on the question continually arose as to the proper place at which to assume commutativity, and, indeed, the precise form the assumption should take. The outcome has been that this book, particularly in its later stages, is often concerned with Quasi-commutative Algebra. By this I mean that non-commutative rings are allowed but the emphasis is on the behaviour of central elements. In fact much that one normally regards as belonging to Commutative Algebra can be accommodated comfortably within this framework. For example, this is true of considerable areas of Multiplicity Theory and the theory of Hilbert Functions. It is also true of the theory of I-adic Completions and, to some extent, the theory of Primary Decompositions, though the latter fact gets only a passing mention in the exercises. Other instances where this observation is valid will doubtless occur to the reader as he proceeds.

It is with pleasure that I take this opportunity to acknowledge many sources of help and information. Since the subject matter of the book is strongly slanted in the direction of Commutative Algebra it was inevitable that the writings of N. Bourbaki, M. Nagata, P. Samuel and O. Zariski should have a persistent influence. Those who are familiar with the literature will also recognize that the chapter dealing with the Koszul Complex owes much to the classic paper on

Codimension and Multiplicity by M. Auslander and D. A. Buchs-
baum. In a similar manner, the chapter describing the properties of
Hilbert Rings is based on papers by O. Goldman and W. Krull.

This book has also profited from the research investigations of
recent postgraduate students at Sheffield University. In particular,
some ideas involving Multiplicity Theory and Hilbert Functions,
made use of here, first appeared in the doctoral theses of K. Black-
burn, D. J. Wright and W. R. Johnstone though they are now more
widely available in standard mathematical journals.

An author is very fortunate if he has someone who is willing to
read his manuscript with care and make detailed comments. D. W.
Sharpe has performed this labour for me with a thoroughness which
is familiar to those who know him well. His observations and con-
structive criticisms ranged from matters of punctuation and assis-
tance with proof-correcting to comments on the organization of whole
chapters. A number of sections have been rewritten to incorporate
improvements which he has suggested. In the later stages, P. Vámos
also helped me in a similar way and the final version has gained by
being modified to take account of his observations.

Finally my thanks go to my secretary, Mrs E. Benson, who typed
the manuscript and remained cheerful when I changed my mind
and asked to have considerable proportions done again. Without
her help this book would have taken very much longer to complete.

D. G. NORTHCOTT

Sheffield
March 1968

SOME NOTES FOR THE READER

There are certain matters which will be quite clear if the text is read consecutively from the beginning, but which require comment if you are primarily interested in particular sections and wish to study them in isolation. For example, it is necessary to know that the term *ring* is always understood to include the existence of an identity element, and the definition of a ring-homomorphism requires that identity element be mapped into identity element. A further point is that a homomorphism $R \to R'$, where R and R' are rings, is called an *epimorphism* only in the case where the mapping is surjective, that is to say when each element of R' is the image of at least one element of R.

Next there are some extensive sections in which only commutative rings are considered and in these the adjective *commutative* is usually suppressed in order to avoid tedious repetition. To discover whether the results of a particular section are established only in the case of commutative rings, it is sufficient to refer to the general remarks at the beginning of the chapter in which the results occur. These remarks contain, among other information, identification of all sections which are subject to this restriction.†

It will be found that the main text provides rather full explanations and for this reason, does not provide opportunities for you to devise your own arguments. To remedy this situation some exercises have been included at the end of each chapter. These are designed to let you test your grasp of basic concepts as well as to add to the information provided by the rest of the book. Certain of the more useful results contained among the exercises are employed at a later stage; but wherever this is the case the result in question is always established in the course of the discussion.

Cross-references are made in the following manner. If there is a reference to (say) Theorem 5 and no chapter or section is specified, then the result quoted is to be found in the chapter where the reference occurs. Where in one chapter it is necessary to recall a result established in some other chapter, the appropriate section is always given. To illustrate this, suppose that you have been referred to Proposition 18, Cor. 1 of section (3.9). Then the result in question is

† Similar use is made of the remarks which precede each set of exercises.

the first corollary of the eighteenth proposition in Chapter 3. To narrow the search, the extra information provided says that it occurs in the ninth main subdivision of that chapter.

The final point concerns notation used in connection with sets. If X and Y are sets and X is contained in Y, then the symbol $X \subseteq Y$ is used to indicate this fact. The advantage gained is that when X is strictly contained in Y, that is to say when $X \subseteq Y$ and $X \neq Y$, it is possible to indicate this by writing $X \subset Y$. Although this conflicts with common practice, it will be found convenient and not a source of confusion.

1

INTRODUCTION TO SOME BASIC IDEAS

1.1 General remarks concerning rings

It is assumed that the reader is familiar with the notions of *group* and *ring*. Nevertheless a few remarks concerning these concepts may help to establish certain conventions about notation and to clarify the attitude adopted in regard to certain extreme situations. Suppose then that R is a ring. The elements of R may be *added* and *multiplied* to give other elements of the same system. With respect to addition, R is a commutative group. The zero element of this group will be denoted by 0 although we may sometimes write 0_R if more than one ring is under consideration and we wish to be quite explicit. So far as multiplication is concerned, if α, β, γ are arbitrary elements of R, then we have the associative law $(\alpha\beta)\gamma = \alpha(\beta\gamma)$ and the two distributive laws $\alpha(\beta+\gamma) = \alpha\beta + \alpha\gamma$ and $(\beta+\gamma)\alpha = \beta\alpha + \gamma\alpha$. The ring R is said to be *commutative* if multiplication is commutative, that is to say if the relation $\alpha\beta = \beta\alpha$ always holds. In the early part of this book a good deal of the theory developed will apply only to modules over these more restricted rings, but in this chapter we shall not require that the commutative law be satisfied.

An element e belonging to R is called an *identity element* if

$$e\alpha = \alpha = \alpha e$$

for every α in R. It is immediately clear that a ring has at most one identity element. We shall confine our attention entirely to those rings in which such an element is present. The reader should therefore note that, *from this point onwards, when we speak of a ring it is to be understood that we always mean a ring with an identity element*. The identity element will be denoted by 1, though we may sometimes embellish this by writing 1_R if other rings are also being considered and we wish to avoid confusion.

Let R be a ring with zero element 0 and identity element 1. If $0 = 1$ and $\alpha \in R$, then $\alpha = \alpha 1 = \alpha 0 = 0$ and therefore 0 is the only element in the ring. In these circumstances, we say that R is a *null ring*.

Although null rings have only a trivial theory, it is convenient not to exclude them from consideration. For in this way certain results become not only more general, but also (and this is more important) easier to formulate.

1.2 Modules

We are now ready to introduce the central concept of our subject. Let R be a ring and M an additive group†. Suppose that, given an element r of R and an element x of M, we have some rule whereby we can form a 'product' rx which is again an element of M. (We can express this more formally in the following way. Let $R \times M$ consist of all ordered pairs (r, x), where $r \in R$ and $x \in M$, so that $R \times M$ is the so-called *Cartesian product* of R and M. Our supposition now amounts to assuming that we have a mapping of $R \times M$ into M together with the convention that the image of the pair (r, x) is to be denoted by rx.) What is now required is that the product rx shall have some natural connection with the ring structure of R and the group structure of M. The precise requirements are set out in the following

Definition. *If the situation is as described above, then M is said to be a 'left R-module' or a 'left module with respect to R' provided that the following four conditions are satisfied:*

(i) $(r_1 + r_2)x = r_1 x + r_2 x$ *whenever r_1, r_2 belong to R and x belongs to M;*

(ii) $r(x_1 + x_2) = rx_1 + rx_2$ *whenever $r \in R$ and x_1, x_2 belong to M;*

(iii) $(r_1 r_2)x = r_1(r_2 x)$ *whenever r_1, r_2 belong to R and x belongs to M;*

(iv) $1x = x$ *for all x in M.*

In (iv), the symbol 1 denotes the identity element of R.

Naturally, there is an analogous concept called a *right R-module*. In this, the product of $r \in R$ and $x \in M$ is written as xr and (i), (ii), (iii) and (iv) are replaced by

(i)' $x(r_1 + r_2) = xr_1 + xr_2$ *when $x \in M$ and $r_1, r_2 \in R$;*

(ii)' $(x_1 + x_2)r = x_1 r + x_2 r$ *when $x_1, x_2 \in M$ and $r \in R$;*

(iii)' $x(r_1 r_2) = (xr_1)r_2$ *when $r_1, r_2 \in R$ and $x \in M$;*

(iv)' $x1 = x$ *for all $x \in M$.*

It will be clear that to every result concerning left R-modules there is a corresponding result for right R-modules and vice versa.

---

† The term 'additive group' is used here to mean a *commutative* group in which the law of composition is written as addition.

Of course, whenever one is concerned *simultaneously* with left modules and right modules, then it is necessary to make careful distinctions. When, however, all our modules have the ring operating on the same side, we shall normally develop the theory in terms of left modules. This will be the case for the remainder of the present chapter. The reader should therefore note that, *for the remainder of Chapter 1, the term 'R-module' will always signify a left R-module*, as defined above, unless there is a definite statement to the contrary.

Let M be an R-module. The zero element of M will usually be denoted by 0. However, it may be necessary, on occasion, to distinguish between the zero element of M and that of R. We then employ the symbols 0_M and 0_R.

Proposition 1. *Let M be an R-module. Then for all $r \in R$ and $x \in M$, we have*

(a) $0_R x = 0_M$;

(b) $r 0_M = 0_M$;

(c) $(-r) x = -(rx) = r(-x)$;

(d) $(-r)(-x) = rx$.

Proof. Since $0_R + 0_R = 0_R$, the definition of an R-module shows that

$$0_R x = (0_R + 0_R) x = 0_R x + 0_R x,$$

whence $0_R x = 0_M$ because M is a group. Next, from $0_M + 0_M = 0_M$ follows

$$r 0_M = r(0_M + 0_M) = r 0_M + r 0_M,$$

whence $r 0_M = 0_M$. Thus (a) and (b) are proved. Again $0_R = r + (-r)$ and therefore, by (a),

$$0_M = 0_R x = (r + (-r)) x = rx + (-r) x$$

which yields $(-r) x = -(rx)$. A similar argument shows that

$$r(-x) = -(rx).$$

Finally, using (c), we obtain

$$(-r)(-x) = -((-r) x) = -(-(rx)) = rx.$$

This completes the proof.

1.3 Homomorphisms and isomorphisms

Let M and N be modules with respect to a ring R and let $f: M \to N$ be a mapping of M *into* N so that with each element x of M there is associated a definite element $f(x)$ of N.

Definition. *The mapping $f: M \to N$ is said to be an ' R-homomorphism'
or a 'homomorphism of R-modules' if it satisfies the following two
conditions*:

(i) $f(x_1 + x_2) = f(x_1) + f(x_2)$ *whenever* x_1, x_2 *belong to* M;

(ii) $f(rx) = rf(x)$ *whenever* $r \in R$ *and* $x \in M$.

An R-homomorphism $f: M \to N$ is also known as an *R-linear*
mapping. Sometimes we speak simply of a homomorphism of M into
N when it is quite clear which is the ring of operators.

Let $f : M \to N$ be a homomorphism of the R-module M into the
R-module N. Since $f(x_1 + x_2) = f(x_1) + f(x_2)$ whenever x_1, x_2 are in M,
we see, in particular, that f is a homomorphism of the additive group
of M into the additive group of N. Certain consequences follow from
this fact alone. To begin with, from $0_M + 0_M = 0_M$, we obtain

$$f(0_M) + f(0_M) = f(0_M)$$

and therefore $\qquad\qquad f(0_M) = 0_N.$ \hfill (1.3.1)

Again, if $x \in M$, then $0_M = x + (-x)$. Applying the mapping f and
making use of (1.3.1), we see that $0_N = f(x) + f(-x)$ whence

$$f(-x) = -f(x). \qquad\qquad (1.3.2)$$

Finally, if $x, y \in M$, then $y - x = y + (-x)$ so that $f(y-x) = f(y) + f(-x)$
which, by virtue of (1.3.2), yields

$$f(y-x) = f(y) - f(x). \qquad\qquad (1.3.3)$$

In certain situations, homomorphisms can be combined. For example, if $f: M \to N$ and $g: N \to P$ are homomorphisms of R-modules,
then we obtain a mapping $h: M \to P$ by first applying f and afterwards applying g. Thus, by definition, $h(x) = g(f(x))$ from which it is
easily verified that $h(x_1 + x_2) = h(x_1) + h(x_2)$ and $h(rx) = rh(x)$ whenever x_1, x_2, x belong to M and r belongs to R. The new mapping is
therefore a homomorphism. It is called the *product* of the original
mappings and is denoted by† gf. Observe that

$$(gf)(x) = g(f(x)) \qquad\qquad (1.3.4)$$

for all $x \in M$.

There are certain kinds of homomorphisms which play particularly
important roles.

Definition. *A homomorphism $f: M \to N$ of R-modules is called a 'monomorphism', an 'injection' or an 'embedding' if distinct elements of M
always have distinct images in N.*

† Note the order of the terms.

Thus the characteristic property of a monomorphism $f: M \to N$ is that $x_1 \neq x_2$ (where $x_1, x_2 \in M$) implies that $f(x_1) \neq f(x_2)$. We shall shortly give a different characterization but first we need another

Definition. *Let $f: M \to N$ be a homomorphism of R-modules. The set of elements of M which are mapped on to the zero element of N is called the 'kernel' of f and is denoted by $\mathrm{Ker} f$.*

It is clear, from (1.3.1), that $\mathrm{Ker} f$ always contains 0_M.

Lemma 1. *Let $f : M \to N$ be a homomorphism of R-modules. Then in order that f should be a monomorphism it is necessary and sufficient that $\mathrm{Ker} f$ contain only the element 0_M.*

Proof. If f is a monomorphism, then 0_M maps into 0_N. Also if $x \in M$ and $x \neq 0_M$, then $f(x) \neq 0_N$. Thus 0_M is the only element of $\mathrm{Ker} f$. Now assume that $\mathrm{Ker} f = \{0_M\}$, i.e. that $\mathrm{Ker} f$ is the set whose only member is 0_M. If, in this situation, $x_1, x_2 \in M$ and $f(x_1) = f(x_2)$, then, by (1.3.3),
$$f(x_2 - x_1) = f(x_2) - f(x_1) = 0_N.$$

and therefore $x_2 - x_1$ belongs to $\mathrm{Ker} f$. Accordingly $x_2 - x_1 = 0_M$ or $x_2 = x_1$. This completes the proof.

Definition. *A homomorphism $f: M \to N$ of R-modules is called an 'epimorphism' or a 'surjection' if each element of N is the image of at least one element of M.*

Thus $f: M \to N$ is an epimorphism if f maps M on to N. The reader will see, as the subject develops, that the notion of an epimorphism is, in a certain sense, dual to that of a monomorphism.

Suppose now that we have a homomorphism $f: M \to N$ which is both a monomorphism and an epimorphism. Then distinct elements of M have distinct images in N and each element of N is the image of at least one (and therefore of exactly one) element of M. In other terms, the homormorphism $f: M \to N$ gives a *one-one* mapping of the set M on to the set N. It follows that there exists a well defined inverse mapping $f^{-1}: N \to M$. *We contend that f^{-1} is also a homomorphism of R-modules.*

For suppose that y_1, y_2 belong to N and that r belongs to R. We can then choose $x_1, x_2 \in M$ so that $f(x_1) = y_1$ and $f(x_2) = y_2$. Then
$$f(x_1 + x_2) = f(x_1) + f(x_2) = y_1 + y_2$$
and therefore
$$f^{-1}(y_1 + y_2) = x_1 + x_2 = f^{-1}(y_1) + f^{-1}(y_2).$$

Again, from $f(rx_1) = rf(x_1) = ry_1$ follows

$$f^{-1}(ry_1) = rx_1 = rf^{-1}(y_1).$$

This establishes the contention.

It is convenient to embody some of these observations in the

Definition. *A homomorphism $f: M \to N$, of R-modules, which is both a monomorphism and an epimorphism is called an 'isomorphism' of M on to N. In such a situation, f will set up a one-one correspondence between the set M and the set N. The uniquely determined inverse mapping $f^{-1}: N \to M$ is then an isomorphism of N on to M.*

If $f: M \to N$ is an isomorphism and $f^{-1}: N \to M$ is the inverse isomorphism, then it is clear that f is the inverse of f^{-1}.

Two R-modules M and N are said to be *isomorphic* if there exists an isomorphism of M on to N. As already observed, this relation is symmetrical for there will then exist an isomorphism of N on to M. The symbol $M \approx N$ is frequently used to indicate that M and N are isomorphic.

We take this opportunity to introduce some simple but useful terminology. If A is a subset of a set B, then we obtain a mapping $j: A \to B$ by putting $j(a) = a$ for every $a \in A$. This mapping is called the *inclusion mapping* of A into B. In the special case where $A = B$, this yields the *identity mapping* of B. It will be convenient to denote the identity mapping of B by i_B. Thus $i_B(b) = b$ for every b in B.

If M is an R-module, then i_M is an isomorphism of M on to itself, hence $M \approx M$. We have already observed that if $M \approx N$ then $N \approx M$. Again, if $f: M \to N$ is an isomorphism of M on to N and $g: N \to P$ is an isomorphism of N on to P, then $gf: M \to P$ is an isomorphism of M on to P. Thus $M \approx N$ and $N \approx P$ together imply that $M \approx P$. This shows that $\approx$ has the properties of an equivalence relation. Indeed, from our point of view, isomorphic R-modules are simply copies of one another and have identical properties.

Lemma 2. *Let $f: M \to N$ and $g: N \to M$ be homomorphisms of R-modules. In order that f should be an isomorphism whose inverse is g, it is necessary and sufficient that $gf = i_M$ and $fg = i_N$.*

Proof. It is clear that if f is an isomorphism and $g = f^{-1}$, then both gf and fg are identity maps. Now suppose that $gf = i_M$, $fg = i_N$. If $y \in N$, then $y = i_N(y) = f(g(y))$ so that y is the image of $g(y)$. Accord-

ingly f is an epimorphism. Now assume that $x, x' \in M$ and $f(x) = f(x')$. Then
$$x = i_M(x) = g(f(x)) = g(f(x')) = i_M(x') = x'.$$
This shows that f is also a monomorphism and hence an isomorphism. It is clear that $g = f^{-1}$.

1.4 Submodules

Let M and N be R-modules and suppose that M is a subset of N. It does not follow (from what has so far been said) that the module structure of M is in any way related to that of N but clearly we shall have an interesting situation if the two happen to be compatible. This leads to the

Definition. *Let the situation be as described above. We say that ' M is a submodule of N' if the inclusion mapping $M \to N$ is a homomorphism of R-modules. If, in addition, $M \neq N$ (so that M is strictly contained in N) then M is called a 'proper submodule' of N.*

For example, since the identity map of N is a homomorphism of R-modules, N is a submodule of itself. Again, if M is a submodule of N and N is a submodule of P, then M must be a submodule of P.

Suppose that M is a submodule of N. Then $M \subseteq N$ and the inclusion mapping is an R-homomorphism. Let x_1, x_2 belong to M and let r be an element of R. Since the image of the sum of x_1 and x_2 is the sum of their separate images, we see that $x_1 + x_2$ is the same whether we regard x_1, x_2 as elements of M or as elements of N. Also, the image of rx_1 is r times the image of x_1. The interpretation of this is that rx_1 is the same element whether we consider x_1 as belonging to M or as belonging to N. A similar observation applies to $x_1 - x_2$. Again, by (1.3.1), $0_M = 0_N$ which means that M and N have their zero element in common.

A submodule of N is, in particular, a non-empty subset of N. However not every non-empty subset of N can be endowed with the structure of a submodule. We have, in fact, the following result.

Proposition 2. *Let N be an R-module and A a non-empty subset of N. Then A can be given the structure of a submodule of N if and only if the following two conditions are both satisfied:*

(i) *whenever a_1 and a_2 belong to A, then $a_1 + a_2$ also belongs to A;*

(ii) *whenever a belongs to A and r belongs to R, then ra belongs to A.*

Proof. If A is a submodule of N, then the remarks made earlier show that the conditions are satisfied. Now suppose that (i) and (ii) hold. By (i), the sum of any two elements of A is again an element of A and, so far as A is concerned, addition is both commutative and associative because these laws hold in the larger system N. Since A is not empty, we can choose $\alpha \in A$ and then $0_R \alpha = 0_N$ belongs to A by virtue of (ii). Thus A contains an element which is neutral for addition. Furthermore, if $a \in A$, then $(-1) a$ belongs to A, by (ii), and we have $a + (-1) a = 0_N$. These remarks amount to a verification that A is an additive abelian group.

Finally, if $a \in A$ and $r \in R$, then, by (ii), the product ra also belongs to A. It is now clear that the module axioms are satisfied by A and that the inclusion mapping $A \to N$ is a homomorphism. Accordingly, A is a submodule of N.

If we take for A the subset $\{0\}$ consisting simply of the zero element of N, then it is immediately clear that (i) and (ii) hold and therefore $\{0\}$ is a submodule. Indeed, modules which have no non-zero elements play an important role in the general theory. Such a module is called a *null module* or a *zero module*.

In the next proposition, we consider an indexed family $\{L_i\}_{i \in I}$ of submodules of a given R-module N. This means that each L_i is a submodule of N and the individual submodules are labelled by means of the elements of a set I, called the *index set*. The index set may be completely arbitrary. In particular it is not required to contain only a finite number of members. If $i \neq i'$ belong to I, it is not assumed that L_i and $L_{i'}$ are necessarily distinct.

Proposition 3. *Let N be an R-module and $\{L_i\}_{i \in I}$ an indexed family of submodules of N. Then their intersection $\bigcap_{i \in I} L_i$ is also a submodule of N.*

Proof. Let $L = \bigcap_i L_i$ so that L consists of those elements which belong to every L_i. Since every submodule of N contains the zero element, we have $0 \in L$ and therefore L is not empty. To show that it is a submodule, we need only verify that conditions (i) and (ii) of Proposition 2 are satisfied. To this end, let x, y belong to L and let r belong to R. Then for each $i \in I$, we have $x \in L_i$ and $y \in L_i$. But L_i is a submodule, hence $x + y$ and rx also belong to L_i. This holds for every $i \in I$. Consequently $x + y \in L$ and $rx \in L$. Thus we have checked that the conditions of Proposition 2 are satisfied and the proof is complete.

Once again, suppose that $\{L_i\}_{i \in I}$ is a family of submodules of an R-module N. For each $i \in I$, choose $x_i \in L_i$ subject to the condition that, *for only finitely many i, shall x_i be different from zero*. We sometimes describe this situation by saying that $x_i = 0$ *for almost all i*.

In these circumstances we are able to form the sum $\sum_i x_i = x$ say, in spite of the fact that the index set I may be infinite. Let L consist of all elements x which can be obtained as sums in this way. *We contend that* L *is a submodule of* N *and that* $L_i \subseteq L$ *for every* i. For let $i_0 \in I$ and take x_{i_0} to be an arbitrary element of L_{i_0}. By putting $x_i = 0$ whenever $i \neq i_0$, we obtain a family $\{x_i\}_{i \in I}$ such that $x_i \in L_i$ and $\sum_i x_i = x_{i_0}$. Thus $x_{i_0} \in L$ and therefore $L_{i_0} \subseteq L$. In particular, L is not empty.

Now suppose that $y, z \in L$ and $r \in R$. To complete the proof we need only show that $y + z$ and ry belong to L. However, we can write $y = \sum_i y_i, z = \sum_i z_i$, where y_i, z_i belong to L_i and only a finite number of the y_i and z_i are non-zero. Then $y + z = \sum_i (y_i + z_i)$ whence, since $y_i + z_i \in L_i$ (L_i is a submodule), we have $y + z \in L$. Likewise $ry = \sum_i ry_i$ and $ry_i \in L_i$. Consequently $ry \in L$ and the proof is complete.

The submodule L, which has just been constructed, is called the *sum* of the L_i and is denoted by $\sum_{i \in I} L_i$. Not only does the sum contain each of the summands L_i, but it is clearly the *smallest submodule* of N which has this property.

Sometimes we are concerned with only a *finite* number of submodules, say $L_1, L_2, \ldots, L_s$. In such a situation it may be more convenient to use the alternative symbols $L_1 \cap L_2 \cap \ldots \cap L_s$ and $L_1 + L_2 + \ldots + L_s$ for their intersection and sum respectively.

Let U be a subset of an R-module M. By Proposition 3, the intersection L of all the submodules of M which contain U is also a submodule. L, *which is the smallest submodule containing* U, is called the *submodule generated by* U. If it happens that the submodule generated by U is M itself, then we say that U *is a system of generators for* M. Note that the submodule generated by the empty set is just $\{0_M\}$.

Proposition 4. *Let* U *be a subset of an* R-*module* M *and let* $x \in M$. *Then* x *belongs to the submodule generated by* U *if and only if we have a relation* $x = r_1 u_1 + r_2 u_2 + \ldots + r_s u_s$, *where the* r_i *are elements of* R *and the* u_i *belong to* U.

Remark. To cover the case where U is the empty set, we adopt the convention that an empty sum has the value zero.

Proof. Let L be the submodule generated by U, and L' the aggregate of all elements which can be written in the form $r_1 u_1 + r_2 u_2 + \ldots + r_s u_s$ with $r_i \in R$ and $u_i \in U$. It is clear that $L' \subseteq L$. Also, if $u \in U$, then

$u = 1u$ belongs to L' and therefore $U \subseteq L'$. Accordingly, if we can show that L' is a submodule of M, then (because L is the smallest submodule containing U) we shall have $L \subseteq L'$ and the proof will be complete.

Let $x, x^* \in L'$ and $r \in R$. By the definition of L, we can write

$$x = r_1 u_1 + \ldots + r_s u_s \quad \text{and} \quad x^* = r_1^* u_1^* + \ldots + r_t^* u_t^*,$$

where $r_i, r_j^* \in R$ and $u_i, u_j^* \in U$. Then

$$x + x^* = r_1 u_1 + \ldots + r_s u_s + r_1^* u_1^* + \ldots + r_t^* u_t^*$$

and
$$rx = (rr_1) u_1 + \ldots + (rr_s) u_s.$$

This shows that $x + x^*$ and rx both belong to L. Accordingly (Proposition 2) L' is a submodule of M and the required result follows.

Definition. *An R-module which can be generated by a finite number of elements is said to be 'finitely generated'. If an R-module can be generated by one element alone, then it is said to be 'singly generated.'*

The elements $x_1, x_2, \ldots, x_s$ generate an R-module M (it is supposed that the x_i belong to M) if and only if every element of the module can be expressed in the form $r_1 x_1 + r_2 x_2 + \ldots + r_s x_s$. Some important problems are connected with questions of finite generation. We must, however, postpone discussion of these for the time being.

Let $f : M \to N$ be a homomorphism of R-modules and suppose that A is a subset of M while B is a subset of N. We recall that $f(A)$ is used to denote the set of all elements of N which are images of elements of A, while $f^{-1}(B)$ denotes the set consisting of those elements of M whose images are contained in B. As is customary, we refer to $f(A)$ as the *image* of A and to $f^{-1}(B)$ as the *inverse image* of B. This provides an opportunity to introduce a concept which is complementary to $\mathrm{Ker} f$.

Definition. *If $f : M \to N$ is a homomorphism, then $f(M)$ is called the 'image' of the mapping f and is denoted by $\mathrm{Im} f$.*

Thus f is an epimorphism if and only if $\mathrm{Im} f = N$. Should $\mathrm{Im} f = \{0_N\}$, then we say that f is a *null homomorphism*.

Certain basic facts, about images and inverse images of submodules, will now be established.

Proposition 5. *Let $f : M \to N$ be a homomorphism of R-modules, let A be a submodule of M and B a submodule of N. Then $f(A)$ and $f^{-1}(B)$ are submodules of N and M respectively. In particular, $\mathrm{Im} f$ is a submodule of N while $\mathrm{Ker} f$ is a submodule of M. Finally, $f^{-1}(f(A)) = A + \mathrm{Ker} f$ and $f(f^{-1}(B)) = B \cap \mathrm{Im} f$.*

Proof. Let u, v belong to $f(A)$ and let r belong to R. We can choose a_1, a_2 in A so that $f(a_1) = u, f(a_2) = v$. Now, because A is a submodule, $a_1 + a_2$ and ra_1 both belong to A. Moreover $f(a_1 + a_2) = u + v$ and $f(ra_1) = ru$. Thus $u + v$ and ru belong to $f(A)$. Consequently $f(A)$ is a submodule of N. By taking $A = M$ it is seen that $\mathrm{Im} f$ is a submodule of N.

Suppose next that x, y belong to $f^{-1}(B)$ and that r belongs to R. Since $f(x)$ and $f(y)$ belong to B and B is a submodule, both

$$f(x + y) = f(x) + f(y) \quad \text{and} \quad f(rx) = rf(x)$$

are in B. Thus $x + y$ and rx are elements of $f^{-1}(B)$ and therefore $f^{-1}(B)$ is a submodule of M. As a special case, we see that $\mathrm{Ker} f = f^{-1}(\{0_N\})$ is a submodule of M.

It is clear that A and $\mathrm{Ker} f$ are contained in the submodule $f^{-1}(f(A))$ of M. Since $A + \mathrm{Ker} f$ is the smallest submodule which contains them both, we obtain $A + \mathrm{Ker} f \subseteq f^{-1}(f(A))$. Now let x belong to $f^{-1}(f(A))$. Then $f(x) \in f(A)$ which means that $f(x) = f(a)$ for some $a \in A$. Thus $f(x - a) = 0$ and therefore $x - a$ is in $\mathrm{Ker} f$. From $x = a + (x - a)$, it now follows that x belongs to $A + \mathrm{Ker} f$. This proves that

$$f^{-1}(f(A)) \subseteq A + \mathrm{Ker} f$$

and allows us to conclude that $f^{-1}(f(A)) = A + \mathrm{Ker} f$.

Finally, since $f(f^{-1}(B)) \subseteq B \cap \mathrm{Im} f$, we need only establish the opposite inclusion. Let u belong to $B \cap \mathrm{Im} f$. Since $u \in \mathrm{Im} f$, we have $u = f(y)$ for a suitable element y in M. But $u \in B$. Consequently $y \in f^{-1}(B)$.

Thus $u = f(y) \in f(f^{-1}(B))$ and the required relation,

$$B \cap \mathrm{Im} f = f(f^{-1}(B)),$$

follows.

Corollary. *Let the assumptions be as in the proposition. Then*

$$f^{-1}(f(A)) = A$$

if and only if $\mathrm{Ker} f \subseteq A$ while $f(f^{-1}(B)) = B$ if and only if $B \subseteq \mathrm{Im} f$.
    This is an immediate consequence of the main result.

Proposition 6. *Let $f: M \to N$ be an epimorphism of R-modules. Then there is a natural one-one correspondence between (a) the submodules A of M which contain $\mathrm{Ker} f$, and (b) the submodules B of N. This correspondence is characterized by the property that, when A and B correspond, $B = f(A)$ and $A = f^{-1}(B)$. Furthermore, inclusion relations are preserved by the correspondence.*

Proof. Let us associate, with each A, the submodule $f(A)$ of N. By the corollary to Proposition 5, $f^{-1}(f(A)) = A$. Consequently, to distinct A's correspond distinct submodules of N. Again, by the same corollary, we always have $f(f^{-1}(B)) = B$. Since $\mathrm{Ker} f \subseteq f^{-1}(B) \subseteq M$, it follows that every B arises from an A. To complete the proof, it is only necessary to observe that $A_1 \subseteq A_2$ and $B_1 \subseteq B_2$ imply

$$f(A_1) \subseteq f(A_2) \quad \text{and} \quad f^{-1}(B_1) \subseteq f^{-1}(B_2).$$

Corollary. *Let $f: M \to N$ be an epimorphism of R-modules. Let $\{A_i\}_{i \,\in\, I}$ be a non-empty family of submodules of M each of which contains $\mathrm{Ker} f$, and let $\{B_i\}_{i \in I}$ be a similarly indexed family of submodules of N. Suppose that, for each $i \in I$, A_i and B_i correspond in the sense of the proposition. Then $\sum_i A_i$ corresponds to $\sum_i B_i$ and $\bigcap_i A_i$ to $\bigcap_i B_i$.*

Proof. If $j \in I$, then $B_j = f(A_j) \subseteq f(\sum_i A_i)$ hence $\sum_i B_i \subseteq f(\sum_i A_i)$. Accordingly

$$f^{-1}(\sum_i B_i) \subseteq f^{-1}(f(\sum_i A_i)) = \sum_i A_i$$

because $\sum_i A_i$ contains $\mathrm{Ker} f$. On the other hand, $f(A_j) = B_j \subseteq \sum_i B_i$ and therefore $A_j \subseteq f^{-1}(\sum_i B_i)$. It follows that $\sum_i A_i \subseteq f^{-1}(\sum_i B_i)$ whence $\sum_i A_i = f^{-1}(\sum_i B_i)$. This proves the first point. Next

$$f^{-1}(\bigcap_i B_i) = \bigcap_i f^{-1}(B_i) = \bigcap_i A_i$$

which establishes the second.

1.5 Factor modules

Let K be a submodule of an R-module M and let x, y be elements of M. If it happens that their difference $x - y$ belongs to K, then it is customary to say that x and y are *congruent modulo K*, and to write $x \equiv y \pmod{K}$. Sometimes, when it is clear from the context which submodule is involved, we simplify the notation and write just $x \equiv y$. Note that $x \equiv x$, because K contains the zero element, and that if $x \equiv y$ then $y \equiv x$. Furthermore, from $x \equiv y$ and $y \equiv z$ follows $x \equiv z$; this is because $x - z = (x - y) + (y - z)$ and each of $x - y$ and $y - z$ is an element of K. The relation $x \equiv y \pmod{K}$ is thus an equivalence relation between the elements of M and therefore it partitions the set M into disjoint classes. These classes are known as the *cosets of K in M* or the *residue classes of M with respect to K*. If $x \in M$, then the residue class or coset, to which it belongs, consists of all elements of the form $x + k$, where $k \in K$. This aggregate is conveniently denoted by $x + K$.

If we consider the different residue classes as the elements of a new set, then the usual symbol for the new set is M/K. A mapping $M \to M/K$ is now obtained by associating, with each $x \in M$, the coset $x + K$ to which it belongs. This particular mapping is called the *natural mapping* of M on to M/K.

Lemma 3. *Let K be a submodule of an R-module M, let x, x', y, y' be elements of M and let r belong to R. If now $x \equiv x' \pmod{K}$ and $y \equiv y' \pmod{K}$, then $x + y \equiv x' + y' \pmod{K}$ and $rx \equiv rx' \pmod{K}$.*

Proof. We have $(x + y) - (x' + y') = (x - x') + (y - y')$ and

$$rx - rx' = r(x - x').$$

Since $x - x'$ and $y - y'$ belong to K and since K is a submodule of M, it follows that $(x + y) - (x' + y')$ and $rx - rx'$ are also elements of K. This is all we need to prove.

The lemma shows that the residue class containing $x + y$ is the same as that containing $x' + y'$. Likewise the residue class of rx coincides with that of rx'. Now suppose that ξ, η belong to M/K. This means that they are residue classes of M with respect to K. Let x, x' be representatives of ξ and y, y' respresentatives of η so that $\xi = x + K = x' + K$ and $\eta = y + K = y' + K$. The lemma shows that $(x + y) + K = (x' + y') + K$. In other terms, *if we form the residue class of $x + y$, then the result will depend only on ξ and η and will be independent of the choice of the representatives x and y.* It is therefore permissible to denote this residue class by $\xi + \eta$. The lemma also shows that $(rx) + K = (rx') + K$. Thus *the residue class of rx is determined solely by r and ξ and is unaffected by the choice of the representative x.* This residue class will be denoted by $r\xi$. We can now 'add' any two elements of M/K and we can 'multiply' an element of M/K by an element of R. It will be seen, in a moment, that M/K has become an R-module.

If $x \in M$, let us use $\bar{x}$ to denote its image under the natural mapping $M \to M/K$. Thus $\bar{x}$ is just a new symbol for $x + K$. If now x, y belong to M, then the above definitions tell us at once that

$$\bar{x} + \bar{y} = \overline{x + y} \tag{1.5.1}$$

and
$$r\bar{x} = \overline{rx}. \tag{1.5.2}$$

Note that, since $M \to M/K$ maps M on to M/K, $\bar{x}$ and $\bar{y}$ are perfectly general elements of M/K.

Proposition 7. *If K is a submodule of an R-module M, then M/K (with the laws of composition defined above) is an R-module and the natural mapping $M \to M/K$ is an epimorphism of R-modules.*

Proof. First it will be shown that M/K is a group with respect to addition. To this end, let x, y, z belong to M. Then, using (1.5.1),

$$(\bar{x} + \bar{y}) + \bar{z} = \overline{x + y} + \bar{z} = \overline{(x + y) + z}$$

$$= \overline{x + (y + z)} = \bar{x} + \overline{y + z} = \bar{x} + (\bar{y} + \bar{z})$$

and $$\bar{x} + \bar{y} = \overline{x + y} = \overline{y + x} = \bar{y} + \bar{x},$$

which shows that addition is associative and commutative. Again

$$\bar{x} + \bar{0} = \overline{x + 0} = \bar{x} \quad \text{and} \quad \bar{x} + \overline{(-x)} = \overline{x + (-x)} = \bar{0}.$$

This proves that $\bar{0}$ is neutral for addition and that every element has a negative. Accordingly, M/K is an additive abelian group.

Next, let r, ρ belong to R. Then, using (1.5.1) and (1.5.2),

$$r(\bar{x} + \bar{y}) = r\overline{(x + y)} = \overline{r(x + y)} = \overline{rx + ry} = \overline{rx} + \overline{ry} = r\bar{x} + r\bar{y},$$

$$(r + \rho)\bar{x} = \overline{(r + \rho)x} = \overline{rx + \rho x} = \overline{rx} + \overline{\rho x} = r\bar{x} + \rho\bar{x},$$

$$r(\rho\bar{x}) = r\overline{(\rho x)} = \overline{(r\rho)x} = (r\rho)\bar{x}$$

and $$1\bar{x} = \overline{1x} = \bar{x}.$$

This proves that M/K has the structure of an R-module. It is now clear, from (1.5.1) and (1.5.2), that $M \to M/K$ is a homomorphism of M on to M/K. Note that an element $x \in M$ belongs to the kernel of the mapping if and only if $\bar{x} = \bar{0}$, that is if and only if $x \equiv 0 \pmod{K}$. Accordingly, *the kernel of $M \to M/K$ is just K itself*.

Definition. *If K is a submodule of an R-module M, then M/K (endowed with the R-module structure described above) is called the 'factor module' or 'residue module' of M with respect to K.*

In connection with this definition, it is worth observing that if $K = 0$ (i.e. K is the zero submodule† of M), then $M/K = M/0$ is none other than M itself.

Let A be a submodule of M which satisfies $K \subseteq A \subseteq M$, and consider the image of A under the natural mapping $M \to M/K$. A typical element $a \in A$ is mapped into $a + K$ from which it is seen that the image of A is none other than A/K. Thus A/K is a submodule of M/K.

---

† Note the new use of the symbol '0'.

Proposition 8. *Let K be a fixed submodule of an R-module M and A a variable submodule satisfying $K \subseteq A \subseteq M$. Then A/K is a submodule of M/K. Furthermore, if B is a given submodule of M/K, then there is one and only one submodule $A (K \subseteq A \subseteq M)$ such that $B = A/K$.*

This is not essentially new. It is, in fact, just Proposition 6 applied to the natural epimorphism $M \to M/K$. The corollary, to the same proposition, yields the next result.

Corollary. *Let $\{A_i\}_{i \in I}$ be a non-empty indexed family of submodules of M with the property that, for each i, $K \subseteq A_i \subseteq M$. Then*

$$(\textstyle\sum_i A_i)/K = \sum_i (A_i/K) \quad and \quad (\textstyle\bigcap_i A_i)/K = \bigcap_i (A_i/K).$$

Suppose that we have a homomorphism $f\colon M \to N$. Let A be a submodule of M, B a submodule of N and assume that $f(A) \subseteq B$. We describe this situation by saying that f *maps the pair* (A, M) *into the pair* (B, N). Now let ξ belong to the factor module M/A and let elements x, x' of M be such that they are both mapped on to ξ by the natural mapping $M \to M/A$. Thus $\xi = x + A = x' + A$ and $x \equiv x'$ $(\mathrm{mod}\, A)$. Then $f(x) - f(x') = f(x - x') \in f(A) \subseteq B$ and therefore

$$f(x) \equiv f(x') \quad (\mathrm{mod}\, B).$$

Accordingly, $f(x)$ and $f(x')$ have the same natural image in N/B. Put differently, we may say that the image of $f(x)$ in N/B depends only on ξ and not on the choice of the representative x. It is therefore reasonable to denote this image by $f^*(\xi)$.

It has now been shown that, when we have a homomorphism $f\colon M \to N$ which maps (A, M) into (B, N), it gives rise to a mapping $f^*\colon M/A \to N/B$. The reader will find it a straightforward matter to verify that f^* is a homomorphism of R-modules. We shall say that f^* is *induced* by f. Note that *if* $f\colon M \to N$ *is an epimorphism, then so is* $f^*\colon M/A \to N/B$.

1.6 Isomorphism theorems

It has previously been remarked that isomorphic modules are essentially copies of each other and therefore have similar properties. For this reason it is both useful and important to be able to recognize when two modules are isomorphic. We shall now establish certain results which are fundamental in this connection.

Theorem 1. *Let $f: M \to N$ be an epimorphism of R-modules, let B be a submodule of N and set $A = f^{-1}(B)$. Then f induces an isomorphism $M/A \approx N/B$. In particular, there is induced an isomorphism $M/\operatorname{Ker} f \approx N$.*

Proof. Since $f(A) \subseteq B$, we have a mapping of the pair (A, M) into the pair (B, N) and this induces a homomorphism $f^*: M/A \to N/B$. *We contend that f^* is actually an isomorphism.* For, since f is an epimorphism so is f^*. It is therefore sufficient to prove that f^* is a monomorphism and we can do this by showing that its kernel contains only the zero element.†

Let $\xi \in \operatorname{Ker} f^*$, so that $f^*(\xi) = 0$, and choose $x \in M$ in such a way that ξ is the natural image of x in M/A. By the definition of f^*, $f^*(\xi)$ is just the natural image of $f(x)$ in N/B. But $f^*(\xi) = 0$. Consequently $f(x) \in B$ and therefore $x \in f^{-1}(B) = A$. It follows that $\xi = 0$. This completes the proof. Note that the special case (referred to in the theorem) comes by taking B to be the zero submodule of N.

Theorem 2. *Suppose that $K \subseteq A \subseteq M$, where M is an R-module and K, A are submodules. Then A/K is a submodule of M/K and the natural mapping $M \to M/K$ induces an isomorphism $M/A \approx (M/K)/(A/K)$.*

Proof. We have only to apply Theorem 1 to the case where f is the natural mapping $M \to M/K$ and B is A/K.

Theorem 3. *Let A, K be submodules of an R-module M. Then $(A + K)/K$ is ismorphic to $A/(A \cap K)$.*

Proof. Let $f: M \to M/K$ be the natural epimorphism. By Proposition 5, $f^{-1}(f(A)) = A + \operatorname{Ker} f = A + K$ whence $f(A + K) = f(A)$. But

$$f(A + K) = (A + K)/K$$

and therefore $f(A) = (A + K)/K$. Thus by restricting f to A there is produced an epimorphism $\phi: A \to (A + K)/K$. Theorem 1 now shows that we have an isomorphism $A/\operatorname{Ker} \phi \approx (A + K)/K$. However, the elements of A which are mapped by ϕ into zero are just the elements of A which belong to K. Accordingly $\operatorname{Ker} \phi = A \cap K$ and the proof is complete.

Theorems 1, 2 and 3 constitute the main isomorphism theorems in the theory of modules. The next result gives a good example of how they are used and, at the same time, provides information which will be needed in the next section.

† See Lemma 1.

Lemma 4. Let $E \subseteq E'$ and $F \subseteq F'$ all be submodules of an R-module M. Then we have the following isomorphisms:

$$(E + E' \cap F')/(E + E' \cap F) \approx (E' \cap F')/(E \cap F' + E' \cap F)$$
$$\approx (F + E' \cap F')/(F + E \cap F').$$

Remark. In interpreting the above expressions, it should be noted that *intersection* takes precedence over *sum*. Thus $E + E' \cap F'$ means $E + (E' \cap F')$ and not $(E + E') \cap F'$. Likewise $E \cap F' + E' \cap F$ stands for $(E \cap F') + (E' \cap F)$.

Proof. The natural mapping $M \to M/E$, when restricted to a submodule A of M, maps that submodule on to† $(A + E)/E$. In particular, the restriction of $M \to M/E$ to $E' \cap F'$ yields an epimorphism

$$E' \cap F' \to (E + E' \cap F')/E. \qquad (1.6.1)$$

Since the elements of M which are mapped into zero are those belonging to E, the elements of $E' \cap F'$ which become zero form the set $E \cap E' \cap F' = E \cap F'$ because $E \subseteq E'$. It now appears that

$$E \cap F' + E' \cap F$$

is a submodule of $E' \cap F'$ containing the kernel $E \cap F'$ of the mapping (1.6.1). To $E \cap F' + E' \cap F$ there will correspond a submodule of $(E + E' \cap F')/E$ of which it is the inverse image (see Proposition 6). This corresponding submodule is the image of $E \cap F' + E' \cap F$ under (1.6.1) or, equivalently, its image under $M \to M/E$. The remark at the beginning of the proof now shows that to $E \cap F' + E' \cap F$ corresponds $(E + E \cap F' + E' \cap F)/E$, that is $(E + E' \cap F)/E$, because $E \cap F' \subseteq E$. By Theorem 1,

$$(E' \cap F')/(E \cap F' + E' \cap F) \approx ((E + E' \cap F')/E)/((E + E' \cap F)/E)$$

which simplifies to

$$(E' \cap F')/(E \cap F' + E' \cap F) \approx (E + E' \cap F')/(E + E' \cap F)$$

by virtue of Theorem 2. This establishes one of the isomorphisms in the lemma and the other follows by symmetry.

1.7 Composition series

Up to this point we have been discussing some very general ideas and it would be possible to continue, in this way, laying broad foundations. We shall, instead, consider some questions, of a more specialized nature, which are concerned with saturated chains of modules, or, as

---

† If this statement is not clear the reader should refer back to the proof of Theorem 3.

they are usually called, composition series. These will provide an example of how the results of the earlier sections may be applied. The theory of composition series leads on to the idea of the length of a module. This notion, which is a generalization of the concept of *dimension* in the theory of vector spaces, will find a number of applications in the later chapters. At the same time, it will serve to introduce further topics of considerable generality with which we must be familiar before embarking on more sophisticated investigations.

Let K be a submodule of an R-module M. By a *chain of submodules from M to K* we mean a (finite) decreasing sequence $E_0, E_1, ..., E_s$ of submodules of M whose first term is M itself and whose last term is K. Thus, in the case of such a chain, we have

$$M = E_0 \supseteq E_1 \supseteq E_2 \supseteq ... \supseteq E_s = K. \tag{1.7.1}$$

Put $C_i = E_{i-1}/E_i \, (1 \leqslant i \leqslant s)$ so that the C_i are factor modules. We shall refer to $C_1, C_2, ..., C_s$ as the *chain factors* of (1.7.1).

Suppose now that we have a second chain

$$M = F_0 \supseteq F_1 \supseteq F_2 \supseteq ... \supseteq F_t = K \tag{1.7.2}$$

which, like the former, begins at M and ends at K. The typical chain factor in this case is $F_{j-1}/F_j = \Gamma_j$ say. We shall say that (1.7.2) is a *refinement* of (1.7.1) if $s \leqslant t$ and if it is possible to find integers $0 \leqslant j_0 < j_1 < ... < j_s \leqslant t$ such that $F_{j_\nu} = E_\nu$ for $0 \leqslant \nu \leqslant s$. In other terms a typical refinement of (1.7.1) is obtained by inserting new submodules, between those already present, in such a way as to make the chain longer.

There is a second concept, relating to chains, which we shall need. Namely (1.7.1) and (1.7.2) are said to be *equivalent* if $s = t$ and there is a rearrangement $\{\rho_1, \rho_2, ..., \rho_s\}$ of $\{1, 2, ..., s\}$ such that $C_\nu \approx \Gamma_{\rho_\nu}$ for $\nu = 1, 2, ..., s$. Thus equivalent chains are characterized by the property that the chain factors of one are (up to isomorphism) simply a rearrangement of those of the other.

The result, which follows, is known as the Jordan–Hölder–Schreier theorem for modules.

Theorem 4. *Let K be a submodule of an R-module M. Then any two chains from M to K possess equivalent refinements.*

Proof. We take (1.7.1) and (1.7.2) to be the two chains under consideration. For the purposes of the proof we may assume that $M \neq K$ so that $s \geqslant 1, t \geqslant 1$. For $1 \leqslant i \leqslant s, 0 \leqslant j \leqslant t$, put

$$E_{i,j} = E_i + (E_{i-1} \cap F_j)$$

while if $0 \leqslant i \leqslant s$, $1 \leqslant j \leqslant t$, let

$$F_{i,j} = F_j + (E_i \cap F_{j-1}).$$

We then have $E_{i-1} = E_{i,0} \supseteq E_{i,1} \supseteq \dots \supseteq E_{i,t} = E_i$ (1.7.3)

and $F_{j-1} = F_{0,j} \supseteq F_{1,j} \supseteq \dots \supseteq F_{s,j} = F_j.$ (1.7.4)

Thus the $E_{i,j}$ and $F_{i,j}$ provide refinements of (1.7.1) and (1.7.2) respectively. *We contend that these extended chains are equivalent.* Indeed Lemma 4 shows that

$$\begin{aligned} E_{i,j-1}/E_{i,j} &= (E_i + E_{i-1} \cap F_{j-1})/(E_i + E_{i-1} \cap F_j) \\ &\approx (F_j + E_{i-1} \cap F_{j-1})/(F_j + E_i \cap F_{j-1}) \\ &= F_{i-1,j}/F_{i,j} \end{aligned}$$

provided only that $1 \leqslant i \leqslant s$ and $1 \leqslant j \leqslant t$. This proves the theorem.

Definition. *An R-module C is said to be 'simple' if* (i) *it is not a zero module and* (ii) *it has no proper submodules other than the zero submodule.*

With the aid of this definition we can introduce a very important kind of chain called a composition series.

Definition. *Let K be a submodule of an R-module M. Then a 'composition series' from M to K is a chain $M = E_0 \supset E_1 \supset E_2 \supset \dots \supset E_s = K$ whose chain factors E_{i-1}/E_i are all simple.*

It should be noted that it is only in special circumstances that a composition series will exist. If we do have a composition series, then, since a simple module is (by definition) non-null, all the inclusions are strict. The assertion that $M = E_0 \supset E_1 \supset \dots \supset E_s = K$ is a composition series means, in effect, that in no case is there a submodule strictly between E_{i-1} and E_i. This is because of the one-one correspondence connecting the modules between E_{i-1} and E_i with the submodules of E_{i-1}/E_i (see Proposition 8). In the case of a composition series, the chain factors are also referred to as *composition factors.*

Theorem 5. *Let K be a submodule of an R-module M and suppose that there exists at least one composition series from M to K. Then every strictly decreasing chain*

$$M = F_0 \supset F_1 \supset F_2 \supset \dots \supset F_t = K \qquad (1.7.5)$$

can be refined into a composition series. Moreover, any two composition series from M to K are equivalent; that is to say they contain the same number of terms, and the same composition factors up to order and isomorphism.

Proof. Suppose that

$$M = E_0 \supset E_1 \supset E_2 \supset ... \supset E_s = K \qquad (1.7.6)$$

is a composition series. By Theorem 4, (1.7.5) and (1.7.6) have equiva-lent refinements. Of course these refinements will, in general, have repeated terms. However, if in both cases we strike out repetitions one at a time, until none remain, then we shall be left with equivalent chains. This is because the sole effect is to remove the chain factors which are null. Now, because (1.7.6) is a composition series, one of the residual chains must be $E_0 \supset E_1 \supset ... \supset E_s$ itself while the other is a refinement of (1.7.5). This latter chain, because it is equivalent to a composition series, will have simple chain factors. In other words, it will be a composition series itself. This proves the first assertion. Furthermore, if (1.7.5) is a composition series to begin with, then the argument just given shows that (1.7.5) and (1.7.6) are equivalent. This proves the theorem.

Suppose now that E is an arbitrary R-module and let 0 denote the zero submodule of E. We propose to define a symbol $L_R(E)$ whose value will always be a non-negative integer or 'plus infinity'. The details are as follows. If $E \neq 0$ and there is no composition series from E to 0, then we put $L_R(E) = \infty$. If $E \neq 0$ and there exists a composition series

$$E = E_0 \supset E_1 \supset E_2 \supset ... \supset E_s = 0,$$

then $L_R(E)$ is defined to be the integer s. By the last theorem, this is the same whichever composition series is chosen. Finally we put $L_R(E) = 0$ when E is a zero module.

Definition. $L_R(E)$ *is called the 'length' of the* R-*module* E.

Note that a simple module is the same as a module of unit length and that $L_R(E) = 0$ if and only if E is a zero module. Again, if F is a submodule of E, then $L_R(F) \leqslant L_R(E)$. For if F is either 0 or E, or if $L_R(E) = \infty$, then there is nothing to prove; while in the remaining case the assertion follows by refining $0 \subset F \subset E$ to a composition series.

Theorem 6. *Let* $K \subseteq N$ *be submodules of an* R-*module* M. *Then*

$$L_R(M/K) = L_R(M/N) + L_R(N/K). \qquad (1.7.7)$$

Remark. In this theorem, it is not necessary that the lengths involved be finite.

Proof. We may suppose that $K \subset N \subset M$ for if either $K = N$ or $N = M$ the assertion is trivial. By Proposition 8, there is a one-one, inclusion-preserving correspondence connecting the modules between M and K with the submodules of M/K. Suppose that $L_R(M/K) < \infty$. Then to a composition series from M/K to its zero submodule will correspond a composition series from M to K. It follows that $M \supset N \supset K$ can be refined to a composition series. Let

$$M = E_0 \supset E_1 \supset \ldots \supset E_s = K$$

with $E_p = N$ be such a refinement. Then $s = L_R(M/K)$ and

$$M = E_0 \supset E_1 \supset \ldots \supset E_p = N$$

and $$N = E_p \supset E_{p+1} \supset \ldots \supset E_s = K$$

are also composition series. Passing to the factor modules M/N and N/K, we see that $L_R(M/N) = p$ and $L_R(N/K) = s - p$. Consequently (1.7.7) holds in this case.

Finally assume that $L_R(M/N)$ and $L_R(N/K)$ are both finite. Then there exists a composition series from M to N and also one from N to K. Putting these together we arrive at a composition series from M to K. Accordingly $L_R(M/K)$ is finite and the desired relation follows as before.

Corollary. *Let M be an R-module of finite length and let N be a submodule. Then $L_R(N) \leqslant L_R(M)$ with equality if and only if $N = M$.*

This follows from the theorem by taking $K = 0$.

1.8 The maximal and minimal conditions

It will be clear that if one is to make use of the theory of lengths, then it is necessary to be able to recognize when the length of a module is finite. In this section, ideas will be developed which throw a good deal of light on this question. In the process, we shall meet concepts which are useful in various other investigations where some kind of finiteness condition is needed.

Definition. *An R-module M is said to satisfy the 'maximal condition' for submodules if every non-empty collection of submodules contains a maximal member.*

Care must be taken to interpret this definition correctly. Let Ω be a non-empty collection of submodules of M. The statement that Ω *contains a maximal member* means that there exists $K \in \Omega$ such that if $K' \in \Omega$ and $K' \supseteq K$, then $K' = K$. Thus K is maximal in the sense that there is no other member of the collection Ω which strictly contains it. *It is not asserted that all the submodules in Ω are contained in K.*

The next definition concerns a related concept.

Definition. *M is said to satisfy the 'ascending chain condition' for submodules if, whenever*

$$K_1 \subseteq K_2 \subseteq K_3 \subseteq \ldots \subseteq K_m \subseteq \ldots$$

is a (weakly) increasing, infinite sequence of submodules, there always exists an integer μ (depending on the sequence) such that $K_m = K_\mu$ for all $m \geqslant \mu$.

In less formal terms, we can say that M satisfies the ascending chain condition if every increasing sequence, of submodules, terminates. The maximal condition and the ascending chain condition are connected with the property of being finitely generated (see section (1.4)). We have, in fact,

Proposition 9. *If M is an R-module, then the following statements are equivalent*:

    (a) *M satisfies the maximal condition for submodules*;

    (b) *M satisfies the ascending chain condition for submodules*;

    (c) *every submodule of M (including M itself) is finitely generated.*

Proof. First, suppose that (a) holds and that we have an ascending sequence $K_1 \subseteq K_2 \subseteq K_3 \subseteq \ldots$ of submodules of M. From among the set of distinct submodules which occur in the sequence, we can select one, say K_μ, which is maximal for that set. Then $K_m = K_\mu$ for $m \geqslant \mu$. On the other hand, if (b) holds and Ω is a non-empty set of submodules of M, then Ω must contain a maximal member. *For assume the contrary.* Choose $K_1 \in \Omega$. Then, since K_1 is not maximal in Ω, we can find $K_2 \in \Omega$ such that $K_1 \subset K_2$. But K_2 is not maximal in Ω; consequently there exists $K_3 \in \Omega$ for which $K_2 \subset K_3$; and so on. In this way a *strictly increasing* sequence $K_1 \subset K_2 \subset K_3 \subset \ldots$ is generated contrary to the assumption that (b) is true. This establishes the equivalence of (a) and (b).

Now assume that (a) is satisfied and let L be a submodule of M. Denote by Ω the collection of all finitely generated submodules of L. Ω is not empty because it contains the zero submodule. From Ω we can select, by virtue of (a), a maximal member L' say. Then $L' \subseteq L$ and L' is finitely generated, say by $u_1, u_2, \ldots, u_p$. *We contend that*

$L' = L$. For if not, then we can choose $v \in L$ so that v does not belong to L'. Denote by L'' the submodule generated by $u_1, \ldots, u_p, v$. Then $L'' \in \Omega$ and L'' strictly contains L' because it contains v. Accordingly, L' is not maximal in Ω and we have a contradiction. This argument shows that (a) implies (c).

To complete the proof, we assume (c) and deduce (b). To this end, let $K_1 \subseteq K_2 \subseteq K_3 \subseteq \ldots$ be an infinite ascending sequence of submodules of M. Denote by K the union of all the K_m so that an element belongs to K if, and only if, it is in K_m for some value of m. We claim that K is a submodule of M. For suppose that $x, y \in K$ and $r \in R$. It is then possible to choose integers m_1, m_2 in such a way that $x \in K_{m_1}$, $y \in K_{m_2}$. If now $m = \max(m_1, m_2)$, then both x and y belong to K_m. Hence $x + y$ and rx belong to K_m and therefore, a fortiori, to K. This completes the verification that K is a submodule.

Since we are assuming (c) to be true, K can be generated by a finite number of elements. Let $u_1, u_2, \ldots, u_p$ generate K and, for each i ($1 \leqslant i \leqslant p$), choose n_i so that $u_i \in K_{n_i}$. If now $\mu = \max(n_1, n_2, \ldots, n_p)$, then all the u_i belong to K_μ and therefore the module which they generate, namely K, is contained in K_μ. Thus $K \subseteq K_\mu \subseteq K_m \subseteq K$ for every $m \geqslant \mu$, whence $K_m = K_\mu$ provided that $m \geqslant \mu$. This completes the proof.

We can now dispose rapidly of two complementary definitions and a complementary proposition. As before, M denotes an R-module.

Definition. *M is said to satisfy the 'minimal condition' for submodules if every non-empty collection of submodules contains a minimal member.*

Thus, if M satisfies the minimal condition and Ω is a non-empty collection of submodules, then there exists $K \in \Omega$ with the following property: whenever $K' \in \Omega$ and $K' \subseteq K$ then $K' = K$.

Definition. *M is said to satisfy the 'descending chain condition' for submodules if, whenever*

$$K_1 \supseteq K_2 \supseteq K_3 \supseteq \ldots \supseteq K_m \supseteq \ldots$$

is a (weakly) decreasing, infinite sequence of submodules of M, there always exists an integer μ such that $K_m = K_\mu$ for all $m \geqslant \mu$.

Of course, the integer μ depends on the sequence under consideration.

Proposition 10. *If M is an R-module, then the following two statements are equivalent:*

  (a) *M satisfies the minimal condition for submodules;*
  (b) *M satisfies the descending chain condition for submodules.*

The proof of this involves only a minor modification of the demonstration that (a) and (b) of Proposition 9 are equivalent. The details will therefore be omitted.

The relevance of these considerations to the results obtained in the last section can be seen from

Theorem 7. *In order that the length $L_R(M)$ of an R-module M be finite, it is necessary and sufficient that M satisfy both the maximal and minimal conditions for submodules.*

Proof. First suppose that M is of finite length and let Ω be a non-empty collection of submodules of M. If $K \in \Omega$, then

$$L_R(K) \leqslant L_R(M) < \infty.$$

We can therefore choose K so that the integer $L_R(K)$ is as large as possible. It is then clear, by the corollary to Theorem 6, that K must be a maximal member of Ω. Similarly Ω contains a minimal member.

Now assume that M satisfies both the maximal and minimal conditions for submodules. Let Ω consist of all the submodules of M which are of finite length. Then Ω is certainly not empty because it contains the zero submodule. Let K be a maximal member of Ω. Then $L_R(K) < \infty$. *We contend that $K = M$.* For if K is strictly contained in M, let Ω' consist of all the submodules of M which strictly contain K. Ω' is not empty (it contains M) and therefore it contains a minimal member K'. By construction, $K' \supset K$ and there is no submodule strictly between them. Thus K'/K is simple and therefore

$$L_R(K'/K) = 1.$$

Finally, by Theorem 6,

$$L_R(K') = L_R(K'/K) + L_R(K) = 1 + L_R(K) < \infty.$$

This shows that $K' \in \Omega$ and now we have the required contradiction, because K' strictly contains K.

We have just seen that, for a module to be of finite length, the maximal and minimal conditions must hold simultaneously. However, for other purposes, the two conditions play independent roles. As a preliminary, to their separate study, we prove

Lemma 5. *Let A, B, K be submodules of an R-module M. If $A \subseteq B$, $A + K = B + K$ and $A \cap K = B \cap K$, then $A = B$.*

Proof. Let b be an element of B. Then, *a fortiori*, b belongs to

$$B + K = A + K$$

and therefore $b = a + k$ for suitable elements $a \in A$ and $k \in K$. Accordingly $b - a$ belongs to B and also to K, whence $b - a$ is in

$$B \cap K = A \cap K \subseteq A.$$

This shows that $b \in A$ and allows us to conclude that $B \subseteq A$. The lemma follows.

Proposition 11. *Let K be a submodule of an R-module M. If M satisfies the minimal condition for submodules, then the same is true of both K and M/K. Conversely, if K and M/K both satisfy the minimal condition, then so does M. The proposition remains true if 'minimal' is replaced by 'maximal' wherever it occurs.*

Proof. Only the statements involving the minimal condition will be considered, because the corresponding results for the maximal condition may be obtained by making trivial modifications to the argument. First suppose that M satisfies the minimal condition. Then K will certainly satisfy the minimal condition as well, because every submodule of K is also a submodule of M. Again, by Proposition 8, there is a one-one correspondence between the submodules of M/K and those submodules of M which contain K. Since the correspondence preserves inclusion relations, it is now clear that M/K must satisfy the minimal condition.

To prove the converse, we assume that K and M/K satisfy the minimal condition. Let

$$A_1 \supseteq A_2 \supseteq A_3 \supseteq \ldots \supseteq A_m \supseteq \ldots$$

be a descending sequence of submodules of M then, by Proposition 10, it will suffice to show that the sequence terminates. Now

$$A_1 \cap K \supseteq A_2 \cap K \supseteq A_3 \cap K \supseteq \ldots$$

and $\quad (A_1 + K)/K \supseteq (A_2 + K)/K \supseteq (A_3 + K)/K \supseteq \ldots$

are descending sequences in K and M/K respectively. By hypothesis, they must both terminate. We can therefore find μ so that, for $m \geqslant \mu$, $A_m \cap K = A_\mu \cap K$ and $(A_m + K)/K = (A_\mu + K)/K$. Thus, when $m \geqslant \mu$, $A_m \cap K = A_\mu \cap K$ and $A_m + K = A_\mu + K$ (see Proposition 8); consequently, since $A_m \subseteq A_\mu$, we have $A_m = A_\mu$ by virtue of Lemma 5. The proof is now complete.

In order to take these ideas a stage further, we must take a closer look at R itself. The first point to notice is that R may be regarded as a left module with respect to itself. For if $r \in R$ and $x \in R$, then the ring structure of R provides a natural meaning for the product rx. Moreover, the ring axioms ensure that, with a self-explanatory notation,

$$(r_1 + r_2)x = r_1 x + r_2 x, \quad r(x_1 + x_2) = rx_1 + rx_2, \quad r_1(r_2 x) = (r_1 r_2)x$$

and $1x = x$. Thus R has a natural structure as a left R-module.

Definitions. *If R is considered as a left module with respect to itself, then its submodules are called the 'left ideals' of R. A left ideal which is different from the whole ring is called a 'proper left ideal'. A proper left ideal, which is not strictly contained by any other proper left ideal, is known as a 'maximal left ideal'.*

Thus, to be explicit, a left ideal is a non-empty subset A of R with the property that, whenever $a_1, a_2 \in A$ and $r \in R$, then $a_1 + a_2$ and ra_1 also belong to A (see Proposition 2).

Once again, let $r \in R$ and let $x \in R$. Since R is a ring, xr also has a natural meaning. If we use this to define the product of r and x, then R becomes a *right module* (see section (1.2)) with respect to itself, and, in this case, the submodules are called *right ideals*. Accordingly, a right ideal is a non-empty subset B of R with the following property: whenever $b_1, b_2 \in B$ and $r \in R$, then $b_1 + b_2 \in B$ and $b_1 r \in B$.

The fact that R is both a left and a right module with respect to itself, produces a new kind of subsystem. For future reference we note the following

Definition. *A subset, of the ring R, which is both a left ideal and a right ideal is called a 'two-sided ideal'.*

Let us return to the consideration of modules. There is a special connection between left ideals and singly generated left modules.†
Before giving the result which sets this in evidence, let us note that if M is a simple module (see section (1·7)), then it is singly generated. For, by definition, $M \neq 0$ and therefore it contains a non-zero element. Such an element must generate M, because M has no submodules other than itself and the zero submodule.

Proposition 12. *Let M be a left R-module generated by a single element u and let $f: R \to M$ be the mapping defined by $f(r) = ru$. Then f is an epimorphism of left R-modules and it induces an isomorphism $R/I \approx M$, where $I = \mathrm{Ker} f$ is a left ideal. Furthermore, M is a simple module if and only if I is a maximal left ideal.*

† In this connection note that, for a left ideal I, the R-module R/I is generated by the image of the identity element.

Proof. Since u generates M, every element of M can be expressed in the form ru. This shows that f maps R on to M. Next, if r_1, r_2, r are in R, then

$$f(r_1+r_2) = r_1u+r_2u = f(r_1)+f(r_2) \quad \text{while} \quad f(rr_1) = r(r_1u) = rf(r_1).$$

Consequently f is a homomorphism of left R-modules. The kernel I, of f, is thus a left ideal of R and, by Theorem 1, there is induced an isomorphism $R/I \approx M$. Finally, for M to be simple we must have (i) $M \neq 0$ which is equivalent to $I \neq R$ (i.e. I is a proper left ideal), and (ii) M must have no submodules other than itself and the zero submodule. Since (ii) is equivalent to there being no left ideals strictly between I and R, the proof is complete.

Theorem 8. *Let R satisfy the minimal condition for left ideals and let M be a finitely generated left R-module. Then M satisfies the minimal condition for submodules. The theorem is also true if 'minimal' is replaced by 'maximal' whenever it occurs.*

Proof. We shall only consider the theorem in so far as it concerns the minimal condition, since the 'maximal' case can be treated similarly. Let M be generated by n elements, say by $u_1, u_2, ..., u_n$. The argument uses induction on n.

If $n = 1$, then M is singly generated and hence (Proposition 12) isomorphic to R/I for a suitable left ideal I. However, R/I satisfies the minimal condition for submodules by virtue of Proposition 11.

We now suppose that $n > 1$ and that the theorem has been proved for R-modules generated by $n-1$ elements. Let K be the submodule generated by u_1 and $f: M \to M/K$ the natural mapping. A typical element $x \in M$ can be expressed in the form $r_1u_1+r_2u_2+...+r_nu_n$. Consequently, since $r_1u_1 \in K$, $f(x) = r_2f(u_2)+...+r_nf(u_n)$. Thus the $n-1$ elements $f(u_2), f(u_3), ..., f(u_n)$ generate M/K. Accordingly (by the inductive hypothesis) M/K satisfies the minimal condition for submodules, while K itself satisfies the condition because it comes under the case $n = 1$. An appeal to Proposition 11, now completes the proof.

By Theorem 7, a necessary and sufficient condition for R (considered as a left R-module) to be of finite length is that it satisfy both the maximal and minimal conditions for left ideals. The next theorem shows that, in this situation, there exist only a finite number of essentially distinct simple R-modules.

Theorem 9. *Let $R = I_0 \supset I_1 \supset \ldots \supset I_s = 0$ be a composition series of left ideals and let M be an arbitrary simple left R-module. Then there exists an integer j $(1 \leqslant j \leqslant s)$ such that I_{j-1}/I_j is isomorphic to M.*

Proof. By Proposition 12, M is isomorphic to R/I', where I' is a suitable maximal left ideal. Theorem 5 now shows that there exists a composition series
$$R = I_0' \supset I_1' \supset \ldots \supset I_s' = 0$$
of left ideals in which $I_1' = I'$. However, by the same theorem, the two composition series from R to its zero ideal are equivalent, hence $R/I' = I_0'/I_1'$ is isomorphic to I_{j-1}/I_j for some j.

We shall, in the later pages, be more concerned with the maximal than the minimal condition. Among the rings whose ideals satisfy this condition, certain polynomial rings figure rather prominently. The fact that these rings have the required property depends on a famous result, due to D. Hilbert, known as the Basis Theorem. Usually the basis theorem is stated only for commutative rings, but, as the reader will observe, this restriction is not necessary. Indeed, considerable advantages may be gained by going further and formulating the theorem for modules rather than for rings. We now prepare the way for a statement of this result by making some observations concerning formal power series and polynomials.

Let X be a new symbol. An infinite *formal* sum $a_0 + a_1 X + a_2 X^2 + \ldots$, where a_0, a_1, a_2 etc. belong to R, is called a *formal power series in X with coefficients in R*. The element a_i, which occurs in the term $a_i X^i$, is called the *coefficient* of X^i. If we have two such power series, say
$$F = a_0 + a_1 X + a_2 X^2 + a_3 X^3 + \ldots$$
and
$$G = b_0 + b_1 X + b_2 X^2 + b_3 X^3 + \ldots,$$
then their *sum* $F + G$ is defined by
$$F + G = (a_0 + b_0) + (a_1 + b_1) X + (a_2 + b_2) X^2 + (a_3 + b_3) X^3 + \ldots.$$
Thus to add two formal power series we simply add their coefficients. The *product* of F and G, which is denoted by FG, is defined to be the series
$$FG = (a_0 b_0) + (a_0 b_1 + a_1 b_0) X + (a_0 b_2 + a_1 b_1 + a_2 b_0) X^2 + \ldots.$$
With these laws of composition, formal power series form a ring having $1 + 0X + 0X^2 + \ldots$ as its identity element. We leave the reader to verify that the ring axioms are satisfied. The usual symbol for this ring is $R[[X]]$. Although power series are written as sums with an infinite number of terms, the reader should note that this is really a notational device. In particular, no concept of 'convergence' is involved.

Now suppose that E is an R-module. Then by a *formal power series in X with coefficients in E* we shall mean an expression

$$e_0 + e_1 X + e_2 X^2 + \ldots,$$

where e_0, e_1, e_2 etc. are arbitrary elements of E. Let $E[[X]]$ denote the set of all these new power series. It is clear that $E[[X]]$ becomes an additive abelian group if we add two of its members by adding the coefficients just as was done in the case of power series with coefficients in R. However we can do more than this for we can turn $E[[X]]$ into an $R[[X]]$-module. In fact if

$$r_0 + r_1 X + r_2 X^2 + r_3 X^3 + \ldots$$

belongs to $R[[X]]$ and

$$e_0 + e_1 X + e_2 X^2 + e_3 X^3 + \ldots$$

to $E[[X]]$, then we achieve the desired result by defining their product to be
$$(r_0 e_0) + (r_0 e_1 + r_1 e_0) X + (r_0 e_2 + r_1 e_1 + r_2 e_0) X^2 + \ldots.$$

Those special elements of $R[[X]]$ which are such that only a finite number of X^i ($i = 0, 1, 2, \ldots$) have non-zero coefficients, are called polynomials. More precisely they are called *polynomials in X with coefficients in R*. It is evident that the sum and product of two of these polynomials are again polynomials. In fact, using $R[X]$ to denote their aggregate, we can easily verify that $R[X]$ is a *ring* which is embedded in $R[[X]]$.

In an entirely similar manner, we can define *polynomials in X with coefficients in the R-module E*. These form an additive group, $E[X]$ say, which is embedded in $E[[X]]$. Moreover the result of multiplying an element of $E[X]$ by an element of $R[X]$ belongs to $E[X]$. Thus we see that $E[X]$ is an $R[X]$-module.

Given a polynomial $e_0 + e_1 X + e_2 X^2 + \ldots$ with (say) coefficients in E, it is possible to find an integer m such that $e_n = 0$ for all $n > m$. In view of this we are able, when it is convenient, to write the 'main part' of the series in reverse order. Thus the above polynomial could also be written as $e_m X^m + e_{m-1} X^{m-1} + \ldots + e_0$. Note that for every r in R and every non-negative integer μ we have

$$(rX^\mu)(e_m X^m + e_{m-1} X^{m-1} + \ldots + e_0)$$
$$= (re_m) X^{\mu+m} + (re_{m-1}) X^{\mu+m-1} + \ldots + (re_0) X^\mu.$$

Finally the term *constant polynomial* is used to describe any polynomial having the property that the coefficient of X^i is zero whenever i is strictly positive. Observe that the constant polynomials with coefficients in R form a ring which is an exact copy of R.

Suppose now that K is an $R[X]$-submodule of $E[X]$ and that $\mu \geqslant 0$ is an integer. We shall denote by $I_\mu(K)$ the set of all elements $e \in E$ for which there is a polynomial of the form $eX^\mu + e_{\mu-1}X^{\mu-1} + \ldots + e_0$ belonging to K. It is a simple matter to verify that $I_\mu(K)$ is a submodule of the R-module E. Moreover if $eX^\mu + e_{\mu-1}X^{\mu-1} + \ldots + e_0$ belongs to K, then so does

$$(1X)(eX^\mu + e_{\mu-1}X^{\mu-1} + \ldots + e_0) = eX^{\mu+1} + e_{\mu-1}X^\mu + \ldots + e_0 X.$$

Thus from $e \in I_\mu(K)$ follows $e \in I_{\mu+1}(K)$ and therefore we have an infinite ascending sequence

$$I_0(K) \subseteq I_1(K) \subseteq I_2(K) \subseteq \ldots \tag{1.8.1}$$

of submodules of E. Again, if L is another $R[X]$-submodule of $E[X]$ and $K \subseteq L$, then $I_\mu(K) \subseteq I_\mu(L)$ for every non-negative integer μ.

Lemma 6. *If* $K \subseteq L$ *are* $R[X]$-*submodules of* $E[X]$ *and* $I_\mu(K) = I_\mu(L)$ *for all* $\mu \geqslant 0$, *then* $K = L$.

Proof. Let $\phi = e_\mu X^\mu + e_{\mu-1}X^{\mu-1} + \ldots + e_0$ belong to L. Then e_μ belongs to $I_\mu(L) = I_\mu(K)$. We can therefore find a polynomial ψ_0 of the form $\psi_0 = e_\mu X^\mu + e'_{\mu-1}X^{\mu-1} + \ldots + e'_0$ which belongs to K. Then

$$\phi - \psi_0 = \xi_{\mu-1}X^{\mu-1} + \xi_{\mu-2}X^{\mu-2} + \ldots + \xi_0 \text{ (say)}$$

belongs to L. Repeating the argument, we find that there exists $\psi_1 \in K$ such that $\phi - \psi_0 - \psi_1$ belongs to L and has the form

$$\eta_{\mu-2}X^{\mu-2} + \eta_{\mu-3}X^{\mu-3} + \ldots + \eta_0.$$

After $\mu + 1$ such steps we obtain polynomials $\psi_0, \psi_1, \ldots, \psi_\mu$, all belonging to K, with the property that $\phi - \psi_0 - \psi_1 - \ldots - \psi_\mu = 0$. Thus $\phi \in K$ and lemma follows.

The next theorem is the *Basis Theorem*.

Theorem 10. *Let the* R-*module* E *satisfy the maximal condition for submodules. Then the* $R[X]$-*module* $E[X]$ *also satisfies the maximal condition for submodules.*

Proof. Let $K_1 \subseteq K_2 \subseteq K_3 \subseteq \ldots$ be an infinite ascending sequence of $R[X]$-submodules of $E[X]$. By Proposition 9, it is enough to show that the sequence becomes constant.

With the notation explained above $I_\mu(K_j)$, where $\mu \geqslant 0$ and $j \geqslant 1$, is an R-submodule of E. From among this doubly indexed set of submodules, choose one, say $I_\nu(K_q)$, which is maximal in this set. If now $\mu \geqslant \nu$ and $m \geqslant q$, then $I_\nu(K_q) \subseteq I_\nu(K_m) \subseteq I_\mu(K_m)$ and therefore

$$I_\mu(K_m) = I_\nu(K_q)$$

by the choice of the latter.

Suppose next that $0 \leqslant \rho < \nu$. Then

$$I_\rho(K_1) \subseteq I_\rho(K_2) \subseteq I_\rho(K_3) \subseteq \ldots;$$

consequently there exists an integer q_ρ such that $I_\rho(K_m) = I_\rho(K_{q_\rho})$ for $m \geqslant q_\rho$. Put $p = \max(q, q_0, q_1, \ldots, q_{\nu-1})$ and suppose that $m \geqslant p$. Then for $0 \leqslant \rho < \nu$ we have

$$I_\rho(K_m) = I_\rho(K_{q_\rho}) = I_\rho(K_p)$$

while if $\nu \leqslant \mu$, then $\quad I_\mu(K_m) = I_\nu(K_q) = I_\mu(K_p)$.

Thus $I_\mu(K_m) = I_\mu(K_p)$ for *all* $\mu \geqslant 0$. Consequently, since $K_p \subseteq K_m$, Lemma 6 shows that $K_m = K_p$. As this holds for every $m \geqslant p$, the proof is now complete.

Corollary. *If R satisfies the maximal condition for left ideals, then so does the polynomial ring $R[X]$.*

This follows at once if we regard R as a left module with respect to itself and then put $E = R$ in the theorem.

1.9 Direct sums

We return to the consideration of some very general ideas, connected with modules, with the intention of illustrating them by means of special problems as soon as we have acquired sufficient familiarity.

Let M be an R-module and $\{M_i\}_{i \in I}$ a family of submodules.

Definition. *If each element x, of M, has a unique representation in the form $x = \sum_i x_i$, where $x_i \in M_i$ and $x_i = 0$ for almost all i, then M is said to be the 'direct sum' or, more explicitly, 'the internal direct sum' of the submodules M_i.*

It is clear that, in such a situation, M is the sum of the M_i in the sense of section (1.4). However, for M to be the (ordinary) sum we only require each x to have at least one representation in the form $x = \sum_i x_i$ and there is no insistence on *uniqueness*. When it is desired to indicate that a particular sum is *direct*, it is customary to write

$$M = \bigoplus_{i \in I} M_i$$

or

$$M = \sum_{i \in I} M_i \text{ (direct sum)}.$$

Should it happen that the family consists of finitely many submodules $M_1, M_2, \ldots, M_s$, then we may use $M = M_1 \oplus M_2 \oplus \ldots \oplus M_s$ as an alternative notation.

Lemma 7. *Let $\{M_i\}_{i \in I}$ be a family of submodules of an R-module M and suppose that $M = \sum_i M_i$. Then in order that the sum be direct, it is necessary and sufficient that the representation of the zero element (in the form $\sum_i x_i$) be unique.*

Proof. Necessity is obvious. To prove sufficiency we assume that the representation of 0_M is unique. Now let x be an arbitrary element of M and suppose that $x = \sum_i x_i = \sum_i x_i'$, where $x_i, x_i' \in M_i$ and each sum has only a finite number of non-zero terms. Then $0_M = \sum_i (x_i - x_i')$ and $x_i - x_i'$ belongs to M_i. But 0_M is also the sum of the zero elements of the various M_i and, by hypothesis, 0_M has only a single representation. Accordingly $x_i - x_i' = 0$ and therefore $x_i = x_i'$ for all i. This establishes sufficiency.

Suppose that $M = \bigoplus_{i \in I} M_i$ and let $j \in I$. For each $x \in M$, we have a unique representation $x = \sum_i x_i$ and therefore we can construct a mapping $M \to M_j$ by associating with x the term x_j which occurs in the sum. $M \to M_j$ is, in fact, an epimorphism of R-modules which we shall refer to as the *projection of M on to the jth summand*. Note that if $M_i \to M$ denotes the inclusion mapping, then a homomorphism $M_i \to M_j$ can be formed by combining $M_i \to M$ with the projection $M \to M_j$. Clearly $M_i \to M_j$ is the identity mapping of M_i if $i = j$ and is a null homomorphism when $i \neq j$.

Let us examine, more closely, the connection between x and the individual terms in the sum $\sum_i x_i$ which represents it. To assist in this, we shall call x_i the *ith component of x*. The components of x form a family $\{x_i\}_{i \in I}$ in which $x_i \in M_i$ for all i and at most a finite number of the members are different from zero. Conversely, given a family $\{y_i\}_{i \in I}$ where $y_i \in M_i$ and $y_i = 0$ for almost all i, there will be just one element of M, namely $y = \sum_i y_i$, which has these as components. Thus we have a one-one correspondence between the elements of M, on the one hand, and the families $\{x_i\}_{i \in I}$ on the other. Moreover, when we add two elements of M the effect is simply to add their components while, if we multiply an element of M by an element r of R, then all the components get multiplied by r. This shows that the *module structure* of M is wholly determined by that of the M_i and suggests that we consider direct sums from a second point of view.

For the new approach, we begin with a family $\{M_i\}_{i \in I}$ of R-modules but do not assume that there is some given module which contains them all. Next we consider families $\{x_i\}_{i \in I}$, where each $x_i \in M_i$ and $x_i = 0$ for almost all i. It is now possible to manufacture a new R-module out of these families. We must, of course, explain how two families are to be 'added' and how a family is to be 'multiplied' by an element of R. The definitions are the obvious ones. If $\{x_i\}$ and $\{y_i\}$ are two families and $r \in R$, then we put

$$\{x_i\} + \{y_i\} = \{x_i + y_i\}, \quad r\{x_i\} = \{rx_i\}.$$

It is left to the reader to verify that, with these laws of composition, the module axioms are satisfied. He will find that the zero element of the new system is just the family formed by the zero elements of the various M_i and that the negative of $\{x_i\}$ is $\{-x_i\}$.

Let us denote the module, which has just been constructed, by M. We say that M is the *external direct sum* of the modules M_i $(i \in I)$ and use the same notation as before namely $M = \bigoplus_{i \in I} M_i$. At first sight, it might appear that internal and external direct sums are distinct concepts rather than the same concept looked at in two different ways. To clarify this point, assume that M is the external direct sum of the modules M_i and let $j \in I$. The families $\{x_i\}_{i \in I}$ with the property that $x_i = 0$ whenever $i \neq j$, form a submodule M_j' (say) of M. Let us associate, with a given element ξ of M_j, the family $\{x_i\}_{i \in I}$ given by $x_j = \xi$, $x_i = 0$ for $i \neq j$. This provides a mapping $\lambda_j \colon M_j \to M_j'$ which is clearly an isomorphism. Thus M contains a family of submodules, namely $\{M_i'\}_{i \in I}$, which are just copies of the original M_i.

Again, let $x = \{x_i\}_{i \in I}$ be an element of M and let j belong to I. On replacing the x_i for which $i \neq j$ by zeros, one is left with an element of M_j'. In this way is obtained a canonical epimorphism $\pi_j \colon M \to M_j'$. Note that if $y \in M_j'$ then $\pi_j(y) = y$, while if $z \in M_i'$ and $i \neq j$, then $\pi_j(z) = 0$.

Since $x_j = 0$ for almost all j, we have $\pi_j(x) = 0$ for almost all j. Also $x = \sum_j \pi_j(x)$. Consequently $M = \sum_i M_i'$. *We contend that this sum is direct.* For suppose that $0 = \sum_i x_i'$, where $x_i' \in M_i'$, is a representation of the zero element. Applying π_j we obtain $0 = \sum_i \pi_j(x_i')$. But

$$\pi_j(x_j') = x_j' \quad \text{and} \quad \pi_j(x_i') = 0 \quad \text{for} \quad i \neq j.$$

Thus $x'_j = 0$ and this holds for every j. That $M = \bigoplus_i M'_i$ now follows from Lemma 7.

To recapitulate, the external direct sum M of the modules $\{M_i\}_{i \in I}$ is also the internal direct sum of the submodules $\{M'_i\}_{i \in I}$ and, for each i, M_i and M'_i are isomorphic. It is customary to identify M_i and M'_i (i.e. we use the same symbol for each of two corresponding elements and leave the context to explain which one is meant) and in this way the distinction between internal and external direct sums disappears.

Proposition 13. *Let $\{M_i\}_{i \in I}$ be a family of submodules of an R-module M and suppose that M is their direct sum. Further, let I be split into two non-overlapping sets I_1 and I_2. Then*

$$(\sum_{i \in I_1} M_i) + (\sum_{i \in I_2} M_i) = M$$

and
$$(\sum_{i \in I_1} M_i) \cap (\sum_{i \in I_2} M_i) = 0.$$

In addition, we have isomorphisms

$$M/(\sum_{i \in I_1} M_i) \approx \sum_{i \in I_2} M_i \quad and \quad M/(\sum_{i \in I_2} M_i) \approx \sum_{i \in I_1} M_i.$$

Proof. $\sum_{i \in I_1} M_i$ contains all the M_i with $i \in I_1$ while $\sum_{i \in I_2} M_i$ contains the remainder. Consequently

$$(\sum_{i \in I_1} M_i) + (\sum_{i \in I_2} M_i)$$

is a submodule which contains them all and therefore it must be equal to M. Note that, so far, we have not used the fact that we are dealing with a *direct* sum. Now let u be an element of

$$(\sum_{i \in I_1} M_i) \cap (\sum_{i \in I_2} M_i).$$

Then $u = \sum_{i \in I_1} y_i = \sum_{i \in I_2} z_i$, where $y_i \in M_i$ $(i \in I_1)$, $z_i \in M_i$ $(i \in I_2)$ and each sum contains only a finite number of non-zero terms. For $i \in I_1$ put $x_i = y_i$ while if $i \in I_2$ let $x_i = -z_i$. Then $x_i \in M_i$ for *all* i and $\sum_{i \in I} x_i = 0$. Since M is the direct sum of the M_i, all the x_i must be zero and hence u also is zero. This proves the second assertion. The statements about isomorphisms follow from Theorem 3.

Corollary. *Let the situation be as described in the proposition. Then $\sum_{i \in I_1} M_i$ is the direct sum of the submodules M_i $(i \in I_1)$ and a similar statement is true of $\sum_{i \in I_2} M_i$.*

Proof. Assume that $\sum\limits_{i\in I_1} x_i = 0$, where $x_i \in M_i$ for $i \in I_1$ and there are only a finite number of non-zero terms in the sum. If we now define x_i to be zero when $i \in I_2$, then $\sum\limits_{i\in I} x_i = 0$. Since M is the direct sum of the M_i, it follows that $x_i = 0$ for all i and, in particular, for $i \in I_1$. This completes the proof.

Proposition 14. *Let $\{M_i\}_{i\in I}$ be a family of submodules of an R-module M and suppose that $M = \sum\limits_{i\in I} M_i$. Then in order that this sum should be direct it is necessary and sufficient that*

$$M_i \cap (\sum_{j\neq i} M_j) = 0 \qquad (1.9.1)$$

for every $i \in I$.

Proof. That the condition is necessary is an immediate consequence of the second assertion in Proposition 13. Now suppose that (1.9.1) holds for every $i \in I$ and assume that $\sum\limits_i x_i = 0$, where $x_i \in M_i$ and there are only a finite number of non-zero terms. Then $x_i = -\sum\limits_{j\neq i} x_j$ and therefore x_i belongs to both M_i and $\sum\limits_{j\neq i} M_j$. It follows that $x_i = 0$ and, as this holds for every i in I, the proof is complete.

Corollary. *Let $\{M_i\}_{i\in I}$ be a family of submodules of an R-module M and suppose that M is their direct sum. In addition let $\{I_\lambda\}_{\lambda\in\Lambda}$ be a family of non-overlapping subsets of I whose union is I itself (i.e. the I_λ partition I into disjoint sets). Then $M = \bigoplus\limits_{\lambda\in\Lambda} (\sum\limits_{i\in I_\lambda} M_i)$.*

Proof. Put $N_\lambda = \sum\limits_{i\in I_\lambda} M_i$, then each M_i is contained in some N_λ and hence in $\sum\limits_\lambda N_\lambda$. Accordingly $M = \sum\limits_\lambda N_\lambda$ and it remains for us to show that this sum is direct.

Let $\rho \in \Lambda$ and denote by J_ρ the complement of I_ρ in I. Then, because the I_λ are disjoint, we have $J_\rho = \bigcup\limits_{\lambda\neq\rho} I_\lambda$.

Given $j \in J_\rho$, it is possible to find $\mu \neq \rho$ so that $j \in I_\mu$ in which case

$$M_j \subseteq \sum_{i\in I_\mu} M_i = N_\mu \subseteq \sum_{\lambda\neq\rho} N_\lambda.$$

This shows that $\sum\limits_{i\in J_\rho} M_i \subseteq \sum\limits_{\lambda\neq\rho} N_\lambda$. On the other hand, if $\lambda \neq \rho$, then

$$N_\lambda = \sum_{i\in I_\lambda} M_i \subseteq \sum_{i\in J_\rho} M_i$$

3-2

because $I_\lambda \subseteq J_\rho$. Thus we also have $\sum_{\lambda \neq \rho} N_\lambda \subseteq \sum_{i \in J_\rho} M_i$. These results can now be combined to give

$$\sum_{i \in J_\rho} M_i = \sum_{\lambda \neq \rho} N_\lambda. \qquad (1.9.2)$$

Next, by Proposition 13,

$$(\sum_{i \in I_\rho} M_i) \cap (\sum_{i \in J_\rho} M_i) = 0$$

that is, using (1.9.2), $N_\rho \cap (\sum_{\lambda \neq \rho} N_\lambda) = 0.$

It is now sufficient to appeal to Proposition 14 to see that, in $M = \sum_\lambda N_\lambda$, the sum is direct.

Sometimes it is necessary to draw a conclusion which goes in the opposite direction. For this we use

Proposition 15. *Let $\{M_i\}_{i \in I}$ be a family of submodules of an R-module M and let $\{I_\lambda\}_{\lambda \in \Lambda}$ be a family of non-overlapping subsets of I whose union is I itself. Suppose that each individual sum $\sum_{i \in I_\lambda} M_i$ is direct and that $M = \bigoplus_{\lambda \in \Lambda} (\sum_{i \in I_\lambda} M_i)$. Then $M = \bigoplus_{i \in I} M_i$.*

Proof. Put $N_\lambda = \sum_{i \in I\lambda} M_i$. Then, by the hypotheses, $N_\lambda = \bigoplus_{i \in I_\lambda} M_i$ and $M = \bigoplus_{\lambda \in \Lambda} N_\lambda$. Let x belong to M. Then x is a finite sum of elements $n_\lambda (n_\lambda \in N_\lambda)$ and each of these n_λ's is a finite sum of elements $x_i (x_i \in M_i)$. It follows that x belongs to $\sum_{i \in I} M_i$. Accordingly $M = \sum_{i \in I} M_i$ and it remains for us to show that the sum is direct.

To this end, we assume that $\sum_i x_i = 0$, where $x_i \in M_i$ and $x_i = 0$ for almost all i in I. Let $j \in I$. By Lemma 7, it will suffice to show that $x_j = 0$.

Writing $y_\lambda = \sum_{i \in I_\lambda} x_i$, we find that $\sum_\lambda y_\lambda = 0$. Consequently (since $M = \bigoplus_\lambda N_\lambda$ and $y_\lambda \in N_\lambda$) $y_\lambda = 0$ for all $\lambda \in \Lambda$. Choose λ so that $j \in I_\lambda$. Since $N_\lambda = \bigoplus_{i \in I_\lambda} M_i$, the zero element has a unique representation as a sum of elements in which there is one representative from each $M_i (i \in I_\lambda)$. However $\sum_{i \in I_\lambda} x_i = y_\lambda = 0$. We may therefore conclude that $x_i = 0$ for all $i \in I_\lambda$. In particular $x_j = 0$ and the proof is complete.

Proposition 16. *Let $\{M_i\}_{i \in I}$ be a family of non-zero submodules of an R-module M and suppose that M is their direct sum. If now M is finitely generated, then the index set I is finite.*

Proof. If x belongs to M, then $x = \sum_i x_i$, where $x_i \in M_i$ and only finitely many terms are different from zero. If the non-zero terms arise from $i = i_1, i_2, \ldots, i_p$, then x belongs to $M_{i_1} + M_{i_2} + \ldots + M_{i_p}$. Now let $u_1, u_2, \ldots, u_n$ be elements which generate M. The preliminary observation shows that, for each μ $(1 \leqslant \mu \leqslant n)$, we can find a finite subset I_μ of I such that $u_\mu \in \sum_{i \in I_\mu} M_i$. Let I^* be the smallest subset of I containing the I_μ. Then I^* is a finite set. Also $u_\mu \in \sum_{i \in I^*} M_i$ for every μ and therefore

$$M = \sum_{i \in I^*} M_i. \tag{1.9.3}$$

Denote by J^* the complement of I^* in I. Then, by Proposition 13,

$$\left(\sum_{i \in I^*} M_i \right) \cap \left(\sum_{i \in J^*} M_i \right) = 0.$$

However (1.9.3) shows that this intersection is just $\sum_{i \in J^*} M_i$ and, by hypothesis, none of the M_i is a zero module. It follows that J^* is empty and hence that $I = I^*$. The proof is now complete.

The next result is more specialized in character, but it is included because it provides a good example of how one can work with direct sums. The information it contains will be useful later.

Proposition 17. *Let $N_1, N_2, \ldots, N_p$ be submodules of an R-module M and suppose that for each i $(1 \leqslant i \leqslant p)$*

$$N_i + (N_1 \cap \ldots \cap N_{i-1} \cap N_{i+1} \cap \ldots \cap N_p) = M.$$

In these circumstances there exists an isomorphism

$$M/(N_1 \cap N_2 \cap \ldots \cap N_p) \approx (M/N_1) \oplus (M/N_2) \oplus \ldots \oplus (M/N_p).$$

Proof. Let $\phi_i \colon M \to M/N_i$ be the natural mapping of M on to the factor module M/N_i. For each $x \in M$, the family $\{\phi_1(x), \phi_2(x), \ldots, \phi_p(x)\}$ is, by definition, an element of the direct sum $(M/N_1) \oplus \ldots \oplus (M/N_p)$. Put $\psi(x) = \{\phi_1(x), \phi_2(x), \ldots, \phi_p(x)\}$ so that we now have a mapping

$$\psi \colon M \to (M/N_1) \oplus \ldots \oplus (M/N_p).$$

This is a homomorphism. For if x, x' belong to M and r belongs to R, then

$$\begin{aligned}
\psi(x+x') &= \{\phi_1(x+x'), \phi_2(x+x'), \ldots, \phi_p(x+x')\} \\
&= \{\phi_1(x) + \phi_1(x'), \phi_2(x) + \phi_2(x'), \ldots, \phi_p(x) + \phi_p(x')\} \\
&= \{\phi_1(x), \ldots, \phi_p(x)\} + \{\phi_1(x'), \ldots, \phi_p(x')\}
\end{aligned}$$

by virtue of the definition of addition in an external direct sum. Thus $\psi(x+x') = \psi(x) + \psi(x')$ and similarly $\psi(rx) = r\psi(x)$.

If the element x is to belong to $\mathrm{Ker}\,\psi$, then we must have $\phi_i(x) = 0$ for all i that is $x \in N_i$ $(i = 1, 2, ..., p)$. Accordingly

$$\mathrm{Ker}\,\psi = N_1 \cap N_2 \cap ... \cap N_p.$$

The required isomorphism will therefore follow from Theorem 1 as soon as we can show that ψ is an epimorphism.

Let $\{\xi_1, \xi_2, ..., \xi_p\}$, where $\xi_i \in M/N_i$, be an arbitrary element of the direct sum. We can choose $x \in M$ so that $\phi_1(x) = \xi_1$; and then, since $N_1 + (N_2 \cap N_3 \cap ... \cap N_p) = M$, x can be expressed in the form $x = u + v$, where $u \in N_1$ and $v \in (N_2 \cap ... \cap N_p)$. Then $\phi_i(v) = 0$ $(2 \leqslant i \leqslant p)$ and therefore

$$\psi(v) = \{\phi_1(v), 0, ..., 0\}$$
$$= \{\phi_1(x - u), 0, ..., 0\}$$
$$= \{\xi_1, 0, ..., 0\}$$

because $\phi_1(u) = 0$. Thus $\{\xi_1, 0, ..., 0\}$ belongs to $\mathrm{Im}\,\psi$ and similarly $\{0, \xi_2, ..., 0\}$, ..., $\{0, 0, ..., \xi_p\}$ also belong to $\mathrm{Im}\,\psi$. On adding these p elements together, it is seen that $\{\xi_1, \xi_2, ..., \xi_p\}$ is the image of an element of M and now the proof is complete.

Proposition 18. *Let M_1, M_2, ..., M_p be submodules of an R-module M and suppose that $M = M_1 \oplus M_2 \oplus ... \oplus M_p$. Then the chain factors of*

$$(M_1 + M_2 + ... + M_p) \supseteq (M_1 + M_2 + ... + M_{p-1}) \supseteq ... \supseteq M_1 \supseteq 0 \quad (1.9.4)$$

are isomorphic to $M_p, M_{p-1}, ..., M_1$. In particular, if the M_i are all simple, then (1.9.4) is a composition series and $L_R(M) = p$.

Proof. Put $E_i = M_1 + M_2 + ... + M_{p-i}$ $(0 \leqslant i \leqslant p-1)$ and $E_p = 0$. The corollary to Proposition 13 shows that, if $1 \leqslant i \leqslant p$, then

$$E_{i-1} = M_1 \oplus M_2 \oplus ... \oplus M_{p-i+1}.$$

Accordingly

$$E_{i-1}/E_i = E_{i-1}/(M_1 + M_2 + ... + M_{p-i}) \approx M_{p-i+1}$$

where, this time, we have made use of Proposition 13 itself.

1.10 The ring of endomorphisms

In this section we shall consider new ways of combining homomorphisms. To explain what is involved, let f and g be homomorphisms of an R-module M into an R-module N and, for each element x in M, put $h(x) = f(x) + g(x)$. It is easily verified that h is also a homomorphism of M into N. This new homomorphism is called the *sum* of f and g and denoted by $f + g$. A straightforward check shows that this method of addition is both commutative and associative; moreover the null homomorphism, see section (1.4), plays the role of a neutral element. Finally, if we put $\phi(x) = -f(x)$, then ϕ is a homomorphism of M into N and $f + \phi$ is the null homomorphism. Gathering together these various observations, we conclude: *the homomorphisms of M into N form an abelian group with respect to addition.*

Suppose now that M, N, P are R-modules. Let f, f_1, f_2 be homomorphisms of M into N and g, g_1, g_2 homomorphisms of N into P. It is now possible to form $g(f_1+f_2)$ and $gf_1 + gf_2$ each of which is a homomorphism of M into P. Another simple verification shows that these are equal. Thus we have

$$g(f_1+f_2) = gf_1 + gf_2 \qquad (1.10.1)$$

and similar considerations lead to

$$(g_1+g_2)f = g_1f + g_2f. \qquad (1.10.2)$$

A homomorphism of M into itself is called an *endomorphism*. Any two endomorphisms can be added and multiplied (in either order) and the results will also be endomorphisms. Using (1.10.1) and (1.10.2) it is easily verified that the endomorphisms form a ring. This is called the *endomorphism ring* of the module. The identity and zero elements of this ring are the identity mapping of M and the null homomorphism (of M into itself) respectively. As we shall see, the study of rings of endomorphisms can lead to important results concerning the structure of rings. First, however, we must develop our vocabulary.

Definition. *A ring R is called a 'division ring' if it is not a null ring (i.e. $1 \neq 0$) and if for each $\alpha \in R$, $\alpha \neq 0$, it is possible to find an element $\beta \in R$ such that $\alpha\beta = \beta\alpha = 1$. A commutative division ring is called a 'field'.*

Let us suppose that R is a division ring. If $\alpha \in R$, $\alpha \neq 0$, then there is only one element β such that $\alpha\beta = \beta\alpha = 1$. For if we also have $\alpha\beta' = \beta'\alpha = 1$, then

$$\beta = 1\beta = (\beta'\alpha)\beta = \beta'(\alpha\beta) = \beta'1 = \beta'.$$

The unique element β is called the *inverse* of α and denoted by α^{-1}. Clearly the inverse of α^{-1} is α itself. With the aid of this new concept we can describe a division ring as a non-null ring in which every non-zero element has an inverse.

Still assuming that R is a division ring, let α be a non-zero element and β an arbitrary element. Then there is a unique element $\gamma \in R$

such that $\alpha\gamma = \beta$. Indeed, if we multiply both sides of this equation (on the left) by α^{-1}, we see that, if there is a solution, then it must be given by $\gamma = \alpha^{-1}\beta$. However $\alpha(\alpha^{-1}\beta) = (\alpha\alpha^{-1})\beta = 1\beta = \beta$ so all is established. In a similar manner it can be shown that the equation $\lambda\alpha = \beta$ has the unique solution $\lambda = \beta\alpha^{-1}$. Thus in a division ring one can always divide (on the left or on the right) by any non-zero element. This is the reason for the name.

Next suppose that we have two rings, say R and R'. A mapping $\phi: R \to R'$ is called a *ring-homomorphism* if it satisfies the following three conditions:

(a) $\phi(\alpha+\beta) = \phi(\alpha)+\phi(\beta)$ for all α, β in R;

(b) $\phi(\alpha\beta) = \phi(\alpha)\phi(\beta)$ for all α, β in R;

(c) $\phi(1_R) = 1_{R'}$.

As in the case of modules, a ring-homomorphism ϕ is called a *monomorphism*, an *injection* or an *embedding* if distinct elements of R always have distinct images in R'. The kernel, Ker ϕ, of ϕ is the set of elements which map on to the zero element of R'. It is easy to see that ϕ is a monomorphism when and only when Ker ϕ contains only the zero element of R. The ring-homomorphism is called an *epimorphism* or a *surjection* if $R' = \phi(R)$, that is if ϕ maps R on to R'.

When $\phi: R \to R'$ is both a monomorphism and an epimorphism, it is called a *ring-isomorphism*. In such a case, the mapping sets up a one-one correspondence between the elements of R and those of R'; and the inverse mapping $\phi^{-1}: R' \to R$ is again a ring-isomorphism. Thus the two rings are involved symmetrically. The reader will appreciate that isomorphic rings are essentially copies of each other and, in the context of our theory, have entirely similar properties.

Since our rings are not assumed to be commutative, there is a second type of one-one correspondence which is of special interest. This is the kind that is addition-true but which reverses the order of products. Such a correspondence is called an *anti-isomorphism*. To be explicit, a one-one mapping $f: R \to R'$ of R on to R' is an anti-isomorphism† if $f(\alpha+\beta) = f(\alpha)+f(\beta)$ and $f(\alpha\beta) = f(\beta)f(\alpha)$. Note that the inverse of such a mapping is an anti-isomorphism of R' on to R.

There is one more ring-theoretic concept that is needed before we return to the study of modules. Let $R_1, R_2, ..., R_p$ be rings and consider the set of all sequences $\{r_1, r_2, ..., r_p\}$, where $r_i \in R_i (1 \leqslant i \leqslant p)$.

† Note that 1_R will necessarily correspond to $1_{R'}$.

If the 'sum' and 'product' of $\{r_1, r_2, ..., r_p\}$ and $\{r_1', r_2', ..., r_p'\}$ are defined by

$$\{r_1, r_2, ..., r_p\} + \{r_1', r_2', ..., r_p'\} = \{r_1 + r_1', r_2 + r_2', ..., r_p + r_p'\}$$

and $\quad \{r_1, r_2, ..., r_p\}\{r_1', r_2', ..., r_p'\} = \{r_1 r_1', r_2 r_2', ..., r_p r_p'\}$

respectively, then the set of these sequences takes on the structure of a ring. (The reader is left to carry out the straightforward verification that the ring axioms really are satisfied.) The new ring, which is called the *direct sum* of $R_1, R_2, ..., R_p$ and denoted by $R_1 \oplus R_2 \oplus ... \oplus R_p$, has $\{0_{R_1}, 0_{R_2}, ..., 0_{R_p}\}$ as zero element and $\{1_{R_1}, 1_{R_2}, ..., 1_{R_p}\}$ as identity element.

Suppose now that $R = R_1 \oplus R_2 \oplus ... \oplus R_p$ and denote by

$$A_i\,(1 \leqslant i \leqslant p)$$

the subset of R which consists of all sequences of the form

$$\{0, ..., r_i, ..., 0\}.$$

Here, of course, r_i belongs to R_i and occurs in the ith place. It is clear that the elements of A_i form a group with respect to addition. Moreover, since

$$\{r_1', ..., r_i', ..., r_p'\}\{0, ..., r_i, ..., 0\} = \{0, ..., r_i' r_i, ..., 0\}$$

and $\quad \{0, ..., r_i, ..., 0\}\{r_1', ..., r_i', ..., r_p'\} = \{0, ..., r_i r_i', ..., 0\},$

it follows that the A_i are left and also right ideals of the ring R. Whichever way we regard them, R is their direct sum and therefore we may write

$$R = A_1 \oplus A_2 \oplus ... \oplus A_p.$$

Next, if $a_i \in A_i$, $a_j \in A_j$ and $i \neq j$, then $a_i a_j = 0$. This is described by saying that these ideals annihilate each other by multiplication. Finally, if we consider the elements of A_i *by themselves*, it is clear that they form a ring which is isomorphic to R_i. It is customary to identify A_i and R_i in which case we can sum up our conclusions as follows: *if R is the direct sum of rings $R_1, R_2, ..., R_p$, then each R_i can be identified (in a natural manner) with a certain two-sided ideal of R. These ideals annihilate each other by multiplication and R is their (internal) direct sum.* It is necessary, however, to add a word of warning. The mapping $R_i \to R$ in which r_i is carried into $\{0, ..., r_i, ..., 0\}$ is *not* a ring-homomorphism according to our definition. This is because the image of the identity element of R_i is $\{0, ..., 1_{R_i}, ..., 0\}$ and this is different from the identity element of R.

After this digression into the terminology of ring theory, we return to the study of modules.

Proposition 19. *The ring of endomorphisms of a simple module is a division ring.*

Proof. Let L be a simple R-module and $\phi: L \to L$ a non-null endomorphism. Then $\phi(L) \neq 0$. Since the only non-zero submodule of L is L itself, it follows that $\phi(L) = L$. Again $\phi(L) \neq 0$ implies $\operatorname{Ker} \phi \neq L$. Consequently (because there are only two possibilities for the submodules of L) $\operatorname{Ker} \phi = 0$. It has now been shown that ϕ is an isomorphic mapping of L on to itself. Let ϕ^{-1} be the inverse isomorphism. Then ϕ^{-1} is an endomorphism and each of $\phi\phi^{-1}$ and $\phi^{-1}\phi$ is the identity mapping i_L of L. Since i_L is the identity element of the ring of endomorphisms, this establishes the proposition.

The proof (given above) that the non-null endomorphism $\phi: L \to L$ is an isomorphism, is readily adapted to yield

Lemma 8. *Let $\phi: L \to L'$ be a non-null homomorphism of a simple R-module L into a simple R-module L'. Then ϕ is an isomorphism of L on to L'.*

There is one consequence of this lemma which should be noted. Namely, if L_1 and L_2 are non-isomorphic simple modules, then the only homomorphism of L_1 into L_2 is the null homomorphism.

In order to carry these ideas a stage further, we consider the ring of endomorphisms of a module M which is expressed as the direct sum of a finite number of submodules; say $M = M_1 \oplus M_2 \oplus \dots \oplus M_n$. Denote by $\lambda_j: M_j \to M$ the inclusion mapping and by $\pi_i: M \to M_i$ the projection mapping of M on to the ith summand. Then from

$$x = x_1 + x_2 + \dots + x_n, \quad \text{where} \quad x_i \in M_i,$$

follows $\pi_i(x) = x_i$, $\sum_i \lambda_i \pi_i(x) = x$ and, when $i \neq j$, $\pi_j \lambda_i \pi_i(x) = 0$.

Suppose now that we have an endomorphism $\phi: M \to M$. Put $\phi_{ij} = \pi_i \phi \lambda_j \, (1 \leqslant i,j \leqslant n)$ so that ϕ_{ij} is a homomorphism of M_j into M_i. It is convenient to think of these n^2 homomorphisms as forming the elements of a matrix

$$\begin{Vmatrix} \phi_{11} & \phi_{12} & \cdots & \phi_{1n} \\ \phi_{21} & \phi_{22} & \cdots & \phi_{2n} \\ \cdot & \cdot & \cdots & \cdot \\ \cdot & \cdot & \cdots & \cdot \\ \cdot & \cdot & \cdots & \cdot \\ \phi_{n1} & \phi_{n2} & \cdots & \phi_{nn} \end{Vmatrix}, \qquad (1.10.3)$$

where the entry in the ith row and jth column is a homomorphism of M_j into M_i. We shall refer to (1.10.3) as *the matrix of the endomorphism ϕ*.

Next let $\psi: M \to M$ be a second endomorphism and suppose that its matrix is $\|\psi_{ij}\|$. Since

$$\pi_i(\phi+\psi)\lambda_j = \pi_i\phi\lambda_j + \pi_i\psi\lambda_j = \phi_{ij}+\psi_{ij},$$

the matrix corresponding to $\phi+\psi$ is the sum of their respective matrices. Again $\lambda_1\pi_1+\lambda_2\pi_2+\ldots+\lambda_n\pi_n$ is the identity mapping of M. Consequently

$$\pi_i(\phi\psi)\lambda_j = \pi_i\phi(\lambda_1\pi_1+\ldots+\lambda_n\pi_n)\psi\lambda_j$$
$$= \sum_k \pi_i\phi\lambda_k\pi_k\psi\lambda_j$$
$$= \sum_k \phi_{ik}\psi_{kj}.$$

This is conveniently expressed by saying that the matrix of the endomorphism $\phi\psi$ is just the product of the matrix corresponding to ϕ with that corresponding to ψ.

Let i_M denote the identity mapping of M. Then

$$i_M = \lambda_1\pi_1+\ldots+\lambda_n\pi_n \quad \text{and} \quad \phi = i_M\phi i_M.$$

Consequently

$$\phi = i_M\phi i_M = (\lambda_1\pi_1+\ldots+\lambda_n\pi_n)\phi(\lambda_1\pi_1+\ldots+\lambda_n\pi_n)$$
$$= \sum_{i,j}\lambda_i\pi_i\phi\lambda_j\pi_j = \sum_{i,j}\lambda_i\phi_{ij}\pi_j.$$

This shows that to distinct endomorphisms of M there correspond distinct matrices.

Finally suppose that we start with a matrix (1.10.3), where

$$\phi_{ij}\,(1 \leqslant i,j \leqslant n)$$

is any homomorphism of M_j into M_i whatsoever. Put

$$\phi^* = \sum_{i,j}\lambda_i\phi_{ij}\pi_j,$$

then ϕ^* is an endomorphism of M. Moreover

$$\pi_\mu\phi^*\lambda_\nu = \sum_{i,j}\pi_\mu\lambda_i\phi_{ij}\pi_j\lambda_\nu$$
$$= \pi_\mu\lambda_\mu\phi_{\mu\nu}\pi_\nu\lambda_\nu$$

because $\pi_j\lambda_\nu$ is a null map if $j \neq \nu$ while $\pi_\mu\lambda_i$ is null if $\mu \neq i$. Thus $\pi_\mu\phi^*\lambda_\nu = \phi_{\mu\nu}$ (note $\pi_\mu\lambda_\mu$ and $\pi_\nu\lambda_\nu$ are identity maps) and this shows that $\|\phi_{\mu\nu}\|$ is the matrix of ϕ^*. Collecting results together we find that we have proved

Proposition 20. *If* $M = M_1 \oplus M_2 \oplus \ldots \oplus M_n$, *then the ring of endomorphisms of* M *is isomorphic to the ring formed by all matrices*

$$\begin{Vmatrix} \phi_{11} & \phi_{12} & \cdots & \phi_{1n} \\ \phi_{21} & \phi_{22} & \cdots & \phi_{2n} \\ \cdot & \cdot & \cdots & \cdot \\ \cdot & \cdot & \cdots & \cdot \\ \phi_{n1} & \phi_{n2} & \cdots & \phi_{nn} \end{Vmatrix},$$

where ϕ_{ij} *denotes an arbitrary homomorphism of* M_j *into* M_i.

It should be noted that the above discussion *proves* that these matrices form a ring though, of course, this can easily be verified directly.

Corollary. *Let* $M = M_1 \oplus M_2 \oplus \ldots \oplus M_n$ *and suppose that, when* $i \neq j$, *the only homomorphism of* M_j *into* M_i *is the null homomorphism. Then the ring of endomorphisms of* M *is isomorphic to* $R_1 \oplus R_2 \oplus \ldots \oplus R_n$, *where* R_i *is the ring of endomorphisms of* M_i.

Proof. If ϕ is an endomorphism of M, then the matrix corresponding to ϕ is a diagonal matrix

$$\begin{Vmatrix} \phi_{11} & 0 & 0 & \cdots & 0 \\ 0 & \phi_{22} & 0 & \cdots & 0 \\ \cdot & \cdot & \cdot & \cdots & \cdot \\ \cdot & \cdot & \cdot & \cdots & \cdot \\ 0 & 0 & 0 & \cdots & \phi_{nn} \end{Vmatrix}.$$

This is because ϕ_{ij} is a null homomorphism whenever $i \neq j$. Let us write $\phi_1, \phi_2, \ldots, \phi_n$ in place of $\phi_{11}, \phi_{22}, \ldots, \phi_{nn}$ and associate with ϕ the sequence $\{\phi_1, \phi_2, \ldots, \phi_n\}$. Since ϕ_i is a homomorphism $M_i \to M_i$ (that is an endomorphism of M_i), it follows that $\{\phi_1, \phi_2, \ldots, \phi_n\}$ belongs to $R_1 \oplus R_2 \oplus \ldots \oplus R_n$. We have, in fact, a one-one correspondence between the endomorphisms of M and the elements of the direct sum. Suppose now that ψ is a second endomorphism and that to ψ corresponds the sequence $\{\psi_1, \psi_2, \ldots, \psi_n\}$. Since diagonal matrices are added and multiplied simply by adding and multiplying their corresponding diagonal elements, we find that to $\phi + \psi$ corresponds $\{\phi_1 + \psi_1, \ldots, \phi_n + \psi_n\}$, which is the sum of $\{\phi_1, \ldots, \phi_n\}$ and $\{\psi_1, \ldots, \psi_n\}$, while to $\phi\psi$ corresponds $\{\phi_1\psi_1, \ldots, \phi_n\psi_n\}$ which is their product. Furthermore, to the identity map of M is matched the sequence formed by the identity maps of the individual M_i. This is just the identity element of the ring $R_1 \oplus R_2 \oplus \ldots \oplus R_n$. Thus the one-one correspondence is indeed a ring-isomorphism and the proof is complete.

Theorem 11. *Let the R-module M be the direct sum of n $(n \geqslant 1)$ submodules each of which is isomorphic to one and the same simple module L. Denote by D the division ring formed by the endomorphisms of L (see Proposition 19). Then the ring of endomorphisms of M is isomorphic to the ring formed by all $n \times n$ matrices with elements in D.*

Proof. Since we have only to determine the ring of endomorphisms of M to within an isomorphism, we may suppose that

$$M = L \oplus L \oplus \ldots \oplus L$$

(n terms). This amounts to assuming that M consists of all sequences formed by n elements of L. The theorem is now an immediate consequence of Proposition 20.

Lemma 9. *Let $M = L_1 \oplus L_2 \oplus \ldots \oplus L_s$, $M' = L_1' \oplus L_2' \oplus \ldots \oplus L_t'$ where $L_1, \ldots, L_s, L_1', \ldots, L_t'$ are simple R-modules. Suppose now that none of $L_1, \ldots, L_s$ is isomorphic to any of $L_1', \ldots, L_t'$. Then the only homomorphism of M into M' is the null homomorphism.*

Proof. We shall assume that $f: M \to M'$ is a non-null homomorphism and derive a contradiction. Choose $x \in M$, say $x = x_1 + x_2 + \ldots + x_s$, where $x_i \in L_i$, such that $f(x) \neq 0$. Then $f(x_1) + f(x_2) + \ldots + f(x_s) \neq 0$ and therefore we can select i so that $f(x_i) \neq 0$. The restriction of f to L_i now yields a non-null homomorphism $\phi: L_i \to M'$.

Choose $y \in L_i$ so that $\phi(y) \neq 0$ and write $\phi(y)$ in the form

$$\phi(y) = x_1' + x_2' + \ldots + x_t', \quad \text{where} \quad x_j' \in L_j'.$$

There will then be an integer j for which $x_j' \neq 0$. Denote by π_j' the projection of M' on to the summand L_j', then $\pi_j' \phi(y) = x_j'$. Thus $\pi_j' \phi$ is a non-null homomorphism of L_i into L_j'. It follows, from Lemma 8, that L_i and L_j' are isomorphic. This is the required contradiction.

Theorem 12. *Let the R-module M be a direct sum of n $(n \geqslant 1)$ simple modules. Then the ring of endomorphisms of M is isomorphic to a ring of the form $R_1 \oplus R_2 \oplus \ldots \oplus R_p$ where (i) each R_i consists of all the $n_i \times n_i$ matrices with elements in a division ring D_i and (ii) $n = n_1 + n_2 + \ldots + n_p$.*

Proof. Suppose that $M = L_1 \oplus L_2 \oplus \ldots \oplus L_n$, where each summand is a simple module. We introduce an equivalence relation between the L_i by putting two of them in the same equivalence class if they are isomorphic. Denote by p the number of equivalence classes and use

$$L_{i1}, L_{i2}, \ldots, L_{in_i} \quad (1 \leqslant i \leqslant p)$$

to describe the modules in the ith class. We then have

$$n = n_1 + n_2 + \ldots + n_p.$$

If now $\qquad M_i = L_{i1} + L_{i2} + \ldots + L_{in_i},$

then this sum is direct (Proposition 13 Cor.) and therefore (Theorem 11) the ring R'_i of endomorphisms of M_i is isomorphic to the ring R_i of all $n_i \times n_i$ matrices with elements in a certain division ring D_i. (D_i may be taken to be the ring of endomorphisms of any of $L_{i1}, L_{i2}, \ldots, L_{in_i}$.) Moreover, Lemma 9 shows that if $i \neq j$, then the only homomorphism of M_i into M_j is the null homomorphism. Also, by Proposition 14 Cor., $M = M_1 \oplus M_2 \oplus \ldots \oplus M_p$. It therefore follows, from the corollary to Proposition 20, that the ring of endomorphisms of M is isomorphic to $R'_1 \oplus R'_2 \oplus \ldots \oplus R'_p$ which itself is isomorphic to

$$R_1 \oplus R_2 \oplus \ldots \oplus R_p.$$

Remarks. The above argument yields rather more information than is actually stated in the theorem. To begin with, Proposition 18 shows that there exists a composition series from M to its zero submodule whose composition factors are isomorphic to $L_1, L_2, \ldots, L_n$; consequently the integer n is the length of M. Next, p is seen to be the number of essentially distinct (i.e. non-isomorphic) composition factors, and if we group together isomorphic composition factors, then $n_1, n_2, \ldots, n_p$ are the numbers of times each of the p types occurs. Finally, the p division rings D_i are obtained (up to isomorphism) by taking one composition factor of each type and forming its ring of endomorphisms.

Theorems 11 and 12 have important applications to the theory which seeks to determine the structure of rings. The possibility of making such applications exists because there is an intimate connection between R and the ring of endomorphisms of R when it is regarded as a left R-module. For let $f : R \to R$ be such an endomorphism. On putting $f(1) = \alpha$ and using the fact that f is R-linear, we see that

$$f(r) = f(r1) = rf(1) = r\alpha$$

for every r in R. Thus the mapping f consists in multiplying each element (of the ring) by α on the *right-hand* side. We have now associated a definite element of R with each endomorphism and this association is such that to different endomorphisms correspond different elements. Next let $\beta \in R$ and for each $r \in R$ put $\phi(r) = r\beta$. If now r, r_1, r_2 are elements of the ring, then

$$\phi(r_1 + r_2) = r_1\beta + r_2\beta = \phi(r_1) + \phi(r_2)$$
and $$\phi(rr_1) = r(r_1\beta) = r\phi(r_1).$$

Thus ϕ is an endomorphism and $\phi(1) = \beta$.

This establishes a one-one correspondence between the endomorphisms $f: R \to R$ and the elements α of R. The correspondence is such that, when f and α are associated, we have $f(r) = r\alpha$ for all $r \in R$. Suppose now that $g: R \to R$ is also an endomorphism and that to g corresponds the element β. Then

$$(f+g)(1) = f(1) + g(1) = \alpha + \beta.$$

Hence *to the sum of two endomorphisms corresponds the sum of the corresponding elements.* Again

$$(fg)(1) = f(g(1)) = f(\beta) = \beta\alpha.$$

This shows that *with the product of f and g is associated the product of the corresponding elements taken in the reverse order.* In other words, R is anti-isomorphic to the ring of endomorphisms. This result is sufficiently important for us to record it as

Theorem 13. *A ring R is anti-isomorphic to its own ring of endomorphisms. Equivalently, if M and R are isomorphic as left R-modules, then the ring of endomorphisms of M is anti-isomorphic to the ring R.*

1.11 Simple and semi-simple rings

The reader will appreciate that the results of the last section give considerable information about those rings which can be expressed as direct sums of simple left ideals. Before attempting to set out this information in an orderly fashion, we shall make a few preliminary observations concerning anti-isomorphisms. In this way we shall avoid becoming submerged in minor details at a critical stage of the discussion.

Given an arbitrary ring R we can form a new ring R^* in the following way. The elements of R^* shall be the same as those of R and addition in R^* shall mean exactly the same as in R itself. Suppose now that α

and β belong to R. They are then both in R^*. The product of α and β, taken in R, will be denoted by $\alpha\beta$. Their product, as elements of R^*, will be written $\alpha * \beta$ in order to avoid confusion. The definition of multiplication in R^* may be expressed by the equation

$$\alpha * \beta = \beta\alpha. \tag{1.11.1}$$

This does, in fact, turn R^* into a ring as the reader can easily verify. The new ring, which is called the *opposite* of R, is characterized by the property that the identity mapping is an anti-isomorphism. The reader should note that the opposite of R^* is the original ring R. He should also note that the opposite of a division ring is again a division ring.

Suppose now that R is a ring and $n\,(n \geqslant 1)$ is an integer. Consider the set of all $n \times n$ matrices with elements in R. Such matrices can be added and multiplied in a familiar manner and thus they give rise to a ring $\mathfrak{M}_n(R)$. This will be called the *full matrix ring* (of order n) over R. If $U = \|u_{ij}\|\,(1 \leqslant i,j \leqslant n)$ belongs to $\mathfrak{M}_n(R)$, then U^T will denote the *transpose* of U, that is the matrix one obtains from U by interchanging its rows and columns.

Lemma 10. *Let R^* denote the opposite of the ring R. Then the mapping $U \to U^T$ is an anti-isomorphism of $\mathfrak{M}_n(R)$ on to $\mathfrak{M}_n(R^*)$.*

Proof. We shall only verify that the mapping has the appropriate property in respect of multiplication. The other details are trivial. Let $U = \|u_{ij}\|$, $V = \|v_{ij}\|$ belong to $\mathfrak{M}_n(R)$ and put $W = UV$. Then the (i,j)th element of W^T (that is the element in its ith row and jth column) is the (j,i)th element of UV or $\sum_k u_{jk}v_{ki} = \sum_k v_{ki} * u_{jk}$. Let $U^T = \|u'_{ij}\|$, $V^T = \|v'_{ij}\|$. Then the (i,j)th element of W^T can now be written as $\sum_k v'_{ik} * u'_{kj}$. However, this is just the (i,j)th element of $V^T * U^T$, where $V^T * U^T$ denotes the product of V^T and U^T when these are regarded as elements of $\mathfrak{M}_n(R^*)$. Thus $W^T = V^T * U^T$ or $(UV)^T = V^T * U^T$. This completes the verification.

Another useful result concerning anti-isomorphisms is

Lemma 11. *Suppose that, for $1 \leqslant i \leqslant p$, the ring R_i is anti-isomorphic to the ring R'_i. Then the direct sum $R_1 \oplus R_2 \oplus ... \oplus R_p$ is anti-isomorphic to $R'_1 \oplus R'_2 \oplus ... \oplus R'_p$.*

Proof. Let σ_i be an anti-isomorphism of R_i on to R'_i. Since a typical element of $R_1 \oplus R_2 \oplus ... \oplus R_p$ is a sequence $\{r_1, r_2, ..., r_p\}$ where $r_i \in R_i$, we obtain a mapping $\sigma : R_1 \oplus ... \oplus R_p \to R'_1 \oplus ... \oplus R'_p$ by putting

$$\sigma(\{r_1, r_2, ..., r_p\}) = \{\sigma_1(r_1), \sigma_2(r_2), ..., \sigma_p(r_p)\}.$$

This, we contend, is an anti-isomorphism. For, with self-explanatory notation,

$$\sigma(\{r_1, \ldots, r_p\}\{r_1', \ldots, r_p'\}) = \sigma(\{r_1 r_1', \ldots, r_p r_p'\})$$
$$= \{\sigma_1(r_1 r_1'), \ldots, \sigma_p(r_p r_p')\}$$
$$= \{\sigma_1(r_1')\,\sigma_1(r_1), \ldots, \sigma_p(r_p')\,\sigma_p(r_p)\}$$
$$= \{\sigma_1(r_1'), \ldots, \sigma_p(r_p')\}\{\sigma_1(r_1), \ldots, \sigma_p(r_p)\}$$

and this may be written as

$$\sigma(\{r_1, \ldots, r_p\}\{r_1', \ldots, r_p'\}) = \sigma(\{r_1', \ldots, r_p'\})\,\sigma(\{r_1, \ldots, r_p\}).$$

Thus σ reverses the order of products. The remaining details of the verification are even more trivial.

We come now to the concepts from which this section gets its heading.

Definition. *A non-null ring R which is expressible as a direct sum of simple left ideals, is said to be ' (left) semi-simple'. If the simple left ideals which occur in such a direct sum representation are mutually isomorphic, then the ring is said to be ' (left) simple'.*

Remarks. There are a number of observations to be made. First, there are other definitions of semi-simple and simple rings which turn out to be equivalent to those given here. The definitions actually chosen are the ones which connect most naturally with the ideas developed in the preceding sections. Next, we can define (right) semi-simple and (right) simple rings by using right ideals in place of left ideals and these will naturally have analogous properties. The reason why 'left' and 'right' have been put inside brackets is because (as we shall see) a ring which is left semi-simple is also right semi-simple and vice versa. Likewise a ring which is left simple is also right simple. Hence, in the long run, there is no need to make a careful distinction, but until these basic facts have been established it is necessary to use a terminology which makes clear which of two (apparently distinct) alternatives we have in mind.

Suppose that the ring R is left semi-simple. Then R is a direct sum of simple left ideals. Now R, considered as a left R-module, is generated by the element 1 because every element of R can be written in the form $r1$. It follows, from Proposition 16, that when R is written as a direct sum of simple left ideals, there can only be a finite number of summands. Let $R = L_1 \oplus L_2 \oplus \ldots \oplus L_n$ where each L_i is a simple left ideal. By Proposition 18, there exists a composition series, from R to the zero ideal, whose composition factors are isomorphic to $L_1, L_2, \ldots, L_n$. It follows, from Theorem 5, that the L_i are uniquely determined up to order and isomorphism. In particular, the number n is uniquely determined because it is equal to $L_R(R)$. Next assume that the ring R is *simple*. Then the L_i are isomorphic to each other. Indeed, by Theorem 9, *any two simple R-modules are isomorphic.*

Theorem 14. *Let R be a (left) simple ring, L a simple R-module, D the division ring of endomorphisms of L and n the length of R considered as left R-module. Then R is ring-isomorphic to $\mathfrak{M}_n(D^*)$, where D^* denotes the division ring opposite to D.*

Proof. Let $R = L_1 \oplus L_2 \oplus \ldots \oplus L_n$, where the L_i are simple left ideals. Then each L_i is isomorphic to L because any two simple modules are isomorphic. Next, by Theorem 11, the ring of endomorphisms of R is isomorphic to $\mathfrak{M}_n(D)$ and hence, by Lemma 10, anti-isomorphic to $\mathfrak{M}_n(D^*)$. On the other hand, Theorem 13 shows that the ring of endomorphisms is anti-isomorphic to R. Finally, by combining the two anti-isomorphisms, we obtain an isomorphism between R and $\mathfrak{M}_n(D^*)$.

Now let Δ be an arbitrary division ring and n a positive integer. It will be shown that $\mathfrak{M}_n(\Delta)$ is both left simple and right simple. When this is taken in conjunction with Theorem 14, we obtain Wedderburn's Theorem namely: *a ring is simple if and only if it is isomorphic to a full matrix ring over a division ring.*

In order to discuss the matrix ring, we introduce some new notation. For $1 \leqslant i, j \leqslant n$, e_{ij} will denote the $n \times n$ matrix which has the identity element (of the division ring Δ) in the ith row and jth column and which has zeros elsewhere. Using a self-explanatory notation, we see that each element of $\mathfrak{M}_n(\Delta)$ has a unique representation in the form $\sum_{i,j} \alpha_{ij} e_{ij}$, where α_{ij} are elements of Δ. Note that, if α, β belong to Δ, then

$$(\beta e_{pq})(\alpha e_{ij}) = \left. \begin{matrix} 0 & (q \neq i) \\ (\beta\alpha)e_{pj} & (q = i) \end{matrix} \right\}. \tag{1.11.2}$$

We first show that the matrix ring is *left* simple.

Theorem 15. *Let Δ be a division ring. Then $\mathfrak{M}_n(\Delta)$ is the direct sum of n mutually isomorphic simple left ideals and hence is a (left) simple ring. If L is any simple left $\mathfrak{M}_n(\Delta)$-module, then the ring of endomorphisms of L is anti-isomorphic to Δ.*

Proof. Let L_j $(1 \leqslant j \leqslant n)$ consists of all those matrices which can be expressed in the form $\sum_i \alpha_i e_{ij}$, where each α_i belongs to Δ. Then L_j is composed of those matrices whose columns, with the possible exception of the jth, are composed of zeros. Clearly L_j is a left ideal. Also if $1 \leqslant j,\, k \leqslant n$, then the mapping $L_j \to L_k$, which carries $\sum_i \alpha_i e_{ij}$ into $\sum_i \alpha_i e_{ik}$, is an isomorphism of left $\mathfrak{M}_n(\Delta)$-modules. Note that the ring is the direct sum of $L_1, L_2, \ldots, L_n$.

Now assume that $z = \sum_i \alpha_i e_{ij}$ is an arbitrary non-zero element of L_j. It is possible to choose s so that $\alpha_s \neq 0$ and then, by (1.11.2),

$$(\beta \alpha_s^{-1} e_{ps}) z = \beta e_{pj} \quad (\beta \in \Delta,\ 1 \leqslant p \leqslant n).$$

Thus βe_{pj} belongs to the submodule generated by z. Since every element of L_j is a sum of elements of the form βe_{pj}, z generates L_j. It follows that L_j has no proper submodules other than the zero submodule. Hence L_j is simple. To complete the proof, we need only show that the ring of endomorphisms of L_j is anti-isomorphic to Δ.

Let $\phi : L_j \to L_j$ be an endomorphism. Then $\phi(e_{1j})$ belongs to L_j, say $\phi(e_{1j}) = \alpha_1 e_{1j} + \ldots + \alpha_n e_{nj}$. For a general element $x = \sum_p \beta_p e_{pj}$ of L_j, we have

$$\phi(x) = \phi\big((\textstyle\sum_p \beta_p e_{p1}) e_{1j}\big)$$

$$= (\textstyle\sum_p \beta_p e_{p1})\, \phi(e_{1j})$$

$$= (\textstyle\sum_p \beta_p e_{p1})\, (\alpha_1 e_{1j} + \ldots + \alpha_n e_{nj})$$

$$= \textstyle\sum_p \beta_p \alpha_1 e_{pj}$$

by virtue of (1.11.2). If therefore we write α in place of α_1, then we can express this relation in the form

$$\phi(x) = x\alpha. \tag{1.11.3}$$

Thus operating with ϕ has the same effect as multiplying on the right by α. Clearly α is uniquely determined by ϕ and each element of Δ, when used as a right multiplier, induces an endomorphism of L_j. Accordingly we have a one-one correspondence between the endomorphisms ϕ of L_j and the elements α of Δ.

Finally, let ϕ, ψ be endomorphisms of L_j and α, β the corresponding elements of Δ. Then

$$(\phi + \psi)(x) = \phi(x) + \psi(x) = x\alpha + x\beta = x(\alpha + \beta),$$

$$(\phi\psi)(x) = \phi(\psi(x)) = \phi(x\beta) = (x\beta)\alpha = x(\beta\alpha).$$

Thus to $\phi + \psi$ corresponds $\alpha + \beta$ while the element associated with $\phi\psi$ is $\beta\alpha$. The correspondence is therefore an anti-isomorphism and the proof is complete.

It will now be clear that $\mathfrak{M}_n(\Delta)$ must be right simple as well as left simple. However, it is misleading to say that the theorem remains true if 'left' is replaced by 'right' whenever it occurs. In order to clear up any possible misunderstanding on this point, we state the facts in full.

Corollary. *Let Δ be a division ring. Then $\mathfrak{M}_n(\Delta)$ can be expressed as the direct sum of n mutually isomorphic simple right ideals and hence is a right simple ring. If Λ is any simple right module over $\mathfrak{M}_n(\Delta)$, then its ring of endomorphisms is isomorphic to Δ.*

The point to note is that, besides replacing 'left' by 'right', we must also change 'anti-isomorphic' to 'isomorphic'. Briefly, if Λ_j consists of all matrices of the form $\sum_i \alpha_i e_{ji}$, then the $\Lambda_j (1 \leqslant j \leqslant n)$ are mutually isomorphic simple right ideals and $\mathfrak{M}_n(\Delta)$ is their direct sum. In addition, if $\phi \colon \Lambda_j \to \Lambda_j$ is an endomorphism, then there is a unique α in Δ such that $\phi(x) = \alpha x$ for all x belonging to Λ_j. It is because ϕ and α occur on the same side of x, in this relation, that we get an isomorphism rather than an anti-isomorphism.

It is now possible to fill in a number of details which previously were missing.

Theorem 16. *Let R be a left simple ring. Then it is also a right simple ring and its length as a left R-module is equal to its length as a right R-module.*

Proof. R is ring-isomorphic† to a matrix ring $\mathfrak{M}_n(\Delta)$, where Δ is a division ring. Since $\mathfrak{M}_n(\Delta)$ is right simple (Theorem 15 Cor.), so is R. Again the length of $\mathfrak{M}_n(\Delta)$ is n whether we regard it as a left module or a right module. Consequently the two lengths are also equal in the case of the ring R. This completes the proof.

From now on we shall speak only of simple rings (rather than left simple and right simple rings) since a possible source of confusion has been ruled out. Theorem 14 tells us that a simple ring can be represented as a matrix ring of a certain kind, but we have not considered in what respects the representation is *unique*. However, this question is readily settled.

Theorem 17. *Let Δ and Δ' be division rings. Then $\mathfrak{M}_n(\Delta)$ is isomorphic to $\mathfrak{M}_p(\Delta')$ if and only if $n = p$ and Δ is isomorphic to Δ'.*

† See Theorem 14.

Proof. Let $R = \mathfrak{M}_n(\Delta)$, $R' = \mathfrak{M}_p(\Delta')$ and suppose that we have a ring-isomorphism of R on to R'. It is clear that if each of the rings is considered as a left module with respect to itself, then $L_R(R) = L_{R'}(R')$. However, by Theorem 15, $L_R(R) = n$ and $L_{R'}(R') = p$. Consequently $n = p$.

Now let L be a simple left ideal of R. The isomorphism between R and R' maps L on to a simple left ideal L' of R'. The correspondence now extends to an isomorphism between the ring of endomorphisms of L and the ring of endomorphisms of L'. By Theorem 15, the first ring of endomorphisms is anti-isomorphic to Δ while the second is anti-isomorphic to Δ'. It follows that Δ and Δ' are isomorphic. One half of the theorem is now proved and the other half is completely trivial.

These results will now be applied to semi-simple rings. First we prove

Lemma 12. *Let $R_1, R_2, ..., R_p$ be left semi-simple rings. Then their direct sum $R_1 \oplus R_2 \oplus ... \oplus R_p$ is also left semi-simple. The lemma also holds if 'left' is everywhere replaced by 'right'.*

Proof. Let $R = R_1 \oplus R_2 \oplus ... \oplus R_p$, then R consists of all sequences $\{r_1, r_2, ..., r_p\}$, where r_i belongs to R_i. If R'_i denotes the set of all sequences of the form $\{0, ..., r_i, ..., 0\}$ where the element r_i occurs in the ith position, then, as was remarked in section (1.10), R'_i is a two-sided ideal of R and it is also ring-isomorphic to R_i.

Suppose now that $\{0, ..., \rho_i, ..., 0\}$ is an arbitrary element of R'_i and $\{r_1, r_2, ..., r_p\}$ is an arbitrary element of R. Then

$$\{r_1, ..., r_i, ..., r_p\}\{0, ..., \rho_i, ..., 0\} = \{0, ..., r_i\rho_i, ..., 0\}$$
$$= \{0, ..., r_i, ..., 0\}\{0, ..., \rho_i, ..., 0\}.$$

It follows that a subset of R'_i is a left ideal of R if and only if it is a left ideal of the ring R'_i. Consequently, *if L' is a simple left ideal of R'_i then it is also a simple left ideal of R.*

Since the ring R'_i is isomorphic to R_i it must be semi-simple. Accordingly

$$R'_i = L'_{i1} \oplus L'_{i2} \oplus ... \oplus L'_{in_i},$$

where each L'_{ij} is a simple left ideal of R'_i and hence a simple left ideal of R. Furthermore, R is the (internal) direct sum $R'_1 \oplus R'_2 \oplus ... \oplus R'_p$. It follows, from Proposition 15, that R is the direct sum of all the L'_{ij} and hence it is semi-simple. The proof is now complete.

Theorem 18. *In order that R should be a left semi-simple ring it is necessary and sufficient that R be isomorphic to the direct sum of a finite number of simple rings.*

Proof. If R is isomorphic to such a direct sum, then it must be left semi-simple by virtue of Lemma 12. On the other hand, if R is a direct sum of simple left ideals then, as we saw earlier, there will only be a finite number of summands. By Theorem 12, the ring $\tilde{R}$ (say) of endomorphisms of R (considered as a left R-module) is isomorphic to a direct sum in which each term has the form $\mathfrak{M}_p(\Delta)$, where Δ is a division ring. It follows, from Lemmas 10 and 11, that $\tilde{R}$ is anti-isomorphic to the direct sum of the rings $\mathfrak{M}_p(\Delta^*)$, where Δ^* denotes the division ring opposite to Δ. Finally, by Theorem 13, $\tilde{R}$ is anti-isomorphic to R. It follows that R is isomorphic to the direct sum of the rings $\mathfrak{M}_p(\Delta^*)$ and, as each of these is simple (Theorem 15), the proof is complete.

Corollary. *If R is a left semi-simple ring, then it is also right semi-simple.*

Proof. R is isomorphic to a direct sum $R_1 \oplus R_2 \oplus \dots \oplus R_p$, where each R_i is simple. Since a simple ring is both left simple and right simple, it follows, from Lemma 12, that $R_1 \oplus R_2 \oplus \dots \oplus R_p$ is right semi-simple. This, however, implies that R is right semi-simple.

Theorem 19. *Let R be a semi-simple ring and $M \neq 0$ a finitely generated R-module. Then M is a finite direct sum of simple submodules.*

Proof. Let $R = L_1 \oplus L_2 \oplus \dots \oplus L_n$, where each L_i is a simple (left) ideal, and let $x_1, x_2, \dots, x_m$ generate the (left) R-module M. For each pair i, j $(1 \leqslant i \leqslant n, \ 1 \leqslant j \leqslant m)$, let $f_{ij}: L_i \to M$ be the mapping which carries $\alpha_i \in L_i$ into $\alpha_i x_j$. This is an R-homomorphism. Put $N_{ij} = \mathrm{Im}\,(f_{ij})$. Then, since $\mathrm{Ker}\,(f_{ij})$ must be either L_i or 0, either (i) N_{ij} is isomorphic to L_i and hence is simple or (ii) $N_{ij} = 0$. Clearly $\sum_{i,j} N_{ij} = M$ and, in the sum, we may strike out any zero modules.

It follows that we can find simple submodules $N_1, N_2, \dots, N_p$ such that $M = N_1 + N_2 + \dots + N_p$. Let us arrange that p is as small as possible. With this extra condition we contend that the *sum*

$$N_1 + N_2 + \dots + N_p$$

is direct. Indeed, by Proposition 14, it is enough to show that, for each integer k $(1 \leqslant k \leqslant p)$,

$$N_k \cap (N_1 + \dots + N_{k-1} + N_{k+1} + \dots + N_p) = 0.$$

But if the intersection is not the zero submodule, then it must be N_k (because N_k is simple) and therefore

$$N_k \subseteq N_1 + \dots + N_{k-1} + N_{k+1} + \dots + N_p.$$

Thus $M = N_1 + \ldots + N_{k-1} + N_{k+1} + \ldots + N_p$ contradicting the choice of p. This completes the proof.

The theory of semi-simple rings is extensive but rather specialized from our present point of view. We shall therefore break off the discussion at this point in order to return to more general ideas.

1.12 Exact sequences and commutative diagrams

Let M, N and P be modules with respect to a ring R, and suppose that we have homomorphisms $\phi: M \to N$ and $\psi: N \to P$. It is convenient to represent this situation by means of the diagram

$$M \overset{\phi}{\to} N \overset{\psi}{\to} P. \tag{1.12.1}$$

Such a diagram will usually be referred to as a *three term sequence*.

Definition. *The three term sequence* (1.12.1) *is said to be 'exact' if* $\operatorname{Im} \phi = \operatorname{Ker} \psi$.

For example, if K is a submodule of M, then

$$K \to M \to M/K$$

is exact, where the first mapping is the inclusion mapping and the second is the natural mapping of M on to its factor module M/K. Note that if 0 denotes a zero module, then there is only one homomorphism of 0 into M and only one homomorphism of M into 0. We may therefore speak of the homomorphisms $0 \to M$ and $M \to 0$ without any possibility of ambiguity.

Now consider a three term sequence of the form

$$0 \to M \overset{\phi}{\to} N. \tag{1.12.2}$$

This is exact if and only if the kernel of ϕ is the image of the mapping $0 \to M$, i.e. it is exact precisely when the kernel of ϕ is the zero submodule of M. Accordingly (1.12.2) *is exact when and only when* ϕ *is a monomorphism.* By contrast

$$M \overset{\psi}{\to} N \to 0 \tag{1.12.3}$$

is exact if and only if $\operatorname{Im} \psi$ is the kernel of $N \to 0$. But $\operatorname{Ker}(N \to 0) = N$. Consequently (1.12.3) *is exact when and only when* ψ *is an epimorphism.* For example, when K is a submodule of M, both $0 \to K \to M$ and $M \to M/K \to 0$ are exact, where $K \to M$ and $M \to M/K$ have their usual meanings.

The term *exact* can also be applied to sequences with more than three terms. Suppose that

$$\ldots \to M_{n+2} \to M_{n+1} \to M_n \to M_{n-1} \to \ldots \qquad (1.12.4)$$

is such a sequence. (It may be finite in length or extend to infinity in one or both directions.) If we say that (1.12.4) is exact, then it is to be understood that each triplet $M_{n+1} \to M_n \to M_{n-1}$ (consisting of three consecutive terms) is exact in the sense already explained. For example, if K is a submodule of M, then

$$0 \to K \to M \to M/K \to 0 \qquad (1.12.5)$$

is exact by virtue of the remarks made earlier.

Let us now consider the most general exact sequence of the form

$$0 \to M \overset{\phi}{\to} N \overset{\psi}{\to} P \to 0. \qquad (1.12.6)$$

Because $0 \to M \to N$ is exact, ϕ is a monomorphism. Consequently if $\phi(M) = M'$, then M' is a submodule of N and one derives (by restricting the range of ϕ) an isomorphism $\phi': M \to M'$ of M on to M'. Let $\alpha: M' \to N$ be the inclusion mapping and i_N the identity mapping of N. Then $\alpha\phi' = \phi = i_N\phi$. The relation $\alpha\phi' = i_N\phi$ may be described by saying that the diagram

$$
\begin{array}{ccc}
M & \overset{\phi}{\longrightarrow} & N \\
{\scriptstyle \phi'}\downarrow & & \downarrow{\scriptstyle i_N} \\
M' & \underset{\alpha}{\longrightarrow} & N
\end{array}
$$

is *commutative*. This idea of a commutative diagram is so useful that we shall take this opportunity to comment upon it.

A rectangular array

$$
\begin{array}{ccc}
A & \overset{h}{\longrightarrow} & B \\
{\scriptstyle f}\downarrow & & \downarrow{\scriptstyle g} \\
X & \underset{k}{\longrightarrow} & Y
\end{array}
$$

of sets and mappings is said to be commutative if $gh = kf$. Likewise the triangular diagram is termed commutative if $\chi = \psi\phi$. The same terminology is also used in more complicated situations, but the reader will find that the more elaborate diagrams are always simple networks of rectangles and triangles. The statement that such a diagram is commutative means always that the *small* component rectangles and triangles are commutative in the above sense. Although this is a somewhat casual explanation, it will be perfectly clear, in each individual case, what is intended. For this reason there is no need to analyse the concept any further.

Let us return to the consideration of the exact sequence (1.12.6). Since $N \to P \to 0$ is exact, ψ is an epimorphism and, since $M \to N \to P$ is exact, $\operatorname{Ker}\psi = \operatorname{Im}\phi = M'$. Accordingly, by Theorem 1, ψ induces an isomorphism $\psi^* : N/M' \to P$ of N/M' on to P. Denote by $\beta : N \to N/M'$ the natural mapping. Then the definition of ψ^* ensures that the diagram

$$
\begin{array}{ccc}
& N/M' & \\
{}^{\beta}\nearrow & & \searrow{}^{\psi^*} \\
N & \xrightarrow[\psi]{} & P
\end{array}
$$

is commutative. In symbols, $\psi^*\beta = \psi$ whence $\beta = \psi^{*-1}\psi$, where ψ^{*-1} denotes the isomorphism inverse to ψ^*. We now have the following configuration of modules and homomorphisms:

$$
\begin{array}{ccccccccc}
0 & \longrightarrow & M & \xrightarrow{\phi} & N & \xrightarrow{\psi} & P & \longrightarrow & 0 \\
& & \downarrow{\phi'} & & \downarrow{i_N} & & \downarrow{\psi^{*-1}} & & \\
0 & \longrightarrow & M' & \xrightarrow{\alpha} & N & \xrightarrow{\beta} & N/M' & \longrightarrow & 0
\end{array}
\qquad (1.12.7)
$$

This is a commutative diagram, because $i_N\phi = \alpha\phi'$ and $\psi^{*-1}\psi = \beta i_N$, in which the two rows are exact and the vertical mappings, namely ϕ', i_N and ψ^{*-1}, are isomorphisms. This may be interpreted as meaning that the two exact sequences

$$
0 \to M \to N \to P \to 0
$$

and
$$
0 \to M' \to N \to N/M' \to 0
$$

are virtually identical so that each can be regarded as a copy of the other. In more concrete terms, if we have to prove something about $0 \to M \to N \to P \to 0$, then, because of the properties of (1.12.7), it normally suffices to prove the corresponding result for the second sequence. Thus when we are concerned with an exact sequence of the form $0 \to M \to N \to P \to 0$ it is usually permissible to treat M as an actual submodule of N and to regard P as the factor module N/M. For example, Theorem 6 tells us that

$$L_R(N) = L_R(M') + L_R(N/M');$$

consequently $\quad\quad L_R(N) = L_R(M) + L_R(P).$

This property of the exact sequence (1.12.6) is generalized in

Theorem 20. *Let* $0 \to E_s \to E_{s-1} \to \ldots \to E_0 \to 0$ *be an exact sequence in which each term is an R-module of finite length. Then*

$$\sum_{\mu=0}^{s} (-1)^{\mu} L_R(E_{\mu}) = 0.$$

Proof. If $s = 0$, then the fact that $0 \to E_0 \to 0$ is exact means simply that E_0 is a zero module. If $s = 1$, then the exactness of

$$0 \to E_1 \to E_0 \to 0$$

shows that $E_1 \to E_0$ is an isomorphism of E_1 on to E_0. In each of these two cases the assertion of the theorem is clearly true. Again, if $s = 2$ then we have an exact sequence

$$0 \to E_2 \to E_1 \to E_0 \to 0$$

and therefore, by our earlier remarks, $L_R(E_1) = L_R(E_0) + L_R(E_2)$ which is precisely what we wish to prove. The remainder of the argument uses induction on s. We therefore suppose that $s > 2$ and that the theorem has been proved for all smaller values of the inductive variable.

Put $E'_{s-1} = \mathrm{Im}\,(E_{s-1} \to E_{s-2})$, then E'_{s-1} has finite length because it is a submodule of E_{s-2}. Again, since

$$\mathrm{Ker}\,(E_{s-2} \to E_{s-3}) = \mathrm{Im}\,(E_{s-1} \to E_{s-2}) = E'_{s-1},$$

the sequence $\quad 0 \to E'_{s-1} \to E_{s-2} \to \ldots \to E_1 \to E_0 \to 0$

is exact, where $E'_{s-1} \to E_{s-2}$ is the inclusion mapping. It therefore follows, from the inductive hypothesis, that

$$\sum_{\mu=0}^{s-2} (-1)^{\mu} L_R(E_{\mu}) + (-1)^{s-1} L_R(E'_{s-1}) = 0. \quad\quad (1.12.8)$$

Again, the mapping $E_{s-1} \to E_{s-2}$ carries E_{s-1} on to E'_{s-1} and has $\mathrm{Im}\,(E_s \to E_{s-1})$ as its kernel. It therefore gives rise to the short exact sequence $\quad\quad 0 \to E_s \to E_{s-1} \to E'_{s-1} \to 0.$

From this we conclude that $L_R(E'_{s-1}) = L_R(E_{s-1}) - L_R(E_s)$ and now the required result follows by substituting for $L_R(E'_{s-1})$ in (1.12.8).

1.13 Free modules

If N is a given R-module, then we can construct new modules by forming direct sums in which each summand is isomorphic to N. In particular, this construction may be carried out using R itself as the given module. This leads to the important concept of a 'free module'. However, there is a different and more convenient way of approaching the same idea. The details are set out below.

Let M be an R-module and $\{x_i\}_{i \in I}$ a family of elements of M.

Definition. *The family* $\{x_i\}_{i \in I}$ *is said to constitute a 'base' for* M *if each element x of M has a unique representation in the form* $x = \sum_i r_i x_i$, *where r_i belongs to R and $r_i = 0$ for almost all i in I.*

Remarks. It is clear that if $\{x_i\}_{i \in I}$ is a base for M, then the submodule generated by the x_i is M itself. Accordingly $\{x_i\}_{i \in I}$ is necessarily a system of generators for the module. The notion of a base goes beyond that of a system of generators in requiring that the representation $x = \sum_i r_i x_i$ always be *unique*. Note that uniqueness has to be understood in the following sense: *if*

$$x = \Sigma r_i x_i = \Sigma r_i' x_i,$$

where r_i, r_i' belong to R and are zero for almost all i, then $r_i = r_i'$ for every i in I. This, of course, is stronger than the assertion that $r_i x_i = r_i' x_i$ for each index i. Note that if the zero element has a unique representation, then no element can be represented in more than one way.

Not every R-module possesses a base.

Definition. *An R-module which possesses a base is said to be 'free'.*

It is convenient to regard a zero module as being free with the empty set forming a base. Note that, considered as left R-module, R itself is free. Indeed the identity element of R is a base in this case.

If $u_1, u_2, \ldots, u_n$ are elements of an R-module M, then the submodule which they generate consists of all elements of the form

$$r_1 u_1 + r_2 u_2 + \ldots + r_n u_n,$$

where the r_i $(1 \leqslant i \leqslant n)$ belong to R. It is sometimes convenient to to denote this submodule by $Ru_1 + Ru_2 + \ldots + Ru_n$. In particular, the submodule generated by a single element u may be denoted by Ru.

Proposition 21. *Let $\{x_i\}_{i \in I}$ be a base of a free R-module M. Then*

$$M = \bigoplus_{i \in I} Rx_i$$

and each summand Rx_i is isomorphic to R considered as a left R-module.

Proof. Since the x_i generate M, we have $M = \sum_i Rx_i$. Denote by f_i the mapping $R \to Rx_i$ which satisfies $f_i(r) = rx_i$. This is an epimorphism of left R-modules and, since the x_i form a base, $f_i(r) = 0$ only when $r = 0$. Accordingly f_i is an isomorphism of R on to Rx_i.

Finally suppose that $\sum_i y_i = 0$, where $y_i \in Rx_i$ and $y_i = 0$ for almost all i. For each i we can find $r_i \in R$ so that $y_i = r_i x_i$. However, as we noted earlier, $r_i x_i = 0$ only when $r_i = 0$. Thus $r_i = 0$ for almost all i and $\sum_i r_i x_i = 0$. The fact that the x_i form a base now shows that $r_i = 0$ for *all* i in I. Accordingly $y_i = 0$ for every i and therefore the sum $\sum_i Rx_i$ is direct.

Proposition 22. *Let $\{N_\lambda\}_{\lambda \in \Lambda}$ be a family of submodules of an R-module M and suppose that M is their direct sum. Suppose further that, for each $\lambda \in \Lambda$, N_λ is free and has the elements $\{x_{i\lambda}\}_{i \in I_\lambda}$ as a base. Then M is also free and has as a base the elements $x_{i\lambda}$ ($\lambda \in \Lambda, i \in I_\lambda$).*

Proof. Proposition 21 shows that, for each λ, N_λ is the direct sum of the modules $Rx_{i\lambda}$ ($i \in I_\lambda$). Since M is the direct sum of the N_λ, it follows, by Proposition 15, that

$$M = \bigoplus_{\lambda \in \Lambda, \, i \in I_\lambda} Rx_{i\lambda}. \tag{1.13.1}$$

Suppose now that $\Sigma r_{i\lambda} x_{i\lambda} = 0$, where $r_{i\lambda}$ belongs to R and the sum contains only a finite number of non-zero terms. By (1.13.1),

$$r_{i\lambda} x_{i\lambda} = 0$$

for all admissible pairs i, λ. Choose one such pair. Since $r_{i\lambda} x_{i\lambda} = 0$ and $x_{i\lambda}$ is a basis element for N_λ, it follows that $r_{i\lambda} = 0$. This establishes the proposition.

Propositions 21 and 22 together show that an R-module is free when and only when it is a direct sum of modules each isomorphic to R. As remarked earlier, this fact might have been used to define free modules. However the method adopted has the advantage of introducing the idea of a base. Clearly it is possible to construct very large free modules. We shall now explore the possibilities in this respect.

Let $\{X_i\}_{i \in I}$ be a family of symbols and denote by M the set of all *formal* sums $\sum_{i \in I} r_i X_i$, where r_i belongs to R and is zero for almost all i. Suppose that $\sum_i r_i X_i$ and $\sum_i r_i' X_i$ are two such sums. Put

$$\sum_i r_i X_i + \sum_i r_i' X_i = \sum_i (r_i + r_i') X_i$$

and, for $r \in R$, define $r(\sum_i r_i X_i)$ by

$$r(\sum_i r_i X_i) = \sum_i (r r_i) X_i.$$

It is easy to verify that these laws of composition turn M into an R-module, the zero element of which is just that formal sum in which every X_i has zero as its coefficient.

Let δ_{ij}, where i and j belong to I, be the *Kronecker symbol*, that is to say δ_{ij} stands for 1_R if $i = j$ and for 0_R if $i \neq j$. Then $\bar{X}_i = \sum_j \delta_{ij} X_j$ is an element of M. Indeed, if r_i belongs to R for all i in I and $r_i = 0$ for almost all i, then the proper sum $\sum_i r_i \bar{X}_i$ coincides with the formal sum $\sum_i r_i X_i$. Thus each element of M has a unique representation in the form $\sum_i r_i \bar{X}_i$ and therefore $\{\bar{X}_i\}_{i \in I}$ is a base for the module. We call M the *free R-module on the symbols* $\{X_i\}_{i \in I}$.

In practice we usually identify $\bar{X}_i$ with X_i, i.e. we use the symbol X_i for both and leave the context to explain which is the appropriate interpretation.† On this understanding, M is a free module with $\{X_i\}_{i \in I}$ as base.

Suppose now that M is a free module with a base $\{x_i\}_{i \in I}$ and N an arbitrary R-module. Let $\{y_i\}_{i \in I}$ be a family of elements of N indexed by the same set I as is used to label the base of M. There is then a unique homomorphism $f : M \to N$ with the property that $f(x_i) = y_i$ for all i. Indeed, if $\sum_i r_i x_i$ denotes a typical element of M, then the homomorphism is given by $f(\sum_i r_i x_i) = \sum_i r_i y_i$. Another way of describing this result is to say that *each mapping of the base into N has a unique extension to a homomorphism of M into N*. The reader should note that this is a special property of a *base* which is not shared by more general kinds of systems of generators.

---

† This identification requires that R should not be a null ring, because otherwise the $\bar{X}_i$ will all be the same element.

Proposition 23. *Given any R-module M, it is always possible to construct an exact sequence* $0 \to K \to F \to M \to 0$

in which F is a free module.

Proof. Let $\{x_i\}_{i \in I}$ be any system of generators for M. (Such a system certainly exists. For example, the family might consist of *all* the elements of M.) Next let $\{X_i\}_{i \in I}$ be a family of new symbols but with the same index set, and let F be the free R-module on these symbols. The remarks made immediately before the statement of the proposition show that there exists a homomorphism $f: F \to M$ for which $f(\bar{X}_i) = x_i$, where $\bar{X}_i$ has the same meaning as before. Thus all the x_i belong to Imf and therefore Im$f = M$. Accordingly $F \to M \to 0$ is exact. Put $K = \text{Ker} f$. Then $0 \to K \to F \to M \to 0$ is exact, where $K \to F$ denotes the inclusion mapping.

Corollary 1. *Every module is isomorphic to a factor module of a free module.*

Corollary 2. *If the R-module M can be generated by n $(n \geqslant 0)$ elements, then it is possible to construct an exact sequence $0 \to K \to F \to M \to 0$, where F is a free module with a base of n elements.*

This follows from the method by which the proposition was established.

Theorem 21. *Let D be a division ring and M a finitely generated D-module. Then M is a free D-module with a finite base. If $u_1, u_2, \ldots, u_n$ is a base of M, then $L_D(M) = n$.*

Proof. Let $v_1, v_2, \ldots, v_m$ be elements which generate M, so chosen that m is as small as possible, and assume that

$$\alpha_1 v_1 + \alpha_2 v_2 + \ldots + \alpha_m v_m = 0,$$

where each α_i belongs to D. *We contend that all the α_i are zero.* For if, for example, $\alpha_m \neq 0$ then, because D is a division ring, α_m has an inverse α_m^{-1}. Accordingly

$$v_m = (-\alpha_m^{-1} \alpha_1) v_1 + \ldots + (-\alpha_m^{-1} \alpha_{m-1}) v_{m-1}.$$

Thus the submodule generated by $v_1, v_2, \ldots, v_{m-1}$ contains v_m and therefore it contains the submodule generated by $v_1, v_2, \ldots, v_{m-1}, v_m$, namely M. It follows that $v_1, v_2, \ldots, v_{m-1}$ is a system of generators of M and this contradicts the choice of m.

It has now been shown that $\alpha_1 v_1 + \alpha_2 v_2 + \dots + \alpha_m v_m = 0$ only when all the α_i are zero. Since the v_i generate M, this proves that they form a base.

Finally, let $u_1, u_2, \dots, u_n$ be a base for M and put

$$E_i = Du_1 + Du_2 + \dots + Du_i.$$

Then
$$E_0 \subseteq E_1 \subseteq E_2 \subseteq \dots \subseteq E_n, \tag{1.13.2}$$

$E_0 = 0$ and $E_n = M$. To complete the proof, it will suffice to show that (1.13.2) is a composition series.

First u_i does not belong to E_{i-1}. For otherwise we should have a relation $u_i = \beta_1 u_1 + \dots + \beta_{i-1} u_{i-1}$ or $\beta_1 u_1 + \dots + \beta_{i-1} u_{i-1} + (-1) u_i = 0$. This however is impossible because, in a division ring, $1 \neq 0$. Accordingly $E_{i-1} \subset E_i$. Next assume that N is a submodule satisfying $E_{i-1} \subset N \subseteq E_i$. The theorem will be established if we can deduce that $N = E_i$.

Choose $y \in N$ so that y is not in E_{i-1}. Then $y = \gamma_1 u_1 + \gamma_2 u_2 + \dots + \gamma_i u_i$ (say) and $\gamma_i \neq 0$ because y is not in E_{i-1}. Thus

$$u_i = \gamma_i^{-1} y - \gamma_i^{-1} \gamma_1 u_1 - \dots - \gamma_i^{-1} \gamma_{i-1} u_{i-1}$$

and therefore u_i belongs to N. This proves that N contains $u_1, u_2, \dots, u_{i-1}$ and u_i and hence it contains E_i. Accordingly $N = E_i$ and all is established.

1.14 Change of ring

Let M be an R-module and denoted by A the set of all elements α, belonging to R, with the property that $\alpha x = 0$ for every x in M. It is clear that 0_R belongs to A and therefore A is not empty. Now suppose that α_1, α_2 belong to A and that r is an arbitrary element of R. Then, for any element x of M, we have

$$(\alpha_1 + \alpha_2) x = \alpha_1 x + \alpha_2 x = 0 + 0 = 0,$$

$$(r\alpha_1) x = r(\alpha_1 x) = r0 = 0,$$

$$(\alpha_1 r) x = \alpha_1 (rx) = 0.$$

Accordingly $\alpha_1 + \alpha_2$, $r\alpha_1$ and $\alpha_1 r$ all belong to A and therefore A is a two-sided ideal.

Definition. *The two-sided ideal consisting of all elements α such that $\alpha M = 0$ is called the 'annihilator' of the module M.*

By αM we mean, of course, the set consisting of all elements of the form αx, where x belongs to M. Note that M is a zero module if and only if its annihilator is the whole ring.

Let us regard the elements of R as operating on the module M, then the elements of the annihilator act very crudely. It is natural to try and 'factor out' this ideal in order to arrive at a situation in which the connection between the ring and the module is more intimate. How this may be accomplished is explained below.

Let A be a two-sided ideal of a ring R. Then R/A has a natural structure as left R-module and also as a right R-module. However, it is possible to go one stage further and turn R/A into a *ring*. To this end let $\phi: R \rightarrow R/A$ be the natural epimorphism and let r, r', ρ, ρ' be elements of R.

Lemma 13. *If* $\phi(r) = \phi(r')$ *and* $\phi(\rho) = \phi(\rho')$, *then* $\phi(r\rho) = \phi(r'\rho')$.

Proof. Since $\phi(r) = \phi(r')$ and $\phi(\rho) = \phi(\rho')$, we have $r' = r + \alpha$, $\rho' = \rho + \beta$, where α, β belong to A. Hence

$$r'\rho' = r\rho + r\beta + \alpha\rho + \alpha\beta = r\rho + \gamma$$

say, and $\gamma = r\beta + \alpha\rho + \alpha\beta$ belongs to A because this ideal is *two-sided*. Thus $\phi(r'\rho') = \phi(r\rho)$ as stated.

We already know that R/A is an abelian group with respect to addition. Let ξ, η belong to R/A. It is possible to choose r, ρ belonging to R so that $\phi(r) = \xi$ and $\phi(\rho) = \eta$. By Lemma 13, the element $\phi(r\rho)$ of R/A depends only on ξ and η and is independent of the choice of their representatives r and ρ. We may therefore define the 'product' of ξ and η by means of the formula $\xi\eta = \phi(r\rho)$. The elements of R/A can now be multiplied together as well as added. It is left to the reader to check that R/A has become a ring and that the natural mapping $R \rightarrow R/A$ is now a ring-epimorphism. He should note, too, that *if R happens to be a commutative ring, then R/A will be commutative as well*.

Definition. *If A is a two-sided ideal of R, then R/A when endowed with the ring-structure described above, is called the 'residue ring of R with respect to A'.*

Let us examine R/A in greater detail. It is not only a ring but also a left R-module and a right R-module. Indeed, if ξ belongs to R/A and r belongs to R, then the connection between these three structures is exhibited in

$$r\xi = \phi(r)\xi, \quad \xi r = \xi\phi(r), \tag{1.14.1}$$

where $\phi: R \rightarrow R/A$ is the natural mapping. Note that ϕ is simultaneously a ring-epimorphism, an epimorphism of left R-modules and an epimorphism of right R-modules. Again a subset of R/A is a left (right) ideal of that ring, if, and only if, it is a submodule of R/A when R/A is considered as a left (right) R-module. It follows that if I is a left (right) ideal of R, then $\phi(I)$ is a left (right) ideal of R/A. In particular, when B is a two-sided ideal of R, $\phi(B)$ is a two-sided ideal of R/A.

Proposition 24. *Let M be a left R-module and A a two-sided ideal contained in the annihilator I of M. Furthermore, let $\phi:R \to R/A$ be the natural mapping. Then M has a natural structure as a left R/A-module. This structure is characterized by the property that, whenever $r \in R$ and $x \in M$, $rx = \phi(r)x$. Finally $\phi(I) = I/A$ is the annihilator of M when it is regarded as a module with respect to R/A.*

Proof. Let ξ belong to R/A and let r, r' be elements of R such that $\phi(r) = \phi(r') = \xi$. Then $r' - r$ belongs to A, which is contained in I, and therefore $(r' - r)x = 0$ for every x in M. Accordingly $rx = r'x$ which means that rx is determined solely by ξ and x and is independent of the element r chosen to represent ξ. We may therefore put $\xi x = rx$. This secures that $rx = \phi(r)x$ for every $r \in R$ and $x \in M$.

To check that M is now an R/A-module, let ξ, ξ_1, ξ_2 belong to R/A and x, x_1, x_2 to M. Choose r, r_1, r_2 so that $\phi(r) = \xi$, $\phi(r_1) = \xi_1$ and $\phi(r_2) = \xi_2$. Then $\phi(r_1 + r_2) = \xi_1 + \xi_2$ and therefore

$$(\xi_1 + \xi_2)x = (r_1 + r_2)x = r_1 x + r_2 x = \xi_1 x + \xi_2 x.$$

Next
$$\xi(x_1 + x_2) = r(x_1 + x_2) = rx_1 + rx_2 = \xi x_1 + \xi x_2,$$

$$\xi_1(\xi_2 x) = r_1(r_2 x) = (r_1 r_2)x = \phi(r_1 r_2)x = (\xi_1 \xi_2)x,$$

and
$$1_{R/A} x = \phi(1_R)x = 1_R x = x.$$

Finally, if ρ belongs to R, then $\rho M = \phi(\rho)M$ and therefore $\phi(\rho)M = 0$ when and only when ρ belongs to I. Thus $\phi(I)$ is the annihilator of M when R/A is taken as the ring of operators.

Corollary 1. *Let the situation be as described in the proposition and let N be a subset of M. Then N is a submodule when M is considered as an R-module if, and only if, it is a submodule when M is considered as an R/A-module.*

Proof. Suppose that N is non-empty and has the property that, whenever y, y' belong to N, so also does $y + y'$. In order that it should be a submodule of the R-module M, we require (in addition) that $ry \in N$ for all $r \in R$ and $y \in N$. For N to be a submodule of the R/A-module M

5

the corresponding condition is that $\phi(r)\,y \in N$ for all $r \in R$ and $y \in N$. However $ry = \phi(r)\,y$ and therefore the two conditions are equivalent.

Corollary 2. *Let the situation be as described in the proposition. Then* $L_R(M) = L_{R/A}(M)$.

This follows immediately from Corollary 1.

It is possible, and also useful, to consider this kind of relationship in a more general setting. To this end let R and R' be rings and suppose that we are given a ring-homomorphism $\psi : R \to R'$. Then any left R'-module M (say) can be turned into an R-module by putting $rx = \psi(r)\,x$. Note that if $\psi(\alpha) = 0$, then $\alpha x = 0$ for every x in M and therefore $\operatorname{Ker}\psi$ is contained in the annihilator of the R-module M. It is clear that any R'-submodule of M must also be an R-submodule though, in general, the converse will not be true. However, *in the special situation where* $\psi : R \to R'$ *is a ring-epimorphism of R on to R', the R'-submodules of M coincide with the R-modules.* It then follows that $L_R(M) = L_{R'}(M)$.

Let us return to the consideration of a general ring-homomorphism $\psi : R \to R'$. Since R' may be regarded as a left R'-module, it acquires (by virtue of the considerations mentioned in the preceding paragraph) the structure of a left R-module. On this understanding, ψ is a homomorphism of left R-modules because, if r, ρ belong to R, then

$$\psi(r\rho) = \psi(r)\,\psi(\rho) = r\psi(\rho).$$

Similarly we may regard R' as a right R-module in which case ψ is also a homomorphism of right R-modules. Moreover every left (right) R'-ideal is a left (right) R-submodule of R' and the converse holds if ψ maps R on to R'.

Now let A be a two-sided ideal of R, A' a two-sided ideal of R' and assume that $\psi(A) \subseteq A'$. As we saw in section (1.5), there is induced a mapping $\psi^* : R/A \to R'/A'$ which is characterized by the fact that the diagram

$$
\begin{array}{ccc}
R & \xrightarrow{\ \psi\ } & R' \\
\downarrow & & \downarrow \\
R/A & \xrightarrow{\ \psi^*\ } & R'/A'
\end{array}
$$

is commutative, where the vertical mappings are the natural epimorphisms. Here ψ^* is not only a homomorphism of left R-modules and right R-modules. It is also a *ring-homomorphism* as the reader may easily verify.

Finally, let us examine in greater detail the case in which $\psi: R \to R'$ is a *ring-epimorphism* so that $\psi(R) = R'$. Since it is then an epimorphism of left R-modules, Proposition 6 and the above remarks show that there is a one-one correspondence between the left ideals I of R which contain $\mathrm{Ker}\,\psi$ and the left ideals I' of R'. Furthermore the correspondence is such that if I and I' correspond, then $I' = \psi(I)$ and $I = \psi^{-1}(I')$. Naturally there is an entirely similar connection between the right R-ideals containing $\mathrm{Ker}\,\psi$ and the right R'-ideals. On combining these observations we obtain: *there is a one-one correspondence between the two-sided ideals A of R which contain $\mathrm{Ker}\,\psi$ and the two-sided ideals A' of R'; the correspondence is such that if A is associated with A', then $\psi(A) = A'$ and $A = \psi^{-1}(A')$.* Note that the two-sided R-ideal which corresponds to the zero ideal of R' is simply $\mathrm{Ker}\,\psi$.

Suppose now that B is a two-sided ideal of R and let us apply these observations to the natural epimorphism $R \to R/B$. This has kernel B. Consequently *if A is any two-sided R-ideal containing B, then A/B is a two-sided ideal of the ring R/B; furthermore, by varying A, we obtain every two-sided ideal of R/B once and once only.*

The final result, of this chapter, is the main isomorphism theorem for rings.

Theorem 22. *Let $\psi: R \to R'$ be a ring-epimorphism of R on to R', A' a two-sided R'-ideal and $A = \psi^{-1}(A')$ the corresponding two-sided R-ideal. Then ψ induces a ring-isomorphism $\psi^*: R/A \to R'/A'$ of R/A on to R'/A'. In particular, there is induced a ring-isomorphism of $R/\mathrm{Ker}\,\psi$ on to R'.*

The proof is essentially the same as that of Theorem 1 and therefore we shall not go into details.

Corollary. *Let B and A, where $B \subseteq A$, be two-sided ideals of the ring R. Then A/B is a two-sided ideal of R/B and the natural mapping $R \to R/B$ induces a ring-isomorphism of R/A on to the residue ring $(R/B)/(A/B)$.*

The corollary is a special case of the theorem. To see this, take ψ to be $R \to R/B$ and replace A' by A/B.

Exercises on Chapter 1

In these exercises R denotes a ring with an identity element. Unless there is a definite statement to the contrary, it is not assumed that R is commutative. The term 'R-module' should always be interpreted as meaning 'left R-module'.

The letter Z is used to denote the ring formed by the integers, that is by the whole numbers 0, ± 1, $\pm 2, \ldots$, where addition and multiplication have their usual meanings.

1. An element γ of a ring R is called a 'central element' if $r\gamma = \gamma r$ for every r in R. Prove that the central elements of R form a ring.

2. M is a commutative group with the law of composition written as addition. For $x \in M$ and k a positive integer, kx is defined to be $x + x + \ldots + x$, where there are k summands. For a negative integer h, let $hx = -(kx)$, where $k = -h$. Finally put $0x = 0_M$. (Thus mx is defined for every integer m and every $x \in M$.) Prove that M now has the structure of a Z-module.

3. The mapping $f : M \to N$ is a homomorphism of R-modules, U is a subset of M, and L the submodule of M generated by U. Prove that the submodule of N generated by $f(U)$ is $f(L)$.

4. Let $\{A_i\}_{i \in I}$ be a family of subsets of an R-module E and let the submodule generated by A_i be X_i. Show that $\sum_i X_i$ is generated by $\bigcup_i A_i$.

5. Let E be an R-module and M_1, M_2 submodules of E. Prove that

$$L_R(M_1 + M_2) + L_R(M_1 \cap M_2) = L_R(M_1) + L_R(M_2)$$

and $\quad L_R(E/(M_1 + M_2)) + L_R(E/(M_1 \cap M_2)) = L_R(E/M_1) + L_R/(E/M_2).$

(It is not assumed that the lengths involved are finite.)

6. Show that if A is a non-zero ideal of the ring Z of integers and d is the smallest strictly positive integer contained in A, then A consists of all the integral multiples of d.

7. If m is a positive integer and p is a prime, prove that mZ/mpZ is a simple Z-module. Hence, or otherwise, find $L_Z(Z/1260Z)$. (In this context, mZ denotes the ideal of the ring Z generated by m.)

8. Show that the ring Z of integers satisfies the maximal but not the minimal condition for ideals.

9. Let Ω denote the set of all rational numbers which can be expressed in the form $m/2^k$, where m and k are integers. Show that Ω is a Z-module having Z itself as a submodule. Establish the following:

(i) each proper submodule of Ω/Z contains only a finite number of elements;

(ii) Ω/Z satisfies the minimal but not the maximal condition for submodules;

(iii) Ω/Z, considered as a Z-module, is not finitely generated.

10. Let the R-module E satisfy the maximal condition for submodules and let X be an indeterminate. Then the $R[[X]]$-module $E[[X]]$ also satisfies the maximal condition for submodules.

11. Let E be an R-module and γ a central element (see Exercise 1) of R. Then the mapping $E \to E$ in which $e \to \gamma e$ is an R-endomorphism of E.

12. Q is a module with respect to the field of real numbers having a base consisting of four elements e, i, j, k. If $q = ae + bi + cj + dk$ and

$$q' = a'e + b'i + c'j + d'k$$

(here $a, b, \ldots, c', d'$ are real numbers), then the product qq' is defined by

$$qq' = (aa' - bb' - cc' - dd')\,e + (ab' + ba' + cd' - dc')\,i + (ac' + ca' + db' - bd')\,j$$
$$+ (ad' + da' + bc' - cb')\,k.$$

Show that Q is a division ring with e as identity element. (This is known as the 'ring of quaternions'.)

13. Prove that a commutative simple ring is a field.

14. Prove that if R is a simple ring, then its only proper two-sided ideal is the zero ideal.

15. The diagram

$$
\begin{array}{ccccccccc}
M_1 & \longrightarrow & M_2 & \longrightarrow & M_3 & \longrightarrow & M_4 & \longrightarrow & M_5 \\
\downarrow{f_1} & & \downarrow{f_2} & & \downarrow{f_3} & & \downarrow{f_4} & & \downarrow{f_5} \\
N_1 & \longrightarrow & N_2 & \longrightarrow & N_3 & \longrightarrow & N_4 & \longrightarrow & N_5
\end{array}
$$

of R-modules and R-homomorphisms is commutative and the two rows are exact. Prove that if f_1 is an epimorphism and f_2, f_4 are monomorphisms, then f_3 is a monomorphism. Prove also that if f_5 is a monomorphism and f_2, f_4 are epimorphisms, then f_3 is an epimorphism. (This is known as the 'Five Lemma'.)

16. Show that the Z-module formed by the rational numbers is not a free module.

2

PRIME IDEALS AND PRIMARY SUBMODULES

General Remarks. In this chapter we shall confine our attention entirely to *commutative* rings. The reader must therefore keep in mind that when, for example, there is a reference to a ring R it is tacitly supposed that R is commutative and possesses an identity element. The null ring, see section (1.1), is not excluded from consideration except where there is a definite statement to this effect.

2.1 Zorn's Lemma

Unless we are prepared to impose certain finiteness conditions on our rings and modules, it is necessary to use 'transfinite methods' in order to prove certain existence theorems. There are various forms which this type of reasoning can take. The one which is embodied in *Zorn's lemma* is by far the most convenient for our purposes. This will now be described.

Let Ω be a non-empty set of objects and suppose that we have a relation which holds between certain pairs of elements of Ω. If x and y belong to Ω and the relation in question holds between them, then we shall write $x \leqslant y$. The relation is said to be (or to define) a *partial order* on Ω if it satisfies the following three conditions:

(i) $x \leqslant x$ *for every* x *in* Ω;

(ii) $x \leqslant y$ *and* $y \leqslant x$ *together imply* $x = y$;

(iii) *whenever* $x \leqslant y$ *and* $y \leqslant z$, *then* $x \leqslant z$.

For example, if A and B are subsets of a set X, let us write $A \leqslant B$ if A is contained in B. Then, denoting by Ω the collection formed by all the subsets of X, it is clear that $\leqslant$ is a partial order on Ω. Note that this example shows that, for certain choices of A and B, it is perfectly possible for neither of the relations $A \leqslant B$ and $B \leqslant A$ to hold.

Returning to the general situation, assume that $\leqslant$ is a partial order on Ω and suppose that Σ is a subset of Ω.

Definition. *An element* u *of* Ω *is said to be an 'upper bound' for the subset* Σ *if* $x \leqslant u$ *for every* x *in* Σ. *A subset which possesses an upper bound is said to be 'bounded above'.*

Another concept, which we shall need, is that of a *totally ordered* set.

Definition. *A relation* $\leqslant$ *on a set* Ω *is called a 'total order' on* Ω *if (a) it is a partial order, and (b) whenever* x *and* y *belong to* Ω *either* $x \leqslant y$ *or* $y \leqslant x$.

For example, the real numbers are totally ordered by the relation '*less than or equal to*'. Again, any partially ordered set consisting of a single element is totally ordered.

Still assuming that $\leqslant$ is a partial order on Ω, suppose that Ω^* is a non-empty subset. If now ξ, η belong to Ω^* let us write (provisionally) $\xi \leqslant^* \eta$ if $\xi \leqslant \eta$ when the elements in question are regarded as elements of Ω. It is clear that $\leqslant^*$ is a partial order on Ω^*. It is called the *induced partial order* and, in practice, we omit the asterisk and simply write $\xi \leqslant \eta$. The reader should note that whenever there is a reference to a partial order on a subset of a partially ordered set, it is always the induced order that we have in mind.

An element μ of Ω is called a *maximal element* if from $x \in \Omega$ and $\mu \leqslant x$ follows $x = \mu$. Note that if μ is a maximal element it does not follow that $y \leqslant \mu$ for every y in Ω. In this respect the use of the term *maximal* is similar to that encountered where we were discussing the maximal condition for submodules.†

It is not always the case that a non-empty partially ordered set possesses a maximal element. (For example, if we take the real numbers with 'less than or equal to' as the ordering relation, then there is no maximal real number.) In fact Zorn's lemma is concerned with an important criterion for the existence of maximal elements. Before stating the criterion it is convenient to give one more

Definition. *A set* Ω, *on which a partial order is given, is called an 'inductive system' if every totally ordered subset is bounded above.*

This brings us finally to

Zorn's Lemma. *Every non-empty inductive system possesses at least one maximal element.*

If the reader has not encountered Zorn's lemma before, it is suggested that he treat it as an axiom. It is in fact, equivalent to the

† See section (1.8).

Axiom of Choice, which we employ without comment, and it is also equivalent to the *Well Ordering Principle*, which we do not require at this stage. It is important to understand clearly what is asserted and to acquire practice, not only in making applications, but also in recognizing situations where the lemma is applicable. So far as Algebra is concerned, the reader will soon be confronted with examples which will help to develop this familiarity. However, Zorn's lemma enshrines such a fundamental logical principle that it finds application in many other fields. We refer the reader to the literature on the Foundations of Mathematics, for an analysis of its connections with related logical ideas. The main feature of Zorn's lemma is its peculiar suitability for the application of these ideas in Mathematics.

Let Ω be a non-empty inductive system and suppose that ω is an element of Ω. By Zorn's lemma, Ω contains a maximal element μ say. This means that if $x \in \Omega$ and $\mu \leqslant x$, then $\mu = x$. It does not secure that $\omega \leqslant \mu$. It is therefore pertinent to ask whether it is possible to choose the maximal element μ in such a way that $\omega \leqslant \mu$. Indeed such a maximal element always exists as is proved in

Lemma 1. *Let ω be an element of an inductive system Ω. Then there exists a maximal element μ of Ω with the property that $\omega \leqslant \mu$.*

Proof. Let Ω^* consist of all x belonging to Ω and satisfying $\omega \leqslant x$. Then Ω^* is not empty because it contains ω. *We contend that Ω^* (with the induced partial order) is an inductive system.* To this end let Σ^* be a non-empty totally ordered subset of Ω^*. We wish to show that Σ^* is bounded above in Ω^*. Since Ω is an inductive system, Σ^* is bounded above in Ω. Let $z \in \Omega$ be such that $y \leqslant z$ for every y in Σ^*. Since Σ^* is not empty, we can find an element y_0 which belongs to it. Then $y_0 \leqslant z$ and also $\omega \leqslant y_0$ because $y_0 \in \Sigma^* \subseteq \Omega^*$. It follows that $\omega \leqslant z$; hence $z \in \Omega^*$. This shows that Σ^* is bounded above in Ω^* (and not merely in Ω) and completes the verification that Ω^* is a non-empty inductive system.

By Zorn's lemma, applied to Ω^*, there exists $\mu^* \in \Omega^*$ such that μ^* is maximal in Ω^*. Since $\mu^* \in \Omega^*$, it follows that $\omega \leqslant \mu^*$. To complete the proof, we show that μ^* is maximal in Ω.

To this end, suppose that $u \in \Omega$ and $\mu^* \leqslant u$. Then $\omega \leqslant \mu^*$ and $\mu^* \leqslant u$. Consequently $\omega \leqslant u$ and hence $u \in \Omega^*$. Since μ^* is maximal in Ω^*, we may now conclude that $\mu^* = u$. It follows that μ^* is maximal in Ω and not merely in Ω^*, which establishes the lemma.

2.2 Prime ideals and integral domains

Let R be a ring† and A a non-empty subset of R. Since R is commutative, A will be a left ideal if and only if it is a right ideal. For this reason, there is no need to keep the distinction and therefore we speak simply of ideals. We recall that A is an ideal provided

(i) *when a_1, a_2 belong to A, then $a_1 + a_2$ also belongs to A*;

(ii) *when a belongs to A and r to R, then ra belongs to A.*

It is further recalled that an ideal is called *proper* if it is not the whole ring. Also a proper ideal which is not strictly contained by any other proper ideal, is known as a *maximal* ideal.

We come now to a new concept. An ideal P is called a *prime* ideal if first it is a proper ideal and, secondly, from $\alpha\beta \in P$ (where α and β belong to R) always follows either $\alpha \in P$ or $\beta \in P$. It should be noted that, because we require prime ideals to be proper, they can only exist in a non-null ring. However, it is not immediately clear that a non-null ring must possess at least one prime ideal though this, in fact, is the case. Indeed one of the reasons for introducing Zorn's lemma was that it will be needed later for the proof of this result.

Lemma 2. *Let P be a prime ideal of R and suppose that a product $\alpha_1 \alpha_2 \ldots \alpha_n$, of elements of R, belongs to P. Then at least one of the α_i belongs to P.*

Proof. We use induction on n. If $n = 1$ or 2 then the assertion is clear. It will therefore be supposed that $n > 2$ and that the lemma has been proved for products with $n - 1$ terms. By hypothesis,

$$(\alpha_1 \alpha_2 \ldots \alpha_{n-1}) \alpha_n$$

belongs to P. Hence, by the definition of a prime ideal, either

$$\alpha_1 \alpha_2 \ldots \alpha_{n-1}$$

belongs to P or α_n belongs to P. In the latter situation there is nothing more to prove and, in the former, the required result follows from the inductive hypothesis.

Definition. *A non-null ring R is called an 'integral domain' if a product $\alpha\beta$ of elements of R is zero only when at least one of the factors is zero.*

Thus an integral domain is a ring (commutative and with an identity element) whose zero ideal is a prime ideal.

---

† It is recalled that, throughout Chapter 2, all rings are assumed to be commutative and to possess identity elements.

Definition. *An element α of a ring R is called a 'zero-divisor' if there exists $\beta \in R$, $\beta \neq 0$, such that $\alpha\beta = 0$.*

Accordingly an integral domain may also be described as a non-null (commutative) ring whose only zero-divisor is the zero element.

In section (1.10) we encountered the concept of a *division ring* and we noted, at that time, that a commutative division ring is known as a *field*. Since we are confining our attention to the commutative case, let us restate this definition in more explicit terms. *A field is a non-null (commutative) ring in which each non-zero element has an inverse.* Note that *if F is a field, then it is also an integral domain.* For F is a non-null ring and from $\alpha\beta = 0$, $\alpha \neq 0$, follows $\beta = 0$ on multiplying by the inverse α^{-1} of α.

Lemma 3. *A non-null ring R is a field if and only if it has no ideals other than the zero ideal and R itself.*

Proof. First suppose that R is a field and let A be a non-zero ideal. It is then possible to choose $\alpha \in A$ so that $\alpha \neq 0$. Now let r be an arbitrary element of R. Then $r = (r\alpha^{-1})\alpha$ belongs to A. This shows that $A = R$.

Next suppose that 0 and R are the only ideals and let β be a non-zero element of R. Then the ideal $R\beta$ must be the whole ring. Thus 1 belongs to $R\beta$ and therefore we can find $\gamma \in R$ so that $\gamma\beta = 1$. Thus β has an inverse and all is proved.

Corollary. *A ring R is a field if and only if its zero ideal is a maximal ideal.*

Now let A be an ideal of a ring R. Since R is commutative, all ideals are two-sided and therefore we can form the residue ring R/A. We know, from the considerations set out in section (1.14), that there is a one-one correspondence, which preserves inclusion relations, between the ideals of R containing A and the ideals of R/A. Indeed if B, where $A \subseteq B$, is an R-ideal, then B/A is the R/A-ideal associated with it.

Proposition 1. *An ideal A is a prime ideal if and only if R/A is an integral domain. It is a maximal ideal if and only if R/A is a field.*

Proof. Let A be a prime ideal. Then A is proper and therefore R/A is a non-null ring. Suppose that $\xi\eta = 0$, where ξ and η belong to R/A. We can choose α, β in R so that $\phi(\alpha) = \xi$, $\phi(\beta) = \eta$, where $\phi: R \to R/A$ is the natural mapping. Then $\phi(\alpha\beta) = \phi(\alpha)\,\phi(\beta) = \xi\eta = 0$, whence $\alpha\beta$

belongs to A. But A is a prime ideal. Consequently either $\alpha \in A$ or $\beta \in A$. It follows that either $\xi = 0$ or $\eta = 0$. Hence R/A is an integral domain.

Now assume that R/A is an integral domain. Then A is certainly a proper ideal. Assume that the product $\alpha'\beta'$ belongs to A. Then $\phi(\alpha')\phi(\beta') = \phi(\alpha'\beta') = 0$. Since R/A is an integral domain, either $\phi(\alpha') = 0$ or $\phi(\beta') = 0$. In the former case $\alpha' \in A$ and in the latter $\beta' \in A$. This shows that A is a prime ideal.

Finally, if A is a maximal ideal, then there are no ideals strictly between A and R. Accordingly R/A is a non-null ring with no proper ideals except the zero ideal. Hence, by Lemma 3, R/A is a field. Conversely, if R/A is a field then A must be a proper ideal. Moreover, since R/A has no non-zero proper ideals, there can be no ideal strictly between A and R.

Corollary. *Every maximal ideal is a prime ideal.*

This follows from the proposition because a field is a special kind of integral domain.

Proposition 2. *Let $\psi: R \to R'$ be a ring-epimorphism of R on to R', A' an ideal of R', and $A = \psi^{-1}(A')$ the corresponding ideal† of the ring R. If now either A or A' is a prime ideal, then so also is the other. Likewise if one is a maximal ideal, then the other is a maximal ideal as well.*

Proof. By Theorem 22 of section (1.14), the residue rings R/A and R'/A' are isomorphic. Hence if one is an integral domain (field), then the other is an integral domain (field). The required result therefore follows from Proposition 1.

Let X be a subset of a ring R. By the *complement* of X in R is meant the set consisting of all elements of R which are not in X. The complement of a prime ideal has an important property. To describe it we first make the

Definition. *A subset S of a ring R is said to be 'multiplicatively closed' if whenever s, s' belong to S, then their product ss' also belongs to S.*

For example, an ideal is a prime ideal if and only if its complement is a non-empty multiplicatively closed set.

This is a convenient place to introduce some new notation. If U and X are sets and U is *not* contained in X, then we shall sometimes

† See section (1.14).

use the symbol $U \nsubseteq X$ to indicate this fact. Likewise, if the object α is *not* in the set X, then, on occasion, we may write $\alpha \notin X$.

We are now ready to tackle certain questions concerning the existence of prime and maximal ideals. There is one piece of reasoning, relevant to these questions, which is likely to be encountered in other contexts. We therefore isolate it in the following lemma.

Lemma 4. *Let a non-empty collection Σ of ideals of a ring R be totally ordered by inclusion. Denote by B the set-theoretic union of the ideals belonging to Σ. Then B is an ideal. Furthermore, if all the members of Σ are proper ideals, then B is also a proper ideal.*

Remarks. The statement that Σ is *totally ordered by inclusion* means that if A, A' are ideals of the collection Σ, then either $A \subseteq A'$ or $A' \subseteq A$. Again, the definition of B amounts to this: an element r of R belongs to B, when and only when there exists $A \in \Sigma$ such that $r \in A$.

Proof. Let β, β' belong to B and r to R. Then there exist A, A' in Σ such that $\beta \in A$ and $\beta' \in A'$. Now either $A \subseteq A'$ or $A' \subseteq A$. For definiteness we shall suppose that the former is correct. In this case, both β and β' belong to A'. Consequently $\beta + \beta'$ and $r\beta$ belong to A' and therefore to B. This proves that B is an ideal.

Finally suppose that every ideal belonging to Σ is proper. Then none of these ideals can contain 1_R because the only ideal which contains the identity element is the whole ring. It follows that $1_R \notin B$ and therefore B is proper.

Proposition 3. *Let S be a non-empty multiplicatively closed subset of a ring R and suppose that S does not contain the zero element of R. Denote by Ω the set of all ideals which do not meet S (i.e. whose intersections with S are empty) and, if A, A' belong to Ω, let $A \leqslant A'$ mean that A is contained in A'. Then Ω, together with the relation $\leqslant$, is a non-empty inductive system and all its maximal elements are prime ideals.*

Proof. It is clear that $\leqslant$ is a partial order on Ω. Also Ω is not empty because it contains the zero ideal. Now let Σ be a non-empty, totally ordered subset of Ω. Then Σ is a collection of ideals totally ordered by inclusion; hence, by Lemma 4, their union B is also an ideal. Let s belong to S. Then s does not belong to any member of Σ and therefore it does not belong to B. Thus B does not meet S; consequently $B \in \Omega$. Clearly B is an upper bound for Σ. The verification that Ω is a non-empty inductive system is now complete.

Let P be a maximal element of Ω. Trivially P is an ideal not meeting S. In particular it is a proper ideal. We shall now prove that P is prime. To this end, suppose that α, β are elements of R whose product $\alpha\beta$ belongs to P. It will suffice to show that at least one of α and β belongs to P. *We shall assume that neither element has this property and derive a contradiction.*

Let C consist of all elements which can be expressed in the form $r\alpha + \pi$, where $r \in R$ and $\pi \in P$. It is a simple matter to check that C is an ideal containing P. Indeed, since $\alpha = 1\alpha + 0$, $\alpha \in C$ and therefore C *strictly* contains P. However P is maximal in Ω. It therefore follows that C meets S. We can therefore find $r_1 \in R$ and $\pi_1 \in P$ so that

$$r_1\alpha + \pi_1 = s_1 \text{ (say)}$$

belongs to S. By entirely similar arguments we can show that there exist $r_2 \in R$ and $\pi_2 \in P$ so that $r_2\beta + \pi_2 = s_2$ (say) also belongs to S. However, because S is multiplicatively closed, it contains

$$s_1 s_2 = r_1 r_2 \alpha\beta + r_1 \alpha\pi_2 + r_2 \beta\pi_1 + \pi_1\pi_2.$$

But $\alpha\beta$, π_1 and π_2 all belong to P; consequently $s_1 s_2 \in P$. Thus P and S have an element in common and we have derived the required contradiction.

Our first application of Proposition 3 is

Theorem 1. *Let S be a non-empty multiplicatively closed subset of a ring R and A an R-ideal which does not meet S. Then there exists a prime ideal P of R which contains A and does not meet S.*

Proof. Since S does not meet A, it cannot contain the zero element. Now let Ω and $\leqslant$ be as in Proposition 3. Then Ω is a non-empty inductive system and, by hypothesis, $A \in \Omega$. By Lemma 1, Ω contains a maximal element P such that $A \leqslant P$. Since $P \in \Omega$, it is an ideal not meeting S. Also, by Proposition 3, P is prime. Of course, the fact that $A \leqslant P$ means simply that A is contained in P.

Theorem 2. *Let A be a proper ideal of a ring R. Then A is contained in a maximal ideal of R.*

Proof. Since A is a proper ideal, R is a non-null ring. Denote by S the set consisting of the identity element of R by itself. Obviously S is multiplicatively closed and does not contain the zero element. Now let Ω and $\leqslant$ be defined as in Proposition 3. Then Ω consists of the proper ideals of R partially ordered by inclusion. In particular, it contains A. Proposition 3 shows that Ω is an inductive system. Applying Lemma 1 we find that there exists a maximal element P such that $A \leqslant P$. The fact that P is maximal in Ω means simply that P is a maximal ideal of the ring, while $A \leqslant P$ signifies that P contains A.

2.3 Minimal prime ideals

In this section we shall establish further results concerning the existence of prime ideals with special properties. First, however, we must set in evidence certain elementary facts which concern ideals in general.

Suppose therefore that $\{A_i\}_{i\in I}$ is a family of ideals of our ring R. This means that each is a submodule of R when R is regarded as a module with respect to itself. Now in section (1.4) we explained how to form the sum of a family of submodules. In the present instance we obtain $\sum_{i\in I} A_i$ and this too is an ideal. Again, the general principles which apply to intersections of submodules† show that $\bigcap_{i\in I} A_i$ is also an ideal of R. Now it is sometimes convenient to extend these ideas to the case of an *empty* family of ideals. In fact, if I is the empty set, then we interpret $\sum_{i\in I} A_i$ as meaning the zero ideal. The appropriate interpretation of $\bigcap_{i\in I} A_i$ when I is empty is less obvious, so we shall digress a little in order to motivate the convention which we shall presently adopt.

Let X be a set. If we intend confining our attention to subsets of X, then we shall indicate this by describing X as the *ambient set*. Suppose now that $\{U_i\}_{i\in I}$ is a family of subsets of X where, in the first instance, I is not empty. Then $\bigcap_{i\in I} U_i$ is a well defined subset of X.

Next, if I_1 is a non-empty subset of I, then $\bigcap_{i\in I_1} U_i$ is also well defined and
$$\bigcap_{i\in I} U_i \subseteq \bigcap_{i\in I_1} U_i.$$

Thus, as the index set is made smaller, the intersection becomes larger. The next step is to extend the meaning of $\bigcap_{i\in I_1} U_i$ so as to cover the case where I_1 is the empty subset of I. The convention which proves most useful says that, in this extreme situation, the intersection is X itself. Thus *the intersection of an empty family of subsets is the ambient set*. Note that it is necessary to know *in advance* which is the ambient set.

† See Proposition 3 of section (1.4).

For example, let us embed X in a larger set Y. Then, when I_1 is the empty set, $\bigcap\limits_{i \in I_1} U_i$ means X if we are taking X to be the ambient set and Y when the latter is so regarded.

In the case of modules, we sometimes consider submodules of a fixed module E which is then referred to as the ambient module. On this understanding, an empty intersection of submodules of E is conventionally equal to E itself. In the case of ideals, it is always understood that the ambient module is the ring to which they belong. *Thus the intersection of an empty set of ideals of R is just R itself.*

Let us take an actual example. Suppose that A is an ideal of a ring R. If A is a proper ideal, then Theorem 2 shows that there exist prime ideals which contain A. It is therefore meaningful to speak of the intersection of all the prime ideals which contain A. This intersection is an ideal called the radical of A about which we shall have more to say later.† However, if A is not a proper ideal then there are no prime ideals which contain A because, by definition, every prime ideal is proper. Thus, this time, the prime ideals containing A form an empty set. Accordingly, by our convention, their intersection is equal to R. Thus the radical of R (considered as an ideal) is itself.

We shall now define the *product* of two ideals, say A and B, of a ring R. Let C denote the set of all elements of R which can be expressed in the form $a_1 b_1 + a_2 b_2 + \ldots + a_n b_n$, where the a_i belong to A and the b_i to B. *Then C is an ideal.* For let α, β belong to C and r to R. Then $\alpha = a_1 b_1 + \ldots + a_n b_n$ and $\beta = a_1' b_1' + \ldots + a_m' b_m'$, where $a_1, \ldots, a_n$, $a_1', \ldots, a_m'$ are in A and $b_1, \ldots, b_n, b_1', \ldots, b_m'$ in B. It follows that

$$\alpha + \beta = a_1 b_1 + \ldots + a_n b_n + a_1' b_1' + \ldots + a_m' b_m'$$

which makes it clear that $\alpha + \beta$ belongs to C. Since

$$r\alpha = (ra_1) b_1 + (ra_2) b_2 + \ldots + (ra_n) b_n$$

and $ra_1, ra_2, \ldots, ra_n$ all belong to A, we also have $r\alpha \in C$. This verifies that C is an ideal. It is, in fact, the ideal generated by all products ab, where $a \in A$ and $b \in B$.

Definition. *If A and B are ideals of the same ring, then the ideal generated by all elements of the form ab, where $a \in A$ and $b \in B$, is called the 'product' of A and B and is denoted by AB.*

Note that, since we are confining our attention to commutative rings, $AB = BA$. Note too that $AR = A$.

---

† When the time comes, we shall give a different definition of the radical of an ideal. However, this will be proved equivalent to the one described above.

Let $A_1, A_2, ..., A_m, B_1, B_2, ..., B_n$ be ideals of a ring R. We contend that

$$(A_1 + A_2 + ... + A_m)(B_1 + B_2 + ... + B_n) = \sum_{i,j} A_i B_j. \qquad (2.3.1)$$

For if $1 \leqslant i \leqslant m$, $1 \leqslant j \leqslant n$, then A_i is contained in $A_1 + A_2 + ... + A_m$ whereas B_j is contained in $B_1 + B_2 + ... + B_n$. Accordingly

$$A_i B_j \subseteq (A_1 + A_2 + ... + A_m)(B_1 + B_2 + ... + B_n)$$

and therefore

$$\sum_{i,j} A_i B_j \subseteq (A_1 + A_2 + ... + A_m)(B_1 + B_2 + ... + B_n).$$

Suppose next that α belongs to $A_1 + A_2 + ... + A_m$ and β to

$$B_1 + B_2 + ... + B_n.$$

Then $\alpha = a_1 + a_2 + ... + a_m$, $\beta = b_1 + b_2 + ... + b_n$, where $a_i \in A_i$ and $b_j \in B_j$. It follows that

$$\alpha\beta = \sum_{i,j} a_i b_j \in \sum_{i,j} A_i B_j.$$

Now every element of the left-hand side of (2.3.1) is a finite sum of elements of the form $\alpha\beta$. Accordingly

$$(A_1 + A_2 + ... + A_m)(B_1 + B_2 + ... + B_n) \subseteq \sum_{i,j} A_i B_j$$

and (2.3.1) is established.

Next let A, B, C be arbitrary ideals of R. It is clear, from the definition, that $(AB)C$ consists of all elements which can be expressed as sums of the form

$$a_1 b_1 c_1 + a_2 b_2 c_2 + ... + a_n b_n c_n,$$

where $a_i \in A_i$, $b_i \in B_i$ and $c_i \in C_i$. But $A(BC)$ also consists of these same elements; consequently.

$$A(BC) = (AB)C. \qquad (2.3.2)$$

Thus multiplication of ideals obeys the associative law. It is therefore possible to form extended products $A_1 A_2 ... A_n$, where $A_1, A_2, ..., A_n$ are arbitrary ideals of R. The elements of $A_1 A_2 ... A_n$ are expressible as finite sums in which each term is a product $\alpha_1 \alpha_2 ... \alpha_n$ with $\alpha_i \in A_i$. Note that

$$A_1 A_2 ... A_n \subseteq A_1 \cap A_2 \cap ... \cap A_n. \qquad (2.3.3)$$

In the proposition which follows $A_1, A_2, ..., A_n$ continue to denote ideals.

Proposition 4. *If $A_1 A_2 ... A_n$ or $A_1 \cap A_2 \cap ... \cap A_n$ is contained in a prime ideal P, then for some i $(1 \leqslant i \leqslant n)$ A_i is contained in P.*

Proof. By (2.3.3), $A_1 A_2 \dots A_n$ is contained in P. *We shall assume that no A_i is contained in P and derive a contradiction.*

For each i choose $a_i \in A_i$ so that $a_i \notin P$. Then $a_1 a_2 \dots a_n$ belongs to $A_1 A_2 \dots A_n$ and hence to P. However, by Lemma 2, this is impossible. This yields the required contradiction.

Sometimes it is necessary to choose elements avoiding certain prime ideals. The next result is fundamental in this connection.

Proposition 5. *Let $P_1, P_2, \dots, P_n$ be prime ideals and let A be an ideal which is not wholly contained in any single one of them. Then A contains an element α which does not belong to any P_i.*

Proof. We may assume that no P_i is contained in any of the remainder. For if, for example, $P_1 \subseteq P_2$ then it suffices to prove the proposition for A and the reduced system $P_2, P_3, \dots, P_n$. Supposing that the extra condition is satisfied, we observe that P_i does not contain any of the ideals $A, P_1, \dots, P_{i-1}, P_{i+1}, \dots, P_n$ and therefore, by Proposition 4, it cannot contain their product. It is therefore possible to choose α_i so that

$$\alpha_i \in A P_1 \dots P_{i-1} P_{i+1} \dots P_n \subseteq A \cap P_1 \cap \dots \cap P_{i-1} \cap P_{i+1} \cap \dots \cap P_n$$

and $\alpha_i \notin P_i$. Let such an element be chosen for each value of i and put $\alpha = \alpha_1 + \alpha_2 + \dots + \alpha_n$. Clearly α belongs to A. However, since $\alpha_1 = \alpha - \alpha_2 - \dots - \alpha_n$ and all of $\alpha_2, \dots, \alpha_n$ belong to P_1, α does not belong to P_1, for the contrary assumption leads to $\alpha_1 \in P_1$ which is not the case. Similarly α does not belong to any of $P_2, P_3, \dots, P_n$. The proof is therefore complete.

At an earlier stage, we had occasion to mention, in a rather informal manner, the notion of the radical of an ideal. This concept will now be discussed in some detail, but our starting point will be somewhat different from that suggested by our earlier comments.

Let A be an ideal of R. By the *radical* of A is meant the set of all those elements α of R which have the property that some positive power of α is contained in A. The radical of A will be denoted by $\operatorname{Rad} A$. Accordingly $\alpha \in \operatorname{Rad} A$ if and only if $\alpha^m \in A$ for some positive integer m. It is clear that

$$A \subseteq \operatorname{Rad} A \tag{2.3.4}$$

and that the radical of the ring R, considered as an ideal, is itself. Also, if A and B are ideals and $A \subseteq B$, then $\operatorname{Rad} A \subseteq \operatorname{Rad} B$. Slightly less obvious is the relation†

$$\operatorname{Rad} (\operatorname{Rad} A) = \operatorname{Rad} A. \tag{2.3.5}$$

† Strictly speaking, this anticipates the result (Proposition 6) that $\operatorname{Rad} A$ is itself an ideal.

To prove this it is only necessary to show that $\mathrm{Rad}\,(\mathrm{Rad}\,A) \subseteq \mathrm{Rad}\,A$. Suppose therefore that $\alpha \in \mathrm{Rad}\,(\mathrm{Rad}\,A)$. Then $\alpha^m = \beta$ (say) belongs to $\mathrm{Rad}\,A$ for a suitable positive integer m. Hence there exists a positive integer n such that $\beta^n = \alpha^{mn}$ belongs to A. Accordingly $\alpha \in \mathrm{Rad}\,A$ and the proof is complete.

Proposition 6. *The radical of an ideal is also an ideal.*

Proof. Let A be an ideal and suppose that α, β belong to $\mathrm{Rad}\,A$. Then there exist positive integers m, n such that α^m, β^n belong to A. Now $(\alpha + \beta)^{m+n}$ can be written as a sum of terms of the form $\alpha^\mu \beta^\nu$, where μ, ν are non-negative integers and $\mu + \nu = m + n$. For such μ, ν either $\mu \geqslant m$ or $\nu \geqslant n$; hence, in any event, $\alpha^\mu \beta^\nu \in A$. This shows that $(\alpha + \beta)^{m+n}$ belongs to A and hence that $\alpha + \beta$ belongs to $\mathrm{Rad}\,A$. Finally, if $r \in R$, then $(r\alpha)^m = r^m \alpha^m$ belongs to A and therefore $r\alpha \in \mathrm{Rad}\,A$. The proof is now complete.

Proposition 7. *Let $A_1, A_2, ..., A_m$ be ideals of R. Then*

$$\mathrm{Rad}\,(A_1 A_2 ... A_m) = \mathrm{Rad}\,(A_1 \cap A_2 \cap ... \cap A_m)$$
$$= \mathrm{Rad}\,A_1 \cap \mathrm{Rad}\,A_2 \cap ... \cap \mathrm{Rad}\,A_m.$$

Proof. It is clear that

$$\mathrm{Rad}\,(A_1 A_2 ... A_m) \subseteq \mathrm{Rad}\,(A_1 \cap A_2 \cap ... \cap A_m) \subseteq \mathrm{Rad}\,A_1 \cap ... \cap \mathrm{Rad}\,A_m.$$

Now suppose that α belongs to $\mathrm{Rad}\,A_1 \cap ... \cap \mathrm{Rad}\,A_m$. For each $i\,(1 \leqslant i \leqslant m)$, it is possible to choose an integer n_i so that $\alpha^{n_i} \in A_i$. Put $n = n_1 + n_2 + ... + n_m$. Then $\alpha^n = \alpha^{n_1} \alpha^{n_2} ... \alpha^{n_m}$ belongs to $A_1 A_2 ... A_m$. Accordingly α belongs to $\mathrm{Rad}\,(A_1 A_2 ... A_m)$ and the proposition is proved.

Since we are able to multiply ideals, we can raise an ideal to a positive power. Taking the ideals in Proposition 7 to be the same, we obtain the

Corollary. *If A is an ideal and m a positive integer, then*

$$\mathrm{Rad}\,(A^m) = \mathrm{Rad}\,A.$$

Before establishing the next result, we shall explain a notation which is commonly used in ideal theory. If $\alpha_1, \alpha_2, ..., \alpha_n$ belong to a ring R, then they will generate an ideal. According to the conventions laid down for modules, it would be reasonable to denote this ideal by $R\alpha_1 + R\alpha_2 + ... + R\alpha_n$. In practice, however, the symbol $(\alpha_1, \alpha_2, ..., \alpha_n)$ is often preferred. Thus the ideal $(\alpha_1, \alpha_2, ..., \alpha_n)$ consists of all elements which can be expressed in the form $r_1 \alpha_1 + r_2 \alpha_2 + \ldots + r_n \alpha_n$, where the r_i belong to R. Note that

$$(\alpha_1, \alpha_2, \ldots, \alpha_n) + (\beta_1, \beta_2, \ldots, \beta_s) = (\alpha_1, \ldots, \alpha_n, \beta_1, \ldots, \beta_s) \qquad (2.3.6)$$

and

$$(\alpha_1, \alpha_2, \ldots, \alpha_n)(\beta_1, \beta_2, \ldots, \beta_s) = (\alpha_1 \beta_1, \ldots, \alpha_i \beta_j, \ldots, \alpha_n \beta_s). \qquad (2.3.7)$$

Proposition 8. *Let A be an ideal and B a finitely generated ideal contained in* Rad A. *Then there exists a positive integer m such that $B^m \subseteq A$.*

Proof. Let $B = (\beta_1, \beta_2, \ldots, \beta_p)$. Then for each i $(1 \leqslant i \leqslant p)$ there exists an integer m_i such that $\beta_i^{m_i} \in A$. Put $m = m_1 + m_2 + \ldots + m_p$. By (2.3.7), B^m is generated by the power-products $\beta_1^{\mu_1} \beta_2^{\mu_2} \ldots \beta_p^{\mu_p}$, where $\mu_1, \mu_2, \ldots,$ μ_p are non-negative integers such that $\mu_1 + \mu_2 + \ldots + \mu_p = m$. Now if $\mu_1, \mu_2, \ldots, \mu_p$ satisfy these conditions, then it is possible to choose i so that $\mu_i \geqslant m_i$. Thus $\beta_1^{\mu_1} \beta_2^{\mu_2} \ldots \beta_p^{\mu_p}$ belongs to $(\beta_i^{m_i}) \subseteq A$ which shows that each of the generators of B^m is contained in A. Accordingly $B^m \subseteq A$ and the proposition is proved.

We shall now develop results which will connect the radical of an ideal A with the prime ideals containing A.

Lemma 5. *Let $\{P_i\}_{i \in I}$ be a non-empty family of prime ideals of a ring R and suppose that the family is totally ordered by inclusion. Then both $\bigcup_i P_i$ and $\bigcap_i P_i$ are prime ideals.*

Remark. The statement that the family is *totally ordered by inclusion* means here that if i, j belong to I, then either $P_i \subseteq P_j$ or $P_j \subseteq P_i$.

Proof. Put $P^* = \bigcup_i P_i$. Then, by Lemma 4, P^* is a proper ideal. Now suppose that $\alpha\beta$ belongs to P^* but α does not. Then there exists i such that $\alpha\beta \in P_i$. Since P_i cannot contain α, we have $\beta \in P_i \subseteq P^*$. This proves that P^* is a prime ideal.

Next put $P^{**} = \bigcap_i P_i$. Clearly P^{**} is a proper ideal. Assume that $\alpha'\beta'$ belongs to P^{**} but α' does not. It is then possible to choose i so that $\alpha' \notin P_i$. This shows that $\beta' \in P_i$. Finally let j be an arbitrary element of I. If $P_i \subseteq P_j$, then $\beta' \in P_j$. On the other hand, if $P_j \subseteq P_i$ then $\alpha'\beta' \in P_j$ while $\alpha' \notin P_j$; consequently $\beta' \in P_j$. Thus in any event $\beta' \in P_j$ and therefore $\beta' \in P^{**}$. Accordingly P^{**} is a prime ideal and the proof is complete.

Definition. *A prime ideal P is called a 'minimal prime ideal' of an ideal A, if it contains A and there is no smaller prime ideal with this property.*

A minimal prime ideal of the zero ideal is also known as a *minimal prime ideal of the ring*. Thus a prime ideal P is a minimal prime ideal of its ring if it does not strictly contain any other prime ideal.

Theorem 3. *Let an ideal A be contained in a prime ideal P. Then P contains a minimal prime ideal of A.*

Proof. Denote by Ω the set of all *prime ideals* which contain A and are contained in P. Then $P \in \Omega$ and therefore Ω is not empty. If P', P'' belong to Ω, then we shall write $P' \leqslant P''$ if $P'' \subseteq P'$. (Note the change in the order of P' and P''.) This gives a partial order on Ω. *We contend that Ω is an inductive system.*

For let Σ be a non-empty totally ordered subset of Ω. By Lemma 5 the intersection of all the members of Σ is a prime ideal $\bar{P}$. This certainly contains A and is contained in P. Consequently $\bar{P} \in \Omega$. Also since $\bar{P} \subseteq P'$ for every $P' \in \Sigma$ we have $P' \leqslant \bar{P}$ for every P' in Σ. Thus $\bar{P}$ is an upper bound for Σ and the contention that Ω is an inductive system has been shown to be correct.

By Zorn's lemma, Ω contains a maximal element P^*. Since $P^* \in \Omega$, it is a prime ideal and $A \subseteq P^* \subseteq P$. Suppose now that P_1 is a prime ideal satisfying $A \subseteq P_1 \subseteq P_1$. Then $P_1 \in \Omega$ and $P^* \leqslant P_1$. Consequently, since P^* is maximal in Ω, $P_1 = P^*$. This shows that P^* is a minimal prime ideal of A and completes the proof.

Corollary 1. *Every proper ideal possesses at least one minimal prime ideal.*

This follows by combining the result just proved with Theorem 2 and the corollary to Proposition 1. Again, if in Theorem 3 we let A be the zero ideal, then we obtain

Corollary 2. *Let P be a prime ideal. Then P contains at least one minimal prime ideal of the ring.*

Theorem 4. *The radical of an ideal A is the intersection of all its minimal prime ideals.*

Proof. If A is the whole ring, then the assertion holds by virtue of the convention regarding the intersection of an empty set of ideals. We shall therefore assume that A is a proper ideal. Now suppose that $\alpha \in \operatorname{Rad} A$ and let P be any prime ideal containing A. Then for a suitable integer m, $\alpha^m \in A \subseteq P$ and hence $\alpha \in P$ by Lemma 2. This shows that Rad A is contained in the intersection of all the minimal prime ideals of A.

Finally assume that $\beta \notin \operatorname{Rad} A$. If we can show that there exists a minimal prime ideal of A which does not contain β, then the proof will be complete.

Denote by S the set of powers $\beta, \beta^2, \beta^3, \ldots$; then S is multiplicatively closed and it does not meet A. Hence, by Theorem 1, there exists a prime ideal P such that $A \subseteq P$ and P does not meet S. In particular, $\beta \notin P$. Lastly, by Theorem 3, there exists a minimal prime ideal P^* of A such that $A \subseteq P^* \subseteq P$. It is clear that $\beta \notin P^*$; consequently all is proved.

There is a special case of the last theorem which is worth noting. Before recording it we make the

Definition. *An element of a ring is said to be 'nilpotent' if some positive power of it is zero.*

Thus the nilpotent elements are just the elements which belong to the radical of the zero ideal. Accordingly if we take A to be the zero ideal in Theorem 4, then we obtain the

Corollary. *An element of a ring is nilpotent if and only if it belongs to all the minimal prime ideals of the ring.*

2.4 Subrings and extension rings

Let R and R' be rings and suppose that each element of R happens also to be an element of R'. Of course, this by itself does not ensure that the ring structure of R has a natural connection with that of R'.

Definition. *If the inclusion mapping $R \to R'$ is a ring-homomorphism, then R is said to be a 'subring' of R' and R' is called an 'extension ring' of R.*

For example, the integers $0, \pm 1, \pm 2, \ldots$ form a ring. This is a subring of the ring consisting of all complex numbers $m + in$, where m, n are integers and $i = \sqrt{-1}$.

Suppose that R is a subring of R'. Then $0_R = 0_{R'}$ and, because of the way we framed the definition of a ring-homomorphism,† we also have $1_R = 1_{R'}$. Again if α, β belong to R, we get the same values for their sum, difference and product whether we regard α, β as belonging to the smaller or the larger ring.

Still assuming that R' is an extension ring of R, let A be an R-ideal. The elements of A will then generate an ideal of R'. This latter ideal is

† See section (1.10).

called the *extension of A to R'* and it is denoted by AR' or $R'A$ depending on which happens to be the more convenient. Note that a typical element of AR' can be expressed in the form $a_1 r'_1 + a_2 r'_2 + ... + a_n r'_n$, where $a_1, a_2, ..., a_n$ belong to A and $r'_1, r'_2, ..., r'_n$ to R'. Note, too, that if $A_1, A_2, ..., A_n$ are R-ideals, then

$$(A_1 + A_2 + ... + A_n) R' = (A_1 R') + (A_2 R') + ... + (A_n R') \quad (2.4.1)$$

and $$(A_1 A_2 ... A_n) R' = (A_1 R') (A_2 R') ... (A_n R'). \quad (2.4.2)$$

Of course, any set of elements which generates an R-ideal is also a set of generators for the extension of that ideal to R'.

Now let A' be an ideal of the extension ring R'. It is an easy matter to verify that $A' \cap R$ is an ideal of the subring R. This latter ideal is called the *contraction of A' in R*. Observe that if A' is proper, then $1_{R'} \notin A'$ and therefore $1_R \notin A' \cap R$. Thus the contraction of a proper ideal is always a proper ideal. However it is not the case that the *extension* of a proper ideal is necessarily proper.

If A is an R-ideal, then AR' is an R'-ideal and therefore we can form its contraction. Note that

$$A \subseteq (AR') \cap R. \quad (2.4.3)$$

On the other hand, if A' is an R'-ideal, then $A' \cap R$ is an R-ideal and we may form its extension. This time we have

$$(A' \cap R) R' \subseteq A'. \quad (2.4.4)$$

Lemma 6. *Let R' be an extension ring of a ring R and let P' be a prime ideal of the larger ring. Then the contraction $P = P' \cap R$ of P' is a prime ideal of R.*

Proof. Since P is the contraction of a proper ideal, it is itself a proper ideal. Now assume that a, b belong to R, $ab \in P$ and $b \notin P$. Then $ab \in P'$, $b \notin P'$. Consequently, since P' is prime, $a \in P'$. Thus

$$a \in P' \cap R = P$$

and the proof is complete.

2.5 Integral extensions

Let R' be an extension ring of R and let α belong to R'. The element α is said to be *integral with respect to R* or *integral over R* if it satisfies a relation of the form

$$\alpha^n + a_1 \alpha^{n-1} + a_2 \alpha^{n-2} + ... + a_n = 0, \quad (2.5.1)$$

where $a_1, a_2, ..., a_n$ are in R and $n \geqslant 1$. If every element of R' is integral with respect to R, then we say that R' is an *integral extension* of R.

Before we begin to develop the theory of integral elements and integral extensions, we shall introduce some additional terminology and notation. If x is a new symbol, then an expression of the form

$$x^m + b_1 x^{m-1} + b_2 x^{m-2} + ... + b_m \quad (b_i \in R)$$

is called a *monic polynomial* in x with coefficients in R. Thus the essential property of a monic polynomial in x is that the highest power of x actually present has the identity element as its coefficient. An element of R' is now seen to be integral with respect to R if it is a root of a monic polynomial with coefficients in R. Observe that if $r \in R$, then $x - r$ is such a monic polynomial and it has r as a root. Thus every element of R, when regarded as an element of R', is integral with respect to R. If no other element of R' is integral with respect to R, then R is said to be *integrally closed* in R'.

Let $\alpha_1, \alpha_2, ..., \alpha_s$ be elements of an extension ring R'. The set of all elements of R' which can be expressed as polynomials in $\alpha_1, \alpha_2, ..., \alpha_s$, with coefficients in R, is denoted by $R[\alpha_1, \alpha_2, ..., \alpha_s]$. Thus an element $\beta \in R'$ belongs to $R[\alpha_1, \alpha_2, ..., \alpha_s]$ if and only if it is a linear combination (with coefficients in R) of a finite number of power-products $\alpha_1^{\mu_1} \alpha_2^{\mu_2} ... \alpha_s^{\mu_s}$. Here $\mu_1, \mu_2, ..., \mu_s$ denote non-negative integers. It is a simple matter to verify that $R[\alpha_1, \alpha_2, ..., \alpha_s]$ is an extension ring of R and a subring of R'. Indeed $R[\alpha_1, \alpha_2, ..., \alpha_s]$ is the smallest subring of R' containing R and the elements $\alpha_1, \alpha_2, ..., \alpha_s$. For this reason it is called the *subring of R' generated by R and the elements $\alpha_i (1 \leqslant i \leqslant s)$*. Suppose for the moment that $s > 1$. Then we can form $R^* = R[\alpha_1, \alpha_2, ..., \alpha_{s-1}]$ and afterwards construct $R^*[\alpha_s]$. Since $R[\alpha_1, \alpha_2, ..., \alpha_s]$ contains both R^* and α_s, it must contain $R^*[\alpha_s]$. On the other hand, $R^*[\alpha_s]$ contains $R[\alpha_1, \alpha_2, ..., \alpha_s]$ because it contains R and all the α_i. Accordingly

$$R[\alpha_1, \alpha_2, ..., \alpha_s] = R^*[\alpha_s]. \tag{2.5.2}$$

One further remark. If $r \in R$ and $\alpha \in R'$, then we can form $r\alpha$ and, in this way, R' becomes an R-module. These various observations are brought together in

Proposition 9. *Let R' be an extension ring of R and α an element of R'. Then α is integral with respect to R if and only if $R[\alpha]$ is a finitely generated R-module.*

Proof. Suppose that α is integral with respect to R. Then we have a relation

$$\alpha^n + a_1 \alpha^{n-1} + a_2 \alpha^{n-2} + ... + a_n = 0, \tag{2.5.3}$$

where the a_i belong to R. We shall show that

$$R[\alpha] = R1 + R\alpha + \ldots + R\alpha^{n-1}.$$

Clearly it will suffice to show that α^m belongs to the right-hand side for all $m \geqslant 0$. For this we use complete induction on m. Note that for $m = 0$ the assertion is trivial.

Now assume that $m > 0$ and that $1, \alpha, \ldots, \alpha^{m-1}$ are all known to belong to $R1 + R\alpha + \ldots + R\alpha^{n-1}$. If $m < n$ then it is obvious that α^m belongs to the module. If, however, $m \geqslant n$ then, by (2.5.3),

$$\alpha^m = (-a_n)\alpha^{m-n} + (-a_{n-1})\alpha^{m-n+1} + \ldots + (-a_1)\alpha^{m-1}.$$

Since, by the induction hypothesis, $\alpha^{m-n}, \alpha^{m-n+1}, \ldots, \alpha^{m-1}$ all belong to $R1 + R\alpha + \ldots + R\alpha^{n-1}$ it immediately follows that α^m belongs to it as well. This proves one half of the proposition. The other half is a special case of

Theorem 5. *Let R' be an extension ring of R and suppose that R' is finitely generated as an R-module. Then every element of R' is integral with respect to R.*

Proof. Since R' is a finitely generated R-module, we can find

$$\omega_1, \omega_2, \ldots, \omega_n \quad \text{so that} \quad R' = R\omega_1 + R\omega_2 + \ldots + R\omega_n.$$

Now let α be an arbitrary element of R'. Then, for each i $(1 \leqslant i \leqslant n)$, we have $\alpha\omega_i \in R'$ and therefore we have a relation

$$\alpha\omega_i = a_{i1}\omega_1 + a_{i2}\omega_2 + \ldots + a_{in}\omega_n \quad (1 \leqslant i \leqslant n),$$

where the a_{ij} are elements of R. Let δ_{ij} $(1 \leqslant i,j \leqslant n)$ be the Kronecker symbol (so that $\delta_{ij} = 1$ if $i = j$ and $\delta_{ij} = 0$ when $i \neq j$). Then

$$\sum_j (\alpha\delta_{ij} - a_{ij})\omega_j = 0.$$

Denote by Δ the value of the nth order determinant $|\alpha\delta_{ij} - a_{ij}|$. Then $\Delta\omega_j = 0$ for every j. Since $R' = R\omega_1 + R\omega_2 + \ldots + R\omega_n$, it follows that $\Delta 1_{R'} = 0$ and hence that $\Delta = 0$. Expanding the determinant

$$\Delta = |\alpha\delta_{ij} - a_{ij}|,$$

we obtain an equation

$$\alpha^n + b_1\alpha^{n-1} + b_2\alpha^{n-2} + \ldots + b_n = 0,$$

where $b_1, b_2, \ldots, b_n$ belong to R. Thus α is integral over R and the proof is complete.

The next lemma will enable us to take full advantage of the results so far obtained.

Lemma 7. *Let R' be an extension ring of R and R'' an extension ring of R'. Then R'' is an extension ring of R. Furthermore, if R' is a finitely generated R-module and R'' a finitely generated R'-module, then R'' is a finitely generated R-module.*

Proof. The first point is clear. Now assume that

$$R' = R\omega_1 + R\omega_2 + \ldots + R\omega_n \quad \text{and} \quad R'' = R'\eta_1 + R'\eta_2 + \ldots + R'\eta_m.$$

Let r'' be an arbitrary element of R''. Then $r'' = r_1'\eta_1 + r_2'\eta_2 + \ldots + r_m'\eta_m$, where the r_j' are suitable elements of R'. Next we have relations

$$r_j' = a_{j1}\omega_1 + a_{j2}\omega_2 + \ldots + a_{jn}\omega_n \quad (1 \leqslant j \leqslant m),$$

where the elements a_{ij} belong to R. It follows that r'' belongs to $\sum_{i,j} R\omega_i\eta_j$. Thus the nm elements $\omega_i\eta_j$ generate R'' as an R-module and the lemma is proved.

Corollary. *Let R' be an extension ring of R and let $\alpha_1, \alpha_2, \ldots, \alpha_n$ be elements of R' which are integral over R. Then $R[\alpha_1, \alpha_2, \ldots, \alpha_n]$ is a finitely generated R-module.*

Proof. For $0 \leqslant i \leqslant n$, put $R_i = R[\alpha_1, \alpha_2, \ldots, \alpha_i]$, where by R_0 we mean R itself. By (2.5.2), $R_i = R_{i-1}[\alpha_i]$ and, since α_i is integral with respect to R, it is *a fortiori* integral with respect to the larger ring R_{i-1}. Proposition 9 now shows that R_i is a finitely generated R_{i-1}-module. The corollary therefore follows by repeated applications of the lemma which has just been proved.

We can now derive the important

Theorem 6. *Let R' be an extension ring of R and denote by $\hat{R}$ the set of all elements which belong to R' and are integral over R. Then $\hat{R}$ is an extension ring of R and a subring of R'.*

Proof. It is clear that $R \subseteq \hat{R} \subseteq R'$ so we need only show that the elements of $\hat{R}$ form a ring. Let α and β belong to $\hat{R}$. Clearly the required result will follow if we can show that $-\alpha$, $\alpha + \beta$, and $\alpha\beta$ also belong to $\hat{R}$. Now α and β are integral with respect to R. Hence, by the corollary to Lemma 7, $R[\alpha, \beta]$ is a finitely generated R-module. It therefore follows, by Theorem 5, that every element of $R[\alpha, \beta]$ is integral with respect to R and hence belongs to $\hat{R}$. However, $-\alpha$, $\alpha + \beta$ and $\alpha\beta$ all belong to $R[\alpha, \beta]$. Consequently the theorem is proved.

Definition. *Let the notation be as in Theorem 6. Then the ring $\hat{R}$ is called the 'integral closure' of R in R'.*

The term *integral closure* suggests that if we repeat the construction using $\hat{R}$ in place of R, then we shall get nothing new. First, however, we shall prove a more general result. This shows that, for rings, the relation of being integrally dependent is *transitive*.

Theorem 7. *Let R' be an integral extension of a ring R and R'' an arbitrary extension ring of R'. If now an element η of R'' is integral with respect to R', then η is integral with respect to R.*

Proof. Since η is integral with respect to R' we have a relation

$$\eta^n + \alpha_1 \eta^{n-1} + \alpha_2 \eta^{n-2} + \dots + \alpha_n = 0, \tag{2.5.4}$$

where $\alpha_1, \alpha_2, \dots, \alpha_n$ all belong to R' and $n \geqslant 1$. Put

$$R^* = R[\alpha_1, \alpha_2, \dots, \alpha_n].$$

Since each α_i is integral with respect to R, the corollary to Lemma 7 shows that R^* is a finitely generated R-module. Next, (2.5.4) tells us that η is integral with respect to R^*. Accordingly, by Proposition 9, $R^*[\eta]$ is a finitely generated R^*-module. It therefore follows, by Lemma 7, that $R^*[\eta]$ is a finitely generated R-module and hence (Theorem 5) that η is integral with respect to R.

Corollary. *Let R' be an arbitrary extension ring of a ring R and let $\hat{R}$ be the integral closure of R in R'. Then the integral closure of $\hat{R}$ in R' is just $\hat{R}$ itself.*

This justifies the use of the expression *integral closure*.

2.6 Prime ideals and integral dependence

Let R' be a ring containing R as subring. If P' is a prime ideal of R', then, by Lemma 6, $P' \cap R$ is a prime ideal of R. In general, not every prime ideal of R will be the contraction of a prime ideal of R' but, as we shall now prove, this is the case when R' is an integral extension of R.

Theorem 8. *Let R' be an integral extension of a ring R and let P be a prime R-ideal. Then R' contains a prime ideal P' such that $P' \cap R = P$.*

Proof. Let S be the complement of P in R. Then S is multiplicatively closed in both R and R'. Denote by Ω the set of all R'-ideals which do not meet S and, if A_1', A_2' belong to Ω, let $A_1' \leqslant A_2'$ mean that A_1' is contained in A_2'. It will be shown that Ω *is a non-empty inductive system every one of whose maximal elements is a prime R'-ideal which contracts to P in R.* The reader is here informed that the statement in italics (which contains more than is actually required for the theorem) will be used again later.

Since S does not meet P, it cannot contain the zero element. It therefore follows, from Proposition 3, that Ω is a non-empty inductive system and all its maximal elements are prime ideals. Let P' be such a maximal element. Since P' does not meet S, it follows that $P' \cap R \subseteq P$. To complete the proof, *we assume that there exists an element* $p \in P$ *which does not belong to P', and derive a contradiction.*

Let B' consist of all elements of R' which can be expressed in the form $r'p + \pi'$, where $r' \in R'$ and $\pi' \in P'$. Then $B' = R'p + P'$ and so B' is an R'-ideal containing both P' and p. Accordingly B' *strictly* contains P'. However, P' is maximal in Ω; consequently $B' \notin \Omega$ that is to say B' meets S. It is therefore possible to choose $r' \in R'$ and $\pi' \in P'$ so that $r'p + \pi' = s$ (say) is an element of S.

Now r' is integral over R and therefore it satisfies a relation

$$r'^n + a_1 r'^{n-1} + a_2 r'^{n-2} + \ldots + a_n = 0,$$

where $a_1, a_2, \ldots, a_n$ belong to R. Multiplying by p^n and putting $r'p = s - \pi'$, we obtain

$$(s - \pi')^n + a_1 p (s - \pi')^{n-1} + a_2 p^2 (s - \pi')^{n-2} + \ldots + a_n p^n = 0$$

whence $s^n = \rho' \pi' + \rho p$, where $\rho' \in R'$ and $\rho \in R$. Thus $s^n - \rho p$ belongs to $P' \cap R \subseteq P$ and therefore $s^n \in P$. But $s^n \in S$ and now we have the required contradiction.

Theorem 9. *Let R' be an integral extension of a ring R, let P'_0 be a prime R'-ideal and $P_0 = P'_0 \cap R$ the contraction of P'_0 in R. If now P is a prime ideal of R such that $P_0 \subseteq P$, then there exists a prime ideal P' of R' such that $P'_0 \subseteq P'$ and $P' \cap R = P$.*

Proof. Let S denote the complement of P in R and define Ω and $\leqslant$ exactly as in the proof of Theorem 8. Then Ω is a non-empty inductive system containing P'_0. Accordingly, by Lemma 1, there exists a maximal element P' of Ω such that $P'_0 \leqslant P'$. As we saw above, the fact that P' is maximal in Ω ensures that P' is a prime ideal whose contraction is P. On the other hand, since $P'_0 \leqslant P', P'_0$ is contained in P'.

Theorem 10. *Let A be a proper R-ideal and let R' be an integral extension of R. Then the extension $R'A$, of A to R', is also a proper ideal.*

Proof. By Theorem 2, there exists a prime ideal P such that $A \subseteq P$ and, by Theorem 8, there exists a prime R'-ideal P' such that

$$P' \cap R = P.$$

Then $R'A \subseteq P'$ which shows that $R'A$ is proper.

Proposition 10. *Let R' be an integral extension of a ring R, let A' be an R'-ideal and P' a prime R'-ideal. If now $P' \subseteq A'$ and $P' \cap R = A' \cap R$, then $A' = P'$.*

Proof. Let α belong to A'. We shall assume that α does not belong to P' and derive a contradiction.

It is possible to find $a_1, a_2, \ldots, a_n$ in R such that $\alpha^n + a_1 \alpha^{n-1} + \ldots + a_n$ belongs to P'. (Indeed we can find elements $b_1, b_2, \ldots, b_m$ of R such that $\alpha^m + b_1 \alpha^{m-1} + \ldots + b_m = 0$.) Let the elements $a_1, a_2, \ldots, a_n$ be so chosen that their number n is as small as possible. Then $n \geqslant 1$. If now

$$\alpha^n + a_1 \alpha^{n-1} + a_2 \alpha^{n-2} + \ldots + a_n = \pi',$$

then $\pi' \in P'$ and, since

$$a_n = \pi' - \alpha(\alpha^{n-1} + a_1 \alpha^{n-2} + \ldots + a_{n-1}),$$

we see that $a_n \in A' \cap R = P' \cap R \subseteq P'$. It follows that

$$\alpha(\alpha^{n-1} + a_1 \alpha^{n-2} + \ldots + a_{n-1})$$

belongs to P'. But α does not belong to P'. Consequently

$$\alpha^{n-1} + a_1 \alpha^{n-2} + \ldots + a_{n-1}$$

is an element of P'. This, however, contradicts the minimal property of the integer n.

Theorem 11. *Let R' be an integral extension of R, P' a prime ideal of R' and $P = P' \cap R$ its contraction in R. Then P' is a maximal ideal of R' if and only if P is a maximal ideal of R.*

Proof. First suppose that P is a maximal ideal of R. By Theorem 2, P' is contained in a maximal R'-ideal A' say. Then

$$P = P' \cap R \subseteq A' \cap R$$

whence, since $A' \cap R$ is a proper ideal, $P = A' \cap R$. That $P' = A'$ now follows from Proposition 10.

Next assume that P' is a maximal ideal. In the ring R it is possible to find a maximal and hence prime ideal P_1 such that $P \subseteq P_1$. Then, by Theorem 9, there exists a prime ideal P'_1 of R' such that $P' \subseteq P'_1$ and $P'_1 \cap R = P_1$. However, P' is a maximal ideal. Consequently $P' = P'_1$ and therefore $P = P_1$. Accordingly P is a maximal ideal of R.

Theorem 12. *Let R and R' be integral domains and suppose that R' is an integral extension of R. If now either R or R' is a field, then so is the other.*

Proof. The zero ideals (0_R) and $(0_{R'})$ of R and R' are prime ideals and the former is the contraction of the latter. Using Theorem 11 and the corollary to Lemma 3, we see that the following four statements are equivalent:

(a) R is a field;
(b) $\{0_R\}$ is a maximal ideal;
(c) $\{0_{R'}\}$ is a maximal ideal;
(d) R' is a field.

This proves the theorem.

2.7 Further operations with ideals and modules

So far we have concerned ourselves mainly with prime ideals. In the next section, we shall begin the study of primary submodules. Here we interpose some general considerations concerning ideals and modules which it is convenient to examine at this stage.

Let R be a ring and E a left R-module. If $r \in R$ and $x \in E$, then rx is defined. Suppose now that we *define* xr by putting $xr = rx$. In the case of a *non-commutative* ring, this would turn E into a right module over the ring which is opposite† to R. However, we have agreed that, in Chapter 2, only commutative rings shall be considered. On this understanding, the above definition of xr turns E into a right R-module. Thus in the commutative case, every left R-module may be regarded as a right R-module and, of course, conversely. There is therefore no point in preserving the distinction between left modules and right modules. Accordingly we shall, in future, speak simply of R-modules.

Let E be an R-module and A an ideal. The submodule of E which is generated by all products ae, where $a \in A$ and $e \in E$, will be denoted by AE. The typical element of AE can therefore be represented as a sum $a_1 e_1 + a_2 e_2 + \dots + a_n e_n$, where $a_i \in A$ and $e_i \in E$. Of course, when E itself is an ideal, AE is just the product of two ideals as defined in section (2.3). Note that if $\{A_i\}_{i \in I}$ is a family of ideals of R and $\{K_j\}_{j \in J}$ a family of submodules of E, then

$$(\sum_i A_i)(\sum_j K_j) = \sum_{i,j} A_i K_j. \tag{2.7.1}$$

This is a generalization of (2.3.1). We leave the verification to the reader.

† See section (1.11).

Again, if A, B are ideals then we can form both $A(BE)$ and $(AB)E$. However each of these is just the submodule of E consisting of those elements which can be expressed as sums $a_1 b_1 e_1 + a_2 b_2 e_2 + \ldots + a_n b_n e_n$, where the notation is self-explanatory. Accordingly

$$A(BE) = (AB)E = B(AE) \qquad (2.7.2)$$

and we may use ABE to denote any one of them. If α is a *fixed* element of R, then the set of all elements αe, where $e \in E$, is a submodule of E which we denote by αE. Of course, $\alpha E = (R\alpha) E$ but the former notation is less clumsy.

Let K be a submodule of E and U an arbitrary subset of E. Denote by $K:U$ the set consisting of all elements α of R such that $\alpha u \in K$ for every $u \in U$. $K:U$ *is an ideal.* For let α_1, α_2 belong to $K:U$ and let $r \in R$. Then for $u \in U$ we have $(\alpha_1 + \alpha_2) u = \alpha_1 u + \alpha_2 u \in K$ while

$$(r\alpha_1) u = r(\alpha_1 u) \in K.$$

Observe that if RU denotes the submodule of E generated by U, then

$$K:U = K:RU. \qquad (2.7.3)$$

Indeed it is clear that $K:RU$ is contained in $K:U$. Now let α be an element of $K:U$. If $x = r_1 u_1 + r_2 u_2 + \ldots + r_n u_n$, where $r_i \in R$ and $u_i \in U$, then $\alpha x = r_1(\alpha u_1) + \ldots + r_n(\alpha u_n)$ which belongs to K because each αu_i belongs to K. Thus α belongs to $K:RU$ and (2.7.3) follows. Note too that if A is an ideal and Γ is a subset of R, then the above construction attaches a meaning to $A:\Gamma$ making it an ideal. (For this we take $E = R$, $K = A$, and $U = \Gamma$.) Naturally $A \subseteq (A:\Gamma) = (A:R\Gamma)$.

Once again, let K be a submodule of E and Γ a *subset* of R. We denote by $K:_E \Gamma$ the set of all $x \in E$ such that $\gamma x \in K$ for every $\gamma \in \Gamma$. It is a simple matter to check that $K:_E \Gamma$ is a submodule of E and that

$$K \subseteq (K:_E \Gamma) = (K:_E R\Gamma). \qquad (2.7.4)$$

Note that $K:\Gamma$ by itself is usually ambiguous because it depends on the module in which K is embedded and we have given no indication as to which this is. For example $K:_K \Gamma = K$. However, if we have a definite ambient module E and are investigating its submodules, then we may use $K:\Gamma$ in place of $K:_E \Gamma$ if there is no risk of confusion.

In the proposition which follows, U is a subset of E and $\{K_i\}_{i \in I}$ a family of submodules of E. Further, Γ denotes a subset of R and $\{A_j\}_{j \in J}$ a family of ideals.

Proposition 11. *With the above notation*

$$\Big(\bigcap_{i \in I} K_i \Big) : U = \bigcap_{i \in I} (K_i : U); \qquad (2.7.5)$$

$$(\bigcap_{i\in I} K_i):_E \Gamma = \bigcap_{i\in I} (K_i:_E \Gamma); \qquad (2.7.6)$$

and if K is a submodule of E, then

$$K:(\sum_{i\in I} K_i) = \bigcap_{i\in I}(K:K_i); \qquad (2.7.7)$$

$$K:_E(\sum_{j\in J} A_j) = \bigcap_{j\in J}(K:_E A_j). \qquad (2.7.8)$$

Proof. Let $\alpha \in R$ and $u\in U$. Then αu belongs to $\bigcap_i K_i$ if and only if $\alpha u \in K_i$ for each i. Keeping α fixed and varying u, we see that α belongs to $(\bigcap_i K_i):U$ if, and only if, α belongs to $K_i:U$ for each i. This establishes (2.7.5) and we can establish (2.7.6) by a similar argument.

If $i_0\in I$ then $K_{i_0} \subseteq \sum_i K_i$ and therefore $(K:\sum_i K_i) \subseteq (K:K_{i_0})$. This shows that $K:\sum_i K_i$ is contained in $\bigcap_i (K:K_i)$. Now let α belong to $\bigcap_i (K:K_i)$ and let $x\in \sum_i K_i$. If we can show that $\alpha x\in K$, then it will follow that α belongs to $K:\sum_i K_i$ and (2.7.7) will be proved. Now $x = \Sigma x_i$, where $x_i\in K_i$ and the sum contains only a finite number of non-zero terms. Hence $\alpha x = \Sigma \alpha x_i\in K$ because, since α is in $K:K_i$, we have $\alpha x_i\in K$ for every i. Thus (2.7.7) is established and a very similar argument yields (2.7.8).

In the proposition which follows, A and B denote ideals of a ring R, E is an R-module and K, N submodules of E.

Proposition 12. *With the notation explained above,*

$$(N:K):A = N:AK = (N:_E A):K \qquad (2.7.9)$$

and $$(N:_E A):_E B = N:_E AB = (N:_E B):_E A. \qquad (2.7.10)$$

Proof. We shall only consider (2.7.9) since (2.7.10) can be treated similarly. First suppose that r belongs to $(N:K):A$ and let k, a denote arbitrary elements of K and A respectively. Then ra is in $N:K$ and therefore $rak\in N$. Now a typical element of AK is a sum of elements of the form ak. It follows that $rx\in N$ for every $x\in AK$ and therefore r belongs to $N: AK$. Thus $(N:K):A$ is contained in $N:AK$.

Next let r be an element of $N:AK$ and let a, k be as before. Then $ark = rak\in N$. Keeping k fixed and varying a, we see that rk belongs to $N:_E A$. As this holds for every k, we may conclude that r is in

$$(N:_E A):K.$$

Accordingly $N:AK$ is contained in $(N:_E A):K$.

Finally let r be in $(N:_E A):K$. Then rk belongs to $N:A$ and therefore $ark \in N$. Here, as before, $a \in A$ and $k \in K$. Keeping a fixed we find that $ar \in (N:K)$ and, as this holds for all a, it follows that r is an element of $(N:K):A$. Thus $(N:_E A):K$ is contained in $(N:K):A$ and all is proved.

It will be recalled† that the *annihilator* of E consists of all the elements $r \in R$ such that $rE = 0$, where in this context 0 denotes the zero submodule of E. Sometimes we use $\text{Ann}_R E$ to denote the annihilator. Thus $\text{Ann}_R E = 0:E$. More generally, if K is a submodule of E then, as is easily verified, $\text{Ann}_R(E/K) = K:E$.

Proposition 13. *Suppose that the R-module E can be generated by n elements and let B be an ideal containing the annihilator $0:E$ of E. If now A is an ideal such that $AE \subseteq BE$, then $A^n \subseteq B$.*

Proof. Let $E = Re_1 + Re_2 + \ldots + Re_n$ and let $\alpha_1, \alpha_2, \ldots, \alpha_n$ belong to A. For each $i \, (1 \leqslant i \leqslant n)$ we have $\alpha_i e_i \in AE \subseteq BE$ and therefore

$$\alpha_i e_i = \sum_j \beta_{ij} e_j,$$

where the β_{ij} are suitable elements of B. Accordingly

$$\sum_j (\beta_{ij} - \alpha_i \delta_{ij}) e_j = 0,$$

where δ_{ij} is the Kronecker symbol. Put

$$\Delta = \begin{vmatrix} \beta_{11} - \alpha_1 & \beta_{12} & \cdots & \beta_{1n} \\ \beta_{21} & \beta_{22} - \alpha_2 & \cdots & \beta_{2n} \\ \cdot & \cdot & \cdots & \cdot \\ \cdot & \cdot & \cdots & \cdot \\ \beta_{n1} & \beta_{n2} & \cdots & \beta_{nn} - \alpha_n \end{vmatrix}.$$

Then $\Delta e_j = 0$ for $1 \leqslant j \leqslant n$. Thus $\Delta E = 0$ and therefore Δ belongs to $0:E \subseteq B$. But, on expanding Δ, we see that this implies that $\alpha_1 \alpha_2 \ldots \alpha_n$ belongs to B. Since each element of A^n is a sum of elements of the form $\alpha_1 \alpha_2 \ldots \alpha_n$, this proves that $A^n \subseteq B$.

Proposition 14. *Let E be an R-module generated by n elements and let α belong to R. Then $[\text{Ann}_R(E/\alpha E)]^n \subseteq \text{Ann}_R E + R\alpha \subseteq \text{Ann}_R(E/\alpha E)$.*

Proof. Put $B = \text{Ann}_R E + R\alpha$ and $A = \text{Ann}_R(E/\alpha E)$. Then $AE \subseteq \alpha E = BE$ hence $A^n \subseteq B$ by the last result. Next from $BE = \alpha E$ follows $B \subseteq \alpha E:E = \text{Ann}_R(E/\alpha E)$. Thus the proof is complete.

† See section (1.14).

2.8 Primary submodules

We come now to one of the central concepts of the present chapter. Let E be an R-module. For the present, we shall concern ourselves with the submodules of E.

Definition. *A submodule N of E is called a 'primary submodule' if*
(a) *N is a proper submodule of E, i.e. $N \neq E$; and*
(b) *from $re \in N$ and $e \notin N$ follows $r^m E \subseteq N$ for some positive integer m.*
It is, of course, understood that, in (b), r denotes an element of R and e an element of E.

If we take $E = R$ and for N an ideal Q, then the above definition tells us what to understand by a primary ideal. However, since $r^m R$ is contained in Q when and only when $r^m \in Q$, the details can be simplified a little.

Definition. *An ideal Q of a ring R (commutative and possessing an identity element) is called a 'primary ideal' if*
(a) *Q is a proper ideal, and*
(b) *from $r\alpha \in Q$ and $\alpha \notin Q$ follows $r^m \in Q$ for some positive integer m.*

Proposition 15. *The radical of a primary ideal is a prime ideal.*

Proof. Let Q be a primary ideal and put $P = \operatorname{Rad} Q$. By Proposition 6, P is an ideal. Furthermore $1 \notin P$. (For otherwise $1 \in Q$ contradicting (a) of the above definition.) Now assume that $\alpha\beta \in P$, $\beta \notin P$, where α, β are elements of R. If m is a large enough positive integer, then

$$\alpha^m \beta^m = (\alpha\beta)^m \in Q.$$

However $\beta^m \notin Q$ for otherwise β would belong to P. Accordingly some power of α^m belongs to Q. This shows that $\alpha \in \operatorname{Rad} Q = P$ and establishes the proposition.

Definition. *If Q is a primary ideal and $P = \operatorname{Rad} Q$, then Q is said to 'belong to P' or to be 'P-primary'.*
For example, every prime ideal is primary and belongs to itself.

Proposition 16. *Let A be a proper ideal of R. Then A is primary if and only if every zero-divisor in R/A is nilpotent.*

Proof. Suppose that the zero-divisors in R/A are nilpotent. If now $\alpha\beta \in A$ and $\beta \notin A$, then $\phi(\alpha)\phi(\beta) = 0$, $\phi(\beta) \neq 0$, where $\phi : R \to R/A$ is the natural mapping. By hypothesis, $\phi(\alpha)$ is nilpotent, say $\phi(\alpha)^m = 0$. Then $\phi(\alpha^m) = 0$ and therefore $\alpha^m \in A$. This proves that A is primary.

The converse is established by reversing the argument and using the fact that ϕ is an epimorphism.

Let $\phi: R \to R'$ be a ring-epimorphism. We saw, in Proposition 2, that there is a one-one correspondence between the prime ideals of R' and the prime ideals of R containing $\operatorname{Ker} \phi$. A similar result holds for primary ideals namely

Proposition 17. *Let $\phi: R \to R'$ be a ring-epimorphism of R on to R', let A' be an R'-ideal, and $\phi^{-1}(A') = A$ (say) the corresponding R-ideal. Then A' is primary if and only if A is primary. Furthermore, if A' is P'-primary then A is P-primary, where $P = \phi^{-1}(P')$.*

Proof. The rings R/A and R'/A' are isomorphic.† Hence if one is non-null and has the property that all its zero-divisors are nilpotent, then the same will be true of the other. This establishes the first point. Now assume that A' is P'-primary. Then $r \in \operatorname{Rad} A$ if and only if $r^m \in A$ for some value of m. This in turn holds if and only if $\phi(r)^m \in A'$ for some value of m. Thus $r \in \operatorname{Rad} A$ when and only when

$$\phi(r) \in \operatorname{Rad} A' = P'.$$

Hence $\operatorname{Rad} A = \phi^{-1}(P')$.

We return to the consideration of primary submodules.

Proposition 18. *Let N be a primary submodule of an R-module E. Then $N:E$ (or equivalently $\operatorname{Ann}_R E/N$) is a primary ideal.*

Proof. Since $E \nsubseteq N$, the ideal $N:E$ is a proper ideal. Now suppose that α, β are elements of R such that $\alpha\beta \in (N:E)$, $\beta \notin (N:E)$. It will suffice to show that $\alpha^m \in (N:E)$, i.e. that $\alpha^m E \subseteq N$ for some positive integer m.

Since $\beta \notin (N:E)$, there exists $e \in E$ such that $\beta e \notin N$. We now have $\alpha(\beta e) \in N$, because $\alpha\beta \in (N:E)$, and $\beta e \notin N$. But N is a primary submodule, consequently $\alpha^m E \subseteq N$ for some integer m. This completes the proof.

Suppose that N is a primary submodule of E. Proposition 18 shows that $N:E$ is a primary ideal. Consequently, by Proposition 15, $P = \operatorname{Rad}(N:E)$ is a prime ideal. The connection between P and N is described in two ways. Sometimes we say that the *primary submodule N belongs to the prime ideal P*. On other occasions, we shall say that *N is P-primary*. Observe that if Q is a P-primary submodule of R, then

† See Theorem 22 of section (1.14).

$P = \operatorname{Rad} Q$. This is because $Q \colon R = Q$. Thus the new terminology is consistent with that introduced earlier for ideals.

The following lemma is often useful in dealing with the prime-primary relationship.

Lemma 8. *Let N be a proper submodule of an R-module E and let Π be an ideal of R. Assume further that the following two conditions are satisfied:*

(a) *if $rx \in N$ and $x \notin N$, then $r \in \Pi$ (here r and x denote elements of R and E respectively);*

(b) *if $r \in \Pi$, then $r^k E \subseteq N$ for some positive integer k.*

Then Π is a prime ideal and N a Π-primary sumbodule of E.

Proof. It follows at once, from (a) and (b), that N is a primary submodule of E. To complete the proof we shall show that $\Pi = \operatorname{Rad}(N \colon E)$.

By (b), Π is contained in $\operatorname{Rad}(N \colon E)$. Now assume that $r \in \operatorname{Rad}(N \colon E)$ and $r \neq 0$. Then $r^s \in (N \colon E)$, that is $r^s E \subseteq N$, for some positive integer s. Choose s to be as small as possible. Then $s \geqslant 1$ and $r^{s-1} E \nsubseteq N$. Accordingly there exists $e \in E$ such that $r^{s-1} e \notin N$. But then $r(r^{s-1} e) \in N$ and $r^{s-1} e \notin N$. Consequently $r \in \Pi$ by (a). This shows that

$$\operatorname{Rad}(N \colon E) \subseteq \Pi$$

and completes the proof.

Proposition 19. *Let N be a P-primary submodule of an R-module E and suppose that $re \in N$, where $r \in R$ and $e \in E$. Then either $r \in P$ or $e \in N$.*

Proof. If $e \notin N$, then $r^m E \subseteq N$ for some positive integer m. Accordingly r belongs to $\operatorname{Rad}(N \colon E) = P$.

Corollary 1. *Let N be a P-primary submodule of the R-module E and suppose that A is an ideal of R and K a submodule of E. If now $AK \subseteq N$, then either $A \subseteq P$ or $K \subseteq N$.*

Corollary 2. *Let N be a P-primary submodule of the R-module E and let D be a subset of R not contained in P. Then $N \colon_E D = N$.*

Proof. Since $N \colon_E D = N \colon_E RD$, where RD denotes the ideal generated by D, we may suppose that D is an ideal. Next $D(N \colon_E D) \subseteq N$ and $D \nsubseteq P$. Hence, by Corollary 1, $N \colon_E D \subseteq N$. However the opposite inclusion is trivial and so the corollary follows.

Proposition 20. *Let N be a P-primary submodule of an R-module E and let K be an arbitrary submodule of E. If $K \nsubseteq N$, then $N \colon K$ is a P-primary ideal. If $K \subseteq N$, then, of course, $N \colon K = R$.*

Proof. We shall make use of Lemma 8. Clearly it is sufficient to consider the case in which $K \nsubseteq N$. This secures that $N:K$ is a proper ideal.

Assume that $r\alpha \in (N:K)$ but $\alpha \notin (N:K)$. Here r, α denote elements of R. Then there exists $k \in K$ such that $\alpha k \notin N$. Accordingly $r(\alpha k) \in N$ but $\alpha k \notin N$. Consequently, by Proposition 19, $r \in P$.

Now assume that $r \in P$. Then $r^m E \subseteq N$ for some positive integer m and therefore $r^m \in (N:K)$. It follows that $r^m R \subseteq N:K$. Applying Lemma 8 we see that $N:K$ is a P-primary submodule of R, that is to say $N:K$ is a P-primary ideal.

The next result is similar to the one just proved.

Proposition 21. *Let N be a P-primary submodule of the R-module E and let A be an ideal of R. If $A \nsubseteq N:E$ then $N:_E A$ is a P-primary submodule of E. If however $A \subseteq N:E$, then, of course, $N:_E A = E$.*

Proof. We need only consider the case in which $A \nsubseteq N:E$. For this situation $N:_E A$ is a proper submodule of E. We shall now apply Lemma 8.

Assume that $rx \in (N:_E A)$ and $x \notin (N:_E A)$, where $r \in R$ and $x \in E$. Choose $\alpha \in A$ so that $\alpha x \notin N$. Then $r(\alpha x) = \alpha(rx) \in N$ and $\alpha x \notin N$. Hence, by Proposition 19, $r \in P$.

Next suppose that $r \in P$. Then there exists a positive integer m such that $r^m E \subseteq N \subseteq N:_E A$. That $N:_E A$ is a P-primary submodule of E now follows at once from Lemma 8.

The next result gives information about the effect of changing the ambient module.

Proposition 22. *Let N be a P-primary submodule of the R-module E and let K be a submodule of E such that $K \nsubseteq N$. Then $K \cap N$ is a P-primary submodule of K.*

Proof. First of all $N \cap K$ is a proper submodule of K. Now assume that $rx \in N \cap K$ and $x \notin N \cap K$, where r is an element of R and x belongs to K. Then $rx \in N$ and $x \notin N$. Consequently $r \in P$.

Next suppose that $r \in P$. Then there exists a positive integer m such that $r^m E \subseteq N$. Accordingly $r^m K \subseteq N \cap K$. The proposition follows by appealing to Lemma 8.

The next proposition concerns the intersection of a finite number of P-primary submodules of a given module.

Proposition 23. *Let $N_1, N_2, ..., N_s (s \geqslant 1)$ be P-primary submodules of an R-module E. Then $N = N_1 \cap N_2 \cap ... \cap N_s$ is also a P-primary submodule of E.*

Proof. Once again we make use of Lemma 8. Clearly N is a proper submodule of E. Now suppose that $rx \in N$, $x \notin N$, where r belongs to R and x belongs to E. We can then choose j so that $x \notin N_j$. We thus have $rx \in N_j$, $x \notin N_j$ whence, by Proposition 19, $r \in P$.

Finally assume that $r \in P$. If k is a large enough positive integer, then $r^k E \subseteq N_i (1 \leqslant i \leqslant s)$ and therefore $r^k E \subseteq N$. That N is a P-primary submodule of E now follows from Lemma 8.

Proposition 24. *Let $\psi : E \to E'$ be an epimorphism of R-modules, N' a submodule of E' and $N = \psi^{-1}(N')$ the corresponding submodule of E. Then N' is a primary submodule of E' if and only if N is a primary submodule of E. When these submodules are primary, they belong to the same prime ideal.*

Proof. Assume that N' is primary and suppose that $rx \in N$, $x \notin N$. Here r denotes an element of R and x an element of E. Then $r\psi(x) \in N'$, $\psi(x) \notin N'$. Hence there exists an integer m such that $r^m E' \subseteq N'$. Now $\psi(r^m E) = r^m E' \subseteq N'$ and therefore $r^m E \subseteq N$. Thus N is a primary submodule of E. By more or less reversing the argument, we can show that N' is primary when N is primary. Finally E/N and E'/N' are isomorphic modules† and therefore they have the same annihilators. Thus $N : E = N' : E'$ and, in consequence, Rad $(N : E) = $ Rad $(N' : E')$. Accordingly when N and N' are primary submodules, they belong to the same prime ideal.

It is frequently necessary to apply these results to the case where ψ is the natural mapping of E on to a factor module. We shall therefore restate the proposition for this case.

Corollary. *Let K and N, where $K \subseteq N$, be submodules of an R-module E. In addition, let P be a prime ideal. Then N is a P-primary submodule of E if and only if N/K is a P-primary submodule of E/K.*

In conclusion, we consider the effect of a change of ring. To this end, let $\phi : R \to R'$ be a ring-epimorphism of R on to R', E an R'-module, and N an R'-submodule of E. Then, as we saw in section (1.14), E may be regarded as an R-module and, when this is done, N will be an R-submodule of E.

Proposition 25. *Let the situation be as described above. Then N is a primary submodule of the R'-module E if and only if it is a primary submodule of the R-module E. If N, considered as R'-submodule, is P'-primary, then it is $\phi^{-1}(P')$-primary when considered as an R-submodule of E.*

† See Theorem 1 of section (1.6).

Proof. Let $r \in R$ and $x \in E$. Then $rx = \phi(r)x$ and, for every positive integer m, $r^m E = \phi(r)^m E$. The first assertion follows from these facts. Assume now that the R'-submodule N is P'-primary. Then, with a self-explanatory notation, $r \in \operatorname{Rad}_R(N:E)$ if and only if

$$\phi(r) \in \operatorname{Rad}_{R'}(N:E) = P'.$$

Accordingly $\operatorname{Rad}_R(N:E) = \phi^{-1}(P')$ and the proposition is proved.

This result will now be restated for a residue ring. Let E be an R-module and A an R-ideal contained in the annihilator of E. Then, as we saw in section (1.14), E has a natural structure as an R/A-module. If now N is an R-submodule of E, then it is also an R/A-submodule of E.

Corollary. *Let the situation be as described above and let P be a prime ideal of R containing A. Then N is a P-primary submodule of E considered as an R-module, if and only if it is a P/A-primary submodule of E considered as an R/A-module.*

2.9 Submodules which possess a primary decomposition

Let E be an R-module and K a submodule. If K can be represented as the intersection of a finite number of primary submodules of E, then we say that K *possesses a primary decomposition in E* or K *is decomposable in E*. A little later, we shall see that when E satisfies the maximal condition for submodules, then all its submodules are decomposable. Note that, since E is an empty intersection of primary submodules, it counts as a decomposable submodule of itself. Again, on taking $E = R$, we are led to the notion of a *decomposable ideal*. Thus an ideal A possesses a primary decomposition if it can be expressed in the form $A = Q_1 \cap Q_2 \cap \ldots \cap Q_s$, where each Q_i is a primary ideal.

Proposition 26. *Let $K = N_1 \cap N_2 \cap \ldots \cap N_s$, where N_i is a P_i-primary submodule of E, and let P be a prime ideal of R. Then*

$$\operatorname{Rad}(K:E) = P_1 \cap P_2 \cap \ldots \cap P_s$$

and P contains $K:E$ if and only if it contains at least one of $P_1, P_2, \ldots, P_s$. It follows that $K:E$ has only finitely many minimal prime ideals and these all occur among $P_1, P_2, \ldots, P_s$. Indeed $P_i \, (1 \leqslant i \leqslant s)$ is a minimal prime ideal of $K:E$ if and only if it does not strictly contain any other member of the set $P_1, P_2, \ldots, P_s$.

Remark. It should be noted that we do not assume that the

$$P_i \, (1 \leqslant i \leqslant s)$$

are distinct.

Proof. By (2.7.5),

$$K:E = (N_1:E) \cap (N_2:E) \cap \ldots \cap (N_s:E).$$

Since $P_i = \mathrm{Rad}\,(N_i:E)$, it follows, from Proposition 7, that

$$K:E \subseteq \mathrm{Rad}\,(K:E) = P_1 \cap P_2 \cap \ldots \cap P_s.$$

Accordingly, if P contains one of the P_i, then it contains $K:E$.

Now assume that $K:E \subseteq P$. It is then obvious, from the definition of the radical of an ideal, that P contains $\mathrm{Rad}\,(K:E) = P_1 \cap \ldots \cap P_s$. Hence, by Proposition 4, P contains P_i for at least one value of i.

Corollary. *Let the submodule K of E have $K = N_1 \cap N_2 \cap \ldots \cap N_s$ as a primary decomposition. Then $K:E = (N_1:E) \cap \ldots \cap (N_s:E)$ and this is a primary decomposition of the ideal $K:E$.*

This is obvious because, by Proposition 18, $N_i:E$ is a primary ideal.

Let S be a non-empty multiplicatively closed subset of R, and K an arbitrary submodule of E. Denote by K^S the set of all elements e of E such that $se \in K$ for at least one element $s \in S$. It is clear that $K \subseteq K^S \subseteq E$. We contend that K^S is a submodule of E. For let e_1, e_2 belong to K^S. There exist s_1, s_2 in S such that $s_1 e_1$ and $s_2 e_2$ are both in K. Put $s = s_1 s_2$. Then s belongs to S and $s(e_1 + e_2) = s_2(s_1 e_1) + s_1(s_2 e_2)$ belongs to K. Thus $e_1 + e_2$ is in K^S. Again if $r \in R$, then

$$s_1(re_1) = r(s_1 e_1) \in K$$

and therefore $re_1 \in K^S$.

Definition. *The submodule K^S is called the 'S-component' of K in E.*

The reader should note that the notation K^S is slightly unsatisfactory because it does not indicate the role played by the ambient module E. However, although the concept of an S-component is an important one, it will appear only occasionally and therefore we shall not elaborate the notation in order to cover this point.

Proposition 27. *Let $K = N_1 \cap N_2 \cap \ldots \cap N_t$, where N_i is a P_i-primary submodule of E, and let S be a non-empty multiplicatively closed subset of R. Suppose that $P_1, P_2, \ldots, P_t$ are so numbered that none of $P_1, P_2, \ldots, P_m$ meets S while each of the remainder does meet S. Then*

$$K^S = N_1 \cap N_2 \cap \ldots \cap N_m.$$

Proof. For each j satisfying $m+1 \leqslant j \leqslant t$, choose $s_j \in S$ so that $s_j \in P_j$. Put $\sigma = s_{m+1} s_{m+2} \dots s_t$. Then $\sigma \in S$ and $\sigma \in P_j$ for $m+1 \leqslant j \leqslant t$. Since $P_j = \mathrm{Rad}\,(N_j : E)$, it is possible to select an integer ν large enough to ensure that $\sigma^\nu \in (N_j : E)$ for $j = m+1, m+2, \dots, t$. If therefore we set $s = \sigma^\nu$, then $s \in S$ and $sE \subseteq N_{m+1} \cap N_{m+2} \cap \dots \cap N_t$. Accordingly for x in $N_1 \cap N_2 \cap \dots \cap N_m$, we have

$$sx \in N_1 \cap \dots \cap N_m \cap N_{m+1} \cap \dots \cap N_t = K$$

which shows that $x \in K^S$. Thus $N_1 \cap \dots \cap N_m \subseteq K^S$.

Now suppose that y belongs to K^S. There exists $s' \in S$ such that $s'y \in K$. If therefore $1 \leqslant i \leqslant m$, then $s'y \in N_i$, $s' \notin P_i$. Hence, by Proposition 19, $y \in N_i$. Thus y is in $N_1 \cap N_2 \cap \dots \cap N_m$. This shows that $K^S \subseteq N_1 \cap N_2 \cap \dots \cap N_m$ and completes the proof.

We shall now consider to what extent primary decompositions are unique. Suppose therefore that

$$K = N_1 \cap N_2 \cap \dots \cap N_t, \qquad (2.9.1)$$

where N_i is a P_i-primary submodule of E. If N_i contains

$$N_1 \cap \dots \cap N_{i-1} \cap N_{i+1} \cap \dots \cap N_t,$$

then N_i is superfluous in (2.9.1) in the sense that the two sides remain equal if N_i is deleted.

Definition. *If in* (2.9.1) *no* N_i *is superfluous, then the representation of* K *is said to be 'irredundant'.*

It is clear that starting with an arbitrary primary decomposition of K we can turn it into one which is irredundant simply by striking out unnecessary terms one at a time. Suppose now that (2.9.1) is an irredundant primary decomposition of K. If a prime ideal P occurs more than once among $P_1, P_2, \dots, P_t$ then, by Proposition 23, we can replace the various N_i which are P-primary by their intersection. In this way we arrive at an irredundant primary decomposition in which the prime ideals, to which the primary components belong, are all distinct. An irredundant primary decomposition with this additional property is called a *normal decomposition* or a *Lasker decomposition*. Note that an arbitrary primary decomposition can be refined into a normal (Lasker) decomposition by first striking out superfluous terms and afterwards grouping those which remain in the manner just explained. It should also be observed that one effect of this is to replace the various prime ideals, associated with the original decomposition, by a subset of them. By Proposition 26, the subset will contain all the minimal prime ideals of $K : E$.

In the theorem which follows, K as usual denotes a submodule of an R-module E.

Theorem 13. *Let* $K = N_1 \cap N_2 \cap \ldots \cap N_p$ *and* $K = N_1' \cap N_2' \cap \ldots \cap N_q'$ *be two normal primary decompositions of* K *as a submodule of* E. *Suppose now that* N_i *is* P_i-*primary and* N_j' *is* P_j'-*primary. Then* $p = q$ *and the prime ideals* $P_1', P_2', \ldots, P_q'$ *are the same as* $P_1, P_2, \ldots, P_p$ *though their order may be different.*

Proof. Let P be any one of the prime ideals $P_1, P_2, \ldots, P_p$. It is sufficient to show that P occurs among $P_1', P_2', \ldots, P_q'$ because the roles of the two sets of prime ideals can then be interchanged. Without loss of generality, we can suppose that the numbering is such that $P = P_m$ and $P_1 \subseteq P, P_2 \subseteq P, \ldots, P_m \subseteq P$; $P_{m+1} \nsubseteq P, P_{m+2} \nsubseteq P, \ldots, P_p \nsubseteq P$ and

$$P_1' \subseteq P, P_2' \subseteq P, \ldots, P_\mu' \subseteq P; \quad P_{\mu+1}' \nsubseteq P, P_{\mu+2}' \nsubseteq P, \ldots, P_q' \nsubseteq P.$$

Denote by S the complement of P in R. Then, by Proposition 27,

$$K^S = N_1 \cap N_2 \cap \ldots \cap N_m = N_1' \cap N_2' \cap \ldots \cap N_\mu'.$$

We shall now assume that P does not occur among $P_1', P_2', \ldots, P_\mu'$ and derive a contradiction.

Since P is not contained by any of $P_1, P_2, \ldots, P_{m-1}$ or $P_1', P_2', \ldots, P_\mu'$, it is possible (see Proposition 5) to find an element $\alpha \in P$ which does not belong to a single one of $P_1, \ldots, P_{m-1}, P_1', \ldots, P_\mu'$. Next

$$P = \mathrm{Rad}\,(N_m : E)$$

and therefore $\alpha^\nu \in (N_m : E)$ provided ν is a large enough positive integer. This† ensures that $N_m : \alpha^\nu = E$. On the other hand it follows, from Proposition 19 Cor. 2, that $N_i : \alpha^\nu = N_i$ for $1 \leqslant i \leqslant m-1$ and

$$N_j' : \alpha^\nu = N_j' \quad \text{for} \quad 1 \leqslant j \leqslant \mu.$$

Thus
$$K^S : \alpha^\nu = (N_1 : \alpha^\nu) \cap \ldots \cap (N_{m-1} : \alpha^\nu) \cap (N_m : \alpha^\nu)$$
$$= N_1 \cap N_2 \cap \ldots \cap N_{m-1}$$

and also
$$K^S : \alpha^\nu = (N_1' : \alpha^\nu) \cap \ldots \cap (N_\mu' : \alpha^\nu)$$
$$= N_1' \cap N_2' \cap \ldots \cap N_\mu'$$
$$= N_1 \cap N_2 \cap \ldots \cap N_m.$$

Accordingly
$$N_m \supseteq N_1 \cap N_2 \cap \ldots \cap N_{m-1}$$
$$\supseteq N_1 \cap \ldots \cap N_{m-1} \cap N_{m+1} \cap \ldots \cap N_t,$$

which contradicts the assumption that $K = N_1 \cap N_2 \cap \ldots \cap N_t$ is an irredundant representation. The theorem is therefore proved.

† To simplify the notation we write $N_m : \alpha^\nu$ rather than $N_m :_E \alpha^\nu$.

Suppose now that the submodule K of E possesses a primary decomposition. This can be refined into a normal decomposition

$$K = N_1 \cap N_2 \cap \ldots \cap N_s,$$

where N_i is (say) a P_i-primary submodule of E. Since the decomposition is normal, it is irredundant and the P_i are distinct. Furthermore, by Theorem 13, $P_1, P_2, \ldots, P_s$ are uniquely determined by K and E. They will be called the *prime ideals belonging to the submodule K of E.* If P_i does not contain any other prime ideal belonging to K, then it is called a *minimal* or *isolated* prime ideal of the submodule. The remaining prime ideals among $P_1, P_2, \ldots, P_s$ are known as the *embedded* prime ideals of K.

Consider, for the moment, the case of an ideal A with a normal decomposition $A = Q_1 \cap Q_2 \cap \ldots \cap Q_s$, where Q_i is a P_i-primary ideal. Since $A : R = A$, it follows (see Proposition 26) that the minimal prime ideals of A in the sense just explained are the same as the minimal prime ideals of A according to the definition used in section (2.3). Again Proposition 26 also yields

Theorem 14. *Let the submodule K of E possess a primary decomposition. Then the minimal prime ideals of the submodule K are the same as the minimal prime ideals of the ideal $K : E$.*

It is worth while noting that if $K = N_1 \cap N_2 \cap \ldots \cap N_s$, where N_i is a P_i-primary submodule of E, is an *irredundant but not necessarily normal* primary decomposition of K, then $P_1, P_2, \ldots, P_s$ is the set of prime ideals belonging to K though, in this instance, there may be repetitions. This is clear if one considers the usual method of refining an irredundant representation to a normal one.

Let K be a submodule of E which possesses a primary decomposition and let $P_1, P_2, \ldots, P_s$ be the prime ideals which belong to K. Suppose now that $1 \leqslant i_1 < i_2 < \ldots < i_\nu \leqslant s$. Then $P_{i_1}, P_{i_2}, \ldots, P_{i_\nu}$ *is called an isolated set of prime ideals belonging to K* if every $P_j (1 \leqslant j \leqslant s)$ which is contained by one of $P_{i_1}, P_{i_2}, \ldots, P_{i_\nu}$ actually occurs as a member of the set $P_{i_1}, P_{i_2}, \ldots, P_{i_\nu}$. For example, *if P is a minimal prime ideal of the submodule K, then P by itself is an isolated set of prime ideals belonging to K.*

Suppose that $P_{i_1}, P_{i_2}, \ldots, P_{i_\nu}$ is such an isolated set. Let

$$K = N_1 \cap N_2 \cap \ldots \cap N_s = N_1' \cap N_2' \cap \ldots \cap N_s'$$

be two normal decompositions of K, where for each $j (1 \leqslant j \leqslant s)$ N_j and N_j' are P_j-primary. *We contend that*

$$N_{i_1} \cap N_{i_2} \cap \ldots \cap N_{i_\nu} = N_{i_1}' \cap N_{i_2}' \cap \ldots \cap N_{i_\nu}'.$$

For let S consist of all elements of R not contained by any of

$$P_{i_1}, P_{i_2}, \ldots, P_{i_\nu}.$$

Then S is multiplicatively closed and, since $1 \in S$, it is not empty. Should P be a prime ideal belonging to K and different from all of $P_{i_1}, P_{i_2}, \ldots, P_{i_\nu}$, then P is not contained in any of them. Hence, by Proposition 5, there exists $\alpha \in P$ such that $\alpha \notin P_{i_1}, \ldots, \alpha \notin P_{i_\nu}$. In other terms, P meets S. It now follows, from Proposition 27, that

$$K^S = N_{i_1} \cap N_{i_2} \cap \ldots \cap N_{i_\nu}.$$

For quite similar reasons, we also have $K^S = N'_{i_1} \cap N'_{i_2} \cap \ldots \cap N'_{i_\nu}$. The proof is therefore complete.

Definition. *With the above notation* $N_{i_1} \cap N_{i_2} \cap \ldots \cap N_{i_\nu}$ *is called the 'isolated component' of K corresponding to the isolated set*

$$P_{i_1}, P_{i_2}, \ldots, P_{i_\nu}$$

of prime ideals.

Corollary. *Let the submodule K of E possess a primary decomposition and let P be a minimal prime ideal belonging to the submodule. Then K has the same P-primary component for all normal decompositions in E.*

Theorem 15. *Let E contain a submodule K which admits a primary decomposition. If now D is a subset of R not contained in any prime ideal belonging to the submodule K, then $K :_E D = K$.*

Proof. Let $K = N_1 \cap N_2 \cap \ldots \cap N_s$, where N_i is P_i-primary, be a normal decomposition. Then, by (2.7.6),

$$K :_E D = (N_1 :_E D) \cap (N_2 :_E D) \cap \ldots \cap (N_s :_E D).$$

However, since D is not contained in P_i, we have $N_i :_E D = N_i$ (see Proposition 19 Cor. 2). Thus $K :_E D = K$.

This result has a partial converse namely

Theorem 16. *Let E contain a submodule K which admits a primary decomposition. If now A is a finitely generated ideal and $K :_E A = K$, then A is not contained in any prime ideal belonging to K.*

Proof. Let $K = N_1 \cap N_2 \cap \ldots \cap N_s$, where N_i is P_i-primary, be a normal decomposition. *We shall assume that* $A \subseteq P_1$ *and derive a contradiction.*

First, by (2.7.10), $K :_E A^{m+1} = (K :_E A^m) :_E A$ whence it follows, since $K :_E A = K$, that $K :_E A^m = K$ for all m. Next,

$$A \subseteq P_1 = \mathrm{Rad}\,(N_1 : E)$$

and A is finitely generated. Hence, by Proposition 8, there exists an integer m such that $A^m \subseteq N_1 \colon E$. It follows that $N_1 \colon_E A^m = E$. Accordingly

$$K = K \colon_E A^m = (N_1 \colon_E A^m) \cap (N_2 \colon_E A^m) \cap \ldots \cap (N_s \colon_E A^m)$$
$$= (N_2 \colon_E A^m) \cap (N_3 \colon_E A^m) \cap \ldots \cap (N_s \colon_E A^m).$$

However, by Proposition 21, $N_i \colon_E A^m$ ($2 \leqslant i \leqslant s$) is either a P_i-primary submodule of E or it coincides with E. It follows that K possesses a primary decomposition not involving P_1. On refining this to a normal decomposition, we see that P_1 does not belong to K. This is the required contradiction.

The remaining theorems of this section give additional information about the prime ideals which belong to a decomposable submodule.

Theorem 17. *Let E contain a submodule K which possesses a primary decomposition and suppose that M is a submodule of E satisfying $K \subseteq M \subseteq E$. Then K, when regarded as a submodule of M, has a primary decomposition. Furthermore, if $P_1, P_2, \ldots, P_s$ resp. $P_1', P_2', \ldots, P_\sigma'$ are the prime ideals belonging to K when it is regarded as a submodule of E resp. M, then $\{P_1', P_2', \ldots, P_\sigma'\}$ is a subset of $\{P_1, P_2, \ldots, P_s\}$. In particular, every minimal prime ideal of $K \colon M$ occurs among $P_1, P_2, \ldots, P_s$.*

Proof. Let $K = N_1 \cap N_2 \cap \ldots \cap N_s$, where N_i is P_i-primary, be a normal decomposition of K in E. Then

$$K = (N_1 \cap M) \cap (N_2 \cap M) \cap \ldots \cap (N_s \cap M). \tag{2.9.2}$$

By Proposition 22, either $N_i \cap M = M$ or $N_i \cap M$ is a P_i-primary submodule of M. If therefore we strike out, from (2.9.2), terms which are superfluous, then we shall arrive at a primary decomposition of K in M. This establishes all but the final assertion of the theorem. To complete the proof, we have only to observe that, by Proposition 26, the minimal prime ideals of $K \colon M$ occur among $P_1', P_2', \ldots, P_\sigma'$ and hence among $P_1, P_2, \ldots, P_s$.

Theorem 18. *Let E contain a submodule K which possesses a primary decomposition and let P be a prime ideal. Then the following two statements are equivalent:*

(1) P belongs to the submodule K;

(2) there exists a submodule M of E such that $K \subseteq M \subseteq E$ and K is a P-primary submodule of M.

Proof. Assume that (1) is true and let $K = N_1 \cap N_2 \cap \ldots \cap N_s$, where N_i is P_i-primary, be a normal decomposition of K in E. We may assume that $P = P_1$. Put $M = N_2 \cap \ldots \cap N_s$. Then $K \subseteq M \subseteq E$ and $M \nsubseteq N_1$

because the representation of K is irredundant. By Proposition 22, $N_1 \cap M = K$ is a P-primary submodule of M. Thus (1) implies (2).

Next assume that (2) is true. Then P belongs to K when K is regarded as a submodule of M. Hence, by Theorem 17, P is one of the prime ideals belonging to K when K is regarded as a submodule of E.

Our final theorem of this section is somewhat more delicate in character. For it, we need a lemma. Suppose that $K = N_1 \cap N_2 \cap \ldots \cap N_t$, where N_i is P_i-primary, is a normal decomposition of the submodule K of E. Set $P = P_1$ so that P is just one of the prime ideals belonging to K, and assume that the numbering of the components has been done in such a way that all of $P_1, P_2, \ldots, P_m$ are contained in P while none of $P_{m+1}, P_{m+2}, \ldots, P_t$ has this property. Put

$$M = N_1 \cap N_2 \cap \ldots \cap N_m.$$

Lemma 9. *Let the notation be as described above and assume that P is finitely generated. Then*

$$(M :_E P) \cap N_{m+1} \cap \ldots \cap N_t$$

strictly contains K and for every submodule L such that

$$K \subset L \subseteq (M :_E P) \cap N_{m+1} \cap \ldots \cap N_t$$

we have $K : L = P$.

Proof. We first note that $M = N_1 \cap N_2 \cap \ldots \cap N_m$ is a normal decomposition of M in E. Consequently P is one of the prime ideals belonging to M and therefore, by Theorem 16, $M :_E P \neq M$. It follows that $M \subset M :_E P$ the inclusion being strict. Next

$$M :_E P = (N_1 :_E P) \cap (N_2 :_E P) \cap \ldots \cap (N_m :_E P).$$

However, for $2 \leqslant i \leqslant m$, $P \nsubseteq P_i$ and therefore $N_i :_E P = N_i$ by Proposition 19 Cor. 2. It follows that

$$M \subset M :_E P = (N_1 :_E P) \cap N_2 \cap \ldots \cap N_m.$$

Note that, by Proposition 21, either $N_1 :_E P = E$ or $N_1 :_E P$ is a P_1-primary submodule of E.

Let S consist of the elements of R which are not in P. Then S is multiplicatively closed. Also it meets none of $P_1, P_2, \ldots, P_m$ but all of $P_{m+1}, \ldots, P_t$. It therefore follows, by Proposition 27, that the S-component of

$$(M :_E P) \cap N_{m+1} \cap \ldots \cap N_t = (N_1 :_E P) \cap N_2 \cap \ldots \cap N_m \cap N_{m+1} \cap \ldots \cap N_t$$

in E is $\qquad (N_1 :_E P) \cap N_2 \cap \ldots \cap N_m = M :_E P.$

On the other hand, the S-component of $K = N_1 \cap \ldots \cap N_t$ in E is $N_1 \cap \ldots \cap N_m = M$. Since the two S-components are different, the modules themselves must be distinct. We may therefore conclude that

$$K \subset (M:_E P) \cap N_{m+1} \cap \ldots \cap N_t$$

the inclusion being strict. This proves the first assertion.

Now let L be as stated. Then $PL \subseteq M \cap N_{m+1} \ldots \cap N_t = K$ and therefore $P \subseteq K:L$. Finally suppose that r belongs to $K:L$. Then $rL \subseteq N_i$ for all i. However L is not contained in $N_1 \cap N_2 \cap \ldots \cap N_m$ for otherwise it would be contained in $N_1 \cap \ldots \cap N_{m+1} \cap \ldots \cap N_t = K$ which is not the case. We can therefore choose i so that $1 \leqslant i \leqslant m$ and $L \nsubseteq N_i$. Select $x \in L$ so that $x \notin N_i$. This yields $rx \in N_i$, $x \notin N_i$; consequently $r \in P_i \subseteq P$ by virtue of Proposition 19. This shows that $K:L \subseteq P$ and completes the proof.

Theorem 19. *Let E contain a submodule K which possesses a primary decomposition and let P be a finitely generated prime ideal. Then P belongs to K if and only if there exists $e \in E$ such that $P = K:e$. When this is the case P also belongs to K when K is considered as a submodule of $K:_E P$.*

Proof. First suppose that P belongs to K. Then, using the notation of Lemma 9, we can choose e in

$$(M:_E P) \cap N_{m+1} \cap \ldots \cap N_t$$

so that $e \notin K$. Taking $L = K + Re$, the lemma shows that

$$P = K:L = (K:K) \cap (K:Re) = K:Re = K:e.$$

Now assume that we have an element $e \in E$ such that $P = K:e$. Then $P = K:(K + Re)$ and therefore, by Theorem 17, P is one of the prime ideals belonging to K when K is considered as a submodule of E. Put $\bar{M} = K:_E P$. Theorem 17 shows that K possesses a primary decomposition in $\bar{M}$. Further $e \in \bar{M}$. Since $K:e = P$, the earlier discussion now shows that P belongs to K when it is regarded as a submodule of $\bar{M}$.

2.10 Existence of primary decompositions

The main purpose of this section is to show that when E satisfies the maximal condition for submodules,† then every submodule possesses a primary decomposition. Actually, when the maximal condition is

---

† See section (1.8).

satisfied, it is possible to say a great deal more, but we shall leave most of the other results until later. For the moment, our aim is to show that, in the last section, we were not working in a vacuum.

Lemma 10. *Let the R-module E satisfy the maximal condition for submodules and let K be a proper submodule of E which is not primary. Then there exists proper submodules K', K" of E which strictly contain K and are such that $K = K' \cap K''$.*

Proof. Since K is not primary in E, there exist $r \in R$ and $x \in E$ such that $rx \in K, x \notin K$ and, for every integer $m, r^m E \nsubseteq K$. Now, by (2.7.10),

$$K:_E Rr^{m+1} = (K:_E Rr^m):_E Rr;$$

consequently we have an ascending sequence

$$K \subset K:_E Rr \subseteq K:_E Rr^2 \subseteq K:_E Rr^3 \subseteq \dots \qquad (2.10.1)$$

of submodules of E the first inclusion being strict because x belongs to $K:_E Rr$ whereas $x \notin K$. Since E satisfies the maximal condition for submodules, it follows, from Proposition 9 of section (1.8), that there exists an integer m such that $K:_E Rr^n = K:_E Rr^m$ for all $n \geq m$. Put $K' = K:_E Rr^m$ and $K'' = K + r^m E$. Then both K' and K'' strictly contain K. Let $e \in K' \cap K''$. To complete the proof we need only show that $e \in K$.

Since $e \in K''$, we may write $e = u + r^m e_1$, where $u \in K$ and $e_1 \in E$. But $e \in K'$ and so it follows that

$$r^{2m} e_1 = r^m e - r^m u \in K$$

whence e_1 is in $K:_E Rr^{2m} = K:_E Rr^m$. Thus $r^m e_1 \in K$ and therefore $e \in K$. This completes the proof.

Theorem 20. *Let the R-module E satisfy the maximal condition for submodules. Then every submodule of E can be expressed as the intersection of a finite number of primary submodules.*

Proof. Denote by Ω the set of all submodules of E which *cannot* be expressed in the required manner. We shall assume that Ω is not empty and derive a contradiction. Before proceeding note that E does not belong† to Ω.

Since E satisfies the maximal condition, it is possible to choose $K \in \Omega$ so that K is maximal in Ω. Then K is a proper submodule and it cannot be primary (for otherwise it would be the intersection of the members of a set consisting of a single primary submodule). It

† Here we use the convention that an empty family of submodules of E has E as its intersection. See section (2.3).

follows, from Lemma 10, that $K = K' \cap K''$, where K', K'' are submodules of E which *strictly* contain K. By the maximal property of K, neither K' nor K'' belongs to Ω and hence each is a finite intersection of primary submodules of E. But in that case $K' \cap K'' = K$ is also an intersection of a finite number of primary submodules and therefore $K \notin \Omega$. This is the required contradiction.

Let $L \subseteq K$ be submodules of an R-module E and suppose that E satisfies the maximal condition for submodules. Then E/L also satisfies the maximal condition for submodules. Accordingly, by Theorem 20, K has a primary decomposition in E and K/L a primary decomposition in E/L. The following corollary shows how these decompositions are related.

Corollary. *Let $N_1, N_2, ..., N_s$ be submodules of E which satisfy*

$$K = N_1 \cap N_2 \cap ... \cap N_s \qquad (2.10.2)$$

and for which therefore we have

$$K/L = (N_1/L) \cap (N_2/L) \cap ... \cap (N_s/L). \qquad (2.10.3)$$

Then (2.10.2) is a primary resp. normal decomposition of K in E if and only if (2.10.3) is a primary resp. normal decomposition of K/L in E/L. Furthermore the prime ideals which belong to the submodule K of E are the same as those which belong to the submodule K/L of E/L.

Proof. Observe first that the intersection representing K will contain no superfluous term if and only if the same holds for the intersection representing K/L. Now let N be any submodule of E containing L and let P be a prime ideal. Then (Proposition 24 Cor.) N is a P-primary submodule of E if and only if N/L is a P-primary submodule of E/L. The assertions of the corollary all follow from these facts.

2.11 Graded rings and modules

Sometimes certain additional structure is imposed on a module. It may then become necessary to know how far the preceding theory combines naturally with the new features. An example of this is provided by graded rings and modules which form the next topic for discussion.

First we need the concept of a grading monoid. In formal technical terms, this is a commutative and associative monoid, with neutral element, which satisfies the cancellation law. The law of composition is written as *addition* and the neutral element is denoted by 0.

Thus if Γ is a grading monoid, then its elements may be 'added' to give further elements of Γ and addition obeys the following axioms:

(i) $\gamma_1 + \gamma_2 = \gamma_2 + \gamma_1$ for all γ_1, γ_2 in Γ;

(ii) $\gamma_1 + (\gamma_2 + \gamma_3) = (\gamma_1 + \gamma_2) + \gamma_3$ for all $\gamma_1, \gamma_2, \gamma_3$ in Γ;

(iii) *there is an element* 0 *(necessarily unique) such that* $\gamma + 0 = \gamma$ *for all* $\gamma \in \Gamma$;

(iv) *if* $\gamma + \gamma_1 = \gamma + \gamma_2$, *where* $\gamma, \gamma_1, \gamma_2$ *are in* Γ, *then* $\gamma_1 = \gamma_2$.

For example, any additively written abelian group is a grading monoid. The non-negatives integers, when the law of composition is ordinary addition, provides another example. More generally, if we consider sequences $(m_1, m_2, \ldots, m_p)$ of p (p is fixed) non-negative integers and define addition by

$$(m_1, m_2, \ldots, m_p) + (m_1', m_2', \ldots, m_p') = (m_1 + m_1', m_2 + m_2', \ldots, m_p + m_p'),$$

then we obtain a third example.

Let Γ be a grading monoid and R a ring (commutative and with an identity element). The elements of R form a group with respect to addition and therefore it makes sense to speak of the subgroups of the additive group of R.

Definition. *A 'Γ-grading on R' is a family* $\{R^{(\gamma)}\}_{\gamma \in \Gamma}$ *of subgroups of the additive group of R such that*

(a) $R = \sum_{\gamma \in \Gamma} R^{(\gamma)}$ *(direct sum)*,

(b) $R^{(\gamma)} R^{(\gamma')} \subseteq R^{(\gamma + \gamma')}$ *for all* γ, γ' *in* Γ.

By way of explanation, it needs to be stated that $R^{(\gamma)} R^{(\gamma')}$ means the set of all elements which can be expressed in the form

$$r_1^{(\gamma)} \rho_1^{(\gamma')} + r_2^{(\gamma)} \rho_2^{(\gamma')} + \ldots + r_s^{(\gamma)} \rho_s^{(\gamma')}, \quad \text{where} \quad r_j^{(\gamma)} \in R^{(\gamma)} \text{ and } \rho_j^{(\gamma')} \in R^{(\gamma')}.$$

Suppose that we have a Γ-grading on R. Each element r of R has a *unique* representation in the form $r = \sum_{\gamma} r^{(\gamma)}$, where $r^{(\gamma)} \in R^{(\gamma)}$ and the sum contains only a finite number of non-zero terms. The elements of $R^{(\gamma)}$ are said to be *homogeneous of degree* γ and $r^{(\gamma)}$ is called the *homogeneous component of r of degree* γ. Note that the zero element is homogeneous of degree γ for every γ in Γ. Also (b) of the above definition tells us that if a homogeneous element of degree γ is multiplied by one of degree γ', then the result is a homogeneous element of degree $\gamma + \gamma'$.

It is worth while observing that given any ring R, we can obtain a Γ-grading on R by putting $R^{(0)} = R$ and $R^{(\gamma)} = 0$ whenever $\gamma \neq 0$. This is called the *trivial* Γ-grading on R.

Theorem 21. *Let R be a Γ-graded ring. Then the identity element of R is homogeneous of degree zero.*

Proof. Let $1_R = \sum\limits_{\gamma} \epsilon^{(\gamma)}$ be the representation of the identity element as a sum of homogeneous elements, and let η be a homogeneous element of degree λ. Then

$$\eta = \eta 1_R = \sum_{\gamma} \eta \epsilon^{(\gamma)}.$$

On comparing terms of degree λ we see that $\eta = \eta \epsilon^{(0)}$ which shows that the effect of multiplying a homogeneous element by $\epsilon^{(0)}$ is to leave it unchanged. However, every element is a sum of homogeneous elements, consequently $r = r\epsilon^{(0)}$ for every $r \in R$. In particular,

$$1_R = 1_R \epsilon^{(0)} = \epsilon^{(0)}.$$

This completes the proof.

Corollary. *$R^{(0)}$ is a subring of R.*

For if α and β belong to $R^{(0)}$, then so do $-\alpha, \alpha + \beta$ and $\alpha\beta$. It follows that $R^{(0)}$ is a ring and, since we have just shown that $1_R \in R^{(0)}$, it is in fact a subring of R.

Let $R = \sum\limits_{\gamma} R^{(\gamma)}$ be a Γ-graded ring and let E be an R-module. The elements of E form a group with respect to addition. This will be referred to as the 'additive group of E'.

Definition. *A 'Γ-grading on E' is a family $\{E^{(\gamma)}\}_{\gamma \in \Gamma}$ of subgroups, of the additive group of E, which satisfies*

(a) $E = \sum\limits_{\gamma} E^{(\gamma)}$ *(direct sum);*

(b) $R^{(\gamma)} E^{(\gamma')} \subseteq E^{(\gamma + \gamma')}$ *for all γ, γ' in Γ.*

In this context, $R^{(\gamma)} E^{(\gamma')}$ means the set of all elements which can be expressed in the form $r_1^{(\gamma)} e_1^{(\gamma')} + r_2^{(\gamma)} e_2^{(\gamma')} + \ldots + r_s^{(\gamma)} e_s^{(\gamma')}$, where

$$r_i^{(\gamma)} \in R^{(\gamma)} \quad \text{and} \quad e_i^{(\gamma')} \in E^{(\gamma')}.$$

Let $E = \sum\limits_{\gamma} E^{(\gamma)}$ be a Γ-graded module over the Γ-graded ring $R = \sum\limits_{\gamma} R^{(\gamma)}$. Then each element $e \in E$ has a unique representation in the form $e = \sum\limits_{\gamma} e^{(\gamma)}$, where $e^{(\gamma)} \in E^{(\gamma)}$ and the sum contains only a finite number of non-zero terms. The elements of $E^{(\gamma)}$ are said to be *homogeneous of degree γ* and $e^{(\gamma)}$ is called the *homogeneous component of e of degree γ*. It follows, from (b), that if a homogeneous element e of degree γ' is multiplied by a homogeneous ring element r of degree γ, then the result is a homogeneous element of E of degree $\gamma + \gamma'$. Note that each $E^{(\gamma)}$ is a module with respect to $R^{(0)}$.

To see that the above definition is not vacuous, let R be any ring and E any R-module. If we now endow R with the trivial Γ-grading and also E with the trivial Γ-grading (i.e. put $E^{(0)} = E$ and $E^{(\gamma)} = 0$ for $\gamma \neq 0$), then E becomes a Γ-graded module over the Γ-graded ring R. We return to the consideration of a general Γ-graded ring $R = \sum_{\gamma} R^{(\gamma)}$. Note that R is a Γ-graded module with respect to itself. Now let $E = \sum_{\gamma} E^{(\gamma)}$ be a general Γ-graded R-module.

Proposition 28. *If K is a submodule of the Γ-graded R-module $E = \sum_{\gamma} E^{(\gamma)}$, then the following statements are equivalent:*

(a) $K = \sum_{\gamma} (E^{(\gamma)} \cap K)$;

(b) *if $y \in K$, then all the homogeneous components of y belong to K*;

(c) K *can be generated, as an R-module, by homogeneous elements.*

Proof. We give a cyclic argument. Suppose that (a) is true and let $y \in K$. Then $y = \sum_{\gamma} y^{(\gamma)}$, where $y^{(\gamma)} \in E^{(\gamma)} \cap K$. It follows that $y^{(\gamma)}$ is the homogeneous component of y of degree γ and, since it belongs to $E^{(\gamma)} \cap K$, it necessarily belongs to K. Thus (a) implies (b).

Next assume that (b) is true. Since each element is the sum of its homogeneous components, each element of K is a sum of homogeneous elements of K. Accordingly the submodule generated by all homogeneous elements belonging to K is K itself. This shows that (b) implies (c).

Finally assume that K is generated by a family $\{x_i\}_{i \in I}$ of elements, where each x_i is homogeneous and of degree γ_i (say). Let $r = \sum_{\gamma} r^{(\gamma)}$ belong to R. Then $r^{(\gamma)} x_i \in E^{(\gamma + \gamma_i)} \cap K$ and therefore $r x_i \in \sum_{\gamma} (E^{(\gamma)} \cap K)$. It follows that $K = \sum_{\gamma} (E^{(\gamma)} \cap K)$. Thus (c) implies (a) and the proposition is proved.

A submodule K of a graded module $E = \sum_{\gamma} E^{(\gamma)}$ which has properties (a), (b) and (c) of Proposition 28, is called a *homogeneous* submodule of E. Of course, as soon as we know that K possesses *one* of the three properties then we can say that it is homogeneous because it necessarily possesses the other two. If, in place of E, we take R itself (regarded as a Γ-graded R-module), then we are led to the notion of a *homogeneous ideal*. For example, an ideal A is homogeneous if and only if it can be generated by homogeneous elements.

Let $E = \sum_\gamma E^{(\gamma)}$ be a Γ-graded module over the Γ-graded ring $R = \sum_\gamma R^{(\gamma)}$ and let K be a homogeneous submodule of E. If we put

$$K^{(\gamma)} = E^{(\gamma)} \cap K, \qquad (2.11.1)$$

then $K^{(\gamma)}$ is a subgroup of the additive group of K. Also, using (a) of Proposition 28, it is easily verified that the family $\{K^{(\gamma)}\}_{\gamma \in \Gamma}$ constitutes a Γ-grading on K. We call this the *induced grading*. Note that the inclusion mapping $K \to E$ injects $K^{(\gamma)}$ into $E^{(\gamma)}$.

Still assuming that K is a homogeneous submodule of E, let

$$\phi : E \to E/K$$

be the natural mapping. Now both E and E/K may be regarded as $R^{(0)}$-modules and on this understanding ϕ is an $R^{(0)}$-homomorphism. Put

$$\phi(E^{(\gamma)}) = (E/K)^{(\gamma)}. \qquad (2.11.2)$$

Then $(E/K)^{(\gamma)}$ is an $R^{(0)}$-submodule of E/K and hence a subgroup of the additive group of E/K. We now contend that $\{(E/K)^{(\gamma)}\}_{\gamma \in \Gamma}$ *is a Γ-grading on E/K*.

To prove this contention, let $x = \sum_\gamma x^{(\gamma)}$, where $x^{(\gamma)} \in E^{(\gamma)}$ and almost all the $x^{(\gamma)}$ are zero. Then $\phi(x) = \sum_\gamma \phi(x^{(\gamma)})$ and $\phi(x^{(\gamma)}) \in (E/K)^{(\gamma)}$. This establishes that $E/K = \sum_\gamma (E/K)^{(\gamma)}$. Clearly $R^{(\gamma)}(E/K)^{(\gamma')} \subseteq (E/K)^{(\gamma+\gamma')}$. It therefore only remains to be shown that E/K is the *direct* sum of the $(E/K)^{(\gamma)}$. For this purpose, assume that $\xi^{(\gamma)} \in (E/K)^{(\gamma)}$, $\xi^{(\gamma)} = 0$ for almost all γ, and $\sum_\gamma \xi^{(\gamma)} = 0$. We wish to prove that $\xi^{(\gamma)} = 0$ for *all* γ.

Choose $x^{(\gamma)} \in E^{(\gamma)}$ so that $\phi(x^{(\gamma)}) = \xi^{(\gamma)}$. This is possible by (2.11.2). In addition, $x^{(\gamma)}$ is to be the zero element whenever $\xi^{(\gamma)} = 0$. This secures that $x^{(\gamma)} = 0$ for almost all γ. Now put $x = \sum_\gamma x^{(\gamma)}$. Then $\phi(x) = \sum_\gamma \xi^{(\gamma)} = 0$ and therefore $x \in K$. However, K is homogeneous and therefore all the homogeneous components of x belong to K. Thus $x^{(\gamma)} \in K$ and hence $\xi^{(\gamma)} = \phi(x^{(\gamma)}) = 0$. This completes the proof that $\{(E/K)^{(\gamma)}\}_{\gamma \in \Gamma}$ *is a Γ-grading on E/K*.

Definition. *The Γ-grading $\{(E/K)^{(\gamma)}\}_{\gamma \in \Gamma}$ on E/K is called the 'factor grading'.*

If E/K is endowed with the factor grading, then (2.11.2) shows that a homogeneous element of E is mapped by ϕ into a homogeneous element of E/K of the same degree. Note that ϕ induces an $R^{(0)}$-epimorphism of $E^{(\gamma)}$ on to $(E/K)^{(\gamma)}$. The kernel of this epimorphism is $E^{(\gamma)} \cap K = K^{(\gamma)}$, see (2.11.1), consequently we have an $R^{(0)}$-isomorphism

$$E^{(\gamma)}/K^{(\gamma)} \approx (E/K)^{(\gamma)} \quad (\gamma \in \Gamma). \tag{2.11.3}$$

Sometimes this is used to identify $(E/K)^{(\gamma)}$ with $E^{(\gamma)}/K^{(\gamma)}$. On the basis of this identification it is often said that E/K is the direct sum of the $R^{(0)}$-modules $E^{(\gamma)}/K^{(\gamma)}$.

Now let $E = \sum\limits_{\gamma} E^{(\gamma)}$ and $F = \sum\limits_{\gamma} F^{(\gamma)}$ be Γ-graded R-modules and suppose that $f\colon E \to F$ is an R-homomorphism.

Definition. *If there exists $\lambda \in \Gamma$ such that $f(E^{(\gamma)}) \subseteq F^{(\lambda+\gamma)}$ for every $\gamma \in \Gamma$, then the mapping f is said to be of 'degree λ'.*

It is clear that only this kind of homomorphism is significant when we are dealing with graded modules.

Lemma 11. *Let E and F be Γ-graded R-modules and let $f\colon E \to F$ be an R-homomorphism of degree λ. Then $\mathrm{Im}\, f$ and $\mathrm{Ker}\, f$ are homogeneous submodules of F and E respectively.*

Proof. With the usual notation, let $x = \sum\limits_{\gamma} x^{(\gamma)}$ be an element of E. Then $f(x) = \sum\limits_{\gamma} f(x^{(\gamma)})$ and each summand $f(x^{(\gamma)})$ is a homogeneous element of $\mathrm{Im}\, f$. Since $f(x)$ is a typical element of $\mathrm{Im}\, f$, it follows that $\mathrm{Im}\, f$ can be generated by homogeneous elements.

Next let $y = \sum\limits_{\gamma} y^{(\gamma)}$ be an element of $\mathrm{Ker}\, f$. Then $\sum\limits_{\gamma} f(y^{(\gamma)}) = 0$. Now the summands $f(y^{(\gamma)})$ are not only homogeneous but have unequal degrees. It follows that $f(y^{(\gamma)}) = 0$ for every γ and therefore each $y^{(\gamma)}$ belongs to $\mathrm{Ker}\, f$. Accordingly, by (b) of Proposition 28, $\mathrm{Ker}\, f$ is a homogeneous submodule of E.

We shall now check that most of the elementary operations possible with modules and ideals produce homogeneous results when the modules and ideals being operated upon are themselves homogeneous.

Proposition 29. *Let $\{K_i\}_{i \in I}$ be a family of homogeneous submodules of a Γ-graded R-module E. Then $\sum\limits_{i} K_i$ and $\bigcap\limits_{i} K_i$ are both homogeneous.*

Proof. Since each K_i possesses a system of homgeneous generators, it is clear that $\sum\limits_{i} K_i$ can be generated by homogeneous elements. This disposes of the first point. Next, with the usual notation, let $x = \sum\limits_{\gamma} x^{(\gamma)}$ be an element of $\bigcap\limits_{i} K_i$. To complete the proof, it will suffice to show

118 PRIME IDEALS AND PRIMARY SUBMODULES

that each $x^{(\gamma)}$ belongs to the intersection. Now x belongs to K_i and K_i is homogeneous; consequently $x^{(\gamma)} \in K_i$. As this holds for every i, we see that $x^{(\gamma)} \in \bigcap_i K_i$. This completes the proof.

Proposition 30. *Let E be a Γ-graded R-module and K a homogeneous submodule of E. Then $K:E$ (or equivalently $\operatorname{Ann}_R E/K$) is a homogeneous ideal.*

Proof. Let $\{\omega_i\}_{i \in I}$ be a family of homogeneous elements which generate E and, with a self-explanatory notation, let $r = \sum_\gamma r^{(\gamma)}$ belong to $K:E$. The proposition will follow if we can show that every $r^{(\gamma)}$ belongs to $K:E$. Now, for each $i \in I$, $r\omega_i = \sum_\gamma r^{(\gamma)} \omega_i$ belongs to K. Moreover, the terms in the sum are homogeneous and have unequal degrees. Since K is homogeneous, this implies that $r^{(\gamma)}\omega_i \in K$. For fixed γ this will hold for all i. Consequently $r^{(\gamma)} E \subseteq K$ and therefore $r^{(\gamma)} \in (K:E)$. The proof is now complete.

Proposition 31. *Let A be a homogeneous ideal, E a Γ-graded R-module and K a homogeneous submodule of E. Then both AK and $K:_E A$ are homogeneous submodules of E.*

Proof. Let $\{\alpha_i\}_{i \in I}$ be a family of homogeneous elements which generate A and $\{k_j\}_{j \in J}$ a family of homogeneous elements which generate K. Then the products $\alpha_i k_j$ are homogeneous and they generate AK which is thus seen to be a homogeneous submodule. The proof that $K:_E A$ is also homogeneous is very similar to the proof of Proposition 30 and therefore it will be omitted.

2.12 Torsionless grading monoids

The reader may have wondered whether the radical of a homogeneous ideal is also homogeneous. For grading monoids in general this is not the case. However, as we shall see later, if the grading monoid Γ is *torsionless*, then this further property holds. First, however, we must define this new concept.

Definition. *The grading monoid Γ is said to be 'torsionless' if from $n\gamma = n\gamma'$, where $n \geqslant 1$ is an integer and γ, γ' belong to Γ, follows $\gamma = \gamma'$.*

By $n\gamma$ we mean, of course, the sum $\gamma + \gamma + \ldots + \gamma$ in which there are n terms.

In order to obtain an insight into the properties of torsionless grading monoids, we introduce a further concept which, to begin with, may appear unrelated.

Definition. *A total order* $\leqslant$, *on a grading monoid* Γ, *is said to be 'compatible with the monoid structure' if from* $\gamma' \leqslant \gamma''$, *where* $\gamma', \gamma'' \in \Gamma$, *follows* $\gamma + \gamma' \leqslant \gamma + \gamma''$ *for all* $\gamma \in \Gamma$.

The lemma which follows lists some of the elementary properties of such a total order.

Lemma 12. *Let* $\leqslant$ *be a total order on* Γ *which is compatible with its monoid structure. If now* γ, γ', γ'', γ^* *all belong to* Γ *and* m *denotes a positive integer, then*

(1) $\gamma + \gamma' \leqslant \gamma + \gamma''$ *if and only if* $\gamma' \leqslant \gamma''$;

(2) $\gamma + \gamma' = \gamma'' + \gamma^*$ *and* $\gamma \leqslant \gamma''$ *together imply* $\gamma^* \leqslant \gamma'$;

(3) $m\gamma \leqslant m\gamma'$ *if and only if* $\gamma \leqslant \gamma'$;

(4) $m\gamma = m\gamma'$ *if and only if* $\gamma = \gamma'$;

(5) *if* $\gamma_i \leqslant \gamma_i'$ $(1 \leqslant i \leqslant n)$ *then* $\gamma_1 + \gamma_2 + \ldots + \gamma_n \leqslant \gamma_1' + \gamma_2' + \ldots + \gamma_n'$ *and there is equality if and only if* $\gamma_i = \gamma_i'$ *for every* i.

Proof. (1) Half the statement follows from the definition. Assume that $\gamma + \gamma' \leqslant \gamma + \gamma''$ and $\gamma'' \leqslant \gamma'$. Then $\gamma + \gamma' \leqslant \gamma + \gamma'' \leqslant \gamma + \gamma'$, whence $\gamma + \gamma' = \gamma + \gamma''$ and therefore $\gamma' = \gamma''$ by the cancellation law. This establishes (1).

(2) Suppose that $\gamma + \gamma' = \gamma'' + \gamma^*$, $\gamma \leqslant \gamma''$ and $\gamma' \leqslant \gamma^*$. Then $\gamma'' + \gamma^* = \gamma + \gamma' \leqslant \gamma + \gamma^* \leqslant \gamma'' + \gamma^*$ which shows that $\gamma + \gamma' = \gamma + \gamma^*$. Accordingly $\gamma' = \gamma^*$ and (2) is proved.

(3) First suppose that $\gamma \leqslant \gamma'$. Then we obtain $m\gamma \leqslant m\gamma'$ by induction on m. For from $n\gamma \leqslant n\gamma'$ follows

$$(n+1)\gamma = n\gamma + \gamma \leqslant n\gamma' + \gamma \leqslant n\gamma' + \gamma' = (n+1)\gamma'.$$

Now assume that $m\gamma \leqslant m\gamma'$. Then either $\gamma \leqslant \gamma'$ or $\gamma' \leqslant \gamma$. In the former case there is nothing to prove. We therefore assume that $\gamma' \leqslant \gamma$. If $m > 1$, then

$$(m-1)\gamma + \gamma' \leqslant (m-1)\gamma + \gamma = m\gamma \leqslant m\gamma' = (m-1)\gamma' + \gamma'$$

whence $(m-1)\gamma \leqslant (m-1)\gamma'$ by (1). We now have

$$(m-1)\gamma \leqslant (m-1)\gamma'$$

and $\gamma' \leqslant \gamma$ so it is possible to repeat the argument provided that $m > 2$. Eventually we obtain $\gamma \leqslant \gamma'$ which is the required relation.

(4) This is an immediate consequence of (3).

(5) First we note that $\gamma_1+\gamma_2+\ldots+\gamma_p \leqslant \gamma_1'+\gamma_2'+\ldots+\gamma_p'$ for $1 \leqslant p \leqslant n$. For this holds when $p = 1$. Also if $1 \leqslant q < n$ and

$$\gamma_1+\gamma_2+\ldots+\gamma_q \leqslant \gamma_1'+\gamma_2'+\ldots+\gamma_q'$$

then

$$\gamma_1+\ldots+\gamma_q+\gamma_{q+1} \leqslant \gamma_1'+\ldots+\gamma_q'+\gamma_{q+1}' \leqslant \gamma_1'+\ldots+\gamma_q'+\gamma_{q+1}'. \quad (2.12.1)$$

Thus the required relation follows by induction.

Next, we observe that if

$$\gamma_1+\ldots+\gamma_q+\gamma_{q+1} = \gamma_1'+\ldots+\gamma_q'+\gamma_{q+1}',$$

then, by (2.12.1),

$$\gamma_1+\ldots+\gamma_q+\gamma_{q+1} = \gamma_1+\ldots+\gamma_q+\gamma_{q+1}' = \gamma_1'+\ldots+\gamma_q'+\gamma_{q+1}'$$

and therefore $\gamma_{q+1} = \gamma_{q+1}'$ and $\gamma_1+\ldots+\gamma_q = \gamma_1'+\ldots+\gamma_q'$. The second assertion in (5) follows from this observation.

It is clear from (4) that if Γ admits a total order compatible with its monoid structure, then Γ is torsionless. The whole of the remainder of this section will be devoted to proving the converse.†

To this end let Γ be a torsionless grading monoid. Suppose that G is a subset of Γ containing 0 and such that $\gamma+\gamma'$ belongs to G whenever γ, γ' belong to G. We can then regard G as a grading monoid in its own right. Such a subset will be referred to as a *submonoid* of Γ.

Consider a pair $(G, \leqslant)$, where G denotes a submonoid of Γ and $\leqslant$ is a total order on G compatible with its structure as a monoid. The set of all such pairs will be denoted by Ω. Note that Ω is not empty because we can take G to be the trivial submonoid consisting of the zero itself.

Now suppose that $(G, \leqslant)$ and $(G^*, \leqslant^*)$ both belong to Ω. We shall write $(G, \leqslant) \ll (G^*, \leqslant^*)$ provided $G \subseteq G^*$ and the total order induced by $\leqslant^*$ on G coincides with $\leqslant$. It is clear that $\ll$ is a partial order on Ω.

The next step is to show that Ω has become an inductive system. For this we assume that we are given a non-empty family $\{(G^i, \leqslant^i)\}_{i \in I}$ of elements of Ω which we assume to be totally ordered by the relation $\ll$. Put $G = \bigcup_i G^i$. Then it is clear that G is a submonoid of Γ. If now g, g' belong to G, then it is possible to find i so that they both belong to G^i. Furthermore, if they also both belong to G^j, then $g \leqslant^i g'$ if and only if $g \leqslant^j g'$. We may therefore indicate this state of affairs by

---

† As the details of the proof are rather lengthy and do not help one to understand the applications, it is suggested that they be omitted at a first reading.

writing $g \leqslant g'$. This gives a total order on G which is easily seen to be compatible with its monoid structure. In addition, $(G^i, \leqslant^i) \ll (G, \leqslant)$ for every i which shows that the family is bounded above in Ω. Accordingly Ω is a non-empty inductive system and therefore, by Zorn's lemma, it contains at least one maximal element.

Let $(G, \leqslant)$ be a maximal member of Ω. It is going to be shown that $G = \Gamma$ which, of course, will establish the result we are trying to prove. The demonstration that $G = \Gamma$ contains a number of stages.

Denote by $\hat{G}$ the set consisting of all elements ξ of Γ for which it is possible to find g, g' in G such that $\xi + g = g'$. Since $g + 0 = g$, we see that $G \subseteq \hat{G}$. Indeed $\hat{G}$ is a submonoid of Γ containing G. It will now be shown that the total order on G can be extended to a total order on $\hat{G}$ in such a way that the extension is compatible with the monoid structure of $\hat{G}$. It will then follow, from the maximal property of $(G, \leqslant)$, that $G = \hat{G}$.

Let ξ_1, ξ_2 belong to $\hat{G}$. We shall suppose that $\xi_1 + g_1 = g_1'$ and $\xi_2 + g_2 = g_2'$, where g_1, g_2, g_1', g_2' all belong to G. Since there will exist other relations of the same kind, we shall also assume that

$$\xi_1 + \tilde{g}_1 = \tilde{g}_1' \text{ and } \xi_2 + \tilde{g}_2 = \tilde{g}_2', \quad \text{where} \quad \tilde{g}_1, \tilde{g}_2, \tilde{g}_1', \tilde{g}_2'$$

belong to G as well. Then $\xi_1 + g_1 + \tilde{g}_1' = \xi_1 + \tilde{g}_1 + g_1'$ and therefore $g_1 + \tilde{g}_1' = \tilde{g}_1 + g_1'$. Similarly $g_2 + \tilde{g}_2' = \tilde{g}_2 + g_2'$. This shows that

$$g_1' + g_2 + \tilde{g}_1 + \tilde{g}_2' = g_1 + g_2' + \tilde{g}_1' + \tilde{g}_2.$$

It now follows, from (2) of Lemma 12, that *if* $g_1' + g_2 \leqslant g_1 + g_2'$ *then* $\tilde{g}_1' + \tilde{g}_2 \leqslant \tilde{g}_1 + \tilde{g}_2'$. Of course, the converse implication holds by symmetry. We may therefore write $\xi_1 \leqslant \xi_2$ when $g_1' + g_2 \leqslant g_1 + g_2'$ and this will give a properly defined relation on $\hat{G}$. Note that when ξ_1, ξ_2 are in G we may take $g_1' = \xi_1, g_2' = \xi_2$ and $g_1 = g_2 = 0$. This shows that the relation on $\hat{G}$ is an extension of the total order on G and therefore no confusion can arise because we have used the same symbol.

It is clear that $\xi_1 \leqslant \xi_1$. Next either $\xi_1 \leqslant \xi_2$ or $\xi_2 \leqslant \xi_1$. If both hold, then $g_1' + g_2 = g_1 + g_2'$ and therefore $\xi_1 + g_1 + g_2 = \xi_2 + g_1 + g_2$ whence $\xi_1 = \xi_2$ by the cancellation law. It is left to the reader to check that the relation on $\hat{G}$ is transitive after which it will follow that $\leqslant$ is a total order on $\hat{G}$. Next assume that $\xi_1 \leqslant \xi_2$ and let ξ be an arbitrary element of $\hat{G}$. We then have $g_1' + g_2 \leqslant g_1 + g_2'$ and there exist g, g' in G such that $\xi + g = g'$. Accordingly $\xi + \xi_1 + g + g_1 = g' + g_1'$, $\xi + \xi_2 + g + g_2 = g' + g_2'$ and $g' + g_1' + g + g_2 \leqslant g + g_1 + g' + g_2'$. This shows that $\xi + \xi_1 \leqslant \xi + \xi_2$. It follows that $(\hat{G}, \leqslant)$ belongs to Ω and indeed

$(G, \leqslant) \ll (\hat{G}, \leqslant)$. However, $(G, \leqslant)$ is, by construction, maximal in Ω. Consequently $G = \hat{G}$ as was asserted earlier.

We recall that our aim is to prove that $G = \Gamma$. Let $\gamma \in \Gamma$. It will suffice to prove that $\gamma \in G$. To this end let H consist of all elements of Γ which can be expressed in the form $g + n\gamma$, where $g \in G$ and $n \geqslant 0$ is an integer. H is a submonoid of Γ and it contains G. We now distinguish two possibilities. Either there exists a positive integer m such that $m\gamma \in G$ or none of $\gamma, 2\gamma, 3\gamma, \ldots$ belongs to G. In the former case we shall show that γ belongs to G and in the latter we shall obtain a contradiction. This will complete the proof.

First suppose that $m\gamma \in G$, where $m \geqslant 1$ is an integer. We then obtain a mapping $\phi : H \to G$ by multiplying each element of H by m. It is clear that $\phi(0) = 0$ and if h_1, h_2 belong to H, then

$$\phi(h_1 + h_2) = \phi(h_1) + \phi(h_2).$$

Also, because Γ is torsionless, distinct elements of H have distinct images. It follows therefore that H and $\phi(H)$ are, in an obvious sense, isomorphic as monoids. Now the total order on G induces a total order on $\phi(H)$ which is compatible with the latter's monoid structure. Let us transfer this compatible total order to H by means of the inverse mapping ϕ^{-1}. Because the restriction of ϕ to G preserves order relations (see (3) of Lemma 12), it follows that the total order now imposed on H extends the original total order on G. Using the same symbol $\leqslant$ to denote the total order on H, we have $(G, \leqslant) \ll (H, \leqslant)$ whence $G = H$ because of the maximal property of $(G, \leqslant)$. But $\gamma \in H$, consequently $\gamma \in G$.

We have now reached the last stage of the argument. Here we assume that none of $\gamma, 2\gamma, 3\gamma, \ldots$ belongs to G and derive a contradiction. First we note that every element of H has a *unique* representation in the form $g + n\gamma$, where $g \in G$ and $n \geqslant 0$. For suppose, with a self-explanatory notation, that $g + n\gamma = g' + n'\gamma$. If we can show that $n = n'$, then it will also follow that $g = g'$. However if (say) $n < n'$, then $g + n\gamma = g' + (n' - n)\gamma + n\gamma$ whence $g = g' + (n' - n)\gamma$ and therefore $(n' - n)\gamma \in \hat{G} = G$. This however contradicts the assumption that no positive multiple of γ belongs to G.

The fact that the representation is unique allows us to define a total order on H. Specifically if $h = g + n\gamma$ and $h^* = g^* + n^*\gamma$ are elements of H, then we put $h \leqslant h^*$ if either (i) $g \leqslant g^*$ and $g \neq g^*$ or (ii) $g = g^*$ and $n \leqslant n^*$. It is clear that this gives a relation on H which extends the total order on G. It is also obvious that either

$h \leqslant h^*$ or $h^* \leqslant h$ and that if both hold then $h = h^*$. Again $h \leqslant h$. The reader is left to complete the verification that we have a total order on H by checking the transitive property.

Finally suppose that $h \leqslant h^*$ and that $\bar{h} = \bar{g} + \bar{n}\gamma$ also belongs to H. Then $h + \bar{h} = (g + \bar{g}) + (n + \bar{n})\gamma$ while $h^* + \bar{h} = (g^* + \bar{g}) + (n^* + \bar{n})\gamma$. If $g \leqslant g^*$, $g \neq g^*$, then $g + \bar{g} \leqslant g^* + \bar{g}$ and $g + \bar{g} \neq g^* + \bar{g}$; consequently $h + \bar{h} \leqslant h^* + \bar{h}$. If however $g = g^*$ and $n \leqslant n^*$, then $g + \bar{g} = g^* + \bar{g}$ and $n + \bar{n} \leqslant n^* + \bar{n}$. Thus again $h + \bar{h} \leqslant h^* + \bar{h}$. Accordingly the total order on H is compatible with its monoid structure. We now have $(G, \leqslant) \ll (H, \leqslant)$ whence $G = H$ and therefore $\gamma \in G$. This is the required contradiction and with it we have completed the proof of

Theorem 22. *A grading monoid can be endowed with a total order compatible with its structure as a monoid if, and only if, it is torsionless.*

In the next section we shall give some applications of this theorem. Before proceeding to these we make one minor observation concerning partial orders. If $\leqslant$ is a partial order on a set Ω and ω, ω' belong to the set, then we write $\omega < \omega'$ if $\omega \leqslant \omega'$, $\omega \neq \omega'$.

2.13 Homogeneous primary decompositions

Throughout the present section, Γ denotes a grading monoid and we continue to study the properties of a Γ-graded ring $R = \sum_{\gamma} R^{(\gamma)}$. However, our results will all be concerned with the case in which Γ is torsionless.

Proposition 32. *Let R be graded by a torsionless monoid Γ and let A be a homogeneous ideal. Then Rad A is also homogeneous.*

Proof. Let $\leqslant$ be a total order on Γ compatible with its monoid structure and let r belong to Rad A. We can write

$$r = r^{(\gamma_1)} + r^{(\gamma_2)} + \ldots + r^{(\gamma_p)},$$

where $r^{(\gamma_i)}$ is homogeneous of degree γ_i and $\gamma_1 < \gamma_2 < \ldots < \gamma_p$. Next there exists an integer m such that $r^m \in A$. Now A is a homogeneous ideal. Consequently the homogeneous component of r^m whose degree is $m\gamma_1$ must belong to A. However this component is just the mth power of $r^{(\gamma_1)}$ so it follows that $r^{(\gamma_1)}$ belongs to Rad A. Accordingly $r^{(\gamma_2)} + r^{(\gamma_3)} + \ldots + r^{(\gamma_p)}$ is in Rad A and we may repeat the argument. In this manner it is seen that all the homogeneous components of r belong to Rad A and this, of course, establishes the proposition.

Lemma 13. *Let R be graded by a torsionless monoid Γ and let P be a homogeneous proper ideal with the following property: whenever $\alpha\beta \in P$ and α, β are both homogeneous elements of R, then either $\alpha \in P$ or $\beta \in P$. In these circumstances, P is a prime ideal.*

In other words, if P behaves like a prime ideal so far as homogeneous elements are concerned, then it is actually a prime ideal.

Proof. Assume the contrary. Then there exist elements r, ρ of R such that $r\rho \in P$ but $r \notin P$, $\rho \notin P$. Let us now introduce a total order on Γ compatible with its structure as a monoid. It is then possible to choose $\gamma_1, \gamma_2, ..., \gamma_p$ in Γ so that $\gamma_1 < \gamma_2 < ... < \gamma_p$ and, with a self-explanatory notation,
$$r = r^{(\gamma_1)} + r^{(\gamma_2)} + ... + r^{(\gamma_p)}$$
and
$$\rho = \rho^{(\gamma_1)} + \rho^{(\gamma_2)} + ... + \rho^{(\gamma_p)}.$$

Not all of $r^{(\gamma_1)}, ..., r^{(\gamma_p)}$ can belong to P. Let $r^{(\gamma_s)}$ be the first which does not. Likewise let $\rho^{(\gamma_t)}$ be the first among $\rho^{(\gamma_1)}, ..., \rho^{(\gamma_p)}$ not to belong to P. Then
$$(r^{(\gamma_s)} + ... + r^{(\gamma_p)})(\rho^{(\gamma_t)} + ... + \rho^{(\gamma_p)}) \in P.$$

Since P is homogeneous, the homogeneous component of degree $\gamma_s + \gamma_t$ of the product on the left-hand side must belong to P. But this is just $r^{(\gamma_s)}\rho^{(\gamma_t)}$. Next, by the hypothesis concerning P, one at least of $r^{(\gamma_s)}$ and $\rho^{(\gamma_t)}$ is in P. This is the required contradiction.

Proposition 33. *Let P be a prime ideal of the Γ-graded ring $R = \sum_{\gamma} R^{(\gamma)}$ and suppose that Γ is torsionless. Denote by P^* the homogeneous ideal generated by all homogeneous elements contained in P. Then P^* is a homogeneous prime ideal.*

Proof. Let r, ρ be *homogeneous* elements of R such that $r\rho \in P^*$, $r \notin P^*$. By Lemma 13, it will suffice to show that $\rho \in P^*$. But $r\rho \in P$ and $r \notin P$. Consequently $\rho \in P$. However ρ is homogeneous and so we have in fact $\rho \in P^*$. This completes the proof.

The lemma which follows forms a companion to Lemma 13. In it $E = \sum_{\gamma} E^{(\gamma)}$ denotes a Γ-graded module with respect to the Γ-graded ring $R = \sum_{\gamma} R^{(\gamma)}$.

Lemma 14. *Let Γ be torsionless and assume that N is a homogeneous proper submodule of E. Furthermore let N satisfy the following condition: whenever a product ρy belongs to N, where ρ is a homogeneous element of R and y a homogeneous element of E, then either $y \in N$ or $\rho \in \mathrm{Rad}\,(N:E)$. In these circumstances, N is a primary submodule of E.*

Proof. As usual we impose a total order on Γ compatible with its monoid structure. Now assume that N is not a primary submodule of E so that there exist $r \in R$ and $x \in E$ such that $rx \in N$, $x \notin N$ and $r \notin \mathrm{Rad}\,(N:E)$.

We express each of r and x as the sum of its *non-zero* homogeneous components. With a self-explanatory notation let these sums be

$$r = r^{(\gamma_1)} + r^{(\gamma_2)} + \ldots + r^{(\gamma_p)},$$
$$x = x^{(\delta_1)} + x^{(\delta_2)} + \ldots + x^{(\delta_q)},$$

where $\gamma_1 < \gamma_2 < \ldots < \gamma_p$ and $\delta_1 < \delta_2 < \ldots < \delta_q$. It will now be supposed that, from among all pairs (r, x) with the properties mentioned above, we have selected one for which the integer q is minimal. Then $x^{(\delta_1)} \notin N$ for otherwise we could replace x by $x - x^{(\delta_1)}$ and so obtain a new pair with a smaller value of q.

Let s $(1 \leqslant s \leqslant p)$ be such that $r^{(\gamma_i)} \in \mathrm{Rad}\,(N:E)$ for $1 \leqslant i \leqslant s-1$ while $r^{(\gamma_s)} \notin \mathrm{Rad}\,(N:E)$. Put $\rho = r^{(\gamma_1)} + r^{(\gamma_2)} + \ldots + r^{(\gamma_{s-1})}$. Then

$$\rho \in \mathrm{Rad}\,(N:E)$$

whence, if k is a large enough positive integer, $\rho^k \in (N:E)$ and therefore $(r - \rho)^k x \in N$. Accordingly

$$(r^{(\gamma_s)} + r^{(\gamma_{s+1})} + \ldots + r^{(\gamma_p)})^k x \in N.$$

The component of degree $k\gamma_s + \delta_1$ of the expression on the left is $(r^{(\gamma_s)})^k x^{(\delta_1)}$. This therefore belongs to N. Since $x^{(\delta_1)} \notin N$, the original hypothesis concerning N shows that $(r^{(\gamma_s)})^k$ belongs to $\mathrm{Rad}\,(N:E)$. Thus $r^{(\gamma_s)} \in \mathrm{Rad}\,(N:E)$ which contradicts the choice of the integer s. The lemma follows.

Theorem 23. *Let the grading monoid Γ be torsionless and let N be a P-primary submodule of the Γ-graded module E. Denote by N^* the (homogeneous) submodule of E generated by all homogeneous elements contained in N, and by P^* the (homogeneous) ideal of R generated by all homogeneous elements contained in P. Then P^* is a prime ideal and N^* is a P^*-primary submodule of E.*

Proof. That P^* is a prime ideal follows from Proposition 33. We next show that N^* is a primary submodule of E. To this end assume that r and y are homogeneous elements of R and E respectively such that $ry \in N^*$ and $y \notin N^*$. If we can show that $r \in \mathrm{Rad}\,(N^*:E)$, then it will follow, from Lemma 14, that N^* is primary. Now $ry \in N$ and $y \notin N$. Hence there exists an integer s such that $r^s E \subseteq N$. It follows that if e is a *homogeneous* element of E, then $r^s e \in N^*$. However, E is generated by homogeneous elements. Consequently $r^s E \subseteq N^*$. Thus r belongs to $\mathrm{Rad}\,(N^*:E)$ as required.

It remains for us to show that N^* belongs to P^*. Since

$$N^*:E \subseteq N:E, \quad \text{we have} \quad \mathrm{Rad}\,(N^*:E) \subseteq \mathrm{Rad}\,(N:E) = P.$$

Further, by Proposition 30, $N^*:E$ is a homogeneous ideal and therefore, by Proposition 32, $\mathrm{Rad}\,(N^*:E)$ is also homogeneous. It follows that $\mathrm{Rad}\,(N^*:E) \subseteq P^*$. To complete the proof it will be sufficient to show that $P^* \subseteq \mathrm{Rad}\,(N^*:E)$ and for this it is enough to show that every homogeneous element of P^* belongs to $\mathrm{Rad}\,(N^*:E)$.

Let $\rho \in P^*$ and be homogeneous. Then $\rho \in P$ and therefore, for some integer q, $\rho^q \in (N:E)$. Accordingly if e is a *homogeneous* element of E, then $\rho^q e \in N^*$. But E is generated by homogeneous elements. Consequently $\rho^q E \subseteq N^*$ and therefore $\rho \in \mathrm{Rad}\,(N^*:E)$. This completes the proof.

It is natural to ask whether a homogeneous decomposable submodule of a graded module possesses a primary decomposition in which all the components are homogeneous. Our final result shows that this is the case when the grading monoid is torsionless.

Theorem 24. *Let R be graded by a torsionless monoid Γ and let E be a Γ-graded R-module. In addition, let K be a homogeneous submodule of E possessing a primary decomposition. Then K has a normal (Lasker) decomposition in which all the primary components are homogeneous. Moreover, the prime ideals belonging to K are necessarily homogeneous.*

Proof. Let $K = N_1 \cap N_2 \cap \ldots \cap N_s$, where N_i is a P_i-primary submodule of E. Denote by N_i^* the homogeneous submodule generated by all homogeneous elements contained in N_i and by P_i^* the homogeneous ideal generated by all homogeneous elements contained in P_i. By Theorem 23, P_i^* is a prime ideal and N_i^* a P_i^*-primary submodule of E. Moreover $K \subseteq N_i^* \subseteq N_i$, whence $K = N_1^* \cap N_2^* \cap \ldots \cap N_s^*$. This is a homogeneous primary decomposition of K in E. If we refine it, in the usual manner, to a normal decomposition, then the components of the result will also be homogeneous (see Proposition 29). Since the prime ideals belonging to K all occur among $P_1^*, P_2^*, \ldots, P_s^*$, the proof is complete.

Exercises on Chapter 2

In these exercises the letters R and R' denote commutative rings each possessing an identity element. Given a ring R, an indeterminate X and a grading monoid Γ, it is possible to construct, in a natural way, a new ring which is here denoted by $R[X;\Gamma]$. The details of the construction are set out in Exercise 13. In certain of the later exercises the notation $R[X;\Gamma]$ is used without further explanation.

1. Let D be a division ring. Use Zorn's lemma to prove that every D-module, whether finitely generated or not, possesses a base.

2. The ring R' is an extension ring of the ring R. Show that if the R'-ideal A' is the extension of an ideal of R, then $R'(A' \cap R) = A'$. Show also that if the R-ideal B is the contraction of an ideal of R', then $R'B \cap R = B$.

3. R' is an extension ring of a ring R and, considered as an R-module, R' is free. Prove that $AR' \cap R = A$ for every ideal A of R.

4. Prove that the ring of integers is integrally closed in the field of rational numbers.

5. K, L are submodules of an R-module E and A is an ideal of R. Prove that

$$\text{(i)} \quad K:(K:_E(K:L)) = K:L;$$

$$\text{(ii)} \quad K:_E(K:(K:_E A)) = K:_E A.$$

6. A finitely generated R-module E is such that $ME = E$ for every maximal ideal M of R. Prove that $E = 0$.

7. N is a P-primary submodule of an R-module E and K is a proper submodule of N. Prove that K is a P-primary submodule of E if and only if it is a P-primary submodule of N.

8. Let $\{N_i\}_{i \in I}$ be a non-empty but otherwise arbitrary family of P-primary submodules of an R-module E. Prove that if there exists a positive integer m such that $P^m E \subseteq \bigcap_i N_i$, then $\bigcap_i N_i$ is a P-primary submodule of E.

9. Show that a non-zero ideal of the ring Z of integers is a prime ideal if and only if it can be generated by a single prime number. Show also that if p is a prime number, then the primary ideals belonging to (p) are simply the ideals (p^k), where k is an arbitrary positive integer. Explain how a normal decomposition of the ideal (m) may be obtained when $m > 1$.

10. Obtain a normal decomposition of the ideal $12600Z$ of the ring Z of integers. Obtain also a normal decomposition of $12600Z$ in the Z-module $225Z$.

11. F is a field and A the ideal $(X_1^2, X_1 X_2)$ of the ring $F[X_1, X_2]$ consisting of polynomials in the indeterminates X_1, X_2 with coefficients in F. Show that

$$\text{(i)} \quad A = (X_1) \cap (X_1^2, X_2),$$

$$\text{(ii)} \quad A = (X_1) \cap (X_1^2, X_1 + X_2),$$

$$\text{(iii)} \quad A = (X_1) \cap (X_1^2, X_1 X_2, X_2^2),$$

are three distinct normal decompositions of A in $F[X_1, X_2]$.

12. Let K be a decomposable submodule of an R-module E, let P be a minimal prime ideal belonging to K, and N the corresponding P-primary component. Show that there exists an element c satisfying $c \in R$, $c \notin P$, and $cN \subseteq K$.

13. Let Γ be a grading monoid and X an indeterminate. We shall call the symbols X^γ, where $\gamma \in \Gamma$, 'formal powers of X with exponents in Γ.' If now R

is a commutative ring with an identity element, let $R[X;\Gamma]$ denote the free R-module having the set $\{X^\gamma\}_{\gamma\in\Gamma}$ of formal powers as base. Next, for any pair

$$\sum_{\gamma\in\Gamma} r_\gamma X^\gamma \quad \text{and} \quad \sum_{\lambda\in\Gamma} \rho_\lambda X^\lambda$$

of elements of this module define their product by means of the relation

$$\left(\sum_{\gamma\in\Gamma} r_\gamma X^\gamma\right)\left(\sum_{\lambda\in\Gamma} \rho_\lambda X^\lambda\right) = \sum_{\gamma,\,\lambda\in\Gamma} r_\gamma \rho_\lambda X^{\gamma+\lambda}.$$

Show that this construction turns $R[X;\Gamma]$ into a commutative ring (with an identity element) which can be Γ-graded by taking as homogeneous elements of degree γ those which have the form rX^γ, where $r\in R$. Show also that R can be identified with the subring consisting of the homogeneous elements of degree zero.

14. Let R be an integral domain and Γ the grading monoid formed by all non-negative rational numbers. Show that the elements of $R[X;\Gamma]$ for which the coefficient of X^0 is zero form a prime ideal P, and that the ideal Q generated by X^1 is P-primary. Show further that no power of P is contained in Q.

15. Let D_p be the monoid whose elements are ordered sets $(m_1, m_2, ..., m_p)$ of p non-negative integers addition being defined by

$$(m_1, m_2, ..., m_p) + (m'_1, m'_2, ..., m'_p) = (m_1 + m'_1, m_2 + m'_2, ..., m_p + m'_p).$$

If $(m_1, m_2, ..., m_p)$ and $(n_1, n_2, ..., n_p)$ belong to D_p write

$$(m_1, m_2, ..., m_p) \leqslant (n_1, n_2, ..., n_p)$$

if the two ordered sets are identical or there is an integer $i\,(1 \leqslant i \leqslant p)$ such that

$$m_1 = n_1, m_2 = n_2, ..., m_{i-1} = n_{i-1}, m_i < n_i.$$

Prove that $\leqslant$ is a total order on D_p compatible with its monoid structure. It is customary to call $\leqslant$ the *lexicographical order* on D_p.

16. Let the monoid D_p be endowed with the lexicographical order (see Exercise 15). Show that if Σ is an arbitrary non-empty subset of D_p, then there exists $(\alpha_1, \alpha_2, ..., \alpha_p)$ in Σ such that

$$(\alpha_1, \alpha_2, ..., \alpha_p) < (\beta_1, \beta_2, ..., \beta_p)$$

for every other element $(\beta_1, \beta_2, ..., \beta_p)$ in Σ.

This result, which asserts that every non-empty subset of D_p has a least element, is usually described by saying that D_p is *well ordered* by the lexicographical order relation.

17. Let K be a field of characteristic 2 (this means that $1_K + 1_K = 0_K$) and let Γ be a cyclic group of order 2 with the law of composition written as addition. Show that the zero ideal of the graded ring $K[X;\Gamma]$ (see Exercise 13) is a homogeneous ideal whose radical is not homogeneous.

18. Let P be a prime ideal of the ring R and Q a primary ideal belonging to it. Show that if Γ is a torsionless grading monoid, then $PR[X;\Gamma]$ is a prime ideal of the ring $R[X;\Gamma]$. Show also that $QR[X;\Gamma]$ is a primary ideal of this ring and that it belongs to $PR[X;\Gamma]$.

19. R is a graded ring, E a graded R-module and K a homogeneous submodule of E. Furthermore there exists an R-submodule L of E such that $E = K \oplus L$. Show that there exists a homogeneous submodule L' of E such that $E = K \oplus L'$.

3

RINGS AND MODULES OF FRACTIONS

General remarks. In all but one of the sections of this chapter, we confine our attention to commutative rings each of which is assumed to possess an identity element. The null ring is not excluded from consideration. The section which is the exception is (3.4). In this we introduce the idea of a *functor* which is extremely useful for describing the properties of certain basic constructions. This notion is remarkably general, a fact which has been slightly stressed by presenting the functor which interests us here against a broader background than applies to the rest of the chapter.

3.1 Formation of fractions

The process whereby the rational numbers are constructed from the integers is very familiar. Nevertheless the corresponding construction for modules over a commutative ring is less well known and therefore we shall deal with its various aspects in a leisurely manner in order to allow the reader time to become properly familiar with all the details.

Let R be a commutative ring with an identity element and S a non-empty multiplicatively closed subset of R. We shall not exclude the possibility that the zero element of R belongs to S but the reader should be aware that, in this event, our conclusions all become trivial.

Now suppose that E is an R-module and consider formal *fractions* $\dfrac{e}{s}$, where the *numerator* e belongs to E and the *denominator* s is an element of S. Such a fraction is simply an ordered pair (written in a special notation) in which one constituent is an element of E and the other an element of S. Sometimes we write e/s in place of $\dfrac{e}{s}$ because the latter consumes more space and is a little inconvenient to print.

The next step is to define an equivalence relation between these formal fractions. To this end consider e_1/s_1 and e_2/s_2, where of course e_1, e_2 belong to E and s_1, s_2 are elements of S. We now write $e_1/s_1 \sim e_2/s_2$ if there exists $s \in S$ such that $ss_2 e_1 = ss_1 e_2$. It is a straightforward exercise, which we leave to the reader, to verify that $\sim$ really is an equivalence relation. The equivalence class to which the fraction $\dfrac{e}{s}$ belongs will be denoted by $\left[\dfrac{e}{s}\right]$ or $[e/s]$. Accordingly

$$[e_1/s_1] = [e_2/s_2]$$

if and only if there exists $s \in S$ such that $ss_2 e_1 = ss_1 e_2$. Observe that if $e \in E$ and s, s' belong to S, then $[e/s] = [s'e/s's]$ and $[se/s] = [s'e/s']$.

Lemma 1. *If, with a self-explanatory notation,* $[e_1/s_1] = [e_1'/s_1']$ *and* $[e_2/s_2] = [e_2'/s_2']$, *then*

$$\left[\frac{s_2 e_1 + s_1 e_2}{s_1 s_2}\right] = \left[\frac{s_2' e_1' + s_1' e_2'}{s_1' s_2'}\right]. \tag{3.1.1}$$

Remark. The reader should note that, because S is multiplicatively closed, the two sides of (3.1.1) do indeed represent classes of fractions of the type under consideration.

Proof. Since $[e_1/s_1] = [e_1'/s_1']$ and $[e_2/s_2] = [e_2'/s_2']$, there exist s, s^* in S such that $ss_1' e_1 = ss_1 e_1'$ and $s^* s_2' e_2 = s^* s_2 e_2'$. It follows that

$$ss^* s_1' s_2'(s_2 e_1 + s_1 e_2) = s^* s_2' s_2 ss_1 e_1' + ss_1' s_1 s^* s_2 e_2' = ss^* s_1 s_2(s_2' e_1' + s_1' e_2')$$

and, since ss^* belongs to S, this proves the lemma.

On the basis of Lemma 1, we can define *addition* of classes of formal fractions by means of the formula

$$\left[\frac{e_1}{s_1}\right] + \left[\frac{e_2}{s_2}\right] = \left[\frac{s_2 e_1 + s_1 e_2}{s_1 s_2}\right]. \tag{3.1.2}$$

Addition is clearly commutative and, since each of

$$\left(\left[\frac{e_1}{s_1}\right] + \left[\frac{e_2}{s_2}\right]\right) + \left[\frac{e_3}{s_3}\right] \quad \text{and} \quad \left[\frac{e_1}{s_1}\right] + \left(\left[\frac{e_2}{s_2}\right] + \left[\frac{e_3}{s_3}\right]\right)$$

is found to be equal to

$$\left[\frac{s_2 s_3 e_1 + s_1 s_3 e_2 + s_1 s_2 e_3}{s_1 s_2 s_3}\right],$$

it is associative as well. Finally if σ is an arbitrary element of S, then

$$\left[\frac{e}{s}\right] + \left[\frac{0}{\sigma}\right] = \left[\frac{\sigma e}{\sigma s}\right] = \left[\frac{e}{s}\right]$$

and

$$\left[\frac{e}{s}\right] + \left[\frac{-e}{s}\right] = \left[\frac{0}{s^2}\right] = \left[\frac{0}{\sigma}\right].$$

If therefore we denote by E_S the set of all classes of formal fractions, then E_S is an abelian group with respect to addition as defined above. The zero element of the group is $[0/\sigma]$ (note that this does not depend on the choice of σ) and $-[e/s] = [(-e)/s]$. A further point to note is that

$$\left[\frac{e}{s}\right] + \left[\frac{e'}{s}\right] = \left[\frac{e+e'}{s}\right].$$

This is because $[(se+se')/s^2] = [(e+e')/s]$.

We can, however, do more than turn E_S into an abelian group. To begin with, let us observe that if $[e/s] = [e'/s']$, then $[re/s] = [re'/s']$ for every element r of R. This allows us to define the *product* of r and $[e/s]$ by means of the formula

$$r\left[\frac{e}{s}\right] = \left[\frac{re}{s}\right] \tag{3.1.3}$$

and suggests that E_S has a natural structure as an R-module. Indeed the verification that the module axioms are satisfied presents no problem. The details will therefore be omitted.

It was observed earlier that if $e \in E$ and $s, s' \in S$, then $[se/s] = [s'e/s']$. This shows that there is a well defined mapping

$$\chi : E \to E_S \tag{3.1.4}$$

which is such that $$\chi(e) = \left[\frac{se}{s}\right]. \tag{3.1.5}$$

It will be convenient to refer to $\chi : E \to E_S$ as the *canonical mapping* of E into E_S.

The canonical mapping $E \to E_S$ is a homomorphism of R-modules. For let e, e' belong to E and r to R. Then

$$\chi(e+e') = \left[\frac{se+se'}{s}\right] = \left[\frac{se}{s}\right] + \left[\frac{se'}{s}\right] = \chi(e) + \chi(e')$$

and $$\chi(re) = \left[\frac{sre}{s}\right] = r\left[\frac{se}{s}\right] = r\chi(e),$$

where in the above equations s may be any element of S. Next, *the kernel of the canonical mapping $E \to E_S$ is the S-component† 0^S of the zero submodule of E.* Indeed if e belongs to 0^S, then it is possible to find $\sigma \in S$ such that $\sigma e = 0$. Consequently

$$\chi(e) = \left[\frac{\sigma e}{\sigma}\right] = \left[\frac{0}{\sigma}\right]$$

† See section (2.9).

and therefore e belongs to $\operatorname{Ker} \chi$. On the other hand, if $e \in \operatorname{Ker} \chi$ then for any $s \in S$ we have $[se/s] = [0/s]$. It follows that $s'se = 0$ for a suitable element s' of S. This shows that e belongs to 0^S and establishes the assertion.

Next assume that E, E' are R-modules and let $f\colon E \to E'$ be a homomorphism. If now $[e_1/s_1] = [e_2/s_2]$, then there exists $s \in S$ such that $ss_2 e_1 = ss_1 e_2$. It follows that

$$ss_2 f(e_1) = f(ss_2 e_1) = f(ss_1 e_2) = ss_1 f(e_2)$$

and therefore $[f(e_1)/s_1] = [f(e_2)/s_2]$. Accordingly there exists a well defined mapping
$$f_S\colon E_S \to E'_S \tag{3.1.6}$$

which is such that
$$f_S\left(\left[\frac{e}{s}\right]\right) = \left[\frac{f(e)}{s}\right]. \tag{3.1.7}$$

A simple verification now shows that f_S is a homomorphism of R-modules. Note that the diagram

$$
\begin{array}{ccc}
E & \xrightarrow{\;f\;} & E' \\
\chi_E \downarrow & & \downarrow \chi_{E'} \\
E_S & \xrightarrow[f_S]{} & E'_S
\end{array}
\tag{3.1.8}
$$

where $\chi_E, \chi_{E'}$ are canonical mappings, is commutative. For if $s \in S$ and $e \in E$, then
$$\chi_{E'} f(e) = \left[\frac{sf(e)}{s}\right]$$

and
$$f_S \chi_E(e) = f_S\left(\left[\frac{se}{s}\right]\right) = \left[\frac{sf(e)}{s}\right]$$
as well.

We have now associated with each R-module E a second R-module E_S; and with each homomorphism $E \to E'$ a homomorphism

$$f_S\colon E_S \to E'_S$$

between the corresponding modules. Before we analyse this situation in greater detail, it is convenient to insert a few remarks about homomorphisms.

If f, g are homomorphisms of an R-module E into an R-module E' then their sum $f + g$ is the homomorphism given by

$$(f+g)(e) = f(e) + g(e).$$

In section (1.10), it was noted that the homomorphisms of E into E' form an abelian group with respect to addition. However, when as in the present instance, the ring involved happens to be commutative, another kind of composition is possible. To see this, let $f:E \to E'$ be an R-homomorphism and suppose that r belongs to R. For $e \in E$ put $h(e) = rf(e)$. Then $e \to h(e)$ defines a mapping $h:E \to E'$ which (bearing in mind that R is commutative) is easily seen to be an R-homomorphism. This new homomorphism is denoted by rf so that we have

$$(rf)(e) = rf(e). \tag{3.1.9}$$

We can therefore 'multiply' a homomorphism of E into E' by an arbitrary element of R. In fact a straightforward verification shows that *these homomorphisms form an R-module* and not merely an abelian group. The details of the verification are left to the reader.

Theorem 1. *Let S be a non-empty multiplicatively closed subset of R and let $f, g:E \to E'$ and $\phi:E' \to E''$ be homomorphisms of R-modules. Then*

(1) *if $i:E \to E$ is the identity mapping of E, then $i_S:E_S \to E_S$ is the identity mapping of E_S;*

(2) $(\phi f)_S = \phi_S f_S;$

(3) $(f+g)_S = f_S + g_S;$

(4) $(rf)_S = r(f_S)$ *for every r belonging to R.*

Proof. The first assertion is an immediate consequence of the definition of i_S. Next let $e \in E$ and $s \in S$. Then

$$(\phi f)_S\left(\left[\frac{e}{s}\right]\right) = \left[\frac{\phi f(e)}{s}\right] = \phi_S\left(\left[\frac{f(e)}{s}\right]\right) = \phi_S f_S\left(\left[\frac{e}{s}\right]\right)$$

and

$$(f+g)_S\left(\left[\frac{e}{s}\right]\right) = \left[\frac{f(e)+g(e)}{s}\right] = \left[\frac{f(e)}{s}\right] + \left[\frac{g(e)}{s}\right]$$

$$= f_S\left(\left[\frac{e}{s}\right]\right) + g_S\left(\left[\frac{e}{s}\right]\right)$$

$$= (f_S + g_S)\left(\left[\frac{e}{s}\right]\right).$$

This disposes of (2) and (3). Finally

$$(rf)_S\left(\left[\frac{e}{s}\right]\right) = \left[\frac{rf(e)}{s}\right] = r\left[\frac{f(e)}{s}\right] = r\left(f_S\left(\left[\frac{e}{s}\right]\right)\right)$$

$$= (rf_S)\left(\left[\frac{e}{s}\right]\right)$$

and so all is proved.

Theorem 2. *Let S be a non-empty multiplicatively closed subset of R and*

$$0 \to E' \overset{\phi}{\to} E \overset{\psi}{\to} E'' \to 0$$

an exact sequence of R-modules. Then the sequence

$$0 \to E'_S \overset{\phi_S}{\to} E_S \overset{\psi_S}{\to} E''_S \to 0$$

is also exact.

Proof. If $e \in E$ and $s \in S$, then $\psi_S([e/s]) = [\psi(e)/s]$. Since $\psi(e)$ may be any element of E'', it follows that ψ_S maps E_S on to E''_S.

Next assume that $[e'/s]$, where $e' \in E'$ and $s \in S$, is mapped by ϕ_S into zero. Then $[\phi(e')/s] = [0/s]$ and therefore there exists $\sigma \in S$ such that $\sigma s \phi(e') = 0$. Accordingly $\phi(\sigma s e') = 0$ whence $\sigma s e' = 0$ because ϕ is a monomorphism. This shows that

$$\left[\frac{e'}{s}\right] = \left[\frac{\sigma s e'}{\sigma s^2}\right] = \left[\frac{0}{s}\right]$$

and proves that ϕ_S is a monomorphism.

Again, $\psi\phi$ is a null homomorphism. Consequently $(\psi\phi)_S = \psi_S\phi_S$ is a null homomorphism. In other terms, $\mathrm{Im}\,\phi_S \subseteq \mathrm{Ker}\,\psi_S$.

Finally assume that $[e/s]$ belongs to $\mathrm{Ker}\,\psi_S$. Then $[\psi(e)/s] = [0/s]$ whence $s^*\psi(e) = 0$ for a suitable element s^* belonging to S. We now have $\psi(s^*e) = 0$ whence $s^*e = \phi(e')$ for some e' in E'. This is because the sequence $0 \to E' \to E \to E'' \to 0$ is exact. Thus

$$\phi_S\left(\left[\frac{e'}{ss^*}\right]\right) = \left[\frac{\phi(e')}{ss^*}\right] = \left[\frac{s^*e}{s^*s}\right] = \left[\frac{e}{s}\right]$$

which shows that $\mathrm{Ker}\,\psi_S \subseteq \mathrm{Im}\,\phi_S$. The proof is now complete.

3.2 Rings of fractions

As before S denotes a non-empty multiplicatively closed subset of R, where R is a commutative ring with an identity element. Since R may be regarded as a module with respect to itself, we can apply the construction described in the last section and so obtain the module R_S. In the lemma which follows, r_1, r_2, r'_1, r'_2 denote elements of R and s_1, s_2, s'_1, s'_2 elements of S.

Lemma 2. *If $[r_1/s_1] = [r'_1/s'_1]$ and $[r_2/s_2] = [r'_2/s'_2]$, then*

$$\left[\frac{r_1 r_2}{s_1 s_2}\right] = \left[\frac{r'_1 r'_2}{s'_1 s'_2}\right].$$

Proof. There exist elements s, σ belonging to S such that $ss_1' r_1 = ss_1 r_1'$ and $\sigma s_2' r_2 = \sigma s_2 r_2'$. It follows that $s\sigma s_1' s_2' r_1 r_2 = s\sigma s_1 s_2 r_1' r_2'$ which proves the lemma.

It is clear that the above result enables us to define multiplication, for pairs of elements of R_S, by means of the formula

$$\left[\frac{r_1}{s_1}\right]\left[\frac{r_2}{s_2}\right] = \left[\frac{r_1 r_2}{s_1 s_2}\right]. \tag{3.2.1}$$

We contend that R_S is now a commutative ring with an identity element. Indeed it is clear that multiplication is both commutative and associative. Further, with a self-explanatory notation,

$$\text{and}\quad \left[\frac{r_1}{s_1}\right]\left(\left[\frac{r_2}{s_2}\right] + \left[\frac{r_3}{s_3}\right]\right) = \left[\frac{s_3 r_1 r_2 + s_2 r_1 r_3}{s_1 s_2 s_3}\right]$$

$$\left[\frac{r_1}{s_1}\right]\left[\frac{r_2}{s_2}\right] + \left[\frac{r_1}{s_1}\right]\left[\frac{r_3}{s_3}\right] = \left[\frac{s_1 s_3 r_1 r_2 + s_1 s_2 r_1 r_3}{s_1^2 s_2 s_3}\right] = \left[\frac{s_3 r_1 r_2 + s_2 r_1 r_3}{s_1 s_2 s_3}\right]$$

which shows that the distributive law holds as well. Finally if $s \in S$, then $[s/s]$ is the identity element of R_S.

The reader should note that if S happens to contain the zero element, then $[r/s] = [0/0]$ for all $r \in R$ and $s \in S$. In other terms, *when S contains the zero element of R, the ring R_S is the null ring.*

Referring back to (3.1.5), we recall that the canonical mapping

$$\chi : R \to R_S$$

is defined by $\chi(r) = [sr/s]$, where s is an arbitrary element of S. Now

$$\chi(rr') = \left[\frac{srr'}{s}\right] = \left[\frac{sr}{s}\right]\left[\frac{sr'}{s}\right] = \chi(r)\,\chi(r')$$

and $\chi(1_R) = [s/s]$ which is the identity element of R_S. Accordingly *the canonical mapping $\chi : R \to R_S$ is a ring-homomorphism.*

Let E be an R-module. In Lemma 3, r, r' denote elements of R, e, e' elements of E and s, s', σ, σ' belong to S.

Lemma 3. *If $[r/\sigma] = [r'/\sigma']$ and $[e/s] = [e'/s']$, then*

$$\left[\frac{re}{\sigma s}\right] = \left[\frac{r'e'}{\sigma' s'}\right].$$

Proof. There exist s_1, s_2 in S such that $s_1 \sigma' r = s_1 \sigma r'$ and $s_2 s' e = s_2 s e'$. Hence $s_1 s_2 \sigma' s' re = s_1 s_2 \sigma s r' e'$ which proves the lemma.

We see, from Lemma 3, that if $[r/\sigma]$ and $[e/s]$ are typical elements of R_S and E_S respectively, then it is possible to define their product unambiguously by means of the equation

$$\left[\frac{r}{\sigma}\right]\left[\frac{e}{s}\right] = \left[\frac{re}{\sigma s}\right]. \qquad (3.2.2)$$

It is now a straightforward matter to check that E_S is a module with respect to the ring R_S. There are, of course, four axioms to be satisfied. We leave the reader to examine the details.

Lemma 4. *If $f\!:\!E' \to E$ is a homomorphism of R-modules, then*

$$f_S\!:\!E'_S \to E_S$$

is a homomorphism of R_S-modules.

Proof. It has already been shown that f_S is compatible with addition. Next, using a notation which explains itself,

$$f_S\left(\left[\frac{r}{\sigma}\right]\left[\frac{e'}{s}\right]\right) = f_S\left(\left[\frac{re'}{\sigma s}\right]\right)$$

$$= \left[\frac{f(re')}{\sigma s}\right]$$

$$= \left[\frac{rf(e')}{\sigma s}\right]$$

$$= \left[\frac{r}{\sigma}\right]\left[\frac{f(e')}{s}\right]$$

$$= \left[\frac{r}{\sigma}\right]f_S\left(\left[\frac{e'}{s}\right]\right).$$

This establishes the lemma.

Now suppose that E' is a submodule of an R-module E. If we apply Lemma 4 and Theorem 2 to the natural exact sequence

$$0 \to E' \to E \to E/E' \to 0,$$

then we see among other things that $E'_S \to E_S$ is a monomorphism of R_S-modules. By identifying E'_S with its image in E_S, we can regard E'_S as an R_S-submodule of E_S. When E'_S is so regarded we shall refer to it as *the extension of E' in E_S*. Thus the extension of E' in E_S consists of all the elements of E_S which can be written in the form $[e'/s]$, where $e' \in E'$ and $s \in S$. In this connection it should be noted that, *when E'_S*

is regarded as a submodule of E_S, there is a natural R_S-isomorphism of E_S/E_S' on to $(E/E')_S$. This is because the exact sequence

$$0 \to E' \to E \to E/E' \to 0$$

induces an exact sequence $0 \to E_S' \to E_S \to (E/E')_S \to 0$ of R_S-modules.

Now consider an R_S-submodule W of E_S. The canonical mapping $\chi: E \to E_S$ is an R-homomorphism; consequently $\chi^{-1}(W)$ is an R-submodule of E. This submodule is known as the *contraction of W in E*. Thus the operation of contraction associates an R-submodule of E with each R_S-submodule of E_S.

Proposition 1. *Let X be an R-submodule of E and let X_S be regarded as a submodule of E_S. Then the contraction of X_S in E is just the S-component† X^S of X in E.*

Proof. Let e belong to X^S. Then there exists $s \in S$ such that $se \in X$. Accordingly $\chi(e) = [se/s]$ belongs to X_S and therefore e belongs to the contraction of X_S.

To prove the converse, we assume that $e \in E$ and $\chi(e) \in X_S$. Then there exists $\sigma \in S$ and $x \in X$ such that $[\sigma e/\sigma] = [x/\sigma]$. Consequently $s'\sigma^2 e = s'\sigma x$ belongs to X for some element $s' \in S$. Since $s'\sigma^2 \in S$, it follows that $e \in X^S$. This completes the proof.

Let us examine what happens if we start with a submodule W of E_S, contract it in E and then extend the contraction in E_S.

Proposition 2. *Let W be an R_S-submodule of E_S and W^c its contraction in E. Then the extension of W^c in E_S coincides with W.*

Proof. If $\chi: E \to E_S$ is the canonical mapping, then $W^c = \chi^{-1}(W)$. Accordingly a typical element of $(W^c)_S$ has the form $[u/s]$, where $\chi(u) \in W$ and $s \in S$. But $[u/s] = [1/s][su/s] = [1/s]\chi(u)$ which belongs to W; consequently $(W^c)_S \subseteq W$.

Now suppose that $[e/\sigma] \in W$, where $e \in E$ and $\sigma \in S$. Then

$$[\sigma^2/\sigma][e/\sigma] = \chi(e)$$

belongs to W and therefore $e \in W^c$. It follows that $[e/\sigma] \in (W^c)_S$. This shows that $W \subseteq (W^c)_S$ and completes the proof.

If we have several R-modules, say E, E', E'' etc., then we have several canonical mappings namely $E \to E_S$, $E' \to E_S'$, $E'' \to E_S''$ and so on. It will be convenient to use χ_E, $\chi_{E'}$, $\chi_{E''}$ etc. to distinguish these. In particular χ_R will denote the canonical mapping $R \to R_S$. Note that if $r \in R$ and $e \in E$, then

$$\chi_E(re) = \chi_R(r)\chi_E(e). \qquad (3.2.3)$$

† See section (2.9).

This is so because if $s \in S$, then

$$\chi_E(re) = \left[\frac{s^2 re}{s^2}\right] = \left[\frac{sr}{s}\right]\left[\frac{se}{s}\right] = \chi_R(r)\,\chi_E(e).$$

Proposition 3. *Let U be a subset of an R-module E and let K be the submodule which it generates. If now $\chi_E : E \to E_S$ is the canonical mapping and K_S is regarded as an R_S-module of E_S, then K_S is generated by $\chi_E(U)$.*

Proof. It is clear that $\chi_E(U) \subseteq K_S$, hence the R_S-submodule $R_S\chi_E(U)$ generated by $\chi_E(U)$ is contained in K_S. Next a typical element of K_S has the form

$$\left[\frac{r_1 u_1 + r_2 u_2 + \dots + r_m u_m}{s}\right],$$

where $r_1, r_2, \dots, r_m$ belong to R, $u_1, u_2, \dots, u_m$ belong to U and s belongs to S. But this can also be written as

$$\left[\frac{r_1}{s}\right]\left[\frac{su_1}{s}\right] + \left[\frac{r_2}{s}\right]\left[\frac{su_2}{s}\right] + \dots + \left[\frac{r_m}{s}\right]\left[\frac{su_m}{s}\right].$$

However $[su_i/s] = \chi_E(u_i) \in \chi_E(U)$ and so $K_S \subseteq R_S\chi_E(U)$.

This completes the proof.

Corollary. *If E is a finitely generated R-module, then E_S is a finitely generated R_S-module.*

Indeed the proposition shows that if $E = Re_1 + Re_2 + \dots + Re_m$, then $E_S = R_S\chi_E(e_1) + \dots + R_S\chi_E(e_m)$.

Proposition 4. *Let E be an R-module and A an R-ideal. Then*

$$(AE)_S = A_S E_S$$

provided that $(AE)_S$ is regarded as a submodule of E_S and A_S as an ideal of R_S.

Proof. The R-module AE is generated by all products ae, where $a \in A$ and $e \in E$. It therefore follows, from Proposition 3, that $(AE)_S$ is the R_S-submodule of E_S generated by all $\chi_E(ae)$, where $\chi_E : E \to E_S$ is the canonical mapping. By (3.2.3), $\chi_E(ae) = \chi_R(a)\chi_E(e)$. Moreover, Proposition 3 also shows that the R_S-ideal generated by the $\chi_R(a)$ is A_S while the R_S-module generated by the $\chi_E(e)$ is E_S.

The proposition follows.

Theorem 3. *If the R-module E satisfies the maximal (minimal) condition for submodules, then so also does the R_S-module E_S.*

Proof. We shall only prove the assertion concerning the maximal condition, since the statement about the minimal condition can be dealt with similarly.

Let $W_1 \subseteq W_2 \subseteq W_3 \subseteq \dots$ be an ascending sequence of submodules of E_S and denote by W_i^c the contraction of W_i in E. Then

$$W_1^c \subseteq W_2^c \subseteq W_3^c \subseteq \dots$$

is an ascending sequence of submodules of E and therefore there exists an integer m such that $W_n^c = W_m^c$ for all $n \geqslant m$. It now follows, from Proposition 2, that $W_n = W_m$ for all $n \geqslant m$. This establishes the theorem.†

We know, from Theorem 7 of section (1.8), that the length $L_R(E)$ of an R-module E is finite if and only if E satisfies both the maximal and the minimal conditions for submodules. It therefore follows, from Theorem 3, that $L_{R_S}(E_S)$ is finite whenever $L_R(E)$ is finite. However, we can easily strengthen this result.

Theorem 4. *If E is an R-module and S a non-empty multiplicatively closed subset of R, then $L_{R_S}(E_S) \leqslant L_R(E)$.*

Remark. *In this result it is not assumed that the lengths involved are finite.*

Proof. Let $W_0 \supset W_1 \supset W_2 \supset \dots \supset W_p$ be a strictly decreasing sequence of R_S-submodules of E_S and let W_i^c denote the contraction of W_i in E. By Proposition 2, these contractions are all distinct; consequently we have a strictly decreasing sequence

$$W_0^c \supset W_1^c \supset W_2^c \supset \dots \supset W_p^c$$

of submodules of E. This shows that $p \leqslant L_R(E)$. The theorem now follows by taking upper bounds when the sequence of W_i's is varied.

3.3 The full ring of fractions

The ring R_S has been constructed for an *arbitrary* non-empty multiplicatively closed subset of R. We now consider a special case of this construction.

Let R be a non-null ring. Assume that α, β belong to R and that neither is a zero-divisor. *We contend that $\alpha\beta$ is not a zero-divisor.* For if $r \in R$ and $\alpha\beta r = 0$, then $\beta r = 0$ (because α is not a zero-divisor) and therefore $r = 0$. This shows that the set S consisting of all elements

† See Proposition 9 of section (1.8).

which are not zero-divisors is multiplicatively closed. Of course, S is not empty because it contains the identity element.

The ring R_S obtained by using this particular multiplicatively closed set is called the *full ring of fractions* of R. Suppose that $[r_1/s_1]$ and $[r_2/s_2]$ belong to the full ring of fractions so that neither s_1 nor s_2 is a zero-divisor. Then $[r_1/s_1] = [r_2/s_2]$ if and only if $ss_2r_1 = ss_1r_2$ for some element s which is not a zero-divisor. But then $s(s_2r_1 - s_1r_2) = 0$ and therefore $s_2r_1 - s_1r_2 = 0$. Accordingly $[r_1/s_1] = [r_2/s_2]$ *if and only if* $s_2r_1 = s_1r_2$. Thus we have a situation very similar to that encountered in the theory of rational numbers. As in that theory, we frequently omit the square brackets and write r/s in place of $[r/s]$. Thus $r_1/s_1 = r_2/s_2$ when and only when $s_2r_1 = s_1r_2$. There is one further point of similarity. This arises from

Lemma 5. *If Σ is the full ring of fractions of R, then the canonical mapping $\chi : R \to \Sigma$ is a monomorphism.*

Proof. Let S consist of all the elements of R which are not zero-divisors and assume that r belongs to $\operatorname{Ker} \chi$. Since $\operatorname{Ker} \chi$ is just the S-component of the zero ideal, it follows that $sr = 0$ for some $s \in S$. This in turn shows that $r = 0$ because s is not a zero-divisor. The lemma follows.

On the basis of this result, we can identify R with its image in Σ. Thus if r denotes an element of R, then the same symbol is also used to denote the fraction $r/1$. In this manner R becomes a subring of its full ring of fractions.

Theorem 5. *If R is an integral domain, then its full ring of fractions is a field.*

Proof. In this case the elements which are not zero-divisors are just the non-zero elements of R. Let $[a/b]$ be a non-zero element of the full ring of fractions, then $a \neq 0$ and $b \neq 0$. It suffices to show that $[a/b]$ has an inverse. But this is obvious because the required inverse is just $[b/a]$.

Definition. *If R is an integral domain, then its full ring of fractions is called the 'quotient field' of R.*

An integral domain is thus a subring of its quotient field. On the other hand, a field is an integral domain and therefore all its subrings are integral domains. These remarks yield the

Corollary. *A ring is an integral domain if and only if it is a subring of a field.*

3.4 Functors

In section (3.3) we considered certain rather special features of the process whereby fractions are formed. In the present one our attitude is quite different. The construction of modules of fractions is examined against a background broad enough to embrace other basic constructions. The technical way of describing our new point of view is to say that fraction formation is regarded as an example of a *functor*. This concept is quite remarkable for the wide variety of situations to which it is applicable. However it is sufficient for our purposes to restrict our attention to functors of modules.

Let R and Λ be rings. It is assumed that each possesses an identity element but it is not necessary to suppose that they are commutative. We shall be concerned with R-modules and Λ-modules. Because we are not restricting ourselves to the commutative case, we have two distinct types of R-module, namely left R-modules and right R-modules. Likewise there are two types of Λ-module. To avoid complicating the terminology, we stipulate that when we speak of a module over the ring R or the ring Λ, we shall always mean a left module. This, however, is only in order that our statements shall be quite explicit. All that really matters is that the term module be used in a *consistent* manner. Thus, to take an example, if we had agreed that 'R-module' should mean 'left R-module' but 'Λ-module' should mean 'right Λ-module', this would not make the slightest difference to the theory which follows.

After these preliminaries, let us suppose that we have a construction, denoted by the symbol T, which accomplishes two things. First, it associates with each R-module E a certain Λ-module $T(E)$. Secondly, whenever we have a homomorphism $f:E_1 \to E_2$ of R-modules, the construction provides a Λ-homomorphism $T(f):T(E_1) \to T(E_2)$ between the corresponding Λ-modules.

Definition. *T is said to be a 'covariant functor' from the category of R-modules to the category of Λ-modules provided it satisfies the following two conditions:*

(a) if i_E is the identity mapping of an R-module E, then $T(i_E)$ is the identity mapping of $T(E)$;

(b) if $f:E_1 \to E_2$ and $g:E_2 \to E_3$ are homomorphisms of R-modules, then $T(gf) = T(g)\,T(f)$.

To explain the significance of the adjective *covariant*, we note in passing that there is another kind of functor known as a *contravariant functor*. The characteristic feature of this is that it reverses the directions of homomorphisms. Thus if U is a contravariant functor from the category of R-modules to the category of Λ-modules and $f: E_1 \to E_2$ is a homomorphism of R-modules, than $U(f)$ maps $U(E_2)$ into $U(E_1)$. However, as we shall only be concerned with covariant functors, we take advantage of this in order to simplify our terminology. Accordingly from now on *the term functor will always mean a covariant functor*.

The present discussion is related to the contents of the earlier sections of this chapter in the following way. Suppose, for the moment, that R is a commutative ring and S a non-empty multiplicatively closed subset of R. If now E is an R-module, then E_S is an R_S-module, and if $f: E \to E'$ is a homomorphism of R-modules then, by Lemma 4, $f_S: E_S \to E'_S$ is a homomorphism of R_S-modules. Indeed Theorem 1 shows that we have a functor from the category of R-modules to the category of R_S-modules. This functor will be called the *fraction functor* associated with S.

Let us return to the general situation and consider an arbitrary functor T from the category of R-modules (R is no longer assumed to be commutative) to the category of Λ-modules.

Lemma 6. *Let* $f: E_1 \to E_2$ *be an isomorphism of R-modules and*

$$g: E_2 \to E_1$$

the inverse isomorphism. Then $T(f): T(E_1) \to T(E_2)$ *is an isomorphism of Λ-modules and* $T(g)$ *is its inverse.*

Proof. Consider gf and fg. These are the identity mappings of E_1 and E_2 respectively. It follows, from the definition of a functor, that

$$T(gf) = T(g)\,T(f)$$

is the identity mapping of $T(E_1)$ and $T(fg) = T(f)\,T(g)$ is the identity mapping of $T(E_2)$. This shows that $T(f)$ is an isomorphism of $T(E_1)$ on to $T(E_2)$ having $T(g)$ as its inverse. Thus the lemma is established.

Let X be a submodule of an R-module E. It is said to be a *direct summand* of E if there exists another R-submodule Y such that $E = X \oplus Y$. The module Y is then referred to as a *supplement of X in E*.

Lemma 7. *Let X be a direct summand of the R-module E. Then the inclusion mapping $X \to E$ gives rise to a monomorphism $T(X) \to T(E)$ of Λ-modules.*

Proof. Let Y be a supplement of X in E so that $E = X \oplus Y$. Denote by $\mu: X \to E$ the inclusion mapping and by $\pi: E \to X$ the projection mapping of E on to the summand X. Then $\pi\mu$ is the identity mapping of X and therefore $T(\pi\mu) = T(\pi) T(\mu)$ is the identity mapping of $T(X)$. This shows, in particular, that $T(\mu)$ is a monomorphism and thereby completes the proof.

We now introduce the notion of an *additive functor*. To this end let E_1 and E_2 be R-modules and suppose that f, g are homomorphisms of E_1 into E_2. Then $f + g$ is also a homomorphism of E_1 into E_2 and $T(f)$, $T(g)$, $T(f+g)$ are all of them Λ-homomorphisms of $T(E_1)$ into $T(E_2)$. The functor T is said to be *additive* if, in the circumstances just described, we always have

$$T(f+g) = T(f) + T(g). \tag{3.4.1}$$

Note that, by (3) of Theorem 1, fraction functors are always additive.

Lemma 8. *Let T be an additive functor. If $f: E_1 \to E_2$ is a null homomorphism of R-modules, then $T(f): T(E_1) \to T(E_2)$ is a null homomorphism of Λ-modules; and if N is a null R-module, then $T(N)$ is a null Λ-module.*

Proof. Since f is a null homomorphism, we have $f + f = f$. It follows that $T(f) + T(f) = T(f)$. However, the Λ-homomorphisms of $T(E_1)$ into $T(E_2)$ form a group with the null homomorphism as neutral element.† It follows that $T(f)$ is the null homomorphism of $T(E_1)$ into $T(E_2)$.

To prove the second assertion, let i be the identity mapping of N. Then $T(i)$ is the identity mapping of $T(N)$. However, since N is a null module, i is also a null homomorphism. It now follows, from the first part of the proof, that the identity mapping of $T(N)$ is a null homomorphism. This shows that $T(N)$ is a null module and completes the proof.

Now suppose that T is an additive functor. Let E be an R-module and assume that $E = X_1 \oplus X_2 \oplus \ldots \oplus X_s$, where the X_i are submodules of E. By the corollary to Proposition 14 in section (1.9),

$$E = X_i \oplus (X_1 + \ldots + X_{i-1} + X_{i+1} + \ldots + X_s).$$

This shows that each X_i is a direct summand of E. It follows, by Lemma 7, that the inclusion mapping $\mu_i : X_i \to E$ induces a monomorphism $\mu_i^* : X_i^* \to E^*$, where we have written $E^* = T(E)$, $X_i^* = T(X_i)$ and $\mu_i^* = T(\mu_i)$. Let $\pi_j : E \to X_j$ be the projection of E

† See section (1.10).

on to the jth summand and put $\pi_j^* = T(\pi_j)$. Then, since $\pi_i \mu_i$ is the identity mapping of X_i, $\pi_i^* \mu_i^*$ is the identity mapping of X_i^*. Again, if $i \neq j$ then $\pi_j \mu_i$ is the null mapping of X_i into X_j and hence $\pi_j^* \mu_i^*$ is the null mapping of X_i^* into X_j^*. Finally, because

$$\mu_1 \pi_1 + \mu_2 \pi_2 + \ldots + \mu_s \pi_s$$

is the identity mapping of E, it follows that $\mu_1^* \pi_1^* + \mu_2^* \pi_2^* + \ldots + \mu_s^* \pi_s^*$ is the identity mapping of E^*.

Let ξ belong to E^*. Then $\xi = \sum_i \mu_i^* \pi_i^*(\xi)$. This shows that

$$E^* = \sum_i \mu_i^*(X_i^*).$$

Now assume that, for $1 \leqslant i \leqslant s$, $\xi_i \in X_i^*$ and suppose that we have $\sum_i \mu_i^*(\xi_i) = 0$. Operating with π_j^*, we find that $\xi_j = 0$ this being true for every j. Accordingly $\mu_i^*(\xi_i) = 0$ for each i and we have shown that

$$E^* = \mu_1^*(X_1^*) \oplus \mu_2^*(X_2^*) \oplus \ldots \oplus \mu_s^*(X_s^*).$$

However, as already observed, $\mu_i^* : X_i^* \to E^*$ is a monomorphism and therefore it may be used to identify X_i^* with $\mu_i^*(X_i^*)$. On this understanding $E^* = X_1^* \oplus X_2^* \oplus \ldots \oplus X_s^*$. Our conclusion is restated in the theorem which follows.

Theorem 6. *Let $E = X_1 \oplus X_2 \oplus \ldots \oplus X_s$, where the X_i are R-submodules of E, and let T be an additive functor. Then the inclusion mapping $X_i \to E$ induces a monomorphism $T(X_i) \to T(E)$ which allows us to identify $T(X_i)$ with a Λ-submodule of $T(E)$. On this understanding, $T(E) = T(X_1) \oplus T(X_2) \oplus \ldots \oplus T(X_s)$.*

We now introduce a new type of functor. A functor T is said to be *exact* if whenever

$$0 \to E_1 \overset{f}{\to} E \overset{g}{\to} E_2 \to 0$$

is an exact sequence of R-modules, the sequence

$$0 \to T(E_1) \overset{T(f)}{\longrightarrow} T(E) \overset{T(g)}{\longrightarrow} T(E_2) \to 0$$

is also exact. Thus, by Theorem 2, all fraction functors are exact. It can happen, however, that a functor has the property in question if one restricts one's attention to a suitable class of R-modules, but fails to possess it if no such restriction is imposed. It is in order to be able to deal with situations of this kind that the following remarks are made.

Let Ω be a non-empty class of R-modules and suppose that the following conditions are satisfied: (i) *whenever a module belongs to Ω then all its submodules and all its images (under R-homomorphisms) belong to Ω as well*; (ii) *the (external) direct sum of a finite number of modules belonging to Ω is also a member of Ω*. To give some examples, Ω might consist of all R-modules. Next, as the reader can easily verify, the class of all R-modules which satisfy the maximal condition for submodules also meets the requirements of (i) and (ii). Again, the aggregate of all R-modules of finite length provides another case in point. However, we do not need to particularize at this stage. We merely suppose, for the remainder of this section, that Ω has the properties described in italics. Note that, because of (i), all null modules belong to Ω. Also, if a module belongs to Ω, then so does any module which is isomorphic to it.

Definition. *The functor T is said to be 'exact on Ω' if whenever*

$$0 \to E_1 \xrightarrow{f} E \xrightarrow{g} E_2 \to 0$$

is an exact sequence of R-modules and E, E_1, E_2 belong to Ω, the sequence

$$0 \to T(E_1) \xrightarrow{T(f)} T(E) \xrightarrow{T(g)} T(E_2) \to 0$$

is also exact.

Observe that when we speak simply of an exact functor without explicit reference to a class Ω, it is always understood that the class in question consists of all R-modules.

Lemma 9. *Let the functor T be exact on Ω. Suppose that $f: E_1 \to E$ is a monomorphism of R-modules and $g: E \to E_2$ an epimorphism of R-modules. If now E, E_1, E_2 belong to Ω, then $T(f)$ is a monomorphism and $T(g)$ an epimorphism.*

Proof. The assumptions concerning Ω ensure that $E/\mathrm{Im}f$ and $\mathrm{Ker}\,g$ both belong to Ω. The lemma now follows by applying T to the exact sequences

$$0 \to E_1 \xrightarrow{f} E \to E/\mathrm{Im}f \to 0$$

and

$$0 \to \mathrm{Ker}\,g \to E \xrightarrow{g} E_2 \to 0,$$

where $E \to E/\mathrm{Im}f$ is the natural mapping and $\mathrm{Ker}\,g \to E$ the inclusion mapping.

Let T be exact on Ω and let E be an R-module belonging to Ω. If now X, Y are submodules of E, then they belong to Ω. Furthermore, by

Lemma 9, the inclusion mapping $X \to E$ gives rise to a monomorphism $T(X) \to T(E)$. *This enables us to regard $T(X)$ as a Λ-submodule of $T(E)$.* Similarly $T(Y)$ can be regarded as a submodule of $T(E)$. Next suppose that $X \subseteq Y$. Since the inclusion mapping $X \to E$ may be obtained by combining the inclusion mappings $X \to Y$ and $Y \to E$, it follows that $T(X) \subseteq T(Y)$ when both are regarded as submodules of $T(E)$.

Lemma 10. *Let the functor T be exact on Ω and let $f: E \to E'$ be a homomorphism of R-modules, where E, E' belong to Ω. If now $T(\mathrm{Ker}f)$ and $T(\mathrm{Im}f)$ are regarded as Λ-submodules of $T(E)$ and $T(E')$ respectively, then $T(\mathrm{Ker}f) = \mathrm{Ker}\,(Tf)$ and $T(\mathrm{Im}f) = \mathrm{Im}\,(Tf)$.*

Remark. To simplify the notation we have written Tf in place of $T(f)$.

Proof. We can express f as a combination of mappings $E \to \mathrm{Im}f \to E'$. Here $E \to \mathrm{Im}f$ is an epimorphism, $\mathrm{Im}f \to E'$ is a monomorphism, and all the modules belong to Ω. $T(f)$ is then a composite mapping

$$T(E) \to T\,(\mathrm{Im}f) \to T(E'),$$

where, by Lemma 9, the first mapping is an epimorphism and the second a monomorphism. It follows that $\mathrm{Im}\,(Tf)$ is the image of $T(\mathrm{Im}f)$ in $T(E')$, that is to say it is $T(\mathrm{Im}f)$ regarded as a submodule of $T(E')$. Thus the second assertion is proved. Next $\mathrm{Ker}\,(Tf)$ is the kernel of $T(E) \to T(\mathrm{Im}f)$. But $0 \to \mathrm{Ker}f \to E \to \mathrm{Im}f \to 0$ is an exact sequence of modules belonging to Ω. Consequently $T(\mathrm{Ker}f) \to T(E) \to T(\mathrm{Im}f)$ is exact. Thus $\mathrm{Ker}(Tf)$ is the image of $T(\mathrm{Ker}f)$ in $T(E)$. In other terms, $\mathrm{Ker}\,(Tf)$ is $T(\mathrm{Ker}f)$ when the latter is regarded as a submodule of $T(E)$. The lemma is therefore proved.

Theorem 7. *Let the functor T be exact on Ω and suppose that $E_1 \overset{f}{\to} E \overset{g}{\to} E_2$ are homomorphisms of R-modules where E, E_1, E_2 all belong to Ω. If now $\mathrm{Im}f \subseteq \mathrm{Ker}\,g$, then $\mathrm{Im}(Tf) \subseteq \mathrm{Ker}\,(Tg)$ and we have a Λ-isomorphism $\mathrm{Ker}\,(Tg)/\mathrm{Im}(Tf) \approx T(\mathrm{Ker}\,g/\mathrm{Im}f)$.*

Proof. The R-modules $\mathrm{Im}f$, $\mathrm{Ker}\,g$ and $\mathrm{Ker}\,g/\mathrm{Im}f$ all belong to Ω. Accordingly the exact sequence

$$0 \to \mathrm{Im}f \to \mathrm{Ker}\,g \to \mathrm{Ker}\,g/\mathrm{Im}f \to 0$$

gives rise to another exact sequence namely

$$0 \to T(\mathrm{Im}f) \overset{\alpha}{\to} T(\mathrm{Ker}\,g) \to T(\mathrm{Ker}\,g/\mathrm{Im}f) \to 0. \qquad (3.4.2)$$

Now consider the commutative diagram

$$
\begin{array}{ccc}
T(\mathrm{Im}f) & \xrightarrow{\ \alpha\ } & T(\mathrm{Ker}g) \\
\beta \downarrow & & \downarrow \gamma \\
T(E) & \longrightarrow & T(E)
\end{array}
$$

where β and γ arise from the inclusion mappings $\mathrm{Im}f \to E$ and $\mathrm{Ker}\,g \to E$ respectively, and $T(E) \to T(E)$ denotes the identity mapping. By Lemmas 9 and 10, β maps $T(\mathrm{Im}f)$ isomorphically on to $\mathrm{Im}(Tf)$ and γ maps $T(\mathrm{Ker}\,g)$ isomorphically on to $\mathrm{Ker}\,(Tg)$. It follows that $\mathrm{Im}\,(Tf) \subseteq \mathrm{Ker}\,(Tg)$ and that the factor module $\mathrm{Ker}\,(Tg)/\mathrm{Im}\,(Tf)$ is isomorphic to $T(\mathrm{Ker}\,g)/\mathrm{Im}\,\alpha$. However, by (3.4.2), the latter module is isomorphic to $T(\mathrm{Ker}\,g/\mathrm{Im}f)$ so all is proved.

Corollary. *Let the functor T be exact on Ω and let $E_1 \to E \to E_2$ be an exact sequence of R-modules. If now E, E_1, E_2, all belong to Ω, then the sequence $T(E_1) \to T(E) \to T(E_2)$ is also exact.*

Proof. With the notation of the theorem we have $\mathrm{Im}f = \mathrm{Ker}\,g$ and this module is in Ω. It therefore follows (Lemma 10) that

$$\mathrm{Im}\,(Tf) = \mathrm{Ker}\,(Tg)$$

which is precisely what we have to prove.

Theorem 8. *Let T be an additive functor and suppose that T is exact on Ω. Further let E be an R-module belonging to Ω and let $X_1, X_2, \ldots, X_s$ be R-submodules of E. Then*

$$T(X_1 + X_2 + \ldots + X_s) = T(X_1) + T(X_2) + \ldots + T(X_s) \quad (3.4.3)$$

and $\quad T(X_1 \cap X_2 \cap \ldots \cap X_s) = T(X_1) \cap T(X_2) \cap \ldots \cap T(X_s). \quad (3.4.4)$

Remark. In (3.4.3) and (3.4.4) it is to be understood that all the Λ-modules which occur are to be regarded as submodules of $T(E)$.

Proof. It is clearly sufficient to prove the theorem for the case $s = 2$. Put $V = X_1 \oplus X_2$ and define R-homomorphisms $g_i : V \to E$ $(i = 1, 2)$ by $g_1(x_1, x_2) = x_1$ and $g_2(x_1, x_2) = x_2$. Then $\mathrm{Im}\,g_1 = X_1$, $\mathrm{Im}\,g_2 = X_2$ and, because

$$(g_1 + g_2)\,(x_1, x_2) = x_1 + x_2,$$

$\mathrm{Im}\,(g_1 + g_2) = X_1 + X_2$. Here, of course, $X_1 + X_2$ denotes the sum of X_1 and X_2 when these are regarded as submodules of E. Now X_1 and X_2

belong to Ω and therefore so does V. Accordingly, by Lemma 10, we have

$$T(X_1 + X_2) = T(\mathrm{Im}\,(g_1 + g_2))$$
$$= \mathrm{Im}\,(T(g_1 + g_2))$$
$$= \mathrm{Im}\,(Tg_1 + Tg_2)$$
$$\subseteq \mathrm{Im}\,(Tg_1) + \mathrm{Im}\,(Tg_2)$$
$$= T(\mathrm{Im}\,g_1) + T(\mathrm{Im}\,g_2)$$
$$= T(X_1) + T(X_2).$$

However, since X_1 and X_2 are both contained in $X_1 + X_2$, the opposite inclusion is obvious. This proves the first assertion.

Next put $U = (E/X_1) \oplus (E/X_2)$ and let $\phi_i \colon E \to E/X_i\,(i = 1, 2)$ denote the natural mapping of E on to E/X_i. We now define R-homomorphisms $f_1 \colon E \to U$ and $f_2 \colon E \to U$ by $f_1(e) = \{\phi_1(e), 0\}$ and $f_2(e) = \{0, \phi_2(e)\}$. Then $\mathrm{Ker}\,f_1 = X_1$, $\mathrm{Ker}\,f_2 = X_2$ and, because

$$(f_1 + f_2)\,(e) = \{\phi_1(e), \phi_2(e)\}, \text{ we have } \mathrm{Ker}\,(f_1 + f_2) = X_1 \cap X_2.$$

Moreover E/X_1 and E/X_2 belong to Ω and so U belongs to Ω as well. We can therefore apply Lemma 10. This gives

$$T(X_1 \cap X_2) = T(\mathrm{Ker}\,(f_1 + f_2))$$
$$= \mathrm{Ker}\,(T(f_1 + f_2))$$
$$= \mathrm{Ker}\,(Tf_1 + Tf_2)$$
$$\supseteq \mathrm{Ker}\,(Tf_1) \cap \mathrm{Ker}\,(Tf_2)$$
$$= T(\mathrm{Ker}\,f_1) \cap T(\mathrm{Ker}\,f_2)$$
$$= T(X_1) \cap T(X_2).$$

But $T(X_1 \cap X_2)$ is contained in both $T(X_1)$ and $T(X_2)$. It follows that $T(X_1 \cap X_2) = T(X_1) \cap T(X_2)$ and this completes the proof of the theorem.

We conclude this section by explaining certain terminology which is useful when one is comparing functors. Suppose that T and U are both functors from the category of R-modules to the category of Λ-modules. It may be that for each R-module E it is possible to construct, in a natural way, a Λ-homomorphism $\mu(E) \colon T(E) \to U(E)$. Assume that this is the case and let these Λ-homomorphisms be such that, whenever $f:E \to E'$ is a homomorphism of R-modules, the diagram

$$
\begin{array}{ccc}
T(E) & \xrightarrow{\mu(E)} & U(E) \\
{\scriptstyle T(f)} \downarrow & & \downarrow {\scriptstyle U(f)} \\
T(E') & \xrightarrow[\mu(E')]{} & U(E')
\end{array}
$$

is commutative. This situation is described by saying that μ is a *natural transformation* of T into U. Should it happen that, for every choice of E, $\mu(E)$ is an isomorphism of $T(E)$ on to $U(E)$, then it is usual to say that μ is an *isomorphism* of T on to U. Naturally, when such an isomorphism exists the functors are said to be *isomorphic*.

If μ is an isomorphism of T on to U, then the inverse isomorphisms $\mu(E)^{-1}:U(E) \to T(E)$ constitute an isomorphism of U on to T. This isomorphism in the reverse direction is denoted by μ^{-1}. Finally suppose that T, U, V are functors such that T is isomorphic to U and U is isomorphic to V. By combining the isomorphisms we see at once that T is isomorphic to V. Thus, for functors, isomorphism is an equivalence relation.

3.5 Further properties of fraction functors

In this section we assume, once again, that the ring R is commutative and possesses an identity element. As before S denotes a multiplicatively closed subset of R and the ring of fractions of R with respect to S is designated by R_S.

Suppose now that E is an R-module. As we have seen, E_S has a natural structure as an R_S-module. Also, if X is an R-submodule of E, then X_S can be regarded as an R_S-submodule of E_S and, when so regarded, is known as the extension of X in E_S.

Let $X^1, X^2, \ldots, X^n$ be submodules of E. Since fraction functors are both additive and exact, it follows from Theorem 8 that

$$(X^1 + X^2 + \ldots + X^n)_S = X_S^1 + X_S^2 + \ldots + X_S^n \tag{3.5.1}$$

and
$$(X^1 \cap X^2 \cap \ldots \cap X^n)_S = X_S^1 \cap X_S^2 \cap \ldots \cap X_S^n. \tag{3.5.2}$$

Here it is to be understood that all the modules of fractions which occur in (3.5.1) and (3.5.2) are considered as submodules of E_S.

Put $X = X^1 + X^2 + \ldots + X^n$. If it happens that this sum is direct so that $X = X^1 \oplus X^2 \oplus \ldots \oplus X^n$, then, by Theorem 6,

$$X_S = X_S^1 \oplus X_S^2 \oplus \ldots \oplus X_S^n. \tag{3.5.3}$$

Naturally it is to be expected that fraction functors will have certain special properties not common to all additive, exact functors. For example, it will now be shown that (3.5.1) and (3.5.3) continue to hold even when the sums involved are not finite.

Proposition 5. *Let $\{X^i\}_{i \in I}$ be a family of submodules of an R-module E. Then*

$$(\sum_i X^i)_S = \sum_i X^i_S, \tag{3.5.4}$$

where both sides are regarded as submodules of E_S.

Proof. Let $\chi: E \to E_S$ be the canonical mapping. Since $\sum_i X^i$ can be generated by the union $\bigcup_i X^i$ of the sets X^i, it follows (see Proposition 3) that $(\sum_i X^i)_S$ is generated by

$$\chi(\bigcup_i X^i) = \bigcup_i \chi(X^i).$$

Next, again by Proposition 3, the R_S-submodule of E_S generated by $\chi(X^i)$ is none other than X^i_S. Consequently the submodule generated by $\bigcup_i \chi(X^i)$ is $\sum_i X^i_S$. Accordingly $(\sum_i X^i)_S = \sum_i X^i_S$ and the proof is complete.

Proposition 6. *Let $\{E^i\}_{i \in I}$ be a family of submodules of E and suppose that E is their direct sum. Then E_S is the internal direct sum of its submodules $\{E^i_S\}_{i \in I}$.*

Proof. By Proposition 5, $E_S = \sum_i E^i_S$. Now suppose that $\sum_{i \in I} \xi^i = 0$, where $\xi^i \in E^i_S$ and the sum contains only a finite number of non-zero terms. The proposition will follow if we can show that all the ξ^i are zero.

There exists a *finite* subset J of I such that $\xi^i = 0$ if $i \notin J$. Put $X = \sum_{j \in J} E^j$. By the corollary to Proposition 13 of section (1.9), X is the direct sum of the $E^j (j \in J)$. It follows, by (3.5.3), that

$$X_S = \bigoplus_{j \in J} E^j_S.$$

Since $\sum_{j \in J} \xi^j = 0$, we may now conclude that $\xi^j = 0$ for every $j \in J$.

The proposition follows.

Corollary. *Let E be a free R-module with base $\{e_i\}_{i \in I}$ and let $\chi: E \to E_S$ be the canonical mapping. Then E_S is a free R_S-module having $\{\chi(e_i)\}_{i \in I}$ as a base.*

Proof. By Proposition 21 of section (1.13), E is the direct sum of the modules Re_i. It therefore follows (from the proposition just proved) that E_S is the direct sum of its submodules $(Re_i)_S$. However, by Proposition 3, $(Re_i)_S$ is the same as $R_S\chi(e_i)$; consequently

$$E_S = \bigoplus_{i \in I} R_S\chi(e_i).$$

To complete the proof, it will suffice to show that if $[r/s]\chi(e_i) = 0$, where $r \in R$ and $s \in S$, then $[r/s]$ is the zero element of R_S.

Now $[r/s]\chi(e_i) = [re_i/s]$. If this is zero, then there exists $\sigma \in S$ such that $\sigma re_i = 0$. But the e_i form a base for E; consequently from $\sigma re_i = 0$ follows $\sigma r = 0$. Finally $[r/s] = [\sigma r/\sigma s] = [0/\sigma s]$ and the corollary is established.

Let Y be a submodule of an R-module E and let $a \in R$. Denote by $\phi: E \to E/Y$ the homomorphism obtained by combining $\alpha: E \to E$, where $\alpha(e) = ae$, with the natural mapping $\beta: E \to E/Y$. Then $\phi = \beta\alpha$ and therefore
$$\operatorname{Ker}\phi = \alpha^{-1}(\operatorname{Ker}\beta) = \alpha^{-1}(Y).$$

But e belongs to $\alpha^{-1}(Y)$ if and only if $ae \in Y$, i.e. if and only if e belongs to $Y:_E a$. Accordingly $\operatorname{Ker}\phi = Y:_E a$.

Next $\phi_S = \beta_S\alpha_S$ whence (using Lemma 10)
$$\operatorname{Ker}(\phi_S) = \alpha_S^{-1}(\operatorname{Ker}(\beta_S)) = \alpha_S^{-1}((\operatorname{Ker}\beta)_S) = \alpha_S^{-1}(Y_S).$$

In addition, $\alpha = ai_E$ (where i_E denotes the identity mapping of E) and therefore, by Theorem 1, $\alpha_S = ai_{E_S}$. Thus if $u \in E_S$, then
$$\alpha_S(u) = au = \chi_R(a)u,$$

where $\chi_R: R \to R_S$ is the canonical mapping. It follows that $\alpha_S^{-1}(Y_S)$ is the same as $Y_S:_{E_S}\chi_R(a)$. Thus $\operatorname{Ker}(\phi_S) = Y_S:_{E_S}\chi_R(a)$ and, since Lemma 10 shows that $\operatorname{Ker}(\phi_S) = (\operatorname{Ker}\phi)_S$, we have proved that

$$(Y:_E a)_S = Y_S:_{E_S}\chi_R(a). \tag{3.5.5}$$

Propositon 7. *Let Y be a submodule of an R-module E and let A be a finitely generated R-ideal. Then*
$$(Y:_E A)_S = Y_S:_{E_S} A_S.$$

Proof. Let A be generated by $a_1, a_2, \ldots, a_n$. Then†
$$Y:A = (Y:a_1) \cap (Y:a_2) \cap \ldots \cap (Y:a_n)$$

† Cf. (2.7.8). In the course of the proof we omit the suffix E from symbols such as $Y:_E A$. Likewise the suffix E_S has also been omitted.

and therefore, by (3.5.2) and (3.5.5),

$$
\begin{aligned}
(Y\!:\!A)_S &= (Y\!:\!a_1)_S \cap (Y\!:\!a_2)_S \cap \dots \cap (Y\!:\!a_n)_S \\
&= (Y_S\!:\!\chi_R(a_1)) \cap \dots \cap (Y_S\!:\!\chi_R(a_n)) \\
&= Y_S\!:\!(\chi_R(a_1), \dots, \chi_R(a_n))R_S \\
&= Y_S\!:\!A_S
\end{aligned}
$$

because, by Proposition 3, the R_S-ideal generated by the elements $\chi_R(a_1), \chi_R(a_2), \dots, \chi_R(a_n)$ is none other than A_S.

Our next proposition is a companion to the one just obtained. As before Y is an R-submodule of E. Let e be a *fixed* element of E. We denote by $\psi : R \to E/Y$ the result of combining the homomorphism $\lambda : R \to E$ defined by $\lambda(r) = re$, with the natural mapping $\mu : E \to E/Y$. Then $\psi = \mu\lambda$ and therefore

$$
\operatorname{Ker}\psi = \lambda^{-1}(\operatorname{Ker}\mu) = \lambda^{-1}(Y) = Y\!:\!e.
$$

Next $\psi_S = \mu_S \lambda_S$ from which we conclude that

$$
\operatorname{Ker}(\psi_S) = \lambda_S^{-1}(\operatorname{Ker}(\mu_S)) = \lambda_S^{-1}((\operatorname{Ker}\mu)_S) = \lambda_S^{-1}(Y_S).
$$

But if $[r/s]$ belongs to R_S, then

$$
\lambda_S\left(\left[\frac{r}{s}\right]\right) = \left[\frac{re}{s}\right] = \left[\frac{r}{s}\right]\left[\frac{se}{s}\right] = \left[\frac{r}{s}\right]\chi(e),
$$

where $\chi : E \to E_S$ is the canonical mapping. It follows that

$$
\lambda_S^{-1}(Y_S) = Y_S\!:\!\chi(e)
$$

and therefore $\operatorname{Ker}(\psi_S) = Y_S\!:\!\chi(e)$. Since $\operatorname{Ker}(\psi_S) = (\operatorname{Ker}\psi)_S$ by Lemma 10, this shows that

$$
(Y\!:\!e)_S = Y_S\!:\!\chi(e). \tag{3.5.6}
$$

Proposition 8. *Let Y be a submodule of a finitely generated R-module E. Then $(Y\!:\!E)_S = Y_S\!:\!E_S$.*

Proof. Let $e_1, e_2, \dots e_n$ be elements which generate E. Then

$$
Y\!:\!E = (Y\!:\!e_1) \cap (Y\!:\!e_2) \cap \dots \cap (Y\!:\!e_n)
$$

and therefore, by (3.5.2) and (3.5.6),

$$
\begin{aligned}
(Y\!:\!E)_S &= (Y\!:\!e_1)_S \cap (Y\!:\!e_2)_S \cap \dots \cap (Y\!:\!e_n)_S \\
&= (Y_S\!:\!\chi(e_1)) \cap \dots \cap (Y_S\!:\!\chi(e_n)) \\
&= Y_S\!:\!(R_S\chi(e_1) + \dots + R_S\chi(e_n)).
\end{aligned}
$$

This completes the proof because, by Proposition 3, the R_S-module generated by $\chi(e_1), \chi(e_2), \dots, \chi(e_n)$ is just E_S.

The reader will appreciate that the results so far obtained in this section could also have been derived, without difficulty, directly from the definitions. Indeed he will probably wonder why this procedure was not followed since it would have made the discussion somewhat shorter. The explanation is that instead of following the most direct route, the opportunity has been taken to introduce a type of reasoning which is often useful when one is dealing with functors. This has the advantage of preparing the way for dealing with more complicated situations, where *ad hoc* arguments are less easy to find.

Now let A be an ideal of the ring R and let us use $\phi : R \to R/A$ to denote the natural ring-epimorphism. If S is a non-empty multiplicatively closed subset of R, then $\phi(S)$ is a non-empty multiplicatively closed subset of R/A. It will be convenient, and also suggestive, to use S/A as an alternative symbol for this subset. Since S/A is multiplicatively closed, we can form a new ring of fractions, namely $(R/A)_{S/A}$. A mapping of R_S into $(R/A)_{S/A}$ will now be constructed. This will enable us to bring out the close connection between the two rings.

To this end assume that $[r_1/s_1] = [r_2/s_2]$, where r_1, r_2 belong to R and s_1, s_2 to S. Then there exists $s \in S$ such that $ss_2 r_1 = ss_1 r_2$. It follows that $\phi(s) \phi(s_2) \phi(r_1) = \phi(s) \phi(s_1) \phi(r_2)$ which shows that

$$[\phi(r_1)/\phi(s_1)] \quad \text{and} \quad [\phi(r_2)/\phi(s_2)]$$

are equal as elements of $(R/A)_{S/A}$. Accordingly we have a well defined mapping of R_S into $(R/A)_{S/A}$ in which $[r/s]$ is mapped into $[\phi(r)/\phi(s)]$.

Proposition 9. *The mapping* $R_S \to (R/A)_{S/A}$ *(in which the image of* $[r/s]$ *is* $[\phi(r)/\phi(s)]$, ϕ *being the natural homomorphism* $R \to R/A$) *is a ring-epimorphism whose kernel is* A_S. *It therefore induces a ring-isomorphism*

$$R_S/A_S \longrightarrow (R/A)_{S/A} \tag{3.5.7}$$

of R_S/A_S *on to* $(R/A)_{S/A}$.

Proof. Every element of $(R/A)_{S/A}$ can be written in the form

$$[\phi(r)/\phi(s)],$$

where $r \in R$ and $s \in S$. Hence R_S is mapped *on to* $(R/A)_{S/A}$. It is clear that the image of the sum (product) of two elements is the sum (product) of their separate images. Again identity element is mapped into identity element. Accordingly we have a ring-epimorphism as was stated. It is obvious that every element of A_S belongs to the kernel.

Finally assume that $[r/s]$ maps into zero. Then there exists $\sigma \in S$ such that $\phi(\sigma)\,\phi(r) = 0$. It follows that $\phi(\sigma r) = 0$ and therefore $\sigma r \in A$. But $[r/s] = [\sigma r/\sigma s]$. Consequently $[r/s]$ belongs to A_S and all is proved.

Consider the diagram

$$(3.5.8)$$

where the mappings of R_S and R_S/A_S into $(R/A)_{S/A}$ are those described in the last proposition and the remaining mappings are the obvious ones. The lower triangle is clearly commutative. Now let r belong to R and s to S. If we apply the combined mapping $R \to R_S \to (R/A)_{S/A}$ to r, then the result is $[\phi(s)\,\phi(r)/\phi(s)]$ which is the same as if we had operated with $R \to R/A \to (R/A)_{S/A}$. Thus (3.5.8) is a commutative diagram.

Still assuming that A is an R-ideal, let E be an R-module and $\psi : E \to E/AE$ the natural mapping. Since A is contained in the annihilator of E/AE, this module may also be regarded† as a module with respect to the ring R/A. Moreover, if $r \in R$ and $e \in E$, then

$$\psi(re) = r\psi(e) = \phi(r)\,\psi(e).$$

As E/AE is an R/A-module, we can form $(E/AE)_{S/A}$ which, of course, is an $(R/A)_{S/A}$-module. The connection between $E_S/A_S E_S$ and $(E/AE)_{S/A}$ will now be investigated in detail.

It is easily verified that there exists a well defined mapping

$$E_S \to (E/AE)_{S/A} \qquad (3.5.9)$$

in which

$$\left[\frac{e}{s}\right] \to \left[\frac{\psi(e)}{\phi(s)}\right]. \qquad (3.5.10)$$

Obviously E_S is mapped on to $(E/AE)_{S/A}$ and it is a simple matter to check that the image of the sum of two elements is the sum of their separate images. Since $\psi(e) = 0$ whenever e belongs to AE, it follows that $(AE)_S$ is contained in the kernel of the mapping. Finally suppose that $[e/s]$ belongs to the kernel. Then $[\psi(e)/\phi(s)] = 0$ and therefore there exists $\sigma \in S$ such that $\phi(\sigma)\,\psi(e) = 0$. But $\phi(\sigma)\,\psi(e) = \psi(\sigma e)$. Consequently $\sigma e \in AE$. Accordingly $[e/s]$, which is the same as $[\sigma e/\sigma s]$, belongs to $(AE)_S$.

† See Proposition 24 of section (1.14).

The above remarks show that the mapping (3.5.9) is an epimorphism whose kernel is $(AE)_S$. However, by Proposition 4, $(AE)_S = A_S E_S$. It follows that *the mapping* (3.5.9) *induces an isomorphism*

$$E_S/A_S E_S \to (E/AE)_{S/A} \qquad (3.5.11)$$

of the additive group of $E_S/A_S E_S$ *on to the additive group of* $(E/AE)_{S/A}$.

Naturally we would prefer to regard (3.5.11) as an isomorphism of *modules*. The immediate difficulty is that $E_S/A_S E_S$ is an R_S/A_S-module whereas $(E/AE)_{S/A}$ is an $(R/A)_{S/A}$-module. We must therefore make use of the ring-isomorphism described in Proposition 9.

Consider the Cartesian product $(R_S/A_S) \times (E_S/A_S E_S)$. This consists of all ordered pairs in which the first component belongs to the ring R_S/A_S while the second belongs to the module $E_S/A_S E_S$. Given such a pair we can obtain an element of $E_S/A_S E_S$ by multiplying the element of the module by the element of the ring. This gives a mapping

$$(R_S/A_S) \times (E_S/A_S E_S) \to E_S/A_S E_S \qquad (3.5.12)$$

which we call a *multiplication mapping*. Likewise we have a multiplication mapping $(R/A)_{S/A} \times (E/AE)_{S/A} \to (E/AE)_{S/A}$. $\qquad (3.5.13)$

Moreover, by combining (3.5.7) and (3.5.11), we obtain (in an obvious manner) a one-one-mapping

$$(R_S/A_S) \times (E_S/A_S E_S) \to (R/A)_{S/A} \times (E/AE)_{S/A}$$

of the first Cartesian product on to the second.

These various mappings can be put together in the single diagram

$$
\begin{array}{ccc}
(R_S/A_S) \times (E_S/A_S E_S) & \longrightarrow & (R/A)_{S/A} \times (E/AE)_{S/A} \\
\downarrow & & \downarrow \\
E_S/A_S E_S & \longrightarrow & (E/AE)_{S/A}
\end{array}
\qquad (3.5.14)
$$

This, we contend, is commutative. To see this let $[r/s]$ resp. $[e/\sigma]$ be an element of R_S resp. E_S and denote by $\overline{[r/s]}$ resp. $\overline{[e/\sigma]}$ its natural image in R_S/A_S resp. $E_S/A_S E_S$. The left vertical mapping changes $(\overline{[r/s]}, \overline{[e/\sigma]})$ into $\overline{[re/s\sigma]}$ which, on applying the lower horizontal mapping, becomes

$$\left[\frac{\psi(re)}{\phi(s\sigma)}\right] = \left[\frac{\phi(r)\,\psi(e)}{\phi(s)\,\phi(\sigma)}\right].$$

On the other hand, the upper horizontal mapping carries $(\overline{[r/s]}, \overline{[e/\sigma]})$ into $([\phi(r)/\phi(s)], [\psi(e)/\phi(\sigma)])$ which becomes $[\phi(r)\,\psi(e)/\phi(s)\,\phi(\sigma)]$

when operated upon with the right vertical mapping. Thus (3.5.14) really is commutative. We sum up our conclusions and put them into slightly informal language in

Proposition 10. *Let A be an ideal of the ring R and let E be an R-module. Further let the rings R_S/A_S and $(R/A)_{S/A}$ be identified by means of the mapping (3.5.7). Then the mapping (3.5.11) is an isomorphism of $E_S/A_S E_S$ on to $(E/AE)_{S/A}$ when both are regarded as modules with respect to the ring $R_S/A_S = (R/A)_{S/A}$.*

There is a further result of the same kind which is sometimes useful. This arises in the following way. As before, A denotes an ideal of R and $\phi : R \to R/A$ the natural mapping. However on this occasion it is necessary to assume that A is *finitely generated*. We continue to use S to denote an arbitrary, non-empty multiplicatively closed subset of R. E denotes a general R-module.

By Proposition 7, $(0 :_E A)_S = 0 :_{E_S} A_S$. This is an R_S/A_S-module because its annihilator contains A_S. On the other hand, $0 :_E A$ is an R/A module and therefore $(0 :_E A)_{S/A}$ is an $(R/A)_{S/A}$-module. We propose to compare the structures of $0 :_{E_S} A_S$ and $(0 :_E A)_{S/A}$.

Since $0 :_{E_S} A_S = (0 :_E A)_S$, a typical element of the former can be represented in the form $[x/s]$, where x belongs to $0 :_E A$ and s belongs to S. This enables us to construct a well defined mapping

$$0 :_{E_S} A_S \to (0 :_E A)_{S/A} \tag{3.5.15}$$

in which
$$\left[\frac{x}{s} \right] \to \left[\frac{x}{\phi(s)} \right]. \tag{3.5.16}$$

It is clear that every element of $(0 :_E A)_{S/A}$ occurs as an image, and that the image of the sum of two elements is the sum of their individual images. Assume that $[x/s]$, where x belongs to $0 :_E A$, is mapped into zero. Then $\phi(\sigma) x = 0$ for some element $\sigma \in S$. But $\phi(\sigma) x = \sigma x$; consequently $\sigma x = 0$ and therefore $[x/s] = [\sigma x / \sigma s] = 0$. This shows that (3.5.15) *is an isomorphism of the additive group of* $0 :_{E_S} A_S$ *on to the additive group of* $(0 :_E A)_{S/A}$.

After these preliminaries we can make use of the mappings (3.5.7) and (3.5.15) to construct a diagram

$$
\begin{array}{ccc}
(R_S/A_S) \times (0 :_{E_S} A_S) & \longrightarrow & (R/A)_{S/A} \times (0 :_E A)_{S/A} \\
\downarrow & & \downarrow \\
0 :_{E_S} A_S & \longrightarrow & (0 :_E A)_{S/A}
\end{array}
\tag{3.5.17}
$$

Here the horizontal mappings are one-one correspondences and the vertical mappings are *multiplications* of the kind encountered in the diagram (3.5.14). It is now claimed that the new diagram is commutative. In order to verify this, assume that r belongs to R, s and σ belong to S, and let x be an element of $0:_E A$. In addition, let us use $\overline{[r/s]}$ to denote the natural image of $[r/s]$ in R_S/A_S. Then the pair $(\overline{[r/s]}, [x/\sigma])$ is a typical element of $(R_S/A_S) \times (0:_{E_S} A_S)$. Its image under the left vertical mapping is $[rx/s\sigma]$ which becomes

$$\left[\frac{rx}{\phi(s)\,\phi(\sigma)}\right] = \left[\frac{\phi(r)\,x}{\phi(s)\,\phi(\sigma)}\right]$$

when it is subjected to the lower horizontal mapping. On the other hand, the upper horizontal mapping turns $(\overline{[r/s]}, [x/\sigma])$ into

$$([\phi(r)/\phi(s)], [x/\phi(\sigma)])$$

and, on applying the right vertical mapping, this also becomes $[\phi(r)\,x/\phi(s)\,\phi(\sigma)]$. This establishes that the diagram (3.5.17) is commutative. We record these results in

Proposition 11. *Let A be a finitely generated R-ideal and E an arbitrary R-module. Further let the rings R_S/A_S and $(R/A)_{S/A}$ be identified by means of the isomorphism (3.5.7). Then the mapping (3.5.15) is an isomorphism of $0:_{E_S} A_S$ on to $(0:_E A)_{S/A}$ when these are regarded as modules with respect to the ring $R_S/A_S = (R/A)_{S/A}$.*

3.6 Fractions and prime ideals

As usual R denotes a commutative ring with an identity element and S a non-empty multiplicatively closed subset of R.

Proposition 12. *Let A be an ideal of R. Then $A_S = R_S$ if and only if S meets A.*

Proof. Suppose first that A contains an element s of S. Then A_S contains $[s/s]$ which is the identity element of R_S. Accordingly $A_S = R_S$.

Now assume that $A_S = R_S$. The contraction of A_S in R is then R itself and therefore it contains the identity element of R. But, by Proposition 1, this contraction is the S-component of A in R; consequently there exists $\sigma \in S$ such that $\sigma 1_R \in A$. Thus A meets S and the proposition is proved.

Proposition 12 has the following generalization:

Propositon 13. *Let K be a submodule of a finitely generated R-module E. Then $K_S = E_S$ if and only if S meets the ideal $K:E$.*

Proof. We have $K_S = E_S$ if and only if $K_S : E_S = R_S$. By Proposition 8, $K_S : E_S = (K:E)_S$ and, by Proposition 12, this equals R_S if and only if S meets $K:E$. This establishes the proposition.

We shall now study the connection between the prime ideals of R and those of R_S.

Proposition 14. *Let P be a prime ideal of R. If P meets S, then $P_S = R_S$. If, however, P does not meet S, then P_S is a prime ideal of R_S whose contraction, in R, is P itself.*

Proof. Proposition 12 shows that P_S is a proper ideal of R_S when and only when P does not meet S. From now on we assume that the intersection of P and S is empty.

Suppose (with a self-explanatory notation) that $[r/s]$ belongs to P_S; say $[r/s] = [p/\sigma]$, where $p \in P$ and $\sigma \in S$. Then there exists $\sigma' \in S$ such that $\sigma'\sigma r = \sigma'sp \in P$. But neither σ' nor σ belongs to P. Consequently, by Lemma 2 of section (2.2), r belongs to P.

Next assume that the product $[r_1/s_1][r_2/s_2]$, that is $[r_1r_2/s_1s_2]$, belongs to P_S. By what has just been said, r_1r_2 belongs to P and therefore either $r_1 \in P$ or $r_2 \in P$. Accordingly either $[r_1/s_1]$ or $[r_2/s_2]$ belongs to P_S. This proves that P_S is a prime ideal.

Finally let r belong to the contraction of P_S in R. By Proposition 1, there exists $\sigma \in S$ such that $\sigma r \in P$. Since however σ does not belong to P, this shows that $r \in P$. Thus the contraction of P_S is contained in P. This establishes the proposition because the opposite inclusion is obvious.

Theorem 9. *There is a one-correspondence between the prime R-ideals P which do not meet S and the prime ideals Π of R_S. This is such that, when P and Π correspond, $\Pi = P_S$ and P is the contraction of Π in R.*

Proof. In view of Proposition 14, it is enough to show that every prime ideal of R_S is the extension of a prime ideal of R.

Let Π be a prime ideal of R_S and denote its contraction in R by P. By Proposition 2, $\Pi = P_S$ and so it is enough to show that P is prime. Note that the relation $\Pi = P_S$ shows that P is a *proper* ideal.

Assume that $r_1r_2 \in P$, where r_1, r_2 are elements of R. If now $\chi: R \to R_S$ is the canonical mapping, then $\chi(r_1r_2) = \chi(r_1)\chi(r_2)$ belongs to Π. However Π is prime; consequently either $\chi(r_1) \in \Pi$ or $\chi(r_2) \in \Pi$. It follows that either $r_1 \in P$ or $r_2 \in P$ because $P = \chi^{-1}(\Pi)$. This establishes the theorem.

The last result of this section shows how the radical of an ideal behaves when it is extended to the ring of fractions.

Proposition 15. *Let A be an ideal of R. Then* $(\mathrm{Rad}\, A)_S = \mathrm{Rad}\,(A_S)$.

Proof. Let $\chi: R \to R_S$ denote the canonical mapping and suppose that $r \in \mathrm{Rad}\, A$. Then $r^m \in A$ for a suitable positive integer m. Since χ is a ring-homomorphism, $\chi(r)^m \in \chi(A) \subseteq A_S$ and therefore $\chi(r) \in \mathrm{Rad}\,(A_S)$. However $(\mathrm{Rad}\, A)_S$ is generated (see Proposition 3) by elements of the form $\chi(r)$. Consequently $(\mathrm{Rad}\, A)_S \subseteq \mathrm{Rad}\,(A_S)$.

Now assume that the fraction $[r/s]$ belongs to $\mathrm{Rad}\,(A_S)$. Then there exists an integer n such that $[r/s]^n = [r^n/s^n]$ belongs to A_S; say $[r^n/s^n] = [a/s']$, where $a \in A$ and $s' \in S$. Accordingly $s^* s' r^n = s^* s^n a \in A$ for some element s^* of S. Put $\sigma = s^* s'$. Then $\sigma r^n \in A$ and therefore $(\sigma r)^n \in A$. It follows that $\sigma r \in \mathrm{Rad}\, A$. Thus $[r/s] = [\sigma r/\sigma s]$ is in $(\mathrm{Rad}\, A)_S$ and we may conclude that $\mathrm{Rad}\,(A_S) \subseteq (\mathrm{Rad}\, A)_S$. This establishes the proposition.

3.7 Primary submodules and fractions

Once again R denotes a commutative ring with an identity element and S a non-empty multiplicatively closed subset of R. E will designate a fixed R-module. If K is a submodule of E, then K_S will be regarded as an R_S-submodule of E_S.

Proposition 16. *Let K be a P-primary† submodule of E. If P meets S, then $K_S = E_S$. On the other hand if P does not meet S, then K_S is a P_S-primary submodule of E_S and its contraction in E is K.*

Proof. First suppose that there is an element s which is common to P and S. Since $P = \mathrm{Rad}\,(K:E)$, there is an integer n such that $s^n = \sigma$ (say) belongs to $K:E$. Accordingly $\sigma E \subseteq K$. Suppose now that $e \in E$ and $s' \in S$. Then $[e/s'] = [\sigma e/\sigma s'] \in K_S$. This shows that $E_S = K_S$ and the first assertion follows.

From here on we assume that P does not meet S. Let e be an element which belongs to the contraction of K_S in E. By Proposition 1, $se \in K$ for a suitable element s of S. Using the facts that K is P-primary in E and $s \notin P$, we conclude, by virtue of Proposition 19 of section (2.8), that $e \in K$. It follows that K is the contraction of K_S in E. This shows, in particular, that K_S is a *proper* submodule of E_S.

† See section (2.8) for the definition.

Assume next that $[r/s][e/\sigma] \in K_S$ and $[e/\sigma] \notin K_S$. (The notation is self-explanatory.) Then $e \notin K$. Multiplying $[re/s\sigma]$ by $[s^2\sigma^2/s\sigma]$, we obtain $[s^2\sigma^2 re/s^2\sigma^2] \in K_S$. Thus re belongs to the contraction of K_S, that is $re \in K$. It now follows that $r^n E \subseteq K$ for a suitable integer n, which in turn implies that $[r/s]^n E_S \subseteq K_S$. This establishes that K_S is a primary submodule of E_S. By Theorem 9 it must be P'_S-primary, where P' is a prime R-ideal not meeting S.

Let $p \in P$. For a suitable integer μ we have $p^\mu E \subseteq K$. Hence if $s \in S$, then $[sp/s]^\mu E_S \subseteq K_S$. It follows that $[sp/s] \in P'_S$ and therefore p belongs to the contraction of P'_S in R. Since (Proposition 14) this contraction is P', the argument shows that $P \subseteq P'$.

Finally let $\pi \in P'$. If now $\sigma \in S$, then $[\sigma\pi/\sigma]^\nu E_S \subseteq K_S$ if ν is properly chosen. Select $e \in E$ so that $e \notin K$. Then

$$\left[\frac{\sigma^{\nu+1}\pi^\nu e}{\sigma^{\nu+1}}\right] = \left[\frac{\sigma\pi}{\sigma}\right]^\nu \left[\frac{\sigma e}{\sigma}\right]$$

belongs to K_S and therefore $\pi^\nu e$ belongs to its contraction which is K. From $\pi^\nu e \in K$, $e \notin K$, follows $\pi^\nu \in P$—see Proposition 19 of section (2.8)—whence $\pi \in P$. This shows that $P' \subseteq P$ and completes the proof.

Theorem 10. *Let P be a prime ideal not meeting S and E an arbitrary R-module. Then there is a one-one correspondence between the P-primary submodules K of E and the P_S-primary submodules W of E_S. This is such that, when K and W correspond, $W = K_S$ and K is the contraction of W in E.*

Proof. Let W be a given P_S-primary submodule of E_S. After Proposition 16 it is enough to show that W is the extension of a P-primary submodule of E.

Let K be the contraction of W in E. By Proposition 2, $K_S = W$. This shows, *inter alia*, that K is a *proper* submodule of E. Now assume that $re \in K$, where $r \in R$, $e \in E$ and $e \notin K$. If s is an arbitrary element of S, then $[sr/s][se/s] = [s^2 re/s^2]$ belongs to $K_S = W$. On the other hand, $[se/s] \notin W$ for the contrary assumption would imply that e belongs to K which is not the case. Since W is a primary submodule of E_S, we now see that there exists an integer m such that $[sr/s]^m E_S \subseteq W$.

Let e' be an arbitrary element of E. Then

$$\left[\frac{s^{m+1}r^m e'}{s^{m+1}}\right] = \left[\frac{sr}{s}\right]^m \left[\frac{se'}{s}\right] \in W,$$

whence $r^m e' \in K$. As this holds for every $e' \in E$, we may conclude that $r^m E \subseteq K$. This proves that K is a primary submodule of E. Let it be P'-primary.

Since $K_S = W$ and $W \neq E_S$, Proposition 16 shows that P' does not meet S. The same proposition shows that $K_S = W$ is P'_S-primary. Thus $P'_S = P_S$ and therefore $P' = P$ by Theorem 9. This completes the proof.

We obtain an interesting special case of the theorem by taking E to be R itself. This is described in the following

Corollary. *Let P be a prime ideal not meeting S. Then there is a one-one correspondence between the P-primary ideals Q of the ring R and the P_S-primary ideals Q' of the ring R_S. This is such that, when Q and Q' correspond, $Q' = Q_S$ and Q is the contraction of Q' in R.*

The final result of this section is concerned with the behaviour of decomposable submodules in relation to the formation of fractions.

Theorem 11. *Suppose that*

$$K = N_1 \cap N_2 \cap \ldots \cap N_t, \qquad (3.7.1)$$

where N_i is a P_i-primary submodule of E, and let the prime ideals $P_1, P_2, \ldots, P_t$ be so numbered that none of $P_1, P_2, \ldots, P_m$ meets S while all of the remainder do meet S. Then, for $1 \leqslant i \leqslant m$, $(N_i)_S$ is a $(P_i)_S$-primary submodule of E_S and

$$K_S = (N_1)_S \cap (N_2)_S \cap \ldots \cap (N_m)_S. \qquad (3.7.2)$$

Furthermore, if (3.7.1) is an irredundant resp. normal decomposition of K in E, then (3.7.2) is an irredundant resp. normal decomposition of K_S in E_S.

Remark. Normal and irredundant primary decompositions were defined in section (2.9).

Proof. The first assertion is an immediate consequence of Proposition 16. By (3.5.2),
$$K_S = (N_1)_S \cap (N_2)_S \cap \ldots \cap (N_t)_S$$

and, again using Proposition 16, $(N_j)_S = E_S$ for $j = m+1, m+2, \ldots, t$. This yields (3.7.2).

Now suppose that (3.7.1) is an irredundant primary decomposition. Then so is (3.7.2). For if, for example, $(N_1)_S$ contains $(N_2)_S \cap \ldots \cap (N_m)_S$, then (contracting in E and remembering that, for $1 \leqslant i \leqslant m$, the contraction of $(N_i)_S$ is N_i by Proposition 16)

$$N_1 \supseteq N_2 \cap \ldots \cap N_m \supseteq N_2 \cap \ldots \cap N_m \cap N_{m+1} \cap \ldots \cap N_t$$

which contradicts the irredundant character of (3.7.1).

Finally assume that (3.7.1) is a normal decomposition of K in E. Then, in particular, it is irredundant and therefore (3.7.2) is irredundant as well. Again the prime ideals $P_1, P_2, \ldots, P_t$ are all distinct and therefore, by Theorem 9, $(P_1)_S$, $(P_2)_S$, $\ldots, (P_m)_S$ are also distinct. This shows that (3.7.2) is a normal decomposition and completes the proof.

3.8 Localization

Let R be a commutative ring with an identity element and E an R-module. If now P is a prime ideal and S its complement in R, then S is a non-empty multiplicatively closed subset of R. We may therefore use S to form the ring R_S of fractions and, at the same time, construct the R_S-module E_S.

Rings and modules of fractions which arise in this way are of such common occurrence that it has proved advantageous to introduce a notation showing directly the connection between the end-products and the prime ideal. The accepted way of doing this is to write R_P and E_P in place of R_S and E_S.

In this connection it is necessary to insert a word of caution. Not only is the complement of P a multiplicatively closed set, but the elements of P, by themselves, form such a system. Thus the general conventions of the theory of fractions already assign meanings to R_P and E_P and the new proposals come into conflict with these. However, since P contains the zero element, the interpretation derived from the general theory would always make R_P a null ring and E_P a null module. Thus although there is an ambiguity associated with the symbols R_P and E_P, one of the possible interpretations is consistently trivial and may therefore be ignored. In practice this causes no confusion.

To recapitulate: *whenever P is a prime ideal, R_P and E_P denote the ring and module obtained by forming fractions with respect to the complement of P in R*. When we pass from R and E to R_P and E_P we shall often describe this by saying that we are *localizing at P*.

If S is the complement of P in R, then the prime ideals which do not meet S are just those which are contained in P. It therefore follows, from Theorem 9, that *the prime ideals of R_P are in one-one correspon-*

dence with the prime ideals of R that are contained in P. Indeed if $P' \subseteq P$ is a prime ideal of R, then the corresponding prime ideal of R_P is its extension. This, however, will be contained in the extension of P. It therefore follows that *the extension of P is the only maximal ideal of R_P.* In order to have a convenient way of describing this situation, we make the following

Definition. *A commutative ring (with an identity element) which has exactly one maximal ideal is called a 'quasi-local† ring'.*

Thus when we localize R at one of its prime ideals, the result is always a quasi-local ring. Localization at the maximal ideals of R plays a particularly important role. The letter M will normally be used to denote a maximal ideal. If E is an R-module, then the results of localizing at M are natually denoted by R_M and E_M.

Proposition 17. *Let E be an R-module and suppose that $E_M = 0$ for every maximal ideal M. Then $E = 0$.*

Thus a module which is locally null at every maximal ideal is necessarily a null module.

Proof. Let e be an element of E. It will suffice to show that $e = 0$. To this end we first fix our attention on one particular maximal ideal M. Considered as an element of E_M, $[e/1]$ is zero by hypothesis. Consequently there exists an element c_M of R such that $c_M \notin M$ and $c_M e = 0$.

We select one such element c_M for each maximal ideal M and denote by I the ideal which these elements generate. Then $Ie = 0$. However $I = R$. For if I were a proper ideal, then, by Theorem 2 of section (2.2), it would be contained in a maximal ideal M' say. But this is impossible because $c_{M'}$ belongs to I but not to M'.

Since $I = R$, 1_R belongs to I and therefore $e = 1_R e = 0$. This completes the proof.

Using similar arguments we can establish

Proposition 18. *Let X be a submodule of an R-module E and, for each maximal ideal M, let X_M be regarded as a submodule of E_M. If now X^M denotes the contraction of X_M in E, then $X = \bigcap_M X^M$.*

Proof. It is clear that $X \subseteq \bigcap_M X^M$. Now let e be an element of the intersection. For each M we can find $c_M \in R$ such that $c_M e \in X$ and $c_M \notin M$. This is because, by Proposition 1, X^M is the S-component of X in E,

† The name 'local ring' is usually given to a quasi-local ring which satisfies the maximal condition for ideals.

where S is the complement of M in R. The ideal generated by the elements c_M is the whole ring R (for the same reasons as were given in the proof of Proposition 17); consequently $Re \subseteq X$. It follows that $e \in X$ and all is proved.

Corollary. *Let X, Y be submodules of an R-module E and, for each maximal ideal M, let X_M and Y_M be considered as R_M-submodules of E_M. If now $X_M = Y_M$ for every M, then $X = Y$.*

This follows immediately from the proposition because X_M and Y_M have the same contraction in E.

Before we go on to consider some applications of the method of localization, we shall establish a useful identity between two rings which are obtained by apparently different constructions. In this respect the situation is similar to that described in Proposition 9.

The setting for the new example is as follows. As usual S denotes a non-empty, multiplicatively closed, subset of R. Now let P be a prime ideal not meeting S. By Proposition 14, P_S is a prime ideal of R_S. We can therefore localize at P_S so obtaining the ring $(R_S)_{P_S}$. It will be shown that $(R_S)_{P_S}$ can be identified with R_P.

First some preparatory observations. Let $[r/s]$ be a typical element of R_S. Should $r \in P$ then $[r/s] \in P_S$; but if $r \notin P$, then $[r/s] \notin P_S$. The latter statement is true because the contraction of P_S in R is P. In what follows the letter c will be used consistently to denote an element of R which does *not* belong to P.

Let $\chi : R \to R_S$ be the canonical mapping of R into R_S. The general element of R_P has the form $[r/c]$, where $r, c \in R$ and $c \notin P$. Now $\chi(c) \notin P_S$ again for the reason that the contraction of P_S is P. It follows that $[\chi(r)/\chi(c)]$ can be regarded as an element of $(R_S)_{P_S}$. Indeed a simple check shows that there exists a well defined mapping

$$R_P \to (R_S)_{P_S} \tag{3.8.1}$$

in which

$$\left[\frac{r}{c}\right] \to \left[\frac{\chi(r)}{\chi(c)}\right]. \tag{3.8.2}$$

The properties of this mapping will now be investigated. In the first place it is obviously a ring-homomorphism. Now let $[r/s], [\rho/\sigma]$ belong to R_S and suppose that $[\rho/\sigma]$ does not belong to P_S. (The notation is self-explanatory.) Then $\rho \notin P$ and $[[r/s]/[\rho/\sigma]]$ is a typical element of $(R_S)_{P_S}$. But in this ring

$$[[r/s]/[\rho/\sigma]] = [[\sigma sr/s]/[\sigma s\rho/\sigma]]$$

because

$$[\sigma s\rho/\sigma][r/s] = [\rho/\sigma][\sigma sr/s].$$

Thus the typical element $[[r/s]/[\rho/\sigma]]$ of $(R_S)_{P_S}$ can also be written as $[\chi(\sigma r)/\chi(s\rho)]$. This shows that (3.8.1) maps R_P on to $(R_S)_{P_S}$.

Now suppose that $[r/c]$ belongs to the kernel of the mapping. Then $[\chi(r)/\chi(c)] = 0$ and therefore $\chi(r)$ becomes zero when multiplied by a suitable element of the complement of P_S in R_S. Let the suitable element be $[c_1/s_1]$, where $c_1 \in R$, $c_1 \notin P$ and $s_1 \in S$. Since $\chi(r) = [s_1 r/s_1]$, this gives $[c_1 s_1 r/s_1^2] = 0$, whence $\sigma c_1 s_1 r_1 = 0$ for some element σ belonging to S. Accordingly $[r/c] = [\sigma c_1 s_1 r/\sigma c_1 s_1 c] = 0$ which shows that (3.8.1) is also a monomorphism. We collect our conclusions in

Proposition 19. *If the prime ideal P does not meet S, then the mapping*

$$R_P \to (R_S)_{P_S} \qquad (3.8.3)$$

in which $[r/c] \to [\chi(r)/\chi(c)]$ *(here* $\chi : R \to R_S$ *is the canonical mapping) is a ring-isomorphism of* R_P *on to* $(R_S)_{P_S}$.

For future reference we note that the diagram

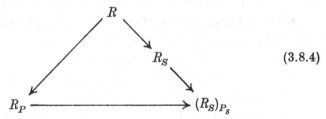

$$(3.8.4)$$

(the lower mapping is the one described in the proposition; the remainder are canonical mappings) is commutative. This is easily verified.

There is, of course, a related result involving modules. Still assuming that P does not meet S, let E be an R-module. We can then form E_S and afterwards localize at P_S to obtain $(E_S)_{P_S}$. It will now be shown that $(E_S)_{P_S}$ can be identified with E_P. In what follows

$$\chi_R : R \to R_S \quad \text{and} \quad \chi_E : E \to E_S$$

denote the usual canonical mappings.

A typical element of E_P has the form $[e/c]$, where $e \in E$, $c \in R$ and $c \notin P$. Next $\chi_E(e)$ belongs to E_S while $\chi_R(c)$ belongs to the complement of P_S in R_S. Accordingly $[\chi_E(e)/\chi_R(c)]$ is an element of $(E_S)_{P_S}$. Indeed, using the relation $\chi_E(re) = \chi_R(r)\,\chi_E(e)$ which we noted earlier in (3.2.3), it is a simple matter to check that there exists a well defined mapping

$$E_P \to (E_S)_{P_S} \qquad (3.8.5)$$

in which

$$\left[\frac{e}{c}\right] \to \left[\frac{\chi_E(e)}{\chi_R(c)}\right]. \qquad (3.8.6)$$

Here we have a situation very similar to that encountered in (3.8.1) and (3.8.2). By adapting the arguments used on the former occasion, it is a straightforward matter to verify that (3.8.5) *is an isomorphism of the additive group of E_P on to the additive group of $(E_S)_{P_S}$*. Note that E_P is an R_P-module while $(E_S)_{P_S}$ is an $(R_S)_{P_S}$-module. The two module structures will now be compared.

To this end we construct the diagram

$$R_P \times E_P \longrightarrow (R_S)_{P_S} \times (E_S)_{P_S}$$
$$\downarrow \qquad\qquad\qquad \downarrow \qquad\qquad\qquad (3.8.7)$$
$$E_P \longrightarrow (E_S)_{P_S}$$

with the aid of the mappings (3.8.3) and (3.8.5). In this the horizontal mappings are one-one correspondences and the vertical mappings represent multiplications. A typical element of $R_P \times E_P$ is a pair $([r/c], [e/\gamma])$, where $r \in R$, $e \in E$ and c, γ belong to the complement of P in R. On first applying the left vertical mapping and then the lower horizontal mapping, the pair becomes

$$\left[\frac{\chi_E(re)}{\chi_R(c\gamma)}\right] = \left[\frac{\chi_R(r)\,\chi_E(e)}{\chi_R(c)\,\chi_R(\gamma)}\right].$$

Now the upper horizontal mapping changes

$$([r/c], [e/\gamma]) \quad \text{into} \quad ([\chi_R(r)/\chi_R(c)], [\chi_E(e)/\chi_R(\gamma)])$$

and from this it is clear that the diagram (3.8.7) is commutative. Accordingly we have proved

Proposition 20. *Let P be a prime ideal not meeting S and let the rings R_P and $(R_S)_{P_S}$ be identified by means of the isomorphism (3.8.3). If now E is an arbitrary R-module, then the mapping $E_P \to (E_S)_{P_S}$ in which $[e/c] \to [\chi_E(e)/\chi_R(c)]$ (where $\chi_R:R \to R_S$ and $\chi_E:E \to E_S$ are the canonical mappings) is an isomorphism of E_P on to $(E_S)_{P_S}$ when these are regarded as modules with respect to the ring $R_P = (R_S)_{P_S}$.*

3.9 Some uses of localization

The next result may be described as the *localization principle* for lengths.

Theorem 12. *Let E be an R-module and let M denote a typical maximal ideal of R. If now $L_R(E)$ denotes the length of the R-module E, then*

$$L_R(E) = \sum_M L_{R_M}(E_M). \qquad (3.9.1)$$

Remark. It should be noted that in (3.9.1) it is not assumed that the lengths involved are finite.

We shall not give a separate proof of this result because it is a special case of the *extension formula* which will be established shortly. First, however, it is necessary to make some preliminary observations.

If F is a *field*, then a module with respect to F is usually known as a *vector space* over F. Suppose that V is a vector space over F. The length of V, considered as an F-module,† is customarily called the *dimension* of the vector space and denoted by $[V:F]$. Thus

$$[V:F] = L_F(V).$$

Suppose now that we have a ring R' which is an integral extension‡ of R. Let M' be a maximal ideal of R'. Then, by Theorem 11 of section (2.6), $M' \cap R$ is a maximal ideal of R. Since R'/M' is an R'-module, it can also be regarded in a natural way as an R-module. However if $\alpha \in M' \cap R$ and $x \in R'/M'$, then $\alpha x = 0$. This shows that $M' \cap R$ is contained in the annihilator of the R-module R'/M' and therefore, by Proposition 24 of section (1.14), R'/M' may be regarded as a module with respect to the residue ring $R/(M' \cap R)$. But, by Proposition 1 of section (2.2), $R/(M' \cap R)$ is a field. We may therefore sum up our conclusions by saying that R'/M' *has a natural structure as a vector space over the field* $R/(M' \cap R)$. Note that

$$L_R(R'/M') = L_{R/(M' \cap R)}(R'/M') = [R'/M':R/(M' \cap R)] \quad (3.9.2)$$

because R'/M' has the same length with respect to R as it has with respect to $R/(M' \cap R)$ by virtue of Proposition 24 Corollary 2 of section (1.14). Of course, there is nothing in our assumptions which will ensure that the dimension $[R'/M':R/(M' \cap R)]$ is *finite*.

Lemma 11. *Let the ring R' be an integral extension of the ring R and let $E' \neq 0$ be an R'-module. Then*

$$\sum_{M'} L_{R_{M'}}(E'_{M'}) [R'/M':R/(M' \cap R)] > 0.$$

Remark. If $L_{R_{M'}}(E'_{M'}) = 0$ and $[R'/M':R/(M' \cap R)] = \infty$, then their product is to be regarded as having the value zero.

Proof. It is clear that $[R'/M':R/(M' \cap R)] > 0$ for every maximal ideal M' of R'. Hence it is sufficient to show that $E'_{M'} \neq 0$ for some M'.

---

† Cf. Theorem 21 of section (1.13).
‡ See section (2.5) for the definition.

However if $E'_{M'} = 0$ for every M', then $E' = 0$ (see Proposition 17) and we have a contradiction.

We are now ready to state and prove the *extension formula* for lengths.

Theorem 13. *Let the ring R' be an integral extension of the ring R, let E' be an R'-module and denote by M' a typical maximal ideal of R'. Then*

$$L_R(E') = \sum_{M'} L_{R'_{M'}}(E'_{M'})\,[R'/M':R/(M' \cap R)]. \qquad (3.9.3)$$

Remark. Theorem 12 is the special case of Theorem 13 obtained by taking $R' = R$. Note that if F is a field, then $[F:F] = 1$. This is because, by Lemma 3 of section (2.2), the only ideals of F are (0) and F itself.

Proof. Suppose that we have a strictly decreasing chain

$$E' = E'^{(0)} \supset E'^{(1)} \supset \ldots \supset E'^{(p)} = 0 \qquad (3.9.4)$$

of R'-modules. Then this is also a chain of R-modules. By successive applications of Theorem 6 of section (1.7), we see that

$$L_R(E') = \sum_{i=1}^{p} L_R(E'^{(i-1)}/E'^{(i)}). \qquad (3.9.5)$$

Next, by localizing at M', we obtain the chain

$$E'_{M'} = E'_{M'}^{(0)} \supseteq E'_{M'}^{(1)} \supseteq \ldots \supseteq E'_{M'}^{(p)} = 0$$

of $R'_{M'}$-modules, whence

$$L_{R'_{M'}}(E'_{M'}) = \sum_{i=1}^{p} L_{R'_{M'}}(E'_{M'}^{(i-1)}/E'_{M'}^{(i)}). \qquad (3.9.6)$$

Again, because fraction functors are exact, the exact sequence

$$0 \to E'^{(i)} \to E'^{(i-1)} \to E'^{(i-1)}/E'^{(i)} \to 0$$

of R'-modules gives rise to the exact sequence

$$0 \to E'^{(i)}_{M'} \to E'^{(i-1)}_{M'} \to (E'^{(i-1)}/E'^{(i)})_{M'} \to 0$$

of $R'_{M'}$-modules. This shows that $(E'^{(i-1)}/E'^{(i)})_{M'}$ and $E'^{(i-1)}_{M'}/E'^{(i)}_{M'}$ are isomorphic. Accordingly, by (3.9.6),

$$L_{R'_{M'}}(E'_{M'}) = \sum_{i=1}^{p} L_{R'_{M'}}\{(E'^{(i-1)}/E'^{(i)})_{M'}\}.$$

If we now multiply by $[R'/M':R/(M' \cap R]$, sum over all M', and rearrange (this is permissible because all terms are non-negative or plus infinity), then it is found that

$$\sum_{M'} L_{R'_{M'}}(E'_{M'})\,[R'/M':R/(M' \cap R)]$$

$$= \sum_{i=1}^{p} \sum_{M'} L_{R'_{M'}}\{(E'^{(i-1)}/E'^{(i)})_{M'}\}\,[R'/M':R/(M' \cap R)]. \qquad (3.9.7)$$

We must now distinguish various cases. First suppose that

$$L_{R'}(E') = \infty.$$

Since (3.9.4) is a chain of R-modules, we see that $L_R(E') \geqslant p$. Also, because $E''^{(i-1)}/E''^{(i)}$ is a non-zero R'-module, Lemma 11 shows that

$$\sum_{i=1}^{p} \sum_{M'} L_{R'_{M'}}\{(E''^{(i-1)}/E''^{(i)})_{M'}\}[R'/M' : R/(M' \cap R)] \geqslant p.$$

However p may be made as large as we please. Consequently both sides of (3.9.3) are infinite and the required result follows in this case.

From now on we suppose that $L_{R'}(E') < \infty$ in which case we may take it that (3.9.4) is a composition series of R'-modules. Comparing (3.9.5) and (3.9.7), it is clear that it is enough to prove the extension formula for each of the modules $E''^{(i-1)}/E''^{(i)}$. However, these are all simple modules. It follows that we need only establish (3.9.3) when E' is a *simple* R'-module. Now, by Proposition 12 of section (1.8), each simple R'-module is isomorphic to a module† of the form R'/I', where I' is a maximal ideal of R'. Accordingly the theorem will follow if we can show that the extension formula is correct when $E' = R'/I'$ and I' is one of the maximal ideals.

Consider this special case. To begin with

$$L_R(R'/I') = L_{R/(I' \cap R)}(R'/I') = [R'/I' : R/(I' \cap R)]. \qquad (3.9.8)$$

Next, from the exact sequence $0 \to I' \to R' \to R'/I' \to 0$, we conclude that $(R'/I')_{M'}$ and $R'_{M'}/I'_{M'}$ are isomorphic as $R'_{M'}$-modules. First suppose that $I' \neq M'$. Then I' cannot be contained in M' because both are maximal ideals. Accordingly I' meets the complement of M' and therefore, by Proposition 14, $I'_{M'} = R'_{M'}$. Thus in this case $(R'/I')_{M'}$ is a null module and therefore $L_{R'_{M'}}\{(R'/I')_{M'}\} = 0$. On the other hand if $I' = M'$, then $I'_{M'}$ is the maximal ideal of $R'_{M'}$ and therefore $R'_{M'}/I'_{M'}$ is a simple $R'_{M'}$-module. It follows that, in this situation

$$L_{R'_{M'}}\{(R'/I')_{M'}\} = 1.$$

Bearing these facts in mind we see that

$$\sum_{M'} L_{R'_{M'}}\{(R'/I')_{M'}\}[R'/M' : R/(M' \cap R)]$$
$$= [R'/I' : R/(I' \cap R)]$$
$$= L_R(R'/I')$$

† It is recalled that a simple module is singly generated. In fact such a module is generated by each of its non-zero elements.

by virtue of (3.9.8). Thus the extension formula is established for simple R'-modules and the general result follows for the reasons previously given.

Before we describe the next application of localization, it will be convenient to establish a lemma.

Lemma 12. *Let R' be an extension ring of R and u, v elements of R' which satisfy $uv = 1$. If now A is a proper ideal of R, then either*

$$AR[u] \ne R[u] \quad or \quad AR[v] \ne R[v].$$

We recall† that $R[u]$ denotes the smallest subring of R' which contains both R and u. It consists of all elements of R' which can be expressed in the form $r_0 + r_1 u + r_2 u^2 + \ldots + r_p u^p$, where $r_0, r_1, \ldots, r_p$ belong to R. By $AR[u]$ is meant, of course, the ideal of $R[u]$ generated by the elements of A. Indeed a typical element of $AR[u]$ can be represented as a sum $a_0 + a_1 u + a_2 u^2 + \ldots + a_m u^m$, where $a_0, a_1, \ldots, a_m$ are elements of A.

Proof. We shall assume that both $AR[u] = R[u]$ and $AR[v] = R[v]$ and then show that this leads to a contradiction. Since the common identity element of R and R' belongs to $AR[u]$, it can be expressed in the form

$$1 = a_0 + a_1 u + a_2 u^2 + \ldots + a_m u^m, \qquad (3.9.9)$$

where the a_i belong to A. Among all such relations we select one for which the integer m is as small as possible. Note that, because $A \ne R$, m must be at least unity.

Next, since the identity element belongs to $AR[v]$, we also have a relation of the form

$$1 = \alpha_0 + \alpha_1 v + \alpha_2 v^2 + \ldots + \alpha_n v^n, \qquad (3.9.10)$$

where $\alpha_0, \alpha_1, \ldots, \alpha_n$ are likewise in A. Here too we select such a relation with n as small as possible. Naturally $n \geqslant 1$.

In what follows we assume that $m \geqslant n$. This is permissible for if $n > m$, then we merely interchange the roles of u and v.

From (3.9.9) is obtained

$$1 - \alpha_0 = (1 - \alpha_0) a_0 + \ldots + (1 - \alpha_0) a_{m-1} u^{m-1} + (1 - \alpha_0) a_m u^m \qquad (3.9.11)$$

while from (3.9.10) follows

$$(1 - \alpha_0) a_m u^m = a_m u^m (\alpha_1 v + \alpha_2 v^2 + \ldots + \alpha_n v^n)$$
$$= \alpha'_{m-1} u^{m-1} + \alpha'_{m-2} u^{m-2} + \ldots + \alpha'_{m-n} u^{m-n},$$

† See section (2.5).

where $\alpha'_j \in A$ for $m-n \leqslant j \leqslant m-1$. Substituting in (3.9.11) and rearranging, we find that

$$1 = a'_0 + a'_1 u + a'_2 u^2 + \ldots + a'_{m-1} u^{m-1},$$

where the coefficients a'_j are elements of A. This however contradicts the minimal property of the integer m and so the lemma follows.

The next theorem is of importance in the theory of valuations and specializations. To explain what is involved, suppose that R is an integral domain and that it is a subring of a field F. Let S be a non-empty, multiplicatively closed subset of R not containing the zero element. Then each element s of S has an inverse s^{-1} in the field F. Furthermore we can map the abstractly defined ring R_S into F by associating with $[r/s]$ the element $s^{-1}r$. This mapping $R_S \to F$ is easily seen to be a ring-homomorphism in which distinct elements of R_S have distinct images. Accordingly R_S is mapped isomorphically on to a subring of F. Let us identify R_S with its image. Then R_S becomes a subring of F and $s^{-1}r$ takes over the role formerly played by $[r/s]$. Note that, after identification, the canonical mapping $R \to R_S$ is just the inclusion mapping. This is because $s^{-1}sr = r$.

Let P be a prime ideal of R. The above remarks (applied to the complement of P) show that R_P is effectively a subring of F its typical element being of the form $c^{-1}r$, where $r, c \in R$ and $c \notin P$. We know that the extension P' (say) of P to R_P is the only maximal ideal of the ring of fractions. Moreover, the contraction of P' in R, that is its inverse image under the canonical mapping $R \to R_P$, is just P. However, in this case, the canonical mapping is the inclusion mapping. Consequently $R \cap P' = P$.

Theorem 14. *Let the integral domain R be a subring of the field F. In addition, let P be a prime ideal of R and α a non-zero element of F. Then at least one of the rings $R[\alpha]$ and $R[\alpha^{-1}]$ contains a prime ideal whose intersection with R is P.*

Proof. Put $R' = R_P$ and let P' be the unique maximal ideal of R'. As already explained, R' may be regarded as a subring of F in which case $P' \cap R = P$.

By Lemma 12, either $P'R'[\alpha] \neq R'[\alpha]$ or $P'R'[\alpha^{-1}] \neq R'[\alpha^{-1}]$. For definiteness we assume the former to be true. Then, by Theorem 2 of section (2.2), there exists a maximal and hence prime ideal Π of $R'[\alpha]$ such that $P'R'[\alpha] \subseteq \Pi$. Next, by Lemma 6 of section (2.4), $\Pi \cap R'$ is a prime ideal of R'. Since however $P' \subseteq \Pi \cap R'$ and P' is the maximal ideal of R', it follows that $\Pi \cap R' = P'$. Again, by the same lemma, $\Pi \cap R[\alpha]$ is a prime ideal of $R[\alpha]$. It has the required property because

$$\Pi \cap R[\alpha] \cap R = \Pi \cap R' \cap R = P' \cap R = P.$$

The theorem has a useful generalization which is contained in the following

Corollary. *Let R, F and P be as in the statement of the theorem and let $\alpha_1, \alpha_2, ..., \alpha_n (n \geqslant 1)$ be elements of F which are not all zero. Then there exists an integer $j (1 \leqslant j \leqslant n)$ such that $\alpha_j \neq 0$ and such that the ring $R[\alpha_1/\alpha_j, \alpha_2/\alpha_j, ..., \alpha_n/\alpha_j]$ contains a prime ideal Π satisfying $\Pi \cap R = P$.*

Remark. In this context it is to be understood that by α_i/α_j is meant the element $\alpha_i \alpha_j^{-1}$ of F. Thus $R[\alpha_1/\alpha_j, \alpha_2/\alpha_j, ..., \alpha_n/\alpha_j]$ is a subring of F and an extension ring of R.

Proof. It is clear that, for the purposes of the proof, we may suppose that *none* of $\alpha_1, \alpha_2, ..., \alpha_n$ is zero. Having made this assumption, we now use induction on n.

If $n = 1$, then there is nothing to prove. When $n = 2$ the assertion is an immediate consequence of the theorem. We shall therefore suppose that $n > 2$ and that the corollary has been established for all smaller values of the inductive variable.

The inductive hypothesis applied to $\alpha_1, \alpha_2, ..., \alpha_{n-1}$ shows that there exists an integer $\nu (1 \leqslant \nu \leqslant n-1)$ such that the ring

$$R[\alpha_1/\alpha_\nu, ..., \alpha_{n-1}/\alpha_\nu] = R' \text{ (say)}$$

contains a prime ideal Π' satisfying $\Pi' \cap R = P$. Next, by the theorem, either $R'[\alpha_n/\alpha_\nu]$ or $R'[\alpha_\nu/\alpha_n]$ contains a prime ideal Π'' for which $\Pi'' \cap R' = \Pi'$ and therefore $\Pi'' \cap R = P$.

Since $R'[\alpha_n/\alpha_\nu] = R[\alpha_1/\alpha_\nu, ..., \alpha_{n-1}/\alpha_\nu, \alpha_n/\alpha_\nu]$, there is no difficulty in the former case. We shall therefore assume that Π'' is a prime ideal of $R'[\alpha_\nu/\alpha_n]$. For $1 \leqslant i \leqslant n-1$, α_i/α_ν belongs to R' and therefore

$$\alpha_i/\alpha_n = (\alpha_i/\alpha_\nu)(\alpha_\nu/\alpha_n) \text{ is in } R'[\alpha_\nu/\alpha_n].$$

This shows that

$$R[\alpha_1/\alpha_n, ..., \alpha_{n-1}/\alpha_n, \alpha_n/\alpha_n] \subseteq R'[\alpha_\nu/\alpha_n].$$

We now put $\Pi = \Pi'' \cap R[\alpha_1/\alpha_n, \alpha_2/\alpha_n, ..., \alpha_n/\alpha_n]$.

By Lemma 6 of section (2.4), Π is a prime ideal of $R[\alpha_1/\alpha_n, ..., \alpha_n/\alpha_n]$. Furthermore, $\Pi \cap R = \Pi'' \cap R = P$ and so the proof is complete.

Exercises on Chapter 3

In these exercises R denotes a commutative ring with an identity element and S a non-empty multiplicatively closed subset of R.

1 Let E be an R-module and W a finitely generated R_S-submodule of E_S. Show that there exists a finitely generated submodule X of E such that $X_S = W$.

2. Let W be an R_S-module. Then, using the canonical ring-homomorphism $R \to R_S$, we may regard W as an R-module. Show that the canonical mapping $W \to W_S$ (of the R-module W into the R_S-module W_S) is, in fact, an isomorphism of R_S-modules.

3. Let Γ be a grading monoid and X an indeterminate. Define $R[X;\Gamma]$ as in Exercise 13 of the Exercises on Chapter 2. If now S is a non-empty multiplicatively closed subset of R, then it can also be regarded as a multiplicatively closed subset of $R[X;\Gamma]$. On this understanding, show that the ring of fractions of $R[X;\Gamma]$ with respect to S is isomorphic to the ring $R_S[X;\Gamma]$.

4. Let Γ, X and $R[X;\Gamma]$ be as in Exercise 3 and, for each R-module E, let $E[X;\Gamma]$ consists of all formal sums $\sum_{\gamma \in \Gamma} e_\gamma X^\gamma$, where $e_\gamma \in E$ and only a finite number of the $e_\gamma (\gamma \in \Gamma)$ are different from zero. Define 'addition' on $E[X;\Gamma]$ in the obvious manner and when

$$\sum_{\alpha \in \Gamma} r_\alpha X^\alpha \text{ resp. } \sum_{\gamma \in \Gamma} e_\gamma X^\gamma$$

belongs to $R[X;\Gamma]$ resp. $E[X;\Gamma]$ define their ' product' by

$$(\sum_\alpha r_\alpha X^\alpha)(\sum_\gamma e_\gamma X^\gamma) = \sum_{\alpha,\gamma} (r_\alpha e_\gamma) X^{\alpha+\gamma}.$$

Show that, with these definitions, $E[X;\Gamma]$ is a module with respect to $R[X;\Gamma]$.

5. Let Γ, X and $R[X;\Gamma]$ be as in Exercise 3. For each R-module E put $T(E) = E[X;\Gamma]$ (see Exercise 4) and for each homomorphism $f:E \to E'$ of R-modules let

$$T(f):T(E) \to T(E')$$

be the mapping in which

$$\sum_\gamma e_\gamma X^\gamma \to \sum_\gamma f(e_\gamma) X^\gamma.$$

Show that T is an additive exact functor from the category of R-modules to the category of $R[X;\Gamma]$-modules.

6. Let S and S' be non-empty multiplicatively closed subsets of R and suppose that S is contained in S'. If now E is an R-module, then we can form both E_S and $E_{S'}$ and each of these may be regarded as an R-module. (In this way we obtain two functors from the category of R-modules to itself.) Show that there exists a homomorphism $E_S \to E_{S'}$ (of R-modules) having the property that the element $[e/s]$ of E_S is mapped into that element of $E_{S'}$ which is also described by the symbol $[e/s]$. Prove that this yields a natural transformation of the fraction functor associated with S into that associated with S'.

7. Let R, S and S' be as in Exercise 6 so that there is a natural transformation of the fraction functor associated with S into that associated with S'. If now, for each element s' in S', the principal ideal Rs' meets S, prove that the natural transformation is an isomorphism of the first functor on to the second.

8. Let Σ be a non-empty multiplicatively closed subset of R_S and denote by S^* the set of all elements η in R such that $[\eta/s] \in \Sigma$ for some $s \in S$. Prove that S^* is a non-empty multiplicatively closed subset of R and show that there is a natural ring-isomorphism of R_{S^*} on to $(R_S)_\Sigma$.

9. Let the notation be as in Exercise 8 and let the rings R_{S^*} and $(R_S)_\Sigma$ be identified by means of the isomorphism referred to in that exercise. Show that if E is an arbitrary R-module, then E_{S^*} and $(E_S)_\Sigma$ are isomorphic modules with respect to the ring $R_{S^*} = (R_S)_\Sigma$.

10. Let E be an R-module, M a maximal ideal of R and N an M-primary submodule of E. Prove that E/N and E_M/N_M are isomorphic as R-modules.

11. Let Q be a quasi-local integral domain with maximal ideal M and quotient field F. Show that if $\alpha \in F$, $\alpha \neq 0$, and $MQ[\alpha] = Q[\alpha]$, then α^{-1} is integral with respect to Q.

12. Let R be an integral domain and, for each maximal ideal M, let R_M be considered as a subring of the quotient field of R. Then $\bigcap_M R_M = R$.

4

NOETHERIAN RINGS AND MODULES

General remarks. This chapter contains a detailed study of *Noetherian modules* that is of modules which satisfy the maximal condition for submodules.† We continue to suppose (for the duration of Chapter 4) that all rings under consideration are commutative and possess identity elements. However, the reader should note that *each of the results proved in section (4.1) holds for any ring with an identity element even if it is not commutative.* Indeed, the proofs given are valid in the more general situation just as they stand. To make this quite precise, we should add that, when employing the wider interpretation, the terms '*R*-module' and 'ideal' should be taken to mean 'left *R*-module' and 'left ideal' respectively. Null rings are not excluded from the discussion.

4.1 Further consideration of the maximal and minimal conditions

Let *R* be a ring and *E* an *R*-module. In section (1.8) we explained what was meant by the statement that *E* satisfies the maximal or minimal condition for submodules. We shall now examine, in greater detail than before, the implications of these conditions. Of course, a substantial number of results in this direction have already been derived. However, as these are scattered among the pages of the preceding chapters, we shall restate certain of them in order to have related results grouped close together. Naturally the proofs of earlier results will not be repeated.

Lemma 1. *Let* $0 \to E' \to E \to E'' \to 0$ *be an exact sequence of R-modules. If E satisfies the maximal condition for submodules, then so do E' and E''. Conversely, if E' and E'' both satisfy the maximal condition for submodules, then E satisfies the condition as well. The lemma remains true if 'maximal' is replaced by 'minimal' wherever it occurs.*

Proof. If the reader refers back to (1.12.7), he will see that there is no loss of generality in supposing that *E'* is a submodule of *E* and

† See section (1.8).

$E'' = E/E'$. However in that case the lemma is just a repetition of Proposition 11 of section (1.8).

Corollary 1. *Let* $E_1, E_2, ..., E_s$ *be* R-*modules. Then the direct sum* $E_1 \oplus E_2 \oplus ... \oplus E_s$ *satisfies the maximal condition for submodules if and only if each* E_i *satisfies the maximal condition for submodules.*

Proof. First suppose that the direct sum satisfies the maximal condition for submodules. Since E_i may be regarded as a submodule of the direct sum, the lemma shows that it too must satisfy the maximal condition.

The converse is easily proved by induction on s. For if $s > 1$, then we have an exact sequence

$$0 \to E_1 \xrightarrow{\alpha} E_1 \oplus E_2 \oplus ... \oplus E_s \xrightarrow{\beta} E_2 \oplus ... \oplus E_s \to 0$$

in which (with a self-explanatory notation)

$$\alpha(e_1) = \{e_1, 0, 0, ..., 0\}$$

and $$\beta(\{e_1, e_2, e_3, ..., e_s\}) = \{e_2, e_3, ..., e_s\}.$$

With the obvious induction hypothesis, both E_1 and $E_2 \oplus ... \oplus E_s$ satisfy the maximal condition for submodules. Hence, by the lemma, the same holds for $E_1 \oplus E_2 \oplus ... \oplus E_s$.

Corollary 2. *Let* $E_1, E_2, ..., E_s$ *be* R-*modules. Then the direct sum* $E_1 \oplus E_2 \oplus ... \oplus E_s$ *satisfies the minimal condition for submodules if and only if each* E_i *satisfies the minimal condition for submodules.*

The proof is identical with that of Corollary 1 save that the word 'maximal' must be replaced by 'minimal' wherever it occurs.

Proposition 1. *Let* $N_1, N_2, ..., N_s$ *be submodules of an* R-*module* E. *Suppose that, for each* i, N_i *satisfies the maximal resp. minimal condition for submodules. Then the sum* $N_1 + N_2 + ... + N_s$ *also satisfies the maximal resp. minimal condition for submodules.*

Proof. We shall confine our attention to the assertion concerning the maximal condition since that relating to the minimal condition may be established similarly. Let

$$\alpha : N_1 \oplus N_2 \oplus ... \oplus N_s \to N_1 + N_2 + ... + N_s$$

be the mapping in which the sequence $\{n_1, n_2, ..., n_s\}$, where $n_i \in N_i$, is mapped into $n_1 + n_2 + ... + n_s$. This is obviously an epimorphism of R-modules. Since (Lemma 1 Cor. 1) $N_1 \oplus N_2 \oplus ... \oplus N_s$ satisfies the maximal condition for submodules, the proposition follows by applying Lemma 1.

Corollary. *Let E be a finitely generated R-module and suppose that R satisfies the maximal (minimal) condition for ideals. Then E satisfies the maximal (minimal) condition for submodules.*

Remark. This is, in fact, identical with Theorem 8 of section (1.8). We give an alternative demonstration.

Proof. There exist elements $e_1, e_2, \ldots, e_s$ of E such that

$$E = Re_1 + Re_2 + \ldots + Re_s.$$

Consider the mapping $R \to Re_i$ in which r, of R, is carried into re_i. This is an epimorphism of R-modules. Consequently, by Lemma 1, Re_i satisfies the maximal (minimal) condition for submodules. The desired result therefore follows from Proposition 1.

Proposition 2. *Let $N_1, N_2, \ldots, N_s$ be submodules of an R-module E. Suppose that, for each i, E/N_i satisfies the maximal resp. minimal condition for submodules. Then $E/(N_1 \cap N_2 \cap \ldots \cap N_s)$ satisfies the maximal resp. minimal condition for submodules.*

Proof. Once again, and for the same reason, we restrict our attention to the assertion concerning the maximal condition. Let $\phi_i : E \to E/N_i$ be the natural mapping and consider the R-homomorphism

$$E \to (E/N_1) \oplus (E/N_2) \oplus \ldots \oplus (E/N_s)$$

in which e is mapped into $\{\phi_1(e), \phi_2(e), \ldots, \phi_s(e)\}$. The kernel of this homomorphism is $N_1 \cap N_2 \cap \ldots \cap N_s$. Hence, by Theorem 1 of section (1.6), $E/(N_1 \cap N_2 \cap \ldots \cap N_s)$ is isomorphic to a submodule of

$$(E/N_1) \oplus \ldots \oplus (E/N_s).$$

Now, according to Corollary 1 of Lemma 1, the direct sum satisfies the maximal condition for submodules. It therefore follows, this time by Lemma 1, that $E/(N_1 \cap N_2 \cap \ldots \cap N_s)$ also satisfies the maximal condition for submodules. This completes the proof.

One of the main methods used for showing (in appropriate situations) that a particular module satisfies the maximal condition for submodules is to invoke the *Basis Theorem*. This was proved in Chapter 1 but, before we can take full advantage of it, the result must be extended. In preparation for this, we make some general observations concerning polynomials.

To this end let R be a ring and let $X_1, X_2, \ldots, X_n$ be new symbols. These will be referred to as *indeterminates*. An expression of the form

$X_1^{\mu_1} X_2^{\mu_2} \ldots X_n^{\mu_n}$, where $\mu_1, \mu_2, \ldots, \mu_n$ are non-negative integers, will be called a *power-product* in the X_i. Using the procedure described in section (1.13), we now construct the free R-module on the set

$$\{X_1^{\mu_1} X_2^{\mu_2} \ldots X_n^{\mu_n}\}$$

of power-products. This free R-module will be denoted by

$$R[X_1, X_2, \ldots, X_n].$$

A typical element of it is uniquely expressible as a sum

$$\Sigma r_{\mu_1 \mu_2 \ldots \mu_n} X_1^{\mu_1} X_2^{\mu_2} \ldots X_n^{\mu_n},$$

where the coefficients $r_{\mu_1 \mu_2 \ldots \mu_n}$ are elements of R and only a finite number of them are non-zero. Naturally we call an element of

$$R[X_1, X_2, \ldots, X_n]$$

a *polynomial in* $X_1, X_2, \ldots, X_n$ *with coefficients in* R.

The next step is to turn $R[X_1, X_2, \ldots, X_n]$ into a *ring*. First we define the product of $r X_1^{\mu_1} X_2^{\mu_2} \ldots X_n^{\mu_n}$ and $\rho X_1^{\nu_1} X_2^{\nu_2} \ldots X_n^{\nu_n}$, where r, ρ are elements of R, to be

$$(r\rho) X_1^{\mu_1 + \nu_1} X_2^{\mu_2 + \nu_2} \ldots X_n^{\mu_n + \nu_n}$$

and then extend the definition (to general polynomials) by linearity in the obvious manner. The result is to provide $R[X_1, X_2, \ldots, X_n]$ with the structure of a ring having $1_R X_1^0 X_2^0 \ldots X_n^0$ as its identity element. Indeed these ideas are very familiar from elementary algebra and the fact that we are using the elements of a general ring as coefficients brings no complication at this stage. We shall therefore not elaborate the arguments. Note that the polynomials of the form $r X_1^0 X_2^0 \ldots X_n^0$, where r belongs to R, form a subring of $R[X_1, X_2, \ldots, X_n]$ which is isomorphic to R. This enables us to identify R with the ring of 'constant' polynomials.

Suppose now that E is an R-module. We have already, in section (1.8), considered polynomials in a single indeterminate with coefficients in E. We extend this idea so as to include polynomials in $X_1, X_2, \ldots, X_n$ with coefficients in E. These have the form

$$\Sigma e_{\sigma_1 \sigma_2 \ldots \sigma_n} X_1^{\sigma_1} X_2^{\sigma_2} \ldots X_n^{\sigma_n},$$

where $\sigma_1, \sigma_2, \ldots, \sigma_n$ are non-negative integers, the coefficients of the power-products belong to E, and at most a finite number of these coefficients are non-zero. It is clear how these new polynomials are to be added and that, with respect to this addition, they form an abelian group. We denote this group by $E[X_1, X_2, \ldots, X_n]$. The final step is to turn $E[X_1, X_2, ..., X_n]$ into an $R[X_1, X_2, ..., X_n]$-module. This is done by defining the product of $rX_1^{\mu_1} X_2^{\mu_2} ... X_n^{\mu_n}$ and $eX_1^{\sigma_1} X_2^{\sigma_2} ... X_n^{\sigma_n}$, where r belongs to R and e to E, to be

$$(re) X_1^{\mu_1+\sigma_1} X_2^{\mu_2+\sigma_2} ... X_n^{\mu_n+\sigma_n}$$

and extending the definition by linearity. In this way we attach a meaning to the product of an element of $E[X_1, X_2, ..., X_n]$ by one of $R[X_1, X_2, ..., X_n]$. Of course it is necessary to check that the axioms for a module are satisfied. However this presents no difficulty so the details will be omitted.

Assume, for the moment, that $n > 1$ and put

$$R' = R[X_1, X_2, ..., X_{n-1}],$$

$E' = E[X_1, X_2, ..., X_{n-1}]$ so that E' is an R'-module. Let

$$f(X_1, X_2, ..., X_n)$$

be an arbitrary polynomial in $X_1, X_2, ..., X_n$ with coefficients in R. Then $f(X_1, X_2, ..., X_n)$ has a unique representation in the form

$$f_0(X_1, X_2, ..., X_{n-1}) + f_1(X_1, X_2, ..., X_{n-1}) X_n + ...$$
$$+ f_p(X_1, X_2, ..., X_{n-1}) X_n^p.$$

Here $f_i(X_1, X_2, ..., X_{n-1})$ is a polynomial in $X_1, X_2, ..., X_{n-1}$ alone, that is to say it belongs to R'. Thus each element of $R[X_1, X_2, ..., X_n]$ has a unique representation as a polynomial in X_n with coefficients in R'. Because of this we write $R[X_1, X_2, ..., X_n] = R'[X_n]$. In a similar sense, we also have $E[X_1, X_2, ..., X_n] = E'[X_n]$.

Theorem 1. *If the R-module E satisfies the maximal condition for submodules, then the same condition is satisfied by $E[X_1, X_2, ..., X_n]$ when this is considered as a module over the ring $R[X_1, X_2, ..., X_n]$.*

Proof. Put $R_0 = R$, $E_0 = E$ and for $1 \leqslant i \leqslant n$, set

$$R_i = R[X_1, X_2, ..., X_i], \quad E_i = E[X_1, X_2, ..., X_i].$$

Then E_i is an R_i-module. Further, by the above remarks,

$$R_{i+1} = R_i[X_{i+1}] \quad \text{and} \quad E_{i+1} = E_i[X_{i+1}].$$

By Theorem 10 of section (1.8), $E_0[X_1] = E_1$ satisfies the maximal condition for submodules when it is regarded as a module with respect to the ring $R_0[X_1] = R_1$. Next, by the same theorem, $E_1[X_2] = E_2$ satisfies the maximal condition for submodules when it is considered as a module over $R_1[X_2] = R_2$. And so on. Finally it is seen that the

R_n-module E_n satisfies the maximal condition for submodules. However $R_n = R[X_1, X_2, ..., X_n]$ and $E_n = E[X_1, X_2, ..., X_n]$. Thus the theorem is proved.

Corollary. *If the ring R satisfies the maximal condition for ideals, then so does the polynomial ring $R[X_1, X_2, ..., X_n]$.*

This follows from the theorem by taking the module E to be R itself.

4.2 Noetherian modules and Artin rings

We recall that if E is an R-module, then $\operatorname{Ann} E$ or (more explicitly) $\operatorname{Ann}_R E$ is used to denote the annihilator of E. Of course $\operatorname{Ann}_R E$ is an ideal and therefore we can form the residue ring $R/\operatorname{Ann}_R E$. It will now be shown that, when E is finitely generated, the behaviour of E is governed to a considerable extent by that of $R/\operatorname{Ann}_R E$. The reader should note that, for the first time in this chapter, we have arrived at a result which requires us to use the fact that multiplication in R is commutative.

Theorem 2. *Let E be a finitely generated R-module and let $A = \operatorname{Ann}_R E$. Then E satisfies the maximal resp. minimal condition for submodules if and only if the residue ring R/A satisfies the maximal resp. minimal condition for ideals.*

Remark. It is worth while noting that the theorem has the following consequence. *If an R-module E satisfies the maximal condition for submodules, then the ring $R/\operatorname{Ann}_R E$ satisfies the maximal condition for ideals.* For, by Proposition 9 of section (1.8), E will be finitely generated and therefore the theorem may be applied.

Proof. Let $E = Re_1 + Re_2 + ... + Re_s$ and put $A_i = \operatorname{Ann}(Re_i)$. Then $A = A_1 \cap A_2 \cap ... \cap A_s$. Next the R-epimorphism $R \to Re_i$ in which $r \to re_i$ has kernel A_i because R is commutative. It therefore follows that R/A_i and Re_i are isomorphic as R-modules. Now suppose that E satisfies the maximal (minimal) condition for submodules. Since R/A_i is isomorphic to the submodule Re_i of E, Lemma 1 shows that R/A_i satisfies the maximal (minimal) condition for submodules. Accordingly, by Proposition 2,

$$R/(A_1 \cap A_2 \cap ... \cap A_s) = R/A$$

satisfies the maximal (minimal) condition for R-submodules. However the R-submodules of R/A are just the ideals of the ring R/A. Thus R/A satisfies the maximal (minimal) condition for ideals.

It will now be supposed that R/A satisfies the maximal (minimal) condition for ideals. By Proposition 24 of section (1.14), E is an R/A-module and as such it is clearly generated by $e_1, e_2, \ldots, e_s$. It therefore follows, from Proposition 1 Cor., that E satisfies the maximal (minimal) condition for R/A-submodules. However the R/A-submodules of E are the same as its R-submodules. Hence all is proved.

Corollary. *Let E be a finitely generated R-module and let $A = \operatorname{Ann}_R E$. Then the length $L_R(E)$ is finite when and only when the ring R/A satisfies both the maximal and minimal conditions for ideals.*

For by Theorem 7 of section (1.8), we have $L_R(E) < \infty$ if and only if E satisfies both the maximal and minimal conditions for submodules.

At this point we introduce two definitions. These will help us to describe, in a comparatively concise manner, certain situations which will occur many times in the sequel.

Definition. *An R-module E which satisfies the maximal condition for submodules is called a 'Noetherian R-module'. If the ring R satisfies the maximal condition for ideals, then it is called a 'Noetherian ring'.*

Definition. *A ring R which satisfies both the maximal and minimal conditions for ideals is called an 'Artin ring'.*

These names have been given in recognition of the pioneer studies of E. Noether and E. Artin. We make a few comments on the definitions.

By Proposition 9 of section (1.8), an R-module E is Noetherian if and only if all its submodules are finitely generated; equivalently (by the same proposition) E is Noetherian if and only if it satisfies the ascending chain condition for submodules. In particular, we have alternative ways of describing a Noetherian ring. We could, for example, define a Noetherian ring as one in which every ideal is finitely generated. Equally well we could say that a ring is Noetherian when it satisfies the ascending chain condition for ideals.

Next, by Theorem 7 of section (1.8), R is an Artin ring if and only if $L_R(R) < \infty$. For example a field F is an Artin ring; indeed, since (0) and F are the only ideals, $L_F(F) = 1$. Actually it is more usual to call a ring an Artin ring if it satisfies just the minimal condition for ideals. However, as we shall prove later, when the minimal condition for ideals holds, then the maximal condition holds as well.

To illustrate the terminology, let E be a *finitely generated* R-module. Then our earlier results show that E *is a Noetherian R-module if and only if $R/\operatorname{Ann}_R E$ is a Noetherian ring; and $L_R(E)$ is finite if and only if*

$R/\text{Ann}_R E$ is an Artin ring. Again, by Theorem 1 Cor., if R is a Noetherian ring, then so is the polynomial ring $R[X_1, X_2, ..., X_n]$. This latter observation is particularly important when taken into conjunction with

Theorem 3. Let $\phi: R \to R'$ be an epimorphism of a ring R on to a ring R'. If now R is a Noetherian resp. Artin ring, then R' is also a Noetherian resp. Artin ring.

This result is obvious from the one-one correspondence between the R'-ideals and the R-ideals containing $\text{Ker}\,\phi$. To show how it may be combined with our results on polynomial rings, we prove

Theorem 4. Let the Noetherian ring R be a subring of a ring R'. If now $\alpha_1, \alpha_2, ..., \alpha_n$ are elements of R', then $R[\alpha_1, \alpha_2, ..., \alpha_n]$ is a Noetherian ring.

Proof. We obtain a ring-homomorphism of the polynomial ring $R[X_1, X_2, ..., X_n]$ on to $R[\alpha_1, \alpha_2, ..., \alpha_n]$ by means of the mapping in which (with a self-explanatory notation)

$$\Sigma r_{\mu_1 \mu_2 \cdots \mu_n} X_1^{\mu_1} X_2^{\mu_2} \cdots X_n^{\mu_n} \to \Sigma r_{\mu_1 \mu_2 \cdots \mu_n} \alpha_1^{\mu_1} \alpha_2^{\mu_2} \cdots \alpha_n^{\mu_n}.$$

Since $R[X_1, X_2, ..., X_n]$ is Noetherian, the required result follows from Theorem 3.

Corollary. Let F be a field and $\alpha_1, \alpha_2, ..., \alpha_n$ elements belonging to some extension ring of F. Then the ring $F[\alpha_1, \alpha_2, ..., \alpha_n]$ is Noetherian.

In contrast to the Noetherian case, extension rings of Artin rings are unlikely to inherit the Artinian property. However, in this connection the following result is sometimes useful.

Theorem 5. Let R' be an extension ring of an Artin ring R and suppose that R', considered as an R-module, is finitely generated. Then R' is also an Artin ring.

Proof. By Proposition 1 Cor., R' satisfies both the maximal and minimal conditions for R-submodules. Now let Ω be a non-empty set of R'-ideals. Then Ω is also a non-empty set of R-submodules of R'. Accordingly Ω contains both a maximal and a minimal member. This proves the theorem.

The behaviour of Noetherian rings and Artin rings in relation to the formation of fractions is easily established. First we need

Lemma 2. Let E be an R-module and S a non-empty multiplicatively closed subset of R. If now E satisfies the maximal (minimal) condition for

R-submodules, then E_S satisfies the maximal (minimal) condition for R_S-submodules.

This is not a new result. In fact we have merely restated Theorem 3 of section (3.2). Taking $E = R$ we obtain

Theorem 6. *Let S be a non-empty multiplicatively closed subset of the ring R. If now R is a Noetherian resp. Artin ring, then R_S is a Noetherian resp. Artin ring.*

Theorems 3, 4 and 6 provide means whereby, starting from a Noetherian ring of a simple kind, it is possible to construct others of a more sophisticated character. The most obvious examples of elementary Noetherian rings are fields and rings having only a finite number of elements. To these we may add the integers. The theorem, which follows, establishes the Noetherian character of this particular ring. Although both the result and its proof are very familiar, they are included for the sake of completeness.

Theorem 7. *Let Z be the ring consisting of the positive, negative and zero whole numbers. Then each ideal of Z can be generated by a single element.*

Proof. It is clear that we may confine our attention to non-zero ideals. Let A be such an ideal. Then we can find $m \in A$ so that $m \neq 0$. Since both m and $-m$ belong to A and one of them is a *positive* integer, it follows that A contains positive integers. Let d be the smallest of these. We shall show that $A = (d)$. Indeed $(d) \subseteq A$ so it is only necessary to prove inclusion in the opposite sense.

Let $n \in A$. Then (by ordinary division of whole numbers) we can find integers q and r so that $n = qd + r$, where $0 \leqslant r < d$. Now $r = n - qd$ belongs to A because n and d belong to A. However d is the smallest strictly positive integer contained in A and $0 \leqslant r < d$. It follows that $r = 0$ and therefore $n = qd \in (d)$. This shows that $A \subseteq (d)$ and completes the proof.

We now begin an investigation of rings whose ideals satisfy the minimal condition. This will lead us to the result that such a ring is an Artin ring according to the definition given above. In the process, we shall acquire other useful information. First, however, it will be convenient to establish a lemma about vector spaces so as not to interrupt the main argument at an awkward stage. We recall that a vector space is just a module with respect to a field and that the length of a vector space is usually known as its dimension.

Lemma 3. *Let E be a vector space with respect to a field F and suppose that E satisfies the minimal condition for subspaces. Then the dimension of E is finite.*

Proof. By Theorem 21 of section (1.13), it will suffice to show that E is finitely generated. *Assume the contrary.* Then it is possible to find an infinite sequence $e_1 (e_1 \neq 0), e_2, e_3, \dots$ of elements of E such that, for each n,

$$e_{n+1} \notin Fe_1 + Fe_2 + \dots + Fe_n. \tag{4.2.1}$$

It follows that if $a_1 e_1 + a_2 e_2 + \dots + a_p e_p = 0$, where the a_i belong to F, then $a_1 = a_2 = \dots = a_p = 0$. (For otherwise let a_q be the last non-zero coefficient. Then, multiplying by a_q^{-1}, we see that e_q belongs to

$$Fe_1 + Fe_2 + \dots + Fe_{q-1}.$$

This, however, contradicts (4.2.1).)

Denote by V_n the subspace of E generated by the elements $e_n, e_{n+1}, \dots$. Then $e_n \in V_n$ and $e_n \notin V_{n+1}$. (For otherwise we should obtain a non-trivial relation between different e_i's and this, as we have seen, is impossible.) Accordingly

$$V_1 \supset V_2 \supset \dots \supset V_n \supset \dots$$

is an infinite strictly decreasing sequence of subspaces. This, however, violates the hypothesis that E satisfies the minimal condition for subspaces. Thus we have reached a contradiction and the proof is complete.

Proposition 3. *Let R be an integral domain and suppose that R satisfies the minimal condition for ideals. Then R is a field.*

Proof. Suppose that $\alpha \in R, \alpha \neq 0$. We wish to show that α has an inverse in R. Since R satisfies the minimal condition, the decreasing sequence

$$(\alpha) \supseteq (\alpha^2) \supseteq (\alpha^3) \supseteq \dots$$

of ideals must terminate.† Accordingly there exists an integer m such that $(\alpha^m) = (\alpha^{m+1})$. Hence $\alpha^m \in (\alpha^{m+1})$ and therefore $\alpha^m = \beta \alpha^{m+1}$ for some $\beta \in R$. However, R is an integral domain and $\alpha^m \neq 0$. Consequently from $\alpha^m (1 - \beta\alpha) = 0$ follows $\beta\alpha = 1$. This establishes the proposition.

Theorem 8. *Let the ring R satisfy the minimal condition for ideals. Then R has only a finite number of prime ideals and each of them is a maximal ideal.*

† See Proposition 10 of section (1.8).

Proof. We may assume that R is not a null ring for otherwise the theorem is vacuous. Let P be a prime ideal. Then R/P is an integral domain and (in view of the correspondence between the ideals of R/P and the R-ideals containing P) it also satisfies the minimal condition. It follows, by Proposition 3, that R/P is a field. Hence, by Proposition 1 of section (2.2), P is a maximal ideal.

Let Ω be the set of all R-ideals which can be obtained as the intersection of a finite number of prime ideals. Then Ω is not empty and therefore it contains a minimal member. Let $P_1 \cap P_2 \cap \ldots \cap P_s$, where the P_i are prime ideals, be such a minimal member.

Now suppose that P is an arbitrary prime ideal of R. Then

$$P \cap P_1 \cap \ldots \cap P_s$$

belongs to Ω and is contained in $P_1 \cap P_2 \cap \ldots \cap P_s$. Consequently, by the minimal property of the latter,

$$P \cap P_1 \cap P_2 \cap \ldots \cap P_s = P_1 \cap P_2 \cap \ldots \cap P_s$$

and therefore $\quad P \supseteq P_1 \cap P_2 \cap \ldots \cap P_s.$

It follows, from Proposition 4 of section (2.3), that there exists $i\,(1 \leqslant i \leqslant s)$ such that $P \supseteq P_i$. However P_i is a maximal ideal of R and so $P = P_i$. This shows that $P_1, P_2, \ldots, P_s$ are the only prime ideals of R and completes the proof.

We recall that an element of a ring is said to be nilpotent if some positive power of it is zero.

Lemma 4. *Let R be a ring satisfying the minimal condition for ideals and let $P_1, P_2, \ldots, P_s$ be its prime ideals. Then every element of*

$$P_1 \cap P_2 \cap \ldots \cap P_s$$

is nilpotent.

Remark. Actually the lemma is an immediate consequence of the corollary to Theorem 4 of section (2.3). However rather than appeal to the general result we give a simple *ad hoc* argument which is applicable to the present very special situation.

Proof. Let α belong to $P_1 \cap P_2 \cap \ldots \cap P_s$. Since R satisfies the minimal condition, the decreasing sequence

$$(\alpha) \supseteq (\alpha^2) \supseteq (\alpha^3) \supseteq \ldots$$

must terminate. It is therefore possible to find an integer m such that $(\alpha^m) = (\alpha^{m+1})$. Then $\alpha^m \in (\alpha^{m+1})$ whence $\alpha^m = r\alpha^{m+1}$ for some $r \in R$. Consider $1 - r\alpha$. It is clear that this cannot belong to any of $P_1, P_2, \ldots, P_s$;

hence the ideal $R(1-r\alpha)$ is not contained in any maximal ideal. Accordingly $R(1-r\alpha) = R$ and therefore $\rho(1-r\alpha) = 1$ for a suitable element ρ of R. Finally, since $\alpha^m(1-r\alpha) = 0$, multiplication by ρ shows that $\alpha^m = 0$. This completes the proof.

Corollary. *Let the assumptions be as in the lemma and put*

$$I = P_1 P_2 \dots P_s.$$

Then there exists an integer m such that $I^m = (0)$.

Proof. Since R satisfies the minimal condition for ideals, the sequence

$$I \supseteq I^2 \supseteq I^3 \supseteq \dots$$

must terminate. Hence there exists an integer m such that $I^n = I^m$ for all $n \geqslant m$. *We shall now assume that $I^m \neq (0)$ and derive a contradiction.*

Let Ω consist of all ideals A such that $I^m A \neq (0)$. Since

$$I^m I^m = I^{2m} = I^m \neq (0),$$

we have $I^m \in \Omega$. Thus Ω is not empty. Among the ideals in the set Ω we select one, B say, which is minimal. Since $I^m B \neq (0)$, there exists $b \in B$ such that $I^m b \neq (0)$. Then $I^m(b) \neq (0)$ and $(b) \subseteq B$; hence, by the minimal property of B, $B = (b)$. Next

$$I^m(IB) = I^{m+1}B = I^m B \neq (0).$$

Accordingly, $IB \in \Omega$. Since $IB \subseteq B$, we may conclude that $B = IB$. We now have
$$(b) = B = IB = I(b).$$

It follows that there exists $c \in I$ such that $b = bc$. Accordingly

$$b = bc = bc^2 = bc^3 = \dots,$$

whence $b = 0$, because, by the lemma, c is nilpotent. This, however, contradicts the fact that $I^m b \neq (0)$. The proof is now complete.

Theorem 9. *Let R be a ring which satisfies the minimal condition for ideals. Then it also satisfies the maximal condition for ideals.*

Proof. We may suppose that R is not a null ring. Clearly the theorem will follow if we show that $L_R(R) < \infty$.

By Theorem 8, R has only a finite number of prime ideals and each of them is a maximal ideal. Let the different prime ideals be

$$P_1, P_2, \dots, P_s.$$

Then, by the corollary to Lemma 4, there exists an integer m such that $P_1^m P_2^m \ldots P_s^m = (0)$. Put $E_0 = R$ and $E_i = P_1^m P_2^m \ldots P_i^m$ for $1 \leqslant i \leqslant s$. We then have

$$R = E_0 \supseteq E_1 \supseteq E_2 \supseteq \ldots \supseteq E_s = (0)$$

and, by Theorem 6 of section (1.7),

$$L_R(R) = \sum_{i=1}^{s} L_R(E_{i-1}/E_i).$$

It is therefore sufficient to prove that $L_R(E_{i-1}/E_i) < \infty$ for each i.

We fix our attention on a particular value of i and put $P = P_i$ in order to simplify the notation. Then

$$E_{i-1} \supseteq P E_{i-1} \supseteq P^2 E_{i-1} \supseteq \ldots \supseteq P^m E_{i-1} = E_i$$

and therefore $L_R(E_{i-1}/E_i) = \sum_{t=1}^{m} L_R(P^{t-1}E_{i-1}/P^t E_{i-1}).$

It will now be shown that $L_R(P^{t-1}E_{i-1}/P^t E_{i-1})$ is finite and this will complete the proof.

First $P^{t-1}E_{i-1}$ is a submodule of R and therefore it satisfies the minimal condition for submodules. It follows, from Lemma 1, that $P^{t-1}E_{i-1}/P^t E_{i-1}$ also satisfies the minimal condition. However the latter module is annihilated by P and R/P is a field because P is a maximal ideal. Thus $P^{t-1}E_{i-1}/P^t E_{i-1}$ is a vector space over R/P and its vector subspaces are the same as its R-submodules. Hence it satisfies the minimal condition for subspaces. Accordingly it follows, from Lemma 3, that the dimension of $P^{t-1}E_{i-1}/P^t E_{i-1}$ (that is its length as an R/P-module) is finite. But the length of $P^{t-1}E_{i-1}/P^t E_{i-1}$ as an R/P-module is the same as its length as an R-module. Thus all is proved.

Theorem 10. *Let $R_1, R_2, \ldots, R_s$ be given rings and let R be their direct sum. If now each R_i is a Noetherian resp. Artin ring, then R is also a Noetherian resp. Artin ring.*

Proof. We have $R = R_1 \oplus R_2 \oplus \ldots \oplus R_s$ and, because each R_i is a commutative ring with an identity element, so too is R. Next, as we saw in section (1.10), each R_i may be regarded (in a natural way) as an ideal of R. This is achieved by identifying the element r_i of R_i with the element $\{0, \ldots, r_i, \ldots, 0\}$ of the direct sum, where naturally r_i occurs in the ith place. Let this identification be made. Then a subset of R_i is an ideal of the ring R if and only if† it is an ideal of the

---

† We have had occasion to make this observation before. See the proof of Lemma 12 in section (1·11), where the argument is given in greater detail.

ring R_i. It follows that the ring R_i satisfies the maximal (minimal) condition for ideals if and only if the R-module R_i satisfies the maximal (minimal) condition for submodules.

Now suppose that each of the rings R_i satisfies the maximal condition for ideals. Then as R-modules they each satisfy the maximal condition. Hence, by Lemma 1 Cor. 1, $R = R_1 \oplus R_2 \oplus ... \oplus R_s$ satisfies the maximal condition for submodules. However this is just another way of saying that R is a Noetherian ring.

The arguments of the last paragraph remain valid if 'maximal' is replaced by 'minimal' and we substitute the second corollary of Lemma 1 for the first. Thus if each R_i satisfies the minimal condition for ideals, then so does R. This establishes the theorem.

Theorem 10 shows that a finite direct sum of Artin rings is again an Artin ring. It will presently be shown that a general Artin ring can be expressed as a direct sum of Artin rings of a comparatively simple kind. First, however, it will be convenient to interpose a few remarks concerning what are called *comaximal* ideals.

To explain this concept, let R be an arbitrary ring (commutative and with identity element) and let A, B be ideals of R.

Definition. *The ideals A, B are said to be 'comaximal' if $A + B = R$.*

Lemma 5. *Let A, B be comaximal ideals. Then $AB = A \cap B$.*

Proof. It is enough to show that $A \cap B \subseteq AB$. Since $A + B = R$, there exist $a \in A$ and $b \in B$ such that $a + b = 1$. Now let x belong to $A \cap B$. Then $x = ax + bx$. But $ax \in AB$ and $bx \in AB$. Accordingly $x \in AB$ and the lemma follows.

Proposition 4. *Let R be an arbitrary ring (commutative and with identity element) and let $A_1, A_2, ..., A_n$ be ideals such that $A_i + A_j = R$ whenever $i \neq j$. Then, for each i $(1 \leqslant i \leqslant n)$,*

$$A_i + (A_1...A_{i-1}A_{i+1}...A_n)$$
$$= A_i + (A_1 \cap ... \cap A_{i-1} \cap A_{i+1} \cap ... \cap A_n) = R. \quad (4.2.2)$$

Furthermore $\quad A_1 A_2 ... A_n = A_1 \cap A_2 \cap ... \cap A_n \quad (4.2.3)$

and the rings

$$R/(A_1 \cap A_2 \cap ... \cap A_n) \quad and \quad R/A_1 \oplus R/A_2 \oplus ... \oplus R/A_n$$

are isomorphic.

Proof. Since $A_1 \ldots A_{i-1} A_{i+1} \ldots A_n$ is contained in

$$A_1 \cap \ldots \cap A_{i-1} \cap A_{i+1} \cap \ldots \cap A_n,$$

the relation (4.2.2) will be established if we can show that

$$A_i + (A_1 \ldots A_{i-1} A_{i+1} \ldots A_n) = R.$$

Naturally it is enough to prove this for the case $i = 1$.

Suppose that $2 \leqslant j \leqslant n$. Then $A_1 + A_j = R$. Hence there exist elements $a_{1j} \in A_1, \alpha_j \in A_j$, such that $a_{1j} + \alpha_j = 1$. It follows that

$$(a_{12} + \alpha_2)(a_{13} + \alpha_3) \ldots (a_{1n} + \alpha_n) = 1$$

and therefore $a_1 + \alpha_2 \alpha_3 \ldots \alpha_n = 1$, where a_1 belongs to A_1. This shows that $A_1 + A_2 A_3 \ldots A_n = R$ and establishes (4.2.2).

The next step will be to prove (4.2.3) by induction on n. When $n = 1$ there is nothing to prove and when $n = 2$ the required result follows from Lemma 5. Now assume that $n > 2$ and that the relation in question has been proved for all smaller values of the inductive variable. By the induction hypothesis,

$$A_1 A_2 \ldots A_{n-1} = A_1 \cap A_2 \cap \ldots \cap A_{n-1};$$

consequently

$$A_1 A_2 \ldots A_{n-1} A_n = (A_1 \cap A_2 \cap \ldots \cap A_{n-1}) A_n.$$

However, by (4.2.2), $A_1 \cap \ldots \cap A_{n-1}$ and A_n are comaximal. Hence, by Lemma 5,

$$(A_1 \cap A_2 \cap \ldots \cap A_{n-1}) A_n = A_1 \cap A_2 \cap \ldots \cap A_{n-1} \cap A_n.$$

Accordingly $A_1 A_2 \ldots A_n = A_1 \cap A_2 \cap \ldots \cap A_n$ and (4.2.3) follows.

Finally let $\phi_i : R \to R/A_i$ be the natural mapping and let

$$\phi : R \to R/A_1 \oplus R/A_2 \oplus \ldots \oplus R/A_n$$

be defined by $\quad \phi(r) = \{\phi_1(r), \phi_2(r), \ldots, \phi_n(r)\}.$

This is clearly a ring-homomorphism. Its kernel is $A_1 \cap A_2 \cap \ldots \cap A_n$. If therefore we can show that ϕ is an epimorphism, then the desired isomorphism will be a consequence of Theorem 22 of section (1.14). However, ϕ is an epimorphism because of (4.2.2). Indeed the reasons why this is so are set out at length in the proof of Proposition 17 of section (1.9). They will therefore not be repeated.

In order to simplify the statement of the next theorem, we make the following

Definition. *An Artin ring which possesses exactly one prime ideal will be called a 'primary' Artin ring.*

Theorem 11. *A non-null Artin ring is isomorphic to the direct sum of a finite number of primary Artin rings.*

Proof. Let R be a non-null Artin ring. By Theorem 8, R has only a finite number of prime ideals and each of them is a maximal ideal. Denote the different prime ideals by $P_1, P_2, \ldots, P_s$. By the corollary to Lemma 4, there exists an integer m such that $P_1^m P_2^m \ldots P_s^m = (0)$. Now when $i \neq j$, P_i^m and P_j^m are comaximal. For if $P_i^m + P_j^m$ were different from R, then it would be contained in a maximal ideal. However the only maximal ideal which contains P_i^m is P_i and the only maximal ideal containing P_j^m is P_j. Since these are different, $P_i^m + P_j^m = R$.

It follows, from Proposition 4, that

$$P_1^m \cap P_2^m \cap \ldots \cap P_s^m = P_1^m P_2^m \ldots P_s^m = (0).$$

Hence, by the same proposition, R and

$$(R/P_1^m) \oplus (R/P_2^m) \oplus \ldots \oplus (R/P_s^m)$$

are isomorphic as rings. By Theorem 3, R/P_i^m is an Artin ring. It only remains for us to show that this ring is primary.

By Proposition 2 of section (2·2), the prime ideals of R/P_i^m correspond to the prime ideals of R which contain P_i^m. But P_i is the only prime R-ideal with this property. Hence P_i/P_i^m is the only prime ideal of R/P_i^m. Thus R/P_i^m is a primary Artin ring and all is proved.

4.3 Noetherian graded modules

We shall now disgress a little in order to consider Noetherian modules in the graded case. The basic definitions relating to grading monoids, graded rings and graded modules are all to be found in section (2.11).

Let Γ be a grading monoid, $R = \sum_{\gamma \in \Gamma} R^{(\gamma)}$ a Γ-graded ring, and

$$E = \sum_{\gamma \in \Gamma} E^{(\gamma)}$$

a Γ-graded module with respect to R. By the corollary to Theorem 21 of section (2.11), $R^{(0)}$ is a subring of R. Clearly each $E^{(\gamma)}$ has a natural structure as an $R^{(0)}$-module.

Theorem 12. *Let $E = \sum_{\gamma} E^{(\gamma)}$ be a Γ-graded module with respect to the Γ-graded ring $R = \sum_{\gamma} R^{(\gamma)}$. If now E is a Noetherian R-module, then, for each $\gamma \in \Gamma$, $E^{(\gamma)}$ is a Noetherian $R^{(0)}$-module.*

Proof. Throughout the proof we shall fix our attention on a particular γ belonging to Γ. Suppose that U is an $R^{(0)}$-submodule of $E^{(\gamma)}$. By Proposition 9 of section (1.8), it will suffice to show that U is finitely generated.

Denote by K the R-submodule of E generated by the elements of U. Since E is a Noetherian R-module, K is finitely generated; say

$$K = Rx_1 + Rx_2 + \ldots + Rx_n.$$

Let x denote any one of $x_1, x_2, \ldots, x_n$. Then x can be expressed in the form

$$x = r_1' u_1' + r_2' u_2' + \ldots + r_p' u_p',$$

where $u_i' \in U$ and $r_i' \in R$. Thus each of $x_1, x_2, \ldots, x_n$ is a linear combination (with coefficients in R) of a *finite number* of elements of U. Accordingly there exists a finite subset $\{u_1, u_2, \ldots, u_s\}$ of U such that, for each $i\, (1 \leqslant i \leqslant n)$, we have a relation

$$x_i = r_{i1} u_1 + r_{i2} u_2 + \ldots + r_{is} u_s,$$

where the r_{ij} are elements of R. It follows that

$$K = Ru_1 + Ru_2 + \ldots + Ru_s.$$

We contend that $U = R^{(0)}u_1 + R^{(0)}u_2 + \ldots + R^{(0)}u_s$. For suppose that $u \in U$. Then $u \in K$ and therefore

$$u = r_1 u_1 + r_2 u_2 + \ldots + r_s u_s \qquad (4.3.1)$$

for suitable elements $r_1, r_2, \ldots, r_s$ belonging to R. Now r_i is uniquely expressible as a sum of homogeneous elements of different degrees; say $r_i = \sum_{\lambda} r_i^{(\lambda)}$. If therefore we compare homogeneous components of degree γ on the two sides of (4.3.1), then we find that

$$u = r_1^{(0)} u_1 + r_2^{(0)} u_2 + \ldots + r_s^{(0)} u_s.$$

This shows that $U \subseteq R^{(0)}u_1 + R^{(0)}u_2 + \ldots + R^{(0)}u_s$ and, as the opposite inclusion is obvious, the theorem is proved.

By letting E coincide with R we at once obtain

Theorem 13. *Let $R = \sum_{\gamma} R^{(\gamma)}$ be a Γ-graded ring. If now R is a Noetherian ring, then $R^{(0)}$ is also a Noetherian ring and each $R^{(\gamma)}$ is a Noetherian $R^{(0)}$-module.*

4.4 Submodules of a Noetherian module

We return now to the study of modules which are not necessarily graded.

Let E be a Noetherian R-module. Put $A = \mathrm{Ann}_R E$. Then, by the remark which immediately follows the statement of Theorem 2, the ring $\bar{R} = R/A$ is Noetherian. Note that E is a Noetherian $\bar{R}$-module as well as a Noetherian R-module.

We shall use ϕ to denote the natural mapping $R \to \bar{R}$ though if r belongs to R we shall also use $\bar{r}$ as an alternative to $\phi(r)$. Thus if $e \in E$, then

$$re = \bar{r}e. \tag{4.4.1}$$

Let K be a submodule of the R-module E, let B be an ideal of R and put $\bar{B} = \phi(B)$ so that $\bar{B}$ is an ideal of $\bar{R}$. It follows from (4.4.1) that

$$K:_E B = K:_E \bar{B} \tag{4.4.2}$$

and $$BE = \bar{B}E. \tag{4.4.3}$$

Since $\phi(B^m) = (\bar{B})^m$ we can immediately generalize (4.4.3) to

$$B^m E = (\bar{B})^m E \tag{4.4.4}$$

which holds for all positive integers m.

Theorem 14. *Let E be a Noetherian R-module and K a submodule of E. Then K can be expressed as the intersection of a finite number of primary submodules of E.*

This is just Theorem 20 of section (2.10) restated with a slight change of language. At the time when that result was established it was remarked that, for modules satisfying the maximal condition, the theory of primary decompositions could be developed considerably beyond the point reached in Chapter 2. We shall now concern ourselves with some aspects of this further development.

Since K admits a primary decomposition in E, it will possess a normal (Lasker) decomposition.† Let

$$K = N_1 \cap N_2 \cap \ldots \cap N_s, \tag{4.4.5}$$

where N_i is a P_i-primary submodule of E, be a normal decomposition. Then the prime ideals $P_1, P_2, \ldots, P_s$, which are uniquely determined by virtue of Theorem 13 of section (2.9), are the prime ideals belonging to K.

† See section (2.9).

It is important to note, because we shall make considerable use of the fact, that each of the prime ideals belonging to K contains the ideal $K:E$ or, what is the same thing, $\mathrm{Ann}_R(E/K)$.

The above assertion is true because

$$K:E \subseteq N_i:E \subseteq P_i.$$

Now $A \subseteq K:E$. *Consequently each of the prime ideals belonging to the submodule K of E contains the annihilator of E.*

Put $\bar{P}_i = P_i/A$ so that $\bar{P}_i$ is a prime ideal of the ring $\bar{R}$. By the corollary to Proposition 25 of section (2.8), N_i is a $\bar{P}_i$-primary submodule of E when E is considered as an $\bar{R}$-module. Accordingly, *if we consider E as an $\bar{R}$-module, then the prime ideals belonging to the $\bar{R}$-submodule K are $\bar{P}_1, \bar{P}_2, ..., \bar{P}_s$, and* (4.4.5) *is still a normal decomposition.*

Corollary 1. *Let K be a submodule of a Noetherian R-module E and let B be an ideal of the ring R. Then $K:_E B = K$ if and only if B is not contained in any of the prime ideals belonging to the submodule K.*

Proof. If B is not contained in any of the prime ideals belonging to K, then $K:_E B = K$ by Theorem 15 of section (2.9).

Now assume that $K:_E B = K$ and let us make use of the notation employed above. By (4.2.2), $K:_E \bar{B} = K$. However $\bar{R}$ is a Noetherian ring, consequently $\bar{B}$ is a finitely generated ideal. Hence, by Theorem 16 of section (2.9), $\bar{B}$ is not contained in any of the prime ideals $\bar{P}_1, \bar{P}_2, ..., \bar{P}_s$ belonging to the $\bar{R}$-submodule K. But then B is not contained in any of $P_1, P_2, ..., P_s$. As these are just the prime ideals belonging to the R-submodule K, this completes the proof.

We note the following special case of the result just proved.

Corollary 2. *Let K be a submodule of a Noetherian R-module E and let α be an element of R. Then $K:_E \alpha = K$ if and only if α is not contained in any of the prime ideals belonging to K.*

By way of proof, it is only necessary to observe that $K:_E \alpha = K:_E R\alpha$. This enables us to apply Cor. 1.

Proposition 5. *Let N be a P-primary submodule of a Noetherian R-submodule E. Then there exists an integer m such that $P^m E \subseteq N$.*

Proof. As before we let A denote the annihilator of the R-module E and write $\bar{R} = R/A$. Then $A \subseteq N:E \subseteq P$. Put $\bar{P} = P/A$ so that N is also a $\bar{P}$-primary submodule of E considered as an $\bar{R}$-module.† Thus $\bar{P}$ is the radical of the $\bar{R}$-ideal $N:E$. Since $\bar{R}$ is a Noetherian ring, $\bar{P}$ is a

---

† See Proposition 25 Cor. of section (2.8).

finitely generated ideal. It follows, by Propostion 8 of section (2.3), that there exists a positive integer m such that $(\bar{P})^m$ is contained in the $\bar{R}$-ideal $N:E$. Accordingly $(\bar{P})^m E \subseteq N$ and therefore, by (4.4.4), $P^m E \subseteq N$. This completes the proof.

For the next result it is convenient to introduce a convention. If we have a finite set of ideals belonging to the same ring, then we know what is meant by the product of the ideals provided the set is not empty. The convention we need is as follows. *If we are concerned with a finite set of ideals of a ring R and the set turns out to be empty, then the product of the ideals is to be interpreted as R.*

Proposition 6. *Let K be a submodule of a Noetherian R-module E and let $P_1, P_2, ..., P_l$ be the minimal prime ideals belonging to K. Then there exists an integer m such that $(P_1 P_2 ... P_l)^m E \subseteq K$.*

Remark. Since the submodule K is decomposable, the concept of the minimal prime ideals of K is applicable.† Indeed if $K \neq E$, then K possesses at least one minimal prime ideal. When $K = E$, the set of minimal prime ideals is empty, but, in that case, the proposition is clearly correct in view of the convention which has just been introduced.

Proof. For the purposes of the proof we may suppose that $K \neq E$. Let $K = N_1 \cap N_2 \cap ... \cap N_s$, where N_i is a P_i-primary submodule of E, be a normal decomposition of K. Without loss of generality we may assume that the numbering of the primary components is done in such a way that $P_1, P_2, ..., P_l$ are the minimal prime ideals of K. Then $P_1 P_2 ... P_l \subseteq P_i$ for $i = 1, 2, ..., s$. By Proposition 5, there exists, for each prime ideal P_i, a positive integer m_i such that $P_i^{m_i} E \subseteq N_i$. Put $m = \max(m_1, m_2, ..., m_s)$. Then

$$(P_1 P_2 ... P_l)^m E \subseteq P_i^m E \subseteq P_i^{m_i} E \subseteq N_i$$

for $i = 1, 2, ..., s$. Thus

$$(P_1 P_2 ... P_l)^m E \subseteq N_1 \cap N_2 \cap ... \cap N_s = K$$

and all is proved.

Proposition 7. *Let K be a submodule of a Noetherian R-module E. Then the following statements are equivalent:*

(a) $L_R(E/K) < \infty$;

(b) *every prime ideal containing $K:E$ is a maximal ideal;*

(c) *there exists a finite set $M_1, M_2, ..., M_n$ ($n \geqslant 0$) of maximal ideals (not necessarily distinct) such that $M_1 M_2 ... M_n E \subseteq K$.*

† See section (2.9) for the definition.

Proof. We shall give a cyclic argument. First suppose that $L_R(E/K)$ is finite. Since $K:E$ is just the annihilator of E/K and E/K is a Noetherian (and therefore finitely generated) R-module, the corollary of Theorem 2 shows that $R/(K:E)$ is an Artin ring. Now let P be a prime ideal of R containing $K:E$. Then $P/(K:E)$ is a prime ideal of the Artin ring $R/(K:E)$ and therefore, by Theorem 8, a maximal ideal of that ring. It now follows, from Proposition 2 of section (2.2), that P is a maximal ideal of R. Thus (a) implies (b).

Now assume that (b) is true and let $P_1, P_2, \ldots, P_l$ be the minimal prime ideals belonging to the submodule K of E. Each of these contains† $K:E$ hence each is a maximal ideal of R. Moreover Proposition 6 shows that there exists an integer m such that $(P_1 P_2 \ldots P_l)^m E \subseteq K$. Accordingly (b) implies (c).

Finally assume that (c) is true. Since we wish to prove (a) we may confine our attention to the case $K \neq E$ in which situation the set $M_1, M_2, \ldots, M_n$ of maximal ideals is not empty. Put $E_0 = E$ and

$$E_i = M_1 M_2 \ldots M_i E \quad \text{for} \quad 1 \leqslant i \leqslant n.$$

If now $0 \leqslant i < n$, then

$$L_R(E_i/E_{i+1}) = L_R(E_i/M_{i+1}E_i) = L_{R/M_{i+1}}(E_i/M_{i+1}E_i)$$

because $E_i/M_{i+1}E_i$ is annihilated by M_{i+1}. Since E is Noetherian, E_i and hence also $E_i/M_{i+1}E_i$ are finitely generated R-modules. It follows that $E_i/M_{i+1}E_i$ is a finitely generated R/M_{i+1}-module. However R/M_{i+1} is a field because M_{i+1} is a maximal ideal. Accordingly, by Theorem 21 of section (1.13), $L_{R/M_{i+1}}(E_i/M_{i+1}E_i)$ is finite. This proves that $L_R(E_i/E_{i+1}) < \infty$. Since

$$L_R(E/M_1 M_2 \ldots M_n E) = \sum_{i=0}^{n-1} L_R(E_i/E_{i+1}) < \infty$$

and $M_1 M_2 \ldots M_n E \subseteq K \subseteq E$, it follows that $L_R(E/K) < \infty$. The proof is now complete.

The special case obtained by taking E to be R and K to be an ideal is worth noting. This is recorded in the

Corollary. *Let A be an ideal of a Noetherian ring R. Then the following statements are equivalent:*

(a) *$L_R(R/A) < \infty$ (equivalently R/A is an Artin ring);*
(b) *every prime ideal containing A is a maximal ideal of R;*
(c) *there exists a finite set $M_1, M_2, \ldots, M_n$ ($n \geqslant 0$) of maximal ideals (not necessarily distinct) such that $M_1 M_2 \ldots M_n \subseteq A$.*

† See the remarks in italics following Theorem 14.

We shall now digress in order to give an application of the above corollary to an interesting result in ring theory.

Let R be an integral domain, F its quotient field† and Λ the integral closure‡ of R in F. If R is Noetherian, then it does not follow that Λ is also Noetherian. However, as we shall show, this is the case when one adds the special assumption that every non-zero prime ideal of R is a maximal ideal. Indeed, in these circumstances, every subring of Λ containing R is Noetherian.

Until the next theorem has been stated we shall assume that *R is a Noetherian integral domain having the property that every non-zero prime ideal is a maximal ideal.* If now A is a non-zero ideal of R then, by the corollary to Proposition 7, $L_R(R/A) < \infty$.

Lemma 6. *Let the integral domain R' be an integral extension of R. Then (assuming R to be as described above) every non-zero prime ideal of R' is a maximal ideal.*

Proof. Let P' be a non-zero prime ideal of R'. The zero ideal $0R'$ is a prime ideal whose contraction in R is the zero ideal of R. Since $0R'$ is strictly contained in P', it follows, from Proposition 10 of section (2.6), that $P' \cap R \neq (0)$. Thus $P' \cap R$ is a non-zero prime ideal of R and therefore (by hypothesis) a maximal ideal. We may now conclude, by appealing to Theorem 11 of section (2.6), that P' is a maximal ideal of R'. This completes the proof.

Let Λ be the integral closure of R in its quotient field F and let R' be a subring of Λ containing R. Thus $R \subseteq R' \subseteq \Lambda$.

Lemma 7. *Let the situation be as described above and suppose that R', considered as an R-module, is finitely generated. Then R' satisfies the same conditions as were assumed to hold for R. Further if $x \in R$, $x \neq 0$, then $L_{R'}(R'/xR') \leqslant L_R(R/xR) < \infty$.*

Proof. R' is Noetherian by Theorem 4 and its non-zero prime ideals are all maximal by virtue of Lemma 6. This establishes the first assertion.

Let x be a non-zero element of R. Since xR is a non-zero ideal of R, $L_R(R/xR) < \infty$. Similarly $L_{R'}(R'/xR')$ is finite.

By hypothesis, R' is a finitely generated R-module. Hence

$$R' = R\xi_1 + R\xi_2 + \ldots + R\xi_p \text{ (say)},$$

† See section (3.3).
‡ This concept is defined in section (2.5).

where each ξ_i belongs to the quotient field of R. Accordingly $\xi_i = a_i/b_i$, where a_i, b_i belong to R and $b_i \neq 0$. Put $c = b_1 b_2 \ldots b_p$. Then $c \in R$, $c \neq 0$, and $cR' \subseteq R$. It follows that, for every positive integer n,

$$cx^n R \subseteq cx^n R' \subseteq cR' \subseteq R.$$

Now $L_{R'}(cR'/cx^n R') \leqslant L_R(cR'/cx^n R')$ because any R'-ideal between cR' and $cx^n R'$ is also an R-ideal. This shows that

$$L_{R'}(cR'/cx^n R') \leqslant L_R(R/cx^n R) = L_R(R/cR) + L_R(cR/cx^n R). \quad (4.4.6)$$

The mapping $R \to cR$ in which $r \to cr$ is an isomorphism of R-modules and it makes $x^n R$ correspond to $cx^n R$. Accordingly $R/x^n R$ and $cR/cx^n R$ are isomorphic R-modules and therefore

$$L_R(cR/cx^n R) = L_R(R/x^n R).$$

Again, from the chain

$$x^n R \subseteq x^{n-1} R \subseteq \ldots \subseteq xR \subseteq R$$

of R-ideals, we conclude that

$$L_R(R/x^n R) = \sum_{i=0}^{n-1} L_R(x^i R/x^{i+1} R).$$

However the mapping $R \to x^i R$ defined by $r \to x^i r$ is also an isomorphism of R-modules and it makes xR correspond to $x^{i+1} R$. Accordingly R/xR and $x^i R/x^{i+1} R$ are isomorphic, whence

$$L_R(x^i R/x^{i+1} R) = L_R(R/xR).$$

It follows that

$$L_R(cR/cx^n R) = L_R(R/x^n R) = nL_R(R/xR)$$

and for entirely similar reasons

$$L_{R'}(cR'/cx^n R') = nL_{R'}(R'/xR').$$

Substituting in (4.4.6) we obtain

$$nL_{R'}(R'/xR') \leqslant L_R(R/cR) + nL_R(R/xR)$$

which holds for all positive integers n. However $L_R(R/cR)$, $L_R(R/xR)$ and $L_{R'}(R'/xR')$ are all finite. Consequently we may conclude that

$$L_{R'}(R'/xR') \leqslant L_R(R/xR)$$

by letting n tend to infinity.

Theorem 15. *Let R be a Noetherian integral domain, with quotient field F, having the property that every non-zero prime ideal is a maximal ideal. Let Λ be the integral closure of R in F and let R^* be any subring of Λ containing R $(R \subseteq R^* \subseteq \Lambda)$. Then R^* is Noetherian and its non-zero prime ideals are maximal ideals.*

Proof. The assertion concerning the prime ideals of R^* follows from Lemma 6. It is therefore only necessary to show that R^* is Noetherian.

Assume the contrary. Then there exists an infinite strictly increasing sequence
$$A_0^* \subset A_1^* \subset A_2^* \subset \dots$$
of proper R^*-ideals and, without loss of generality, we may suppose that $A_0^* \neq (0)$. Select a non-zero element of A_0^*. This belongs to F and therefore it can be expressed in the form x/c, where x, c belong to R and neither is zero. Put $p = L_R(R/xR)$.

For each $i\,(i = 1, 2, \dots, p)$ there exists $\alpha_i \in A_i^*$ such that $\alpha_i \notin A_{i-1}^*$. Since α_i belongs to Λ it is integral with respect to R and therefore, by Lemma 7 Cor. of section (2.5), $R[\alpha_1, \alpha_2, \dots, \alpha_p] = R'$ (say) is a finitely generated R-module. Now $R \subseteq R' \subseteq R^* \subseteq \Lambda$. We can therefore apply Lemma 7 which shows that
$$L_{R'}(R'/xR') \leqslant L_R(R/xR) = p.$$
However, by construction,
$$A_0^* \cap R' \subset A_1^* \cap R' \subset \dots \subset A_p^* \cap R' \subset R'$$
is a strictly increasing sequence of R'-ideals and $xR' \subseteq A_0^* \cap R'$. Accordingly $L_{R'}(R'/xR') \geqslant p+1$. This gives a contradiction and completes the proof.

The theorem just proved has a useful extension. Before this is given, it is necessary to make a few preliminary observations.

Let R and R', where R is a subring of R', be integral domains. Denote the quotient field of R' by F' and regard R' as being embedded in F'. Then the quotient field F of R can be identified with the subfield of F' consisting of all elements of the form a/b, where a, b belong to R and $b \neq 0$. After this identification, F' may be regarded as a vector space over the field F and therefore the dimension $[F':F]$ is defined.

Theorem 16. *Let R be a Noetherian integral domain with the property that every non-zero prime ideal is a maximal ideal. Let R' be both an integral domain and an integral extension of R. If now $[F':F] < \infty$, where F and F' are the quotient fields of R and R' respectively, then R' is a Noetherian ring and all its non-zero prime ideals are maximal ideals.*

Proof. Since $[F':F] < \infty$, F' is finitely generated over the field F and therefore, by Theorem 21 of section (1.13), F' is a free F-module with a base $\eta_1, \eta_2, \dots, \eta_n$ of $n = [F':F]$ elements. Each η_i belongs to the quotient field of R' and therefore it can be expressed in the form

$\eta_i = u_i/v_i$, where u_i, v_i belong to R' and $v_i \ne 0$. Put $w = v_1 v_2 \dots v_n$. Then $w \in R'$, $w \ne 0$, and $\eta_i = \lambda_i/w$ ($1 \le i \le n$), where $\lambda_i \in R'$.

We contend that $\lambda_1, \lambda_2, \dots, \lambda_n$ is a base (with respect to F) of

$$F\lambda_1 + F\lambda_2 + \dots + F\lambda_n.$$

For if $f_1\lambda_1 + f_2\lambda_2 + \dots + f_n\lambda_n = 0$, where $f_i \in F$, then

$$f_1\eta_1 + f_2\eta_2 + \dots + f_n\eta_n = 0$$

and therefore $f_1 = f_2 = \dots = f_n = 0$ because the η_i constitute a base for F'. Accordingly, by Theorem 21 of section (1.13),

$$[F\lambda_1 + \dots + F\lambda_n : F] = n = [F' : F].$$

Bearing in mind that $F\lambda_1 + F\lambda_2 + \dots + F\lambda_n \subseteq F'$ it now follows, from Theorem 6 Cor. of section (1.7), that

$$F\lambda_1 + F\lambda_2 + \dots + F\lambda_n = F'. \tag{4.4.7}$$

Put $R'_0 = R[\lambda_1, \lambda_2, \dots, \lambda_n]$. By Theorem 4, R'_0 is Noetherian and, since $R \subseteq R'_0 \subseteq R'$, it is an integral extension of R. It therefore follows, from Lemma 6, that every non-zero prime ideal of R'_0 is maximal.

Again, since $R \subseteq R'_0 \subseteq F'$, the quotient field of R'_0 is contained in F' and it contains both F and the elements $\lambda_1, \lambda_2, \dots, \lambda_n$. Hence, by (4.4.7), the quotient field of R'_0 is F'. Of course, R' is an integral extension of R'_0 because it is already an integral extension of R. Accordingly R' is between R'_0 and the integral closure of R'_0 in its quotient field F'. We can therefore apply Theorem 15 with R'_0, R' and F' taking over the roles of R, R^* and F respectively. This yields the desired result.

4.5 Lengths and primary submodules

In this section we obtain results which make possible further applications of the concept of length. Usually these applications will require that the Noetherian condition be satisfied but the next proposition, which contains preparatory material, embodies a very general idea not dependent on this particular hypothesis.

Proposition 8. *Let N be a P-primary submodule of an R-module E and let K be a proper submodule of N. Then K is a P-primary submodule of E if and only if it is a P-primary submodule of N.*

Proof. First suppose that K is a P-primary submodule of E. Since K does not contain N, $K \cap N = K$ is a P-primary submodule of N by virtue of Proposition 22 of section (2.8).

Now suppose that K is a P-primary submodule of N. Assume that $re \in K$, $e \notin K$, where $r \in R$ and $e \in E$. We shall distinguish two possibilities namely either (i) $e \notin N$, or (ii) $e \in N$. In the former case we have $re \in N$, $e \notin N$. Hence, by Proposition 19 of section (2.8), $r \in P$. In the latter case we have $e \in N$, $re \in K$, $e \notin K$. Since K is a P-primary submodule of N, we may apply the same proposition as before and conclude that r belongs to P. Thus, in either event, $r \in P$.

Again $P = \operatorname{Rad}(N:E) = \operatorname{Rad}(K:N)$. Consequently there exist positive integers m, n such that $r^m E \subseteq N$ and $r^n N \subseteq K$. It follows that $r^{m+n} E \subseteq K$. This proves that K is a primary submodule of E. Let it be P'-primary. Then, by Proposition 22 of section (2.8), K is a P'-primary submodule of N. It follows that $P' = P$ and so all is proved.

Proposition 9. *Let E be a Noetherian R-module, M a maximal ideal of R and N a proper submodule of E. Then N is an M-primary submodule of E if and only if N contains $M^k E$ for some positive integer k. It follows that if N is an M-primary submodule of E, then so is every proper submodule of E containing N.*

Proof. Assume that $M^k E \subseteq N$. We shall show that N is an M-primary submodule of E. The assertion of the proposition which goes in the reverse direction follows immediately from Proposition 5.

Let P be an arbitrary prime ideal belonging to the submodule N. Then $M^k \subseteq N:E \subseteq P$, whence $M \subseteq P$. But M is a maximal ideal; consequently $M = P$. Thus M is the only prime ideal belonging to N, i.e. N is M-primary. This completes the proof.

Corollary. *Let R be a Noetherian ring, M a maximal ideal and A a proper ideal. Then A is an M-primary ideal if and only if A contains a power of M.*

This follows from the proposition by taking $E = R$.

We take this opportunity to digress a little in order to put on record another useful property of primary submodules which belong to a maximal ideal.

Proposition 10. *Let E be an R-module and N an M-primary submodule of E, where M is a maximal ideal. Then E/N and E_M/N_M are isomorphic R-modules.*

Remark. By E_M, of course, we mean the result of localizing E at M. This is not only an R_M-module but also an R-module. Indeed the two structures are connected in the following way. Let $\chi : R \to R_M$ be the canonical ring-homomorphism.† Then, if $r \in R$ and $x \in E_M$, we have $rx = \chi(r)\,x$. On this understanding, N_M is both an R_M-submodule and an R-submodule of E_M. In particular, E_M/N_M has a natural structure as an R-module and it is this which is referred to in the statement of the proposition.

Proof. Consider the composite mapping $E \to E_M \to E_M/N_M$, where $E \to E_M$ is the canonical and $E_M \to E_M/N_M$ the natural mapping. As the latter are R-homomorphisms, $E \to E_M/N_M$ is also an R-homomorphism.

The kernel of $E \to E_M/N_M$ is the inverse image of N_M with respect to $E \to E_M$, that is to say it is the contraction of N_M in E. However, by Proposition 16 of section (3.7), this contraction is just N. It follows that to complete the proof it is sufficient to show that $E \to E_M/N_M$ is an epimorphism.

To this end let $[e/c]$, where $e \in E$, $c \in R$ and $c \notin M$, be an arbitrary element of E_M. Since $M = \mathrm{Rad}\,(N:E)$, any prime ideal containing $N:E$ must also contain M. But M is a maximal ideal; consequently M is the only prime ideal containing $N:E$. Now c does not belong to M. It follows that there is no maximal ideal containing $(N:E) + Rc$ and therefore $(N:E) + Rc = R$.

Choose $\alpha \in (N:E)$ and $r \in R$ so that $\alpha + rc = 1$. Then $e = \alpha e + cre$ and therefore

$$\left[\frac{e}{c}\right] = \left[\frac{\alpha e}{c}\right] + \left[\frac{cre}{c}\right] = \left[\frac{\alpha e}{c}\right] + \left[\frac{ce'}{c}\right],$$

where $e' = re$. From $\alpha \in (N:E)$ we deduce that $[\alpha e/c]$ belongs to N_M. Thus $[e/c]$ and $[ce'/c]$ have the same image in E_M/N_M. Moreover $[ce'/c]$ is just the image of e' when one applies the canonical mapping $E \to E_M$. Accordingly the element e' of E and the element $[e/c]$ of E_M have the same image in E_M/N_M under the appropriate mappings. Since $[e/c]$ was an entirely arbitrary element of E_M, it now follows that $E \to E_M/N_M$ is an epimorphism. This completes the proof.

Let E be a Noetherian R-module and P a prime ideal of R. Put $R' = R_P$. Then R' has a unique maximal ideal P' (P' is the extension of P) and, by Lemma 2, $E' = E_P$ is a Noetherian R'-module.

By Theorem 10 of section (3.7), there is a one-one correspondence between the P-primary submodules N of E and the P'-primary submodules N' of E'. This is such that, when N and N' correspond, N' is the extension of N in E' and N is the contraction of N' in E. Naturally the correspondence preserves inclusion relations.

† See section (3.2).

Let N' be a P'-primary submodule of E'. Since E' is a Noetherian R'-module and P' is the maximal ideal of R', it follows (Proposition 9) that every proper R'-submodule of E' containing N' is P'-primary. Let N be the P-primary submodule of E corresponding to N'. The above remarks then show that the one-one correspondence, described in the last paragraph, induces *a one-one correspondence between the P-primary submodules of E containing N and the proper R'-submodules of E' containing N'.*

Again $P' = \text{Rad}\,(N':E')$. Hence the only prime ideal containing $N':E'$ is the maximal ideal P'. It therefore follows, from Proposition 7, that $L_{R'}(E'/N') < \infty$.

Next we see that to each chain

$$N = N_1 \subset N_2 \subset \ldots \subset N_q \subset E, \qquad (4.5.1)$$

where N_i is a P-primary submodule of E, corresponds a chain

$$N' = N_1' \subset N_2' \subset \ldots \subset N_q' \subset E' \qquad (4.5.2)$$

of R'-submodules of E' and vice versa. If (4.5.2) happens to be a composition series of R'-modules, and such a composition series exists because $L_{R'}(E'/N')$ is finite, then we shall say that (4.5.1) is a *P-primary composition series from N to E.* In order to discuss this situation, we introduce some new terminology.

Definition. *Let E be an R-module. By a 'maximal P-primary submodule of E' will be meant a P-primary submodule of E which is not strictly contained by any other P-primary submodule of E.*

Definition. *Let $N_1 \subset N_2$ be P-primary submodules of an R-module E. Then we shall say that N_1 and N_2 are 'adjacent in E' if there is no P-primary submodule of E strictly between them.*

Assume that $N_1 \subset N_2$ are P-primary submodules of E. Then, by Proposition 8, N_1 is a P-primary submodule of N_2. Furthermore, by the same proposition, N_1 and N_2 are adjacent in E if and only if N_1 is a maximal P-primary submodule of N_2.

Once again let E be a Noetherian R-module and in (4.5.1) let each N_i be a P-primary submodule of E. Then (4.5.1) is a *P-primary composition series from N to E* if, and only if, N_{i-1} is adjacent to N_i in E $(2 \leqslant i \leqslant q)$ and N_q is a maximal P-primary submodule of E.

Proposition 11. *Let E be a Noetherian R-module and N a P-primary submodule of E. Suppose that*

$$N = N_1 \subset N_2 \subset \ldots \subset N_q \subset E, \qquad (4.5.3)$$

*where each N_i is a P-primary submodule of E. Then $q \leqslant L_{R_P}(E_P/N_P)$.
Moreover if it is not possible to lengthen (4.5.3) by inserting new P-primary submodules of E between the terms of the chain, then*

$$q = L_{R_P}(E_P/N_P).$$

This is clear from the properties of the associated chains (4.5.1) and (4.5.2). The following corollary is an immediate consequence of the proposition.

Corollary. *Let E be a Noetherian R-module and N a P-primary submodule of E. Then any chain*

$$N = N_1 \subset N_2 \subset \ldots \subset N_q \subset E,$$

where each N_i is a P-primary submodule of E, can be refined to a P-primary composition series from N to E by the addition of a finite number of P-primary submodules of E.

Let E be a Noetherian R-module and N a P-primary submodule of E. The number of P-primary submodules in a P-primary composition series from N to E will be denoted by $\lambda_E(N;P)$. Thus, by Proposition 11,

$$\lambda_E(N;P) = L_{R_P}(E_P/N_P). \tag{4.5.4}$$

Theorem 17. *Let E be a Noetherian R-module and let K, N, where $K \subseteq N$, be P-primary submodules of E. Then $\lambda_E(K;P) \geqslant \lambda_E(N;P)$ and there is equality if and only if $K = N$. If $K \neq N$, then K is a P-primary submodule of the Noetherian R-module N and*

$$\lambda_E(K;P) = \lambda_N(K;P) + \lambda_E(N;P).$$

Proof. By (4.5.4),

$$\lambda_E(K;P) = L_{R_P}(E_P/K_P) = L_{R_P}(E_P/N_P) + L_{R_P}(N_P/K_P);$$

consequently

$$\lambda_E(K;P) = \lambda_E(N;P) + L_{R_P}(N_P/K_P). \tag{4.5.5}$$

It follows that $\lambda_E(K;P) \geqslant \lambda_E(N;P)$ with equality if and only if $K_P = N_P$. However, if $K_P = N_P$ then, taking contractions in E, we see that $K = N$. This disposes of the first part.

Now suppose that $K \neq N$. Then K is a proper submodule of N and we may conclude, from Proposition 8, that K is a P-primary submodule of N. It follows, from (4.5.5), that

$$\lambda_E(K;P) = \lambda_E(N;P) + \lambda_N(K;P).$$

This completes the proof.

Proposition 12. *Let E be a Noetherian R-module and N a P-primary submodule. Then the following two statements are equivalent:*

(a) *N is a maximal P-primary submodule of E;*

(b) *for each submodule L of E satisfying $N \subset L \subseteq E$, we have $N:L = P$ and $L:E \supset P$ (strict inclusion).*

Proof. We begin by assuming that N is a maximal P-primary submodule of E. By Theorem 14 Cor. 1, $N:_E P$ strictly contains N. On the other hand, by Proposition 21 of section (2.8), either $N:_E P = E$ or else $N:_E P$ is a P-primary submodule of E. However the latter situation is impossible because of the maximal property of N. Accordingly $N:_E P = E$ and therefore $P \subseteq N:E$. Since $P = \text{Rad}\,(N:E)$, this shows that $P = N:E$.

Now suppose that L is a submodule of E satisfying $N \subset L \subseteq E$. Then $P = N:E \subseteq N:L$ and $N:L \subseteq \text{Rad}\,(N:L) = P$ because, by Proposition 22 of section (2.8), N is a P-primary submodule of L. Accordingly $N:L = P$.

Again $P = N:E \subseteq L:E$. However we cannot have $P = L:E$ for then, by Proposition 26 of section (2.9), P would be one of the prime ideals belonging to L considered as a submodule of E. But in that case there would exist a P-primary submodule of E containing L and therefore strictly containing N which is impossible. Thus $L:E$ strictly contains P.

From now on we shall assume that (b) is true. It will be convenient to put $R' = R_P$ and to denote by P' the extension of P in R' so that P' is the unique maximal ideal of the ring of fractions. Let us now choose $e \in E$ so as to satisfy $e \notin N$. Then $N \subset N + Re \subseteq E$ and therefore, by hypothesis, $N:(N + Re) = P$ and $(N + Re):E \supset P$.

In this paragraph we shall only use the facts† that $P(Re) \subseteq N$ and $e \notin N$. It is important that this be noted because our conclusions will be used again in the proof of Lemma 8. By (3.5.1),

$$(N + Re)_P = N_P + (Re)_P = N' + R'e',$$

where we have put $N' = N_P$ and e' denotes the image of e under the canonical mapping $E \to E_P$. (Here we have made use of Proposition 3 of section (3.2).) Now $N' \subset N' + R'e'$ (strict inclusion) because the contraction of $N' + R'e'$ in E contains e whereas the contraction of N', namely N, does not. Again, by Proposition 4 of section (3.2), the extension of $P(Re)$ in E_P is $P'(R'e')$. Hence, since $P(Re) \subseteq N$, $P'(R'e') \subseteq N'$.

---

† The general assumptions that E is a Noetherian R-module and N a P-primary submodule of E are made tacitly.

Consider the mappings

$$R' \to N' + R'e' \to (N' + R'e')/N',$$

where $R' \to N' + R'e'$ carries r' into $r'e'$ and $N' + R'e' \to (N' + R'e')/N'$ is the natural mapping. Their combined effect is to produce an epimorphism $R' \to (N' + R'e')/N'$ and, because $P'(R'e') \subseteq N'$, the kernel contains P'. However P' is a maximal ideal and the module

$$(N' + R'e')/N'$$

is not a null module. It therefore follows that the kernel is precisely P'. Accordingly $(N' + R'e')/N'$ and R'/P' and isomorphic R'-modules. This in turn implies that

$$(N' + R'e')/N' = (N + Re)_P/N_P$$

is a simple R_P-module. It is, in fact, this conclusion which is used again in the proof of Lemma 8.

For the last step we observe that, since $(N + Re):E$ strictly contains P,

$$((N + Re):E)_P = R_P.$$

Accordingly, by Proposition 8 of section (3.5), $(N + Re)_P : E_P = R_P$ or, changing the notation and writing E' for E_P, $(N' + R'e'):E' = R'$. Thus $E' = N' + R'e'$. The remarks of the last paragraph now show that E'/N' is a simple R'-module ; consequently there can be no P-primary submodule of E which strictly contains N. Thus N is a maximal P-primary submodule of E and the proposition is proved.

Corollary. *Let E be a Noetherian R-module and let N_1, N_2, where $N_1 \subset N_2$, be adjacent P-primary submodules of E. If now L is a submodule of E such that $N_1 \subset L \subseteq N_2$, then $N_1 : L = P$ and $L : N_2 \supset P$ (strict inclusion).*

Proof. N_2 is a Noetherian R-module and† N_1 is a maximal P-primary submodule of N_2. The corollary therefore follows from the proposition.

The lemma, which follows, is sometimes useful when primary composition series are being discussed. This is a convenient place to record it for future reference.

Lemma 8. *Let E be a Noetherian R-module and N a non-maximal P-primary submodule of E. Suppose now that e is an element of E such that $P(Re) \subseteq N$ and $e \notin N$. Then $N + Re$, considered as a submodule of E, has P as a minimal prime ideal. Furthermore, if N_1 is the P-primary component of $N + Re$ in E, then $N \subset N_1$ and N, N_1 are adjacent P-primary submodules of E.*

† See the remarks immediately following the definition of adjacent primary submodules.

Remark. As soon as it has been established that the submodule $N + Re$ of E has P as a minimal prime ideal, we shall know, by the corollary following Theorem 14 of section (2.9), that $N + Re$ has the same P-primary component for all normal decompositions. It is this component that is denoted by N_1.

Proof. Referring back to the fifth paragraph of the proof of Proposition 12, we see that $(N + Re)_P/N_P$ is a simple R_P-module. However E_P/N_P is not a simple R_P-module because otherwise N would be a maximal P-primary submodule of E. It follows that

$$N_P \subset (N + Re)_P \subset E_P \qquad (4.5.6)$$

the inclusions being strict. Let us denote the maximal ideal of R_P by P'. It then follows, by virtue of Proposition 9, that $(N + Re)_P$ is a P'-primary submodule of E_P. However we know, from Theorem 11 of section (3.7), that the prime ideals belonging to the R_P-submodule $(N + Re)_P$ of E_P are the extensions of those prime ideals belonging to $N + Re$ (considered as a submodule of E) that are contained in P. Putting these two observations together, we see first that P is one of the prime ideals belonging to $N + Re$ and, secondly, that all the other prime ideals belonging to it meet the complement of P in R. In other words, P is a minimal prime ideal of the submodule $N + Re$ of E. Let N_1 be its P-primary component. A second application of Theorem 11 of section (3.7) now shows that $(N + Re)_P = (N_1)_P$. Accordingly $(N_1)_P/N_P$ is a simple R_P-module and therefore there are no R_P-submodules of E_P strictly between N_P and $(N_1)_P$. However this implies that there are no P-primary submodules of E strictly between N and N_1. Thus all is proved.

4.6 The Intersection Theorem

The following very useful result is frequently referred to as the *Intersection Theorem*.

Theorem 18. *Let E be a Noetherian R-module and A an ideal of R. Then an element e of E belongs to $\bigcap_{n=1}^{\infty} A^n E$ if and only if $e = \alpha e$ for some element $\alpha \in A$.*

Proof. Put $K = \bigcap_n A^n E$ so that K is a submodule of E. Then, since E is a Noetherian R-module, AK has a decomposition

$$AK = N_1 \cap N_2 \cap \ldots \cap N_s,$$

where N_i $(i = 1, 2, ..., s)$ is a primary submodule of E. Let N_i belong to the prime ideal P_i.

Suppose now that $1 \leqslant i \leqslant s$. If $A \subseteq P_i$, then (Proposition 5) there exists an integer n such that

$$A^n E \subseteq P_i^n E \subseteq N_i.$$

It follows that $K \subseteq N_i$. On the other hand, if $A \nsubseteq P_i$, then from $AK \subseteq N_i$ and $A \nsubseteq P_i$ it again follows† that $K \subseteq N_i$. Accordingly

$$K \subseteq N_1 \cap N_2 \cap ... \cap N_s = AK$$

which shows that $K = AK$.

Next, because E is Noetherian, K is finitely generated, say

$$K = Ru_1 + Ru_2 + ... + Ru_m.$$

Since $u_i \in K = AK$, we must have m relations of the form

$$u_i = \sum_{j=1}^{m} \alpha_{ij} u_j,$$

where the coefficients α_{ij} belong to A. Accordingly $\sum_j (\delta_{ij} - \alpha_{ij}) u_j = 0$ for $1 \leqslant i \leqslant m$ where, of course, δ_{ij} is the familiar Kronecker symbol. If now Δ denotes the determinant $|\delta_{ij} - \alpha_{ij}|$, then $\Delta u_j = 0$ for $1 \leqslant j \leqslant m$ and therefore $\Delta K = 0$. On the other hand, expanding the determinant we find that $\Delta = 1 - \alpha$, where $\alpha \in A$. Thus if $e \in K$, then $(1 - \alpha)e = 0$ or $e = \alpha e$.

To complete the proof we have only to observe that the converse is trivial. For if $e = \alpha e$, where $e \in E$ and $\alpha \in A$, then

$$e = \alpha e = \alpha^2 e = \alpha^3 e = ...$$

whence e belongs to $\bigcap_n A^n E$ as required.

Of particular interest are those situations in which the intersection of the submodules $A^n E$ $(n = 1, 2, 3, ...)$ is the zero submodule of E. Before we proceed to prove a useful result in this direction, it is convenient to make the following

Definition. *The 'Jacobson radical' of a commutative ring is the intersection of all its maximal ideals.*

Theorem 19. *Let A be an ideal of a ring R. Then the following two statements are equivalent:*

(a) *A is contained in the Jacobson radical of R;*

(b) *for every Noetherian R-module E, $\bigcap_{n=1}^{\infty} A^n E = 0$.*

† See Proposition 19 Cor. 1 of section (2.8).

Proof. First, assume that A is contained in the Jacobson radical. Let E be a Noetherian R-module and e an element of $\bigcap_n A^n E$. By Theorem 18, $e = \alpha e$ for some element α belonging to A. Now $(1-\alpha)R$ is not a proper ideal because, since α is contained in every maximal ideal, there is no maximal ideal which contains $1-\alpha$. Accordingly

$$(1-\alpha)R = R$$

and therefore there exists $\beta \in R$ such that $(1-\alpha)\beta = 1$. Multiplying $(1-\alpha)e = 0$ by β, we find that $e = 0$. This shows that (a) implies (b).

Now assume that (b) is true. Since we wish to deduce (a), we may assume that R is a non-null ring. Let M be a maximal ideal and put $E = R/M$. Then E is a simple R-module and, *a fortiori*, a Noetherian R-module. Accordingly $\bigcap_n A^n E = 0$. However, because E is simple, $E \neq 0$ and either $AE = E$ or $AE = 0$. But if $AE = E$, then $A^n E = E$ for $n = 1, 2, \dots$ contradicting the fact that the intersection of the $A^n E$ is zero. It follows that $AE = 0$ and therefore $A \subseteq M$. Since M was prescribed arbitrarily, it follows that A is contained in the Jacobson radical and all is proved.

Corollary. *Let the ideal A be contained in the Jacobson radical of the ring R. If now K is a submodule of a Noetherian R-module E, then*

$$\bigcap_{n=1}^{\infty} (K + A^n E) = K.$$

Proof. Put $E/K = F$. Then $K + A^n E$ is the inverse image of $A^n F$ with respect to the natural epimorphism $E \to F$. Accordingly, by the corollary to Proposition 6 of section (1.4), $\bigcap_n (K + A^n E)$ is the inverse image of $\bigcap_n A^n F$. However F is a Noetherian R-module and therefore the latter intersection is the zero submodule of F by virtue of Theorem 19. It follows that

$$\bigcap_{n=1}^{\infty} (K + A^n E) = K$$

as required.

4.7 The Artin–Rees Lemma

In this section we shall establish a beautiful result known as the *Artin–Rees Lemma* after E. Artin and D. Rees who discovered it independently. However, because of its great importance† we shall

---

† One of the main applications of this result is to be found in the theory of filtered rings and modules. See Chapter 9.

classify it as a theorem rather than as a lemma. The proof, which makes ingenious use of the theory of graded rings and modules, requires some preliminary observations.

Let R be a ring, $A = (\gamma_1, \gamma_2, ..., \gamma_s)$ an ideal of R generated by s elements, and E an R-module. We now introduce $s + 1$ indeterminates which we designate by $X_1, X_2, ..., X_s$ and T. Using these we form the ring $R[X_1, X_2, ..., X_s]$ of polynomials in $X_1, X_2, ..., X_s$ with coefficients in R and the additive abelian group $E[T]$ of polynomials in T with coefficients in E.

The next step is to turn $E[T]$ into an $R[X_1, X_2, ..., X_s]$-module. This is accomplished by defining the product of $\Sigma r_{\mu_1 \mu_2 ... \mu_s} X_1^{\mu_1} X_2^{\mu_2} ... X_s^{\mu_s}$, in $R[X_1, X_2, ..., X_s]$, and $\Sigma e_m T^m$, in $E[T]$, to be

$$\Sigma \Sigma (\gamma_1^{\mu_1} \gamma_2^{\mu_2} ... \gamma_s^{\mu_s} r_{\mu_1 \mu_2 ... \mu_s} e_m) T^{m+\mu_1+\cdots+\mu_s}.$$

Of course, in the double sum, it is the sequence $\{\mu_1, \mu_2, ..., \mu_s\}$ which varies and also the non-negative integer m. A straightforward check shows that the axioms for a module are satisfied. The details are left to the reader.

We now establish a connection between the $R[X_1, X_2, ..., X_s]$-modules $E[T]$ and $E[X_1, X_2, ..., X_s]$. (The latter module was described in detail in section (4.1).) To this end we introduce the mapping

$$\phi : E[X_1, X_2, ..., X_s] \rightarrow E[T]$$

in which (with self-explanatory notation)

$$\Sigma e_{\nu_1 \nu_2 ... \nu_s} X_1^{\nu_1} X_2^{\nu_2} ... X_s^{\nu_s} \rightarrow \Sigma \gamma_1^{\nu_1} \gamma_2^{\nu_2} ... \gamma_s^{\nu_s} e_{\nu_1 \nu_2 ... \nu_s} T^{\nu_1 + \nu_2 + \cdots + \nu_s}.$$

Another straightforward verification shows that ϕ is a homomorphism of $R[X_1, X_2, ..., X_s]$-modules.

Consider the image of the homorphism ϕ. The coefficient of T^n in

$$\Sigma \gamma_1^{\nu_1} \gamma_2^{\nu_2} ... \gamma_s^{\nu_s} e_{\nu_1 \nu_2 ... \nu_s} T^{\nu_1 + \nu_2 + \cdots + \nu_s}$$

is

$$\sum_{\nu_1 + \cdots + \nu_s = n} \gamma_1^{\nu_1} \gamma_2^{\nu_2} ... \gamma_s^{\nu_s} e_{\nu_1 \nu_2 ... \nu_s}.$$

But the set of products $\gamma_1^{\nu_1} \gamma_2^{\nu_2} ... \gamma_s^{\nu_s}$ for which $\nu_1, \nu_2, ..., \nu_s$ are non-negative integers with sum n generates the ideal A^n. Consequently the coefficient in question may be any element of $A^n E$. Thus $\operatorname{Im} \phi$ consists of all polynomials $e_0 + e_1 T + e_2 T^2 + ...$ for which $e_n \in A^n E$ $(n = 0, 1, 2, ...)$. Here it is understood that by $A^0 E$ we mean E itself. Let us use $\Sigma (A^n E) T^n$ to denote the module formed by these special polynomials so that our conclusion may be written as

$$\operatorname{Im} \phi = \Sigma (A^n E) T^n.$$

We now claim that *if E is a Noetherian R-module, then $\Sigma(A^n E) T^n$ is a Noetherian $R[X_1, X_2, ..., X_s]$-module.* For, by Theorem 1,

$$E[X_1, X_2, ..., X_s]$$

is a Noetherian $R[X_1, X_2, ..., X_s]$-module and we have just seen that ϕ gives rise to an $R[X_1, X_2, ..., X_s]$-linear mapping of $E[X_1, X_2, ..., X_s]$ on to $\Sigma(A^n E) T^n$. The contention therefore follows by Lemma 1.

The elements of $R[X_1, X_2, ..., X_s]$ which can be expressed in the form

$$\sum_{\mu_1 + ... + \mu_s = n} r_{\mu_1 \mu_2 ... \mu_s} X_1^{\mu_1} X_2^{\mu_2} ... X_s^{\mu_s}$$

will be said to be *homogeneous of degree n*. Let their aggregate be denoted by $R[X]^{(n)}$. Then $R[X]^{(n)}$ is an additive abelian group and

$$R[X_1, X_2, ..., X_s] = R[X]^{(0)} \oplus R[X]^{(1)} \oplus R[X]^{(2)} \oplus \dots.$$

Further the product of a homogeneous polynomial of degree n and one of degree p is homogeneous of degree $n+p$. Thus we have endowed $R[X_1, X_2, ..., X_s]$ with the structure of a ring graded by the non-negative integers. Note that $E[T]$ can now be regarded as a graded module with respect to the graded ring $R[X_1, X_2, ..., X_s]$. To accomplish this we have to specify which elements of $E[T]$ are to be regarded as being homogeneous of degree m, where m is an arbitrary non-negative integer. This is done by taking them to be the elements of the form eT^m. It is easily checked that this meets all the requirements for a graded module.

One final remark. When $E[T]$ is regarded as a graded module with respect to the ring $R[X_1, X_2, ..., X_s]$ its submodule $\Sigma(A^n E) T^n$ is homogeneous. Accordingly $\Sigma(A^n E) T^n$ has an induced structure as a graded $R[X_1, X_2, ..., X_s]$-module.

We are now ready to state and prove the theorem of Artin–Rees. The particular proof offered here has been chosen because, with only a minor modification, it suffices to establish a more general result that will be needed in Chapter 7.

Theorem 20. *Let E be a Noetherian R-module, K a submodule of E, and A an ideal of the ring R. Then there exists an integer $q \geqslant 0$ such that*

$$A^n E \cap K = A^{n-q}(A^q E \cap K)$$

for all $n \geqslant q$.

Proof. We begin by showing that, without loss of generality, we may suppose that A is finitely generated. By the remark following the statement of Theorem 2, the ring $R/\text{Ann}_R E$ is Noetherian and therefore the ideal $(A + \text{Ann}_R E)/\text{Ann}_R E$ of this ring is finitely generated. Choose $\gamma_1, \gamma_2, \ldots, \gamma_s$ in A so that their images in $R/\text{Ann}_R E$ generate the ideal in question and put $A_0 = (\gamma_1, \gamma_2, \ldots, \gamma_s)$. Then

$$A^n E \cap K = A_0^n E \cap K \quad \text{(all } n \geqslant 0\text{)}$$

and for every non-negative integer q we have

$$A^{n-q}(A^q E \cap K) = A_0^{n-q}(A_0^q E \cap K)$$

provided that $n \geqslant q$. This shows that it will suffice to prove the theorem with A_0 in place of A. Equally well it shows that we may assume that A is finitely generated.

From now on we shall suppose that $A = (\gamma_1, \gamma_2, \ldots, \gamma_s)$. This enables us (with the aid of the construction described above) to regard $\Sigma(A^n E) T^n$ as a graded module with respect to the graded ring $R[X_1, X_2, \ldots, X_s]$. Moreover our previous discussion shows that this is a Noetherian module. Consider the elements $\epsilon_0 + \epsilon_1 T + \epsilon_2 T^2 + \ldots$ of $E[T]$ which satisfy $\epsilon_j \in (A^j E \cap K)$ for all $j \geqslant 0$. A simple verification shows that they form a homogeneous $R[X_1, X_2, \ldots, X_s]$-submodule N (say) of $\Sigma(A^n E) T^n$. Since $\Sigma(A^n E) T^n$ is Noetherian, N is a finitely generated $R[X_1, X_2, \ldots, X_s]$-module. Let it be generated by $u_1, u_2, \ldots, u_p$. Each u_i is the sum of its non-zero homogeneous components and each of these components belongs to N. If therefore we replace the u_i by the non-zero homogeneous components to which they give rise, then we shall obtain a finite number of homogeneous elements which generate N. In what follows we shall suppose that $u_1, u_2, \ldots, u_p$ have been chosen so that they are already homogeneous. Specifically we shall assume that $u_i = \omega_i T^{n_i}$, where $\omega_i \in (A^{n_i} E \cap K)$. Thus

$$N = \sum_{i=1}^{p} R[X_1, X_2, \ldots, X_s] (\omega_i T^{n_i}).$$

Put $q = \max(n_1, n_2, \ldots, n_p)$ and suppose that $n \geqslant q$. We shall show that $A^n E \cap K \subseteq A^{n-q}(A^q E \cap K)$ and, as the opposite inclusion is obvious, this will complete the proof.

Suppose therefore that $\eta \in (A^n E \cap K)$. Then ηT^n belongs to N and so it can be expressed in the form

$$\eta T^n = \sum_{i=1}^{p} g_i(X_1, X_2, \ldots, X_s)(\omega_i T^{n_i}), \tag{4.7.1}$$

where the $g_i(X_1, X_2, \ldots, X_s)$ belong to $R[X_1, X_2, \ldots, X_s]$. Each

$$g_i(X_1, X_2, \ldots, X_s)$$

can be expressed as a finite sum of homogeneous polynomials of unequal degrees. If we express them in this way and then, in (4.7.1), compare terms of degree n on the two sides, we find that we can arrange that, for $1 \leqslant i \leqslant p$, $g_i(X_1, X_2, ..., X_s)$ is homogeneous of degree $n - n_i$.

We now claim that $g_i(X_1, X_2, ..., X_s)(\omega_i T^{n_i})$ belongs to

$$A^{n-q}(A^q E \cap K) T^n.$$

For let $r X_1^{\nu_1} X_2^{\nu_2} ... X_s^{\nu_s}$, where $\nu_1 + \nu_2 + ... + \nu_s = n - n_i$, be a typical term of $g_i(X_1, X_2, ..., X_s)$. Then, so far as our present claim is concerned, it will suffice to show that $(r X_1^{\nu_1} X_2^{\nu_2} ... X_s^{\nu_s})(\omega_i T^{n_i})$ is of the form $c T^n$, where c is in $A^{n-q}(A^q E \cap K)$. Now $X_1^{\nu_1} X_2^{\nu_2} ... X_s^{\nu_s}$ can be expressed as a product

$$X_1^{\sigma_1} X_2^{\sigma_2} ... X_s^{\sigma_s} X_1^{\tau_1} X_2^{\tau_2} ... X_s^{\tau_s},$$

where the σ_i and τ_i are non-negative integers satisfying

$$\sigma_1 + ... + \sigma_s = n - q \quad \text{and} \quad \tau_1 + ... + \tau_s = q - n_i.$$

Hence

$$(r X_1^{\nu_1} X_2^{\nu_2} ... X_s^{\nu_s})(\omega_i T^{n_i}) = \gamma_1^{\sigma_1} \gamma_2^{\sigma_2} ... \gamma_s^{\sigma_s}(\gamma_1^{\tau_1} \gamma_2^{\tau_2} ... \gamma_s^{\tau_s} r \omega_i) T^n.$$

But, since $\omega_i \in (A^{n_i} E \cap K)$,

$$\gamma_1^{\sigma_1} \gamma_2^{\sigma_2} ... \gamma_s^{\sigma_s}(\gamma_1^{\tau_1} \gamma_2^{\tau_2} ... \gamma_s^{\tau_s} r \omega_i) \in \gamma_1^{\sigma_1} \gamma_2^{\sigma_2} ... \gamma_s^{\sigma_s}(A^q E \cap K)$$

which itself is contained in $A^{n-q}(A^q E \cap K)$. Thus

$$g_i(X_1, X_2, ..., X_s)(\omega_i T^{n_i})$$

belongs to $A^{n-q}(A^q E \cap K) T^n$ as we claimed and therefore, by (4.7.1), the same is true of ηT^n. This however means that η is in $A^{n-q}(A^q E \cap K)$ and so we have shown that

$$A^n E \cap K \subseteq A^{n-q}(A^q E \cap K).$$

As previously observed, the theorem follows.

As a first application we shall establish

Proposition 13. *Let E be a Noetherian R-module, A an ideal of R and β and element of R. Then there exists an integer $q \geqslant 0$ such that*

$$A^n E :_E \beta = A^{n-q}(A^q E :_E \beta) + (0 :_E \beta)$$

for all $n \geqslant q$.

Proof. By Theorem 20, there exists an integer $q \geqslant 0$ such that

$$A^n E \cap \beta E = A^{n-q}(A^q E \cap \beta E)$$

for all $n \geqslant q$. Now $A^n E \cap \beta E = \beta(A^n E :_E \beta)$. Consequently, when $n \geqslant q$, $\beta(A^n E :_E \beta) = A^{n-q}(A^q E \cap \beta E) = \beta A^{n-q}(A^q E :_E \beta)$.

Let $e \in (A^n E :_E \beta)$ and suppose that $n \geqslant q$. Then βe belongs to

$$\beta(A^n E :_E \beta)$$

and therefore $\beta e = \beta e'$, where $e' \in A^{n-q}(A^q E :_E \beta)$. Since $\beta(e - e') = 0$, it follows that $e - e' \in (0 :_E \beta)$. Thus $e = e' + (e - e')$ belongs to
$$A^{n-q}(A^q E :_E \beta) + (0 :_E \beta).$$
Accordingly $\quad A^n E :_E \beta \subseteq A^{n-q}(A^q E :_E \beta) + (0 :_E \beta)$
and, as the opposite inclusion is obvious, this completes the proof.

Corollary. *Let the assumptions be as in the proposition. Then there exists an integer $q \geqslant 0$ such that*

for all $n \geqslant q$. $\quad A^n E :_E \beta \subseteq A^{n-q} E + (0 :_E \beta)$

Another useful application is contained in

Proposition 14. *Let $K_1, K_2, \ldots, K_s$ be submodules of a Noetherian R-module E and let A be an ideal of R. Then there exists an integer $q \geqslant 0$ such that*
$$A^n K_1 \cap A^n K_2 \cap \ldots \cap A^n K_s = A^{n-q}(A^q K_1 \cap A^q K_2 \cap \ldots \cap A^q K_s)$$
for all $n \geqslant q$.

Proof. Since K_i is a submodule of the Noetherian R-module E, it is itself Noetherian. Put
$$L = K_1 \oplus K_2 \oplus \ldots \oplus K_s.$$
Then, by Lemma 1 Cor. 1, L is also a Noetherian R-module. Now a typical element of L is a sequence of the form $\{k_1, k_2, \ldots, k_s\}$, where $k_i \in K_i$. Denote by J the subset of L which consists of all sequences $\{x, x, \ldots, x\}$ of s elements, where x belongs to $K_1 \cap K_2 \cap \ldots \cap K_s$. Clearly J is an R-submodule of L and therefore by Theorem 20, there exists an integer $q \geqslant 0$ such that
$$A^n L \cap J = A^{n-q}(A^q L \cap J)$$
whenever $n \geqslant q$. Assume now that $n \geqslant q$ and let y be an element of $A^n K_1 \cap A^n K_2 \cap \ldots \cap A^n K_s$. Then $\{y, y, \ldots, y\}$ is in
$$A^n L \cap J = A^{n-q}(A^q L \cap J)$$
and therefore it can be represented in the form
$$\{y, y, \ldots, y\} = \sum_{j=1}^{p} \beta_j \{z_j, z_j, \ldots, z_j\},$$
where $\beta_j \in A^{n-q}$ and $\{z_j, z_j, \ldots, z_j\} \in (A^q L \cap J)$. Thus $y = \Sigma \beta_j z_j$ and z_j belongs to $A^q K_1 \cap A^q K_2 \cap \ldots \cap A^q K_s$. Accordingly
$$y \in A^{n-q}(A^q K_1 \cap A^q K_2 \cap \ldots \cap A^q K_s)$$
which shows that
$$A^n K_1 \cap A^n K_2 \cap \ldots \cap A^n K_s \subseteq A^{n-q}(A^q K_1 \cap A^q K_2 \cap \ldots \cap A^q K_s).$$
However the opposite inclusion is obvious and so the proof is complete.

4.8 Rank and dimension

Let R be a commutative ring with an identity element and P one of its prime ideals. We consider finite strictly decreasing sequences of prime ideals which begin with P itself.

Definition. *Let $n \geqslant 0$ be an integer. The prime ideal P is said to be of 'rank n', and we write* rank $P = n$, *if there exists a strictly decreasing sequence*

$$P \supset P_1 \supset P_2 \supset \ldots \supset P_n$$

of $n+1$ prime ideals beginning with P itself and if there is no similar sequence with more than $n+1$ terms.

Should it happen that for every non-negative integer n there exists at least one sequence

$$P \supset P_1 \supset P_2 \supset \ldots \supset P_n$$

of $n+1$ prime ideals, then we say that the rank of P is infinite and write rank $P = \infty$. Thus every prime ideal has a definite rank which is either a non-negative integer or infinity. As immediate consequences of the definition we have

   (i) *a prime ideal P is of rank zero if and only if it is a minimal prime ideal of the ring*†;

  (ii) *if $P \subset P'$ (strict inclusion) are prime ideals, then* rank $P \leqslant$ rank P' *and there is strict inequality if* rank P *is finite.*

The notion of rank can be extended to ideals which are not necessarily prime. To this end, let A be a proper ideal and put

$$\operatorname{rank} A = \inf_{P \supseteq A} \operatorname{rank} P. \qquad (4.8.1)$$

Thus rank A is the lower bound of the ranks of all the prime ideals which contain A and, in view of (ii), we recover the original definition when A itself happens to be prime. Now, by Theorem 3 of section (2.3), every prime ideal which contains A also contains a minimal prime ideal of A. It follows that *the value of the infimum in* (4.8.1) *remains unchanged if we restrict P so that it only ranges over the minimal prime ideals of A*. In the Noetherian case this can be a great advantage because A has then only a finite number of minimal prime ideals. Another point to note is that *if A and B are proper ideals and $A \subseteq B$, then* rank $A \leqslant$ rank B.

---

† See section (2.3) for the definition.

Suppose now that S is a non-empty multiplicatively closed subset of R. We know, from Theorem 9 of section (3.6), that there is a one-one correspondence between the prime ideals of R which do not meet S and the prime ideals of R_S. The theorem just quoted sets out in detail how the correspondence works. In particular we see that it preserves inclusion relations. Accordingly, if P is a prime ideal of R not meeting S and P_S is the corresponding prime ideal of R_S, then there is an inclusion-preserving one-one correspondence between the prime R-ideals contained in P and the prime R_S-ideals contained in P_S. It follows that

$$\operatorname{rank} P = \operatorname{rank} P_S, \qquad (4.8.2)$$

i.e. the rank of P does not change when we pass to the ring of fractions.

In this section we shall establish a number of important results concerning the ranks of ideals in Noetherian rings. The reader may be wondering whether one obtains a useful concept by considering increasing rather than decreasing sequences of prime ideals. In fact this leads to the notion of *dimension* about which we shall have something to say at a later stage.

The key result in the theory of rank is known as the *Principal Ideal Theorem*. Like the *Intersection Theorem* it is due to W. Krull. The lemma which follows disposes of most of the technicalities which occur in the proof of this result.

Lemma 9. *Let R be a Noetherian integral domain with a single maximal ideal P. Assume that P is a minimal prime ideal of a principal ideal αR. If now $\beta \in R$, $\beta \neq 0$, then $\alpha^q R \subseteq \beta R$ for some positive integer q.*

Proof. Clearly we may assume that $\alpha \neq 0$. By Theorem 20, there exists an integer $q \geqslant 0$ such that

$$\alpha^n R \cap \beta R = \alpha^{n-q}(\alpha^q R \cap \beta R)$$

for all $n \geqslant q$ so that, in particular,

$$\alpha^{q+1} R \cap \beta R = \alpha(\alpha^q R \cap \beta R).$$

Next, since P is a minimal prime ideal of αR and the only maximal ideal of R, there is no other prime ideal which contains $\alpha^{q+1}R$. It therefore follows, from the corollary to Proposition 7, that $L_R(R/\alpha^{q+1}R) < \infty$. Again, by Theorem 3 of section (1.6), the R-modules

$$(\alpha^{q+1} R + \beta R)/\alpha^{q+1} R$$

and $\qquad \beta R/(\alpha^{q+1} R \cap \beta R) = \beta R/\alpha(\alpha^q R \cap \beta R)$

are isomorphic. Consequently†

$$L_R\{(\alpha^{q+1}R+\beta R)/\alpha^{q+1}R\} = L_R\{\beta R/\alpha(\alpha^q R \cap \beta R)\}$$
$$= L_R\{\beta R/\alpha\beta R\} + L_R\{\alpha\beta R/\alpha(\alpha^q R \cap \beta R)\}$$

$$(4.8.3)$$

and all the lengths involved are finite.

The mapping $\beta R \to \alpha\beta R$, obtained by multiplying the elements of βR by α, is an isomorphism of R-modules because R is an integral domain and $\alpha \neq 0$. Moreover this isomorphism maps $\alpha^q R \cap \beta R$ on to $\alpha(\alpha^q R \cap \beta R)$. Hence, by Theorem 1 of section (1·6), there results an isomorphism $\quad \beta R/(\alpha^q R \cap \beta R) \approx \alpha\beta R/\alpha(\alpha^q R \cap \beta R)$.

On the other hand, by Theorem 3 of section (1.6), $\beta R/(\alpha^q R \cap \beta R)$ is isomorphic to $(\alpha^q R + \beta R)/\alpha^q R$. We may therefore conclude that

$$L_R\{\alpha\beta R/\alpha(\alpha^q R \cap \beta R)\} = L_R\{(\alpha^q R + \beta R)/\alpha^q R\}. \qquad (4.8.4)$$

Again the mapping $R \to \beta R$, in which $r \to \beta r$, is an isomorphism of R-modules. With its aid we see at once that $R/\alpha R$ is isomorphic to $\beta R/\alpha\beta R$. Thus $\quad L_R\{\beta R/\alpha\beta R\} = L_R\{R/\alpha R\}$,

whence, using (4.8.4), the relation (4.8.3) becomes

$$L_R\{(\alpha^{q+1}R+\beta R)/\alpha^{q+1}R\} = L_R\{R/\alpha R\} + L_R\{(\alpha^q R + \beta R)/\alpha^q R\}.$$

However $\alpha^q R/\alpha^{q+1}R$ is isomorphic to $R/\alpha R$ as may be seen by considering the isomorphism $R \to \alpha^q R$ produced by multiplication by α^q. Accordingly $\quad L_R\{R/\alpha R\} = L_R\{\alpha^q R/\alpha^{q+1}R\}$.

Thus

$$L_R\{(\alpha^{q+1}R+\beta R)/\alpha^{q+1}R\} = L_R\{\alpha^q R/\alpha^{q+1}R\} + L_R\{(\alpha^q R + \beta R)/\alpha^q R\}$$
$$= L_R\{(\alpha^q R + \beta R)/\alpha^{q+1}R\}. \qquad (4.8.5)$$

But $(\alpha^{q+1}R+\beta R)/\alpha^{q+1}R$ is contained in $(\alpha^q R+\beta R)/\alpha^{q+1}R$ and the lengths occurring in (4.8.5) are finite. It therefore follows, from the corollary to Theorem 6 of section (1.7), that

$$(\alpha^{q+1}R+\beta R)/\alpha^{q+1}R = (\alpha^q R+\beta R)/\alpha^{q+1}R$$

and hence that $\alpha^{q+1}R+\beta R = \alpha^q R+\beta R$. Thus α^q belongs to $\alpha^{q+1}R+\beta R$ and therefore $\alpha^q = \alpha^{q+1}r+\beta r'$ for suitable elements r, r' of R. This shows that $\alpha^q(1-\alpha r) \in \beta R$. However, since $\alpha \in P$, $1-\alpha r$ does not belong to P. But P is the only maximal ideal, consequently $(1-\alpha r)R = R$. It follows that $\quad \alpha^q R = \alpha^q(1-\alpha r)R \subseteq \beta R$

and now the lemma is proved.

† The properties of lengths, as required here, are developed in section (1.7). See, in particular, Theorem 6 in that section.

Corollary. *Let R, P and α be as described in the lemma. Then* rank $P \leqslant 1$.

Proof. Assume the contrary. Then there exists a prime ideal P' such that $(0) \subset P' \subset P$. Choose $\beta \in P'$ so that $\beta \neq 0$. Then, by the lemma, $\alpha^q \in \beta R \subseteq P'$ for a suitable positive integer q. Thus $\alpha \in P'$ and therefore $\alpha R \subseteq P'$. This, however, is impossible because, by hypothesis, P is a minimal prime ideal of αR.

We are now ready to establish the *Principal Ideal Theorem* itself.

Theorem 21. *Let R be a Noetherian ring and $R\alpha$ a proper principal ideal generated by an element α. If now P is a minimal prime ideal of $R\alpha$, then* rank $P \leqslant 1$.

Proof. Put $R' = R_P$ and denote by P' the extension of P in R'. Then, by Theorem 6, R' is a Noetherian ring with P' as its only maximal ideal and, by (4.8.2), rank $P = $ rank P'. Furthermore, by Theorem 11 of section (3.7), P' is a minimal prime ideal of $(R\alpha)_P = R'\alpha'$, where α' denotes the image of α under the canonical mapping $R \to R_P$. It will suffice to show that rank $P' \leqslant 1$. The same observation may be reformulated as follows: *for the purposes of the proof it may be assumed that R has only one maximal ideal and that this is the prime ideal P occurring in the statement of the theorem.*

It will now be assumed that rank $P > 1$ and from this a contradiction will be derived. Since rank $P > 1$, there exist prime ideals P_1, P_2 such that $P \supset P_1 \supset P_2$ the inclusions being strict. Put $\bar{R} = R/P_2$, $\bar{P} = P/P_2$, $\bar{P}_1 = P_1/P_2$ and denote by $\bar{\alpha}$ the image of α under the natural mapping $R \to \bar{R}$. By Theorem 3, $\bar{R}$ is a Noetherian integral domain. In addition, $\bar{P}$ is a maximal ideal of $\bar{R}$ and moreover it is the only maximal ideal of this ring.† Now $\bar{\alpha}$ belongs to $\bar{P}$ and there can be no other prime ideal of $\bar{R}$ containing $\bar{\alpha}$ because, by hypothesis, P is a minimal prime ideal of $R\alpha$. It therefore follows, from the corollary to Lemma 9, that rank $\bar{P} \leqslant 1$. However $\bar{P} \supset \bar{P}_1 \supset (0)$ is a strictly decreasing sequence of prime ideals and so we have the desired contradiction.

Theorem 22. *Let A be a proper ideal of a Noetherian ring R and suppose that A can be generated by n ($n \geqslant 0$) elements. If now P is a minimal prime ideal of A, then* rank $P \leqslant n$.

Proof. Let $A = (\alpha_1, \alpha_2, ..., \alpha_n)$. The proof employs induction on n. For $n = 0$ the assertion means that a minimal prime ideal of R has zero rank (which is obvious) while if $n = 1$ it reduces to the principal ideal

† This follows from Proposition 2 of section (2.2).

theorem. It will therefore be assumed that $n > 1$ and that the theorem has been proved for all smaller values of the inductive variable.

Put $R' = R_P$ and $A' = A_P$. Then R' is a Noetherian ring whose only maximal ideal is the extension P' of P and we have

$$A' = R'\alpha'_1 + R'\alpha'_2 + \ldots + R'\alpha'_n,$$

where α'_i is the image of α_i under the canonical mapping $R \to R'$. Furthermore, by Theorem 11 of section (3.7), P' is a minimal prime ideal of A'. Since rank $P' = $ rank P, we need only show that rank $P' \leqslant n$. This means that, for the purposes of the proof, we may assume that R has P as its one and only maximal ideal. This assumption is made in the following discussion.

It will now be assumed that rank $P > n$ and from this a contradiction will be derived. Since rank $P > n$, there exists at least one sequence
$$P \supset P_1 \supset P_2 \supset \ldots \supset P_{n+1}$$
containing $n + 2$ prime ideals. Among all such sequences we now select one with the property that P_1 contains as many of $\alpha_1, \alpha_2, \ldots, \alpha_n$ as possible. It cannot contain them all because otherwise P would not be a minimal prime ideal of $(\alpha_1, \alpha_2, \ldots, \alpha_n)$. For definiteness we shall suppose that $\alpha_1 \notin P_1$.

We next observe that if P^* is a prime ideal such that†
$$(P_1, \alpha_1) \subseteq P^* \subseteq P,$$
then $P = P^*$. (For otherwise we could replace P_1 by P^* in the sequence considered above thereby obtaining a contradiction to the statement that P_1 contains the greatest possible number of the α_i.) Since P is the only maximal ideal of R, it now follows that P is the only prime ideal which contains (P_1, α_1). Thus (P_1, α_1) is P-primary. Now $\alpha_1, \alpha_2, \ldots, \alpha_n$ all belong to P. Consequently there exists an integer t such that $\alpha_i^t \in (P_1, \alpha_1)$ for $1 \leqslant i \leqslant n$. Accordingly we have relations of the form $\alpha_i^t = r_i \alpha_1 + \beta_i$, where (for $2 \leqslant i \leqslant n$) $r_i \in R$ and $\beta_i \in P_1$. Put $B = (\beta_2, \beta_3, \ldots, \beta_n)$. Then $B \subseteq P_1$. However rank $P_1 \geqslant n$ and so, by the inductive hypothesis, P_1 cannot be a minimal prime ideal of B. Accordingly there exists a prime ideal Q such that $B \subseteq Q \subset P_1 \subset P$.

Any prime ideal which contains (Q, α_1) must contain α_i^t and hence α_i for $i = 1, 2, \ldots, n$. But P is a minimal prime ideal of $(\alpha_1, \alpha_2, \ldots, \alpha_n)$. It must therefore be a minimal prime ideal of (Q, α_1).

Put $\bar{R} = R/Q$, $\bar{P} = P/Q$ and let $\bar{\alpha}_1$ denote the image of α_1 in $\bar{R}$ under the natural mapping $R \to R/Q$. Then $\bar{R}$ is a Noetherian integral domain

† We use (P_1, α_1) as an alternative for $P_1 + R\alpha_1$.

and, since P is a minimal prime ideal of (Q, α_1), $\bar{P}$ is a minimal prime ideal of $\bar{\alpha}_1\bar{R}$. It therefore follows, from Theorem 21, that rank $\bar{P} \leqslant 1$. However

$$\bar{P} = P/Q \supset P_1/Q \supset (0)$$

is a strictly decreasing sequence of prime ideals and this provides the desired contradiction.

Since all the ideals in a Noetherian ring are finitely generated, the last theorem shows that *in such a ring every proper ideal has finite rank*. Indeed if a proper ideal A in a Noetherian ring can be generated by n elements, then rank $A \leqslant n$.

Our next result is a useful companion to Theorem 22.

Proposition 15. *Let R be a Noetherian ring and A a proper ideal of R. Put $n = $ rank A. Then there exists a sequence $\alpha_1, \alpha_2, ..., \alpha_n$ of n elements of A such that rank $(\alpha_1, \alpha_2, ..., \alpha_i) = i$ for $i = 1, 2, ..., n$.*

Proof. The elements $\alpha_1, \alpha_2, ..., \alpha_n$ will be found in succession. Suppose that $0 \leqslant j < n$ and assume that $\alpha_1, \alpha_2, ..., \alpha_j$ have already been obtained and possess the necessary properties. Let the minimal prime ideals of $(\alpha_1, \alpha_2, ... \alpha_j)$ be $P_1, P_2, ..., P_s$. (It is to be understood that if $j = 0$, then $(\alpha_1, \alpha_2, ..., \alpha_j)$ is the zero ideal.) If now $1 \leqslant \mu \leqslant s$, then

$$j = \text{rank} \, (\alpha_1, \alpha_2, ..., \alpha_j) \leqslant \text{rank} \, P_\mu$$

and, by Theorem 22, rank $P_\mu \leqslant j$. Thus all of $P_1, P_2, ..., P_s$ have rank equal to j. Now $A \nsubseteq P_\mu$ for otherwise we should have

$$\text{rank} \, A \leqslant \text{rank} \, P_\mu = j < n.$$

Accordingly, by Proposition 5 of section (2.3), there exists $\alpha_{j+1} \in A$ such that $\alpha_{j+1} \notin P_1, ..., \alpha_{j+1} \notin P_s$.

Let P' be a minimal prime ideal of $(\alpha_1, \alpha_2, ..., \alpha_{j+1})$. By Theorem 22, rank $P' \leqslant j+1$. On the other hand, P' contains $(\alpha_1, \alpha_2, ..., \alpha_j)$ and therefore it contains one of its minimal prime ideals. Let it contain P_μ. Then $P_\mu \subset P'$ (strict inclusion) because α_{j+1} belongs to P' but not to P_μ. Accordingly rank $P' > $ rank $P_\mu = j$. It follows that rank $P' = j+1$. Thus every minimal prime ideal of $(\alpha_1, \alpha_2, ..., \alpha_{j+1})$ has its rank equal to $j+1$ and therefore rank $(\alpha_1, \alpha_2, ..., \alpha_{j+1}) = j+1$. This shows that the sequence $\alpha_1, \alpha_2, ..., \alpha_j$ can be extended in accordance with the stipulated conditions. The proof is now complete.

Proposition 16. *Let R be a Noetherian ring, A an ideal of R generated by n elements and P a prime ideal containing A. Then*

$$\text{rank} \, P \leqslant n + \text{rank} \, (P/A).$$

Remark. By rank (P/A) is meant the rank of the prime ideal P/A in the ring R/A.

Proof. Put $k = \text{rank}\,(P/A)$, $\bar{R} = R/A$ and $\bar{P} = P/A$. If $k = 0$, then P is a minimal prime ideal of A and therefore the required result is a consequence of Theorem 22. From now on it will be assumed that $k > 0$.

By Proposition 15, there exist k elements $\beta_1, \beta_2, ..., \beta_k$ in $\bar{P}$ such that $\beta_1\bar{R} + \beta_2\bar{R} + ... + \beta_k\bar{R}$ has rank equal to k. This implies that $\bar{P}$ is a minimal prime ideal of $\beta_1\bar{R} + \beta_2\bar{R} + ... + \beta_k\bar{R}$. For each i ($1 \leqslant i \leqslant k$) choose $b_i \in P$ so that β_i is the natural image of b_i in R/A. Let $A = (\alpha_1, \alpha_2, ..., \alpha_n)$. Then P is a minimal prime ideal of $(\alpha_1, ..., \alpha_n, b_1, ..., b_k)$ because $\bar{P}$ is a minimal prime ideal of $\beta_1\bar{R} + ... + \beta_k\bar{R}$. It follows, from Theorem 22, that rank $P \leqslant n + k$. This completes the proof.

Proposition 17. *Let A be a proper ideal of a Noetherian ring R and suppose that A contains an element α which is not a zero-divisor. Then* rank $(A/\alpha R) = $ rank $A - 1$.

Proof. Let P be a typical minimal prime ideal of A. Then $P/\alpha R$ is a typical minimal prime ideal of $A/\alpha R$. If therefore we can show that rank $(P/\alpha R) = $ rank $P - 1$, then the proposition will follow at once. It is seen from this that, for the rest of the proof, we may assume that A itself is a prime ideal. This assumption will be made and, to keep it in mind, we shall employ the letter P in place of A.

Put rank $(P/\alpha R) = q$ and rank $P = k$. Then, by Proposition 16, $k \leqslant q + 1$ so we need only prove the opposite inequality. Next, because rank $(P/\alpha R) = q$, there exist prime ideals $P_1, P_2, ..., P_q$ such that

$$P \supset P_1 \supset P_2 \supset ... \supset P_q \supseteq \alpha R.$$

Now P_q cannot be a minimal prime ideal of the zero ideal. (For otherwise, by Theorem 14 Cor. 2, $0R : \alpha$ would strictly contain $0R$ and this contradicts the assumption that α is not a zero-divisor.) Accordingly P_q strictly contains a minimal prime ideal P' of the zero ideal. We thus have a sequence

$$P \supset P_1 \supset ... \supset P_q \supset P'$$

which shows that rank $P \geqslant q + 1$ and completes the proof.

Before closing this section, we shall introduce the concept of dimension to which a casual reference has already been made.

Let P be a prime ideal of a ring R and let $n \geqslant 0$ be an integer.

We shall say that P is of *dimension n* and write $\dim P = n$ if

(i) *there exists a strictly increasing sequence*

$$P \subset P_1 \subset P_2 \subset \ldots \subset P_n \qquad (4.8.6)$$

of $n + 1$ prime ideals whose first term is P, and

(ii) *there is no such sequence having more than $n + 1$ terms.*

If it happens that for each $n \geqslant 0$ there exists at least one sequence (4.8.6), then P is said to be of *infinite dimension* and we write

$$\dim P = \infty.$$

Thus $\dim P$ is defined for every prime ideal P and is either a non-negative integer or infinity. It is important to note that if $\dim P = \infty$, then it does not follow that there exists an infinite strictly increasing sequence of prime ideals having P as its first term. All that can be said in this case is that there exist arbitrarily long *finite* sequences of the type shown in (4.8.6).

The following immediate consequences of the definition should be noted:

(a) $\dim P = 0$ if and only if P is a maximal ideal;

(b) if $P \subset P'$ (strict inclusion) are prime ideals, then

$$\dim P \geqslant \dim P';$$

moreover $\dim P > \dim P'$ should $\dim P'$ be finite.

The concept of dimension, like that of rank, can be extended so that it applies to all proper ideals. To be explicit, in the case of a proper ideal A we put

$$\dim A = \sup_{P \supseteq A} \dim P, \qquad (4.8.7)$$

where the supremum is taken over all the prime ideals P which contain A. Accordingly $\dim A$ is either a non-negative integer or infinity. We recall that every prime ideal which contains A contains a minimal prime ideal of A. It now follows that *the value of the supremum in (4.8.7) is not changed if P is restricted so that it only ranges over the minimal prime ideals of A.* We take this opportunity to observe that *if A and B are proper ideals and $A \subseteq B$, then $\dim A \geqslant \dim B$.* This follows immediately from (4.8.7).

Definition. *The 'dimension' of a non-null ring R is defined to be the dimension of its zero ideal. The dimension of R will be denoted by* $\mathrm{Dim}\, R$.

Thus if $\mathrm{Dim}\, R = n$, then there exists at least one strictly increasing sequence

$$P_0 \subset P_1 \subset P_2 \subset \ldots \subset P_n \qquad (4.8.8)$$

containing $n+1$ prime ideals but there is no longer sequence of the same kind. Should however Dim $R = \infty$, then there exists a sequence such as (4.8.8) for every value of n. We note that, regardless of whether Dim R is finite or infinite,

$$\text{Dim } R = \sup_{M} \text{rank } M, \qquad (4.8.9)$$

where M ranges over all the maximal ideals of R.

The dimension of a non-null Noetherian ring need not be finite. However should the ring contain only a finite number of maximal ideals, then its dimension will be finite by virtue of (4.8.9). Noetherian rings whose maximal ideals are finite in number form the subject of the next section.

4.9 Local and semi-local rings

We have already encountered a number of situations in which a problem has been greatly simplified by localizing it at a suitable prime ideal. The advantage gained by this procedure arises from the fact that, after localization, we find ourselves dealing with a very special kind of ring. These matters will now be made the subject of a more systematic investigation. We begin by making the

Definition. *A non-null Noetherian ring is called a 'semi-local ring' if it possesses only a finite number of maximal ideals; a Noetherian ring which has exactly one maximal ideal is called a 'local ring'.*

It will be remembered† that a ring with exactly one maximal ideal was called a *quasi-local* ring. Accordingly a local ring is just a Noetherian quasi-local ring. There is another way of describing a local ring which is useful. Before giving this we make a further

Definition. *An element α of a non-null ring R is called a 'unit' of R if $\alpha R = R$.*

Thus if R is a non-null ring, then an element $\alpha \in R$ is a unit if and only if there exists $\beta \in R$ such that $\alpha\beta = 1$. In this case β is uniquely determined. (For if $\alpha\beta = \alpha\beta' = 1$, then $\beta' = \beta'\alpha\beta = 1\beta = \beta$.) This unique element is called the *inverse* of α and denoted by α^{-1}. Thus the units of R are simply those elements which have inverses. Note that if α is a unit, then so is α^{-1} and $(\alpha^{-1})^{-1} = \alpha$.

Now suppose that R is a quasi-local ring and let M be its unique maximal ideal. If $\alpha \in R$ then, since every proper ideal is contained in

† See section (3.8).

a maximal ideal, we have $\alpha R = R$ if and only if $\alpha \notin M$. Thus, in this case, the elements of M are precisely the elements of R which are *not* units.

Next suppose that R is a non-null ring in which the non-units form an ideal M' say. (Observe that there are elements of R which are not units. For instance the zero element is a case in point.) Since a proper ideal is composed entirely of non-units, we see that every proper ideal must be contained in M'. Thus M' is a maximal ideal and, indeed, it is the only maximal ideal which it possesses. Thus R is a quasi-local ring.

The observations of the last two paragraphs tell us that a non-null ring is a quasi-local ring if and only if its non-units form an ideal. Hence *a local ring may be described as a non-null Noetherian ring whose non-units form an ideal.* Note that if R is an arbitrary Noetherian ring and P any one of its prime ideals, then R_P is a local ring.

Suppose now that $\phi: R \to R'$ is a ring-epimorphism of R on to a non-null ring R'. *If R is a semi-local ring, then R' is also semi-local; while if R is a local ring, then R' will be a local ring as well.* This follows from Theorem 3 if we remember† that $\phi^{-1}(M')$ is a maximal ideal of R whenever M' is a maximal ideal of R'.

Proposition 18. *Let R be a semi-local ring and R' an extension ring of R which, when considered as an R-module, is finitely generated. Then R' is also a semi-local ring.*

Proof. By Theorem 4, R' is Noetherian and, by Theorem 5 of section (2.5), it is an integral extension of R. It therefore follows, from Theorem 11 of section (2.6), that a prime ideal of R' is a maximal ideal of that ring if and only if it contracts to a maximal ideal in R. Let $M_1, M_2, ..., M_s$ be the maximal ideals of R and suppose that $1 \leqslant i \leqslant s$. The proof will be complete if we show that there are only a finite number of prime R'-ideals which contract to M_i in R.

By Theorem 10 of section (2.6), $R'M_i$ is a proper ideal of R'. Let P' be any prime ideal of R' which contains $R'M_i$. Then $P' \cap R$ is a proper ideal of R containing the maximal ideal M_i. It follows that $P' \cap R = M_i$. Thus *the prime ideals of R' which contract to M_i are just the prime ideals which contain $R'M_i$.* Furthermore we see, from the argument just given, that every prime ideal containing $R'M_i$ must be a maximal ideal of R'. Hence if P' is a prime ideal containing $R'M_i$, there can be no prime ideal strictly contained in P' which also con-

† See Proposition 2 of section (2.2).

tains $R'M_i$. Accordingly P' is a *minimal prime ideal* of $R'M_i$ and therefore, since R' is a Noetherian ring, such prime ideals are finite in number. The proposition is therefore proved.

It is worthwhile noting that should R happen to be a *local ring* while R' satisfies the hypotheses of the proposition, we still can only assert that R' is *semi-local*. It is this fact which makes it desirable to consider both semi-local and local rings rather than to restrict our attention to the latter.

Let R be a semi-local ring and $M_1, M_2, ..., M_s$ its maximal ideals.

Definition. *An ideal B of R will be called an 'ideal of definition' of R if*

$$(M_1 \cap M_2 \cap ... \cap M_s)^k \subseteq B \subseteq M_1 \cap M_2 \cap ... \cap M_s$$

for some integer k.

Bearing in mind that a prime ideal contains B if and only if it contains $\operatorname{Rad} B$, we see, using Propositions 4 and 7 of section (2.3), that the following statements are equivalent:

(a) B is an ideal of definition of R;

(b) $\operatorname{Rad} B = M_1 \cap M_2 \cap ... \cap M_s$;

(c) the minimal prime ideals of B are $M_1, M_2, ..., M_s$.

Note that, since $M_i + M_j = R$ if $i \neq j$, we have, from Proposition 4,

$$M_1 M_2 ... M_s = M_1 \cap M_2 \cap ... \cap M_s.$$

This ideal is, of course, the Jacobson radical of R. Note also that *if R is a local ring with M as its ideal of non-units, then an ideal of definition is the same as an M-primary ideal.*

The name 'ideal of definition' needs a few words of explanation. For certain purposes it is convenient to regard R as a topological ring in which the powers of $M_1 \cap M_2 \cap ... \cap M_s$ form a fundamental system of neighbourhoods of the zero element. This we may call the *natural topology* of the ring. However if we take any ideal B whatsoever, then R can be regarded as a topological ring in which the powers of B form a fundamental system of neighbourhoods of the zero element. Now it is easy to see that the topology arising from B coincides with the natural topology if and only if B is an ideal of definition in the sense explained above. Thus it would be more explicit to speak of an 'ideal of definition for the natural topology of R'. Since, however, we are not concerned with topological questions at the moment, we have preferred to use the shorter, if less suggestive, description.

We saw in (4.8.9) that the dimension of a ring is equal to the upper bound of the ranks of its maximal ideals. It follows that the dimension of a semi-local ring is necessarily finite. We now contend that *a zero-dimensional semi-local ring is the same as a non-null Artin ring.* For suppose first that R is a semi-local ring satisfying $\text{Dim } R = 0$. Then every prime ideal must be a maximal ideal and therefore, by Proposition 7 Cor. applied to the zero ideal, $L_R(R) < \infty$. This shows that R is an Artin ring. Next assume that R is a non-null Artin ring. Then R is Noetherian and, by Theorem 8, it has only a finite number of prime ideals and each of these is a maximal ideal. Hence R is a semi-local ring and, since no prime ideal can contain any other prime ideal, we also have $\text{Dim } R = 0$. Note that *if R is a zero-dimensional semi-local ring, then the zero ideal is an ideal of definition.*

It will now be shown that the dimension of a semi-local ring may be characterized as the smallest number of elements required to generate an ideal of definition.

Theorem 23. *Let R be a semi-local ring of dimension d ($d \geqslant 0$). Then it is possible to find d elements which generate an ideal of definition of R. Furthermore no ideal of definition can be generated by fewer than d elements.*

Remark. In order that the case $d = 0$ may be covered, we adopt the usual convention that the ideal generated by the empty set is the zero ideal.

Proof. First suppose that $B = (b_1, b_2, ..., b_p)$ is an ideal of definition and let $M_1, M_2, ..., M_s$ be the maximal ideals of R. Then, for $1 \leqslant i \leqslant s$, M_i is a minimal prime ideal of B and therefore, by Theorem 22, $\text{rank } M_i \leqslant p$. It follows that

$$\text{Dim } R = \max_{1 \leqslant i \leqslant s} \text{rank } M_i \leqslant p.$$

Consequently no ideal of definition can be generated by fewer than $\text{Dim } R$ elements.

In order to prove the assertion that there exists an ideal of definition generated by d elements, we use induction on d. There is, of course, no problem when $d = 0$. It will therefore be assumed that $d > 0$ and also that the result in question has already been proved for semi-local rings whose dimensions are smaller than d.

Let $M_1, M_2, ..., M_s$ be the maximal ideals of R. Since every prime ideal of rank zero is a minimal prime ideal of the zero ideal, the total number of such prime ideals is finite. It may be that there are some prime ideals of rank zero among the M_i though, since

$$\max (\text{rank } M_i) = d > 0,$$

not all of them can have this property. Without loss of generality we may suppose that $\operatorname{rank} M_i > 0$ for $1 \leqslant i \leqslant t$ and $\operatorname{rank} M_j = 0$ for $t+1 \leqslant j \leqslant s$. Here t is an integer satisfying $1 \leqslant t \leqslant s$. Besides $M_{t+1}, \ldots,$ M_s there will be other prime ideals of rank zero. These we shall denote by $P_1, P_2, \ldots, P_q$. Since M_1 contains a minimal prime ideal of the zero ideal but does not contain any of $M_{t+1}, \ldots, M_s$, it follows that $q \geqslant 1$.

The product $M_1 M_2 \ldots M_s$ is not contained by any of $P_1, P_2, \ldots, P_q$. Accordingly, by Proposition 5 of section (2.3), there exists

$$\alpha_1 \in M_1 M_2 \ldots M_s$$

such that $\alpha_1 \notin P_1, \ldots, \alpha_1 \notin P_q$.

Suppose now that $1 \leqslant i \leqslant t$ and $M_i \supseteq P' \supseteq R\alpha_1$, where P' is a prime ideal. P' must contain a prime ideal of rank zero but it cannot contain any of $M_{t+1}, \ldots, M_s$ because $M_1, M_2, \ldots, M_s$ are distinct maximal ideals. Accordingly it contains one of $P_1, P_2, \ldots, P_q$. However, α_1 does not belong to any of these. It follows that P' *strictly contains* some other prime ideal, hence

$$\operatorname{rank}(M_i/R\alpha_1) < \operatorname{rank} M_i \quad (1 \leqslant i \leqslant t)$$

from which we see that

$$\operatorname{rank}(M_i/R\alpha_1) \leqslant d-1 \quad (1 \leqslant i \leqslant s). \tag{4.9.1}$$

Put $\bar{R} = R/R\alpha_1$, $\bar{M}_i = M_i/R\alpha_1$. Then $\bar{R}$ is a semi-local ring with $\bar{M}_1, \bar{M}_2, \ldots, \bar{M}_s$ as its maximal ideals and, by (4.9.1), $\operatorname{Dim} \bar{R} = \delta - 1$ (say), where $\delta \leqslant d$. The inductive hypothesis now shows that there exist elements $\bar{\alpha}_2, \ldots, \bar{\alpha}_\delta$ in $\bar{R}$ such that $(\bar{\alpha}_2, \ldots, \bar{\alpha}_\delta)$ is an ideal of definition of that ring. Choose α_i $(i = 2, 3, \ldots, \delta)$ in R so that $\bar{\alpha}_i$ is the image of α_i when the natural mapping $R \to \bar{R}$ is applied. Then, since

$$\bar{M}_1, \bar{M}_2, \ldots, \bar{M}_s$$

are the minimal prime ideals of $(\bar{\alpha}_2, \ldots, \bar{\alpha}_\delta)$, we see that $M_1, M_2, \ldots, M_s$ are the minimal prime ideals of $(\alpha_1, \alpha_2, \ldots, \alpha_\delta)$. This shows that $(\alpha_1, \alpha_2, \ldots, \alpha_\delta)$ is an ideal of definition of R. But $\delta \leqslant d$ and, as we saw earlier, no ideal of definition of R can be generated by fewer than d elements. Thus $\delta = d$ and all is proved.

Definition. *If R is a d-dimensional $(d \geqslant 0)$ semi-local ring, then any set of d elements which generates an ideal of definition of R is called a 'system of parameters' of R.*

By Theorem 23, systems of parameters always exist. Note that *if R is a d-dimensional local ring having M as its ideal of non-units, then a system of parameters is just a set of d elements which generates an M-primary ideal.*

Proposition 19. *Let R be a semi-local ring and* $\alpha_1, \alpha_2, ..., \alpha_k$ *elements each of which belongs to all the maximal ideals of R. Then* $R/(\alpha_1, ..., \alpha_k)$ *is a semi-local ring and*

$$\text{Dim } R \geqslant \text{Dim } R/(\alpha_1, ..., \alpha_k) \geqslant \text{Dim } R - k.$$

Furthermore $\text{Dim } R/(\alpha_1, ..., \alpha_k) = \text{Dim } R - k$ *if and only if* $\alpha_1, \alpha_2, ..., \alpha_k$ *are distinct elements belonging to some system of parameters.*

Proof. Let $M_1, M_2, ..., M_s$ be the maximal ideals of R. Then

$$\text{rank } M_i \geqslant \text{rank } (M_i/(\alpha_1, ..., \alpha_k))$$

and, by Proposition 16,

$$\text{rank } (M_i/(\alpha_1, ..., \alpha_k)) \geqslant \text{rank } M_i - k.$$

It follows that $\bar{R} = R/(\alpha_1, ..., \alpha_k)$ is a semi-local ring whose dimension satisfies

$$\text{Dim } R \geqslant \text{Dim } \bar{R} \geqslant \text{Dim } R - k.$$

Next assume that $\text{Dim } \bar{R} = d - k$, where $d = \text{Dim } R$. By Theorem 23, it is possible to find elements $\alpha_{k+1}, ..., \alpha_d$ in R such that if

$$\bar{\alpha}_j \quad (k+1 \leqslant j \leqslant d)$$

denotes the natural image of α_j in $\bar{R}$, then $(\bar{\alpha}_{k+1}, ..., \bar{\alpha}_d)$ is an ideal of definition of $\bar{R}$ and so has $M_1/(\alpha_1, ..., \alpha_k), ..., M_s/(\alpha_1, ..., \alpha_k)$ as its minimal prime ideals. It follows that $(\alpha_1, ..., \alpha_k, \alpha_{k+1}, ..., \alpha_d)$ is an ideal of definition of R. We may now conclude that the α_i are distinct (otherwise R would have an ideal of definition generated by fewer than $\text{Dim } R$ elements, which is impossible by Theorem 23) and that the set $\alpha_1, ..., \alpha_k$ can be enlarged to a system of parameters.

Finally let $\alpha_1, ..., \alpha_k, \alpha_{k+1}, ..., \alpha_d$ be a system of parameters of R. Then

$$(\alpha_1, ..., \alpha_k, \alpha_{k+1}, ..., \alpha_d)/(\alpha_1, ..., \alpha_k)$$

is an ideal of definition of $\bar{R}$ and it can be generated by $d - k$ elements, namely the images of $\alpha_{k+1}, \alpha_{k+2}, ..., \alpha_d$. It therefore follows, by Theorem 23, that $\text{Dim } \bar{R} \leqslant d - k$. However we have already established the opposite inequality. Thus $\text{Dim } \bar{R} = d - k$ and all is proved.

Proposition 20. *Let R be a semi-local ring and* α *an element which belongs to all the maximal ideals of R. If* α *is not a zero-divisor, then*

$$\text{Dim } R/(\alpha) = \text{Dim } R - 1.$$

228 NOETHERIAN RINGS AND MODULES

Proof. Let $M_1, M_2, \ldots, M_s$ be the maximal ideals of R. By Proposition 17, rank $(M_i/(\alpha))$ = rank $M_i - 1$. Hence

$$\operatorname{Dim} R/(\alpha) = \max_{1 \leqslant i \leqslant s} \operatorname{rank} (M_i/(\alpha)) = \operatorname{Dim} R - 1$$

as required.

For the remainder of this section we shall study the properties of local and quasi-local rings. If R is a quasi-local or local ring, then it has a unique maximal ideal M. Since M is a maximal ideal, R/M is a field. This field will be referred to as the *residue field* of R.

Proposition 21. *Let R be a quasi-local ring and M its maximal ideal. Suppose that E is an R-module and that K, L are submodules of E. If now K is finitely generated and $K \subseteq L + MK$, then $K \subseteq L$.*

Proof. Let $x \in K$. Then $x = y + z$, where $y \in L$ and $z \in MK$. Accordingly $y = x - z$ belongs to $K \cap L$ which shows that $K \subseteq (K \cap L) + MK$. It follows that, for the remainder of the proof, we may assume that L is contained in K. On this understanding $K = L + MK$ whence, by applying the natural mapping $K \to K/L$, we find that $K/L = M(K/L)$.

Put $E' = K/L$. It will suffice to prove that $E' = 0$ and this will follow if we show that $0: E' = R$. Assume the contrary. Then $0: E' \subseteq M$, because M is the only maximal ideal, and

$$RE' = K/L = M(K/L) = ME'.$$

Now, since K is finitely generated, E' is also finitely generated. We may therefore apply Proposition 13 of section (2.7). This shows that $R^n \subseteq M$ for some positive integer n and now we have the required contradiction because M is a proper ideal.

Theorem 24. *Let R be a quasi-local ring with maximal ideal M and E a finitely generated R-module. Suppose that $e_1, e_2, \ldots, e_n$ belong to E and let $\bar{e}_1, \bar{e}_2, \ldots, \bar{e}_n$ denote their images under the natural mapping $E \to E/ME$. Then $e_1, e_2, \ldots, e_n$ generate the R-module E if and only if $\bar{e}_1, \bar{e}_2, \ldots, \bar{e}_n$ generate the R-module E/ME. Should it happen that $e_1, e_2, \ldots, e_n$ generate E but no proper subset of them has this property, then $n = [E/ME : R/M]$.*

Remark. Since M annihilates E/ME, we may regard E/ME as a vector space over the residue field $F = R/M$. Note that $\bar{e}_1, \bar{e}_2, \ldots, \bar{e}_n$ generate E/ME as an R-module if and only if they generate it as an F-space. By $[E/ME : R/M]$ we mean, of course, the dimension of the vector space.

Proof. If $e_1, e_2, ..., e_n$ generate E, then it is clear that $\bar{e}_1, \bar{e}_2, ..., \bar{e}_n$ generate E/ME. Now suppose that $\bar{e}_1, \bar{e}_2, ..., \bar{e}_n$ generate the R-module E/ME. Let e belong to E and denote by $\bar{e}$ its natural image in E/ME. Then $\bar{e} = r_1\bar{e}_1 + r_2\bar{e}_2 + ... + r_n\bar{e}_n$. Thus e and $r_1 e_1 + r_2 e_2 + ... + r_n e_n$ have the same image in E/ME and therefore their difference belongs to ME. This shows that e belongs to $(Re_1 + ... + Re_n) + ME$ from which it follows that $E \subseteq (Re_1 + ... + Re_n) + ME$.

By hypothesis, E is finitely generated. Hence, by Proposition 21, $E \subseteq Re_1 + Re_2 + ... + Re_n$. However the opposite inclusion is obvious and so we see that $E = Re_1 + Re_2 + ... + Re_n$. This establishes the first assertion.

Now assume that $e_1, e_2, ..., e_n$ generate E but no proper subset of the e_i has this property. Then $E/ME = F\bar{e}_1 + F\bar{e}_2 + ... + F\bar{e}_n$. Hence if we can show that $\bar{e}_1, \bar{e}_2, ..., \bar{e}_n$ form a base, for the F-space E/ME, it will follow† that $n = [E/ME : F]$. We shall therefore suppose that

$$\alpha_1 \bar{e}_1 + \alpha_2 \bar{e}_2 + ... + \alpha_n \bar{e}_n = 0,$$

where $\alpha_1, \alpha_2, ..., \alpha_n$ belong to F, and seek to show that all the α_i are zero.

Assume the contrary. For definiteness we suppose that $\alpha_1 \neq 0$. Then $\bar{e}_1 = -\alpha_1^{-1}(\alpha_2 \bar{e}_2 + ... + \alpha_n \bar{e}_n)$. It follows that $\bar{e}_1, \bar{e}_2, ..., \bar{e}_n$ all belong to $F\bar{e}_2 + ... + F\bar{e}_n$ and therefore

$$E/ME = F\bar{e}_2 + ... + F\bar{e}_n = R\bar{e}_2 + ... + R\bar{e}_n.$$

But now the first part of the proof shows that $e_2, e_3, ..., e_n$ generate E by themselves; and thus we have the required contradiction.

The following lemma has several applications.

Lemma 10. *Let R be a quasi-local ring with maximal ideal M and let $P_1, P_2, ..., P_q$ be prime ideals all different from M. If now*

$$A = (\alpha_1, \alpha_2, ..., \alpha_n)$$

is an ideal of R not contained by any of $P_1, P_2, ..., P_q$, then it is possible to find elements $\beta_1, \beta_2, ..., \beta_n$ in R such that

(a) $A = (\alpha_1, \alpha_2, ..., \alpha_n) = (\beta_1, \beta_2, ..., \beta_n)$;

(b) *for each i $(1 \leqslant i \leqslant n), \beta_i \notin P_1, \beta_i \notin P_2, ..., \beta_i \notin P_q$*;

(c) $\alpha_i \equiv \beta_i \pmod{AM}$ *for* $1 \leqslant i \leqslant n$.

Remark. We recall that $\alpha_i \equiv \beta_i \pmod{AM}$ means that $\alpha_i - \beta_i$ belongs to AM. This terminology was introduced in section (1.5) but it has not been used for some time.

_____

† See Theorem 21 of section (1.13).

Proof. Without loss of generality we may assume that none of the prime ideals P_i is contained by any of the others. For if say P_1 were contained by one of $P_2, P_3, ..., P_q$ it would suffice to prove the assertion for the reduced set $P_2, P_3, ..., P_q$.

By renumbering $P_1, P_2, ..., P_q$ we can suppose that α_1 belongs to none of $P_1, P_2, ..., P_t$ and to all of $P_{t+1}, ..., P_q$. Next, if $t+1 \leqslant j \leqslant q$, then $AMP_1 ... P_t$ is not contained in P_j because none of $A, M, P_1, ..., P_t$ is contained in P_j. It therefore follows, by Proposition 5 of section (2.3), that there exists $u \in AMP_1 ... P_t$ such that $u \notin P_{t+1}, ..., u \notin P_q$. Put $\beta_1 = \alpha_1 + u$. Then $\alpha_1 \equiv \beta_1 \pmod{AM}$ and β_1 does not belong to any of the $P_\nu (1 \leqslant \nu \leqslant q)$. (In the case where $t = q$, we take $\beta_1 = \alpha_1$. If $t = 0$, then it is to be understood that u is chosen to be an element of AM which is not in any of $P_1, P_2, ..., P_q$.)

By entirely similar arguments, we can show that for each i $(1 \leqslant i \leqslant n)$ there exists β_i satisfying $\alpha_i \equiv \beta_i \pmod{AM}$ and $\beta_i \notin P_1, ..., \beta_i \notin P_q$.

From $\alpha_i \equiv \beta_i \pmod{AM}$ it follows that α_i belongs to

$$(\beta_1, \beta_2, ..., \beta_n) + AM.$$

Hence $\quad A \subseteq (\beta_1, \beta_2, ..., \beta_n) + AM$

and therefore, by Proposition 21, $A \subseteq (\beta_1, \beta_2, ..., \beta_n)$. However $\beta_i = (\beta_i - \alpha_i) + \alpha_i$ belongs to A and so we also have the opposite inclusion. Accordingly $A = (\beta_1, \beta_2, ..., \beta_n)$ and the lemma is proved.

Theorem 25. *Let R be a local ring, E a finitely generated R-module and A, B ideals of R. If now $0:_E A = 0$, then there exists an integer q such that*

$$B^n E:_E A = B^{n-q}(B^q E:_E A)$$

for all $n \geqslant q$.

Proof. If either $E = 0$ or $A = R$, then the assertion is trivial. We therefore exclude both of these possibilities. Before proceeding note that E is a Noetherian R-module.†

Let M denote the maximal ideal of R. Then $A \subseteq M$. Now denote by $P_1, P_2, ..., P_s$ the prime ideals which belong to the zero submodule of E. Since $0:_E A = 0$ and $A \subseteq M$, it follows that A is not contained by any P_i and also that M does not occur among $P_1, P_2, ..., P_s$. Accordingly, by Lemma 10, we can find elements $\alpha_1, \alpha_2, ..., \alpha_t$ in R such that

$$A = (\alpha_1, \alpha_2, ..., \alpha_t)$$

† See Proposition 1 Cor.

and for each $i(1 \leqslant i \leqslant t)$ α_i is not contained by any of $P_1, P_2, ..., P_s$. This secures that $0 :_E \alpha_i = 0$.

By Proposition 13, there exists an integer $\nu_i \geqslant 0$ such that

$$B^n E :_E \alpha_i = B^{n-\nu_i}(B^{\nu_i} E :_E \alpha_i)$$

for $n \geqslant \nu_i$. Put $\nu = \max(\nu_1, \nu_2, ..., \nu_t)$. Then for $n \geqslant \nu$

$$B^n E :_E \alpha_i = B^{n-\nu} B^{\nu-\nu_i}(B^{\nu_i} E :_E \alpha_i)$$

$$\subseteq B^{n-\nu}(B^{\nu} E :_E \alpha_i).$$

Accordingly when $n \geqslant \nu$

$$B^n E :_E A = (B^n E :_E \alpha_1) \cap ... \cap (B^n E :_E \alpha_t)$$

$$\subseteq B^{n-\nu}(B^{\nu} E :_E \alpha_1) \cap ... \cap B^{n-\nu}(B^{\nu} E :_E \alpha_t).$$

Next, by Proposition 14, there exists $\mu \geqslant 0$ such that

$$B^m(B^{\nu} E :_E \alpha_1) \cap ... \cap B^m(B^{\nu} E :_E \alpha_t)$$

$$= B^{m-\mu}\{B^{\mu}(B^{\nu} E :_E \alpha_1) \cap ... \cap B^{\mu}(B^{\nu} E :_E \alpha_t)\}$$

$$\subseteq B^{m-\mu}\{(B^{\mu+\nu} E :_E \alpha_1) \cap ... \cap (B^{\mu+\nu} E :_E \alpha_t)\}$$

$$= B^{m-\mu}(B^{\mu+\nu} E :_E A)$$

provided that $m \geqslant \mu$. Put $q = \mu + \nu$. Then for $n \geqslant q$

$$B^n E :_E A \subseteq B^{n-\nu}(B^{\nu} E :_E \alpha_1) \cap ... \cap B^{n-\nu}(B^{\nu} E :_E \alpha_t)$$

$$\subseteq B^{n-q}(B^q E :_E A)$$

$$\subseteq B^n E :_E A.$$

This establishes the theorem.

Exercises on Chapter 4

In these exercises R denotes a ring with an identity element. For Exercises 1–23 (inclusive) it is assumed that R is commutative, but in the exercises which come later R is allowed to be a non-commutative ring.

1. Let $A_i (1 \leqslant i \leqslant m)$ and $B_j (1 \leqslant j \leqslant n)$ be ideals of a ring R. Show that if for each $i(1 \leqslant i \leqslant m)$ and each $j(1 \leqslant j \leqslant n)$ A_i and B_j are comaximal, then $A_1 A_2 ... A_m$ and $B_1 B_2 ... B_n$ are also comaximal.

2. Let R be a Noetherian ring and $A = (a_1, a_2, ..., a_s)$ an ideal containing at least one element which is not a zero-divisor. Show that there exist elements $\lambda_2, \lambda_3, ..., \lambda_s$ in R such that $a_1 + \lambda_2 a_2 + ... + \lambda_s a_s$ is not a zero-divisor.

3. Prove that if every prime ideal of a ring R is finitely generated, then R is a Noetherian ring. (*Hint*. Show that the set Ω of ideals which are not finitely generated forms an inductive system with respect to the partial order determined by inclusion. Assuming that Ω is not empty, use the fact that a maximal member of Ω cannot be prime to derive a contradiction.) This result is known as *Cohen's Theorem* after I. S. Cohen.

4. Let I be an ideal of the ring R. Show that the direct sum

$$\bar{R} = R/I \oplus I/I^2 \oplus I^2/I^3 \oplus \dots$$

can be given the structure of a ring in such a way that the product of ξ in I^s/I^{s+1} and η in I^t/I^{t+1} is the image of xy in I^{s+t}/I^{s+t+1}, where x is a representative of ξ and y a representative of η. (This ring is called the *form ring of R with respect to I*. The form ring can obviously be graded by means of the monoid of non-negative integers so that I^s/I^{s+1} becomes the set of homogeneous elements of degree s.) Show also that if I is a finitely generated ideal and the residue ring R/I is Noetherian, then $\bar{R}$ is a Noetherian ring. (*Hint.* If I can be generated by n elements, show that $\bar{R}$ is a homomorphic image of the polynomial ring

$$(R/I)[X_1, X_2, \dots, X_n].)$$

5. Let the ideal I of R satisfy $\bigcap\limits_{n=1}^{\infty} I^n = (0)$ and suppose that the form ring of R, with respect to I, is an integral domain. Deduce that R is also an integral domain.

6. I and B are ideals of the ring R, $\bigcap\limits_{n=1}^{\infty}(B + I^n) = B$, and

$$\bar{B} = \bigoplus_{s \geq 0} (B \cap I^s + I^{s+1})/I^{s+1}$$

is a prime ideal of the form ring $\quad \bar{R} = \bigoplus\limits_{s \geq 0} I^s/I^{s+1}$

of R with respect to I. Prove that B is a prime ideal of R. (*Hint.* Apply Exercise 5 to the form ring of $R' = R/B$ with respect to $I' = (B+I)/B$.)

7. Let K be a submodule of a Noetherian R-module E, and $P_1, P_2, \dots, P_s$ the prime ideals which belong to the submodule K. Further, let

$$K = N_{i1} \cap N_{i2} \cap \dots \cap N_{is} \quad (1 \leq i \leq s)$$

be s normal decompositions of K in E, where, for each pair $i, j (1 \leq i, j \leq s)$, N_{ij} is a P_j-primary submodule of E. Prove that

$$K = N_{11} \cap N_{22} \cap \dots \cap N_{ss}$$

and show that this is also a normal decomposition of K in E.

8. Let E be a Noetherian R-module and P a prime ideal of R which contains $\text{Ann}_R E$. Prove that $PE : E = P$. Hence, or otherwise, show that there exists a submodule of E which is P-primary.

9. Let E be a Noetherian R-module and P a prime ideal of R. Prove that P belongs to the zero submodule of E if and only if $P = 0 : e$ for some element e of E. (*Hint.* Use Theorem 19 of section (2.9).)

10. Let E be a Noetherian R-module, P a prime ideal belonging to its zero submodule, and A an ideal contained in P. Prove that P belongs to the zero submodule of $0 :_E A$.

11. Let $E \neq 0$ be a Noetherian R-module, P a prime ideal belonging to the zero submodule of E, and N an arbitrary submodule of E. Show that either P belongs to the submodule N of E, or else it belongs to the zero submodule of N.

12. Let $E_1, E_2, \dots, E_s$ $(s \geq 1)$ be Noetherian R-modules and put

$$E = E_1 \oplus E_2 \oplus \dots \oplus E_s.$$

Show that a prime ideal P of R belongs to the zero submodule of E if and only if it belongs to the zero submodule of E_i for some i satisfying $1 \leq i \leq s$.

13. Let E be a Noetherian R-module and K a submodule of E. Further let

$$K = N_1 \cap N_2 \cap \dots \cap N_p,$$

where N_i is P_i-primary, be a normal decomposition of K in E. Show that the prime ideals which belong to the submodule K of N_i are precisely

$$P_1, ..., P_{i-1}, P_{i+1}, ..., P_p.$$

14. Let $E \neq 0$ be a Noetherian R-module and $P_1, P_2, ..., P_s$ the prime ideals that belong to its zero submodule. Let the integer t satisfy $1 \leqslant t < s$. Prove that there exists a submodule K, of E, such that (i) $P_1, P_2, ..., P_t$ are precisely the prime ideals belonging to the zero submodule of K, whereas (ii) $P_{t+1}, P_{t+2}, ..., P_s$ are precisely the prime ideals that belong to the submodule K of E.

15. Let $E \neq 0$ be a Noetherian R-module. Show that there exists a strictly decreasing sequence
$$E = E_0 \supset E_1 \supset E_2 \supset ... \supset E_s = 0$$
of submodules of E such that, for each $i (1 \leqslant i \leqslant s)$, E_{i-1}/E_i is isomorphic to R/P_i, where $P_1, P_2, ..., P_s$ are prime ideals. Show also that, for every such sequence, the set $\{P_1, P_2, ..., P_s\}$ of prime ideals includes every prime ideal that belongs to the zero submodule of E.

16. Let E and F be R-modules and denote by $\mathrm{Hom}_R(E, F)$ the set of all R-homomorphisms of E into F. Show that $\mathrm{Hom}_R(E, F)$ can be endowed with the structure of an R-module in which addition has the same meaning as in section (1.10) and, for r in R and u in $\mathrm{Hom}_R(E, F)$, ru is defined by

$$(ru)(e) = r(u(e))$$

for all e in E. Show also that if E is finitely generated and F is a Noetherian R-module, then $\mathrm{Hom}_R(E, F)$ is a Noetherian R-module.

17. Let R be a Noetherian ring, E a finitely generated R-module, F an arbitrary R-module, and S a non-empty multiplicatively closed subset of R. Then, with the notation of Exercise 16,

$$\{\mathrm{Hom}_R(E, F)\}_S \quad \text{and} \quad \mathrm{Hom}_{R_S}(E_S, F_S)$$

are isomorphic R_S-modules.

18. Let R be a Noetherian ring and A, B ideals of R. Show that there exists a positive integer m and ideals A', B' such that $A' \supseteq A^m, B' \supseteq B^m$, and

$$AB = A' \cap B = A \cap B'.$$

19. Let K, L, where $K \subseteq L$, be submodules of a Noetherian R-module E. Show that if α belongs to the Jacobson radical of R, $L \subseteq \alpha E + K$, and $L :_E \alpha = L$, then $K = L$.

20. Let R be a ring graded by the non-negative integers and, for each non-negative integer s, let $R^{(s)}$ denote the set of all homogeneous elements of degree s. Show that every proper homogeneous ideal is contained in a maximal ideal of the form

$$M + \sum_{s=1}^{\infty} R^{(s)},$$

where M is a maximal ideal of the ring $R^{(0)}$. (The intersection of all such maximal ideals of R is called the *graded radical* of R. Note that the graded radical is a homogeneous ideal.)

21. Let E be a Noetherian graded R-module, where both E and R are graded by the non-negative integers. Show that if a homogeneous ideal B is contained in the graded radical (see Exercise 20) of R, then $\bigcap_{n=1}^{\infty} B^n E = 0$.

22. Show that if E is a Noetherian R-module, B an ideal of R, and e an element of E, then there exists an integer $q \geqslant 0$ such that

$$B^n E : e = B^{n-q}(B^q E : e) + (0 : e)$$

for all $n \geqslant q$.

234   NOETHERIAN RINGS AND MODULES

23. The non-null ring R is Noetherian and has only a finite number of prime ideals. Prove that $\operatorname{Dim} R \leqslant 1$.

In the exercises that follow *it is not necessary for the ring R to be commutative*, but it is assumed that R possesses an identity element.

24. Let R be a (not necessarily commutative) ring with an identity element. Prove that the intersection J of all the maximal left ideals coincides with the intersection of all the maximal right ideals. (J is called the *Jacobson radical* of R. Note that it is a two-sided ideal.)

25. Let R be a ring and x an element belonging to its Jacobson radical. Show that there exists an element y in R such that $(1+x)y = y(1+x) = 1$.

26. E is a finitely generated left R-module and J is the Jacobson radical of R. Prove that if $JE = E$, then $E = 0$.

27. The ring R has the property that the set of all elements x such that $Rx \neq R$ is a left ideal. Prove that this ideal is the Jacobson radical J of R and that R/J is a division ring. (This result can be used to generalize the concept of a quasi-local ring.)

5

THE THEORY OF GRADE

General remarks. Throughout this chapter we confine our attention to commutative rings each of which is assumed to possess an identity element. The results developed here were originally discovered by means of the relatively sophisticated techniques of what is known as 'homological algebra'. However further study has revealed that a substantial part of the theory can be derived by simpler and more direct methods. It is with this simplified presentation that we shall now be concerned.

5.1 The concept of grade

Let R be a ring (commutative and possessing an identity element), E an R-module, and $\alpha_1, \alpha_2, \ldots, \alpha_s$ a sequence of elements of R.

Definition. *The sequence $\alpha_1, \alpha_2, \ldots, \alpha_s$ will be called an ' R-sequence on E' provided that*

(a) $(\alpha_1, \alpha_2, \ldots, \alpha_s)E \neq E$, *and*

(b) $(\alpha_1, \alpha_2, \ldots, \alpha_{i-1})E :_E \alpha_i = (\alpha_1, \alpha_2, \ldots, \alpha_{i-1})E$ *for* $1 \leqslant i \leqslant s$.

Naturally when $i = 1$ it is to be understood that (b) asserts that $0 :_E \alpha_1 = 0$. If $\alpha_1, \alpha_2, \ldots, \alpha_s$ is an R-sequence on R, then we say simply that it is an *R-sequence*. It is convenient to introduce some further terminology in order to facilitate the discussion of R-sequences on modules.

Definition. *An element γ of the ring R is said to be a 'zero-divisor' on the R-module E if there exists $e \in E$, $e \neq 0$, such that $\gamma e = 0$.*

It is important to note that, when E is a *Noetherian* R-module, γ is a zero-divisor on E when and only when† it is contained in one of the prime ideals belonging to the zero submodule of E. Again, returning to the general case, an element $\alpha \in R$ is *not* a zero-divisor on E if and only if $0 :_E \alpha = 0$.

Now suppose that $(\alpha_1, \alpha_2, \ldots, \alpha_s)E \neq E$. Then $\alpha_1, \alpha_2, \ldots, \alpha_s$ is an

† See Theorem 14 Cor. 2 of section (4.4).

R-sequence on E precisely when, for each $i\,(1 \leqslant i \leqslant s)$, α_i is not a zero-divisor on $E/(\alpha_1, ..., \alpha_{i-1})E$. Further, for $p+1 \leqslant i \leqslant s$, we have
$$(\alpha_{p+1}, ..., \alpha_i)\,(E/(\alpha_1, ..., \alpha_p)E) = (\alpha_1, ..., \alpha_p, ..., \alpha_i)E/(\alpha_1, ..., \alpha_p)E.$$
It follows that *if* $\alpha_1, \alpha_2, ..., \alpha_s$ *is an R-sequence on E, then*
$$\alpha_{p+1}, \alpha_{p+2}, ..., \alpha_s$$
is an R-sequence on $E/(\alpha_1, \alpha_2, ..., \alpha_p)E$.

Lemma 1. *Let* $\alpha_1, \alpha_2, ..., \alpha_{s-1}, \alpha_s$ *be an R-sequence on E and suppose that* $s \geqslant 2$. *Then*
$$(\alpha_1, ..., \alpha_{s-2}, \alpha_s)E :_E \alpha_{s-1} = (\alpha_1, ..., \alpha_{s-2}, \alpha_s)E.$$
It follows that $\alpha_1, ..., \alpha_{s-2}, \alpha_s, \alpha_{s-1}$ *is an R-sequence on E if and only if*
$$(\alpha_1, ..., \alpha_{s-2})E :_E \alpha_s = (\alpha_1, ..., \alpha_{s-2})E.$$

Proof. Let e belong to $(\alpha_1, ..., \alpha_{s-2}, \alpha_s)E :_E \alpha_{s-1}$. Then
$$\alpha_{s-1}e = \alpha_1 e_1 + ... + \alpha_{s-2}e_{s-2} + \alpha_s e_s$$
for suitable elements $e_1, ..., e_{s-2}, e_s$ in E. Accordingly
$$e_s \in ((\alpha_1, ..., \alpha_{s-2}, \alpha_{s-1})E_E :_E \alpha_s) = (\alpha_1, ..., \alpha_{s-2}, \alpha_{s-1})E$$
and therefore we may express e_s in the form
$$e_s = \alpha_1 e_1' + ... + \alpha_{s-2}e_{s-2}' + \alpha_{s-1}e_{s-1}'.$$
It follows that $\quad \alpha_{s-1}e - \alpha_s \alpha_{s-1} e_{s-1}' \in (\alpha_1, ..., \alpha_{s-2})E.$
Thus $\quad e - \alpha_s e_{s-1}' \in (\alpha_1, ..., \alpha_{s-2})E :_E \alpha_{s-1} = (\alpha_1, ..., \alpha_{s-2})E$
and therefore $e \in (\alpha_1, ..., \alpha_{s-2}, \alpha_s)E$. This proves that
$$(\alpha_1, ..., \alpha_{s-2}, \alpha_s)E :_E \alpha_{s-1} \subseteq (\alpha_1, ..., \alpha_{s-2}, \alpha_s)E$$
and, as the opposite inclusion is obvious, the lemma follows.

Lemma 2. *Let E be an R-module and A an ideal of R. Further, let α and α' be elements of A which are not zero-divisors on E. Then*
$$(\alpha E :_E A)/\alpha E \quad and \quad (\alpha' E :_E A)/\alpha' E$$
are isomorphic as R-modules.

Proof. Let S denote the set of all elements of R which are not zero-divisors on E. Then S, as is easily verified, is multiplicatively closed and, by hypothesis, both α and α' belong to it. We now form the module E_S of fractions and, in what follows, regard it as an R-module.

The mapping $\phi : (\alpha E :_E A) \to E_S$ defined by $\phi(x) = [x/\alpha]$, where x is an arbitrary element of $\alpha E :_E A$, is a homomorphism of R-modules. If $\phi(x) = 0$, then $sx = 0$ for some $s \in S$ and therefore $x = 0$. Thus ϕ is a monomorphism.

Let $\chi\colon E \to E_S$ be the canonical mapping†. We contend that

$$\mathrm{Im}\,\phi = \chi(E)\!:_{E_s}\!A. \tag{5.1.1}$$

For let x belong to $\alpha E\!:_E\!A$ and $a \in A$. Then $a\phi(x) = [ax/\alpha]$. But $ax = \alpha e'$ for some $e' \in E$ and therefore $[ax/\alpha] = [\alpha e'/\alpha]$ which belongs to $\chi(E)$. Thus $a\phi(x) \in \chi(E)$, whence $\phi(x)$ belongs to $\chi(E)\!:_{E_s}\!A$.

Now suppose that ξ is an element of $\chi(E)\!:_{E_s}\!A$. Then $\alpha\xi \in \chi(E)$, say $\alpha\xi = \chi(e)$. Let $a \in A$. Then

$$\chi(ae) = a\chi(e) = a\alpha\xi \in \alpha\chi(E)$$

and so $\chi(ae) = \chi(ae^*)$ for some $e^* \in E$. But, as was shown in section (3.1), the kernel of χ is the S-component of the zero submodule of E. Hence $s(ae - \alpha e^*) = 0$ for some $s \in S$ and therefore $ae = \alpha e^* \in \alpha E$ because s is not a zero-divisor on E. It follows that e belongs to $\alpha E\!:_E\!A$. Furthermore

$$\alpha\xi = \chi(e) = \left[\frac{\alpha e}{\alpha}\right] = \alpha\left[\frac{e}{\alpha}\right] = \alpha\phi(e).$$

But E_S is an R_S-module as well as an R-module. We can therefore multiply $[\alpha/1]\xi = [\alpha/1]\phi(e)$ by $[1/\alpha]$ to obtain $\xi = \phi(e) \in \mathrm{Im}\,\phi$. Thus (5.1.1) is established.

This result, combined with the fact that ϕ is a monomorphism, shows that ϕ induces an isomorphism of the R-module $\alpha E\!:_E\!A$ on to the R-module $\chi(E)\!:_{E_s}\!A$. Moreover αE is carried on to $\chi(E)$. Accordingly there is induced an isomorphism

$$(\alpha E\!:_E\!A)/\alpha E \approx (\chi(E)\!:_{E_s}\!A)/\chi(E).$$

Since we have a similar isomorphism in which α' replaces α, the lemma follows.

Corollary. *Let E be an R-module, K a submodule of E and A an ideal of R. If now $\alpha, \alpha' \in A$ and $K\!:_E\!\alpha = K = K\!:_E\!\alpha'$, then*

$$\{(\alpha E + K)\!:_E\!A\}/(\alpha E + K) \quad and \quad \{(\alpha' E + K)\!:_E\!A\}/(\alpha' E + K)$$

are isomorphic R-modules.

Proof. The hypotheses show that neither α nor α' is a zero-divisor on E/K. Also

$$\alpha(E/K) = (\alpha E + K)/K$$

and

$$\alpha(E/K)\!:_{E/K}\!A = \{(\alpha E + K)\!:_E\!A\}/K$$

with similar relations involving α'. The corollary therefore follows from the lemma.

238   THE THEORY OF GRADE

Theorem 1. *Let E be a Noetherian R-module and A an ideal of R. Assume that $\alpha_1, \alpha_2, ..., \alpha_k$ and $\beta_1, \beta_2, ..., \beta_k$ are two R-sequences on E each consisting of k elements belonging to the ideal A. In these circumstances there is an isomorphism*

$$\{(\alpha_1, \alpha_2, ..., \alpha_k)E :_E A\}/(\alpha_1, \alpha_2, ..., \alpha_k)E$$
$$\approx \{(\beta_1, \beta_2, ..., \beta_k)E :_E A\}/(\beta_1, \beta_2, ..., \beta_k)E$$

between the two sides considered as R-modules.

Proof. The modules $(\alpha_1, \alpha_2, ..., \alpha_i)E$ $(0 \leqslant i \leqslant k-1)$ and $(\beta_1, \beta_2, ..., \beta_j)E$ $(0 \leqslant j \leqslant k-1)$ are all of them proper submodules of the Noetherian R-module E and therefore to each one of them belongs a finite number of prime ideals. Denote by $P_1, P_2, ..., P_s$ the complete set of prime ideals which arise in this way.

Consider P_1. It belongs to one of the set of $2k$ submodules. Suppose for definiteness that it belongs to $(\alpha_1, \alpha_2, ..., \alpha_i)E$, where $0 \leqslant i \leqslant k-1$. Since

$$(\alpha_1, \alpha_2, ..., \alpha_i)E :_E \alpha_{i+1} = (\alpha_1, \alpha_2, ..., \alpha_i)E$$

we see that $\alpha_{i+1} \notin P_1$ and therefore $A \nsubseteq P_1$. It follows, for similar reasons, that A is not contained by any of $P_1, P_2, ..., P_s$.

By Proposition 5 of section (2.3), there exists $\gamma_k \in A$ such that

$$\gamma_k \notin P_1, \gamma_k \notin P_2, ..., \gamma_k \notin P_s.$$

Accordingly

$$(\alpha_1, \alpha_2, ..., \alpha_i)E :_E \gamma_k = (\alpha_1, \alpha_2, ..., \alpha_i)E \quad (0 \leqslant i \leqslant k-1) \quad (5.1.2)$$

and

$$(\beta_1, \beta_2, ..., \beta_i)E :_E \gamma_k = (\beta_1, \beta_2, ..., \beta_i)E \quad (0 \leqslant i \leqslant k-1). \quad (5.1.3)$$

In particular

$$(\alpha_1, \alpha_2, ..., \alpha_{k-1})E :_E \alpha_k = (\alpha_1, \alpha_2, ..., \alpha_{k-1})E = (\alpha_1, \alpha_2, ..., \alpha_{k-1})E :_E \gamma_k$$

and therefore, by the Cor. to Lemma 2, we have an isomorphism

$$\{(\alpha_1, ..., \alpha_{k-1}, \alpha_k)E :_E A\}/(\alpha_1, ..., \alpha_{k-1}, \alpha_k)E$$
$$\approx \{(\alpha_1, ..., \alpha_{k-1}, \gamma_k)E :_E A\}/(\alpha_1, ..., \alpha_{k-1}, \gamma_k)E. \quad (5.1.4)$$

It is clear that $\alpha_1, ..., \alpha_{k-1}, \gamma_k$ is an R-sequence on E. Further, by (5.1.2) and Lemma 1, we may conclude that $\alpha_1, ..., \alpha_{k-2}, \gamma_k, \alpha_{k-1}$ is also an R-sequence on E. Indeed another application of these results shows that $\alpha_1, ..., \alpha_{k-3}, \gamma_k, \alpha_{k-2}, \alpha_{k-1}$ is again an R-sequence on E and, proceeding in this way, we establish that $\gamma_k, \alpha_1, \alpha_2, ..., \alpha_{k-1}$ is an R-sequence on E. Note that (5.1.4) may be rewritten as

$$\{(\alpha_1, ..., \alpha_k)E :_E A\}/(\alpha_1, ..., \alpha_k)E$$
$$\approx \{(\gamma_k, \alpha_1, ..., \alpha_{k-1})E :_E A\}/(\gamma_k, \alpha_1, ..., \alpha_{k-1})E. \quad (5.1.5)$$

In exactly the same manner, we deduce that $\gamma_k, \beta_1, \beta_2, ..., \beta_{k-1}$ is an R-sequence on E and, as a companion to (5.1.5), we have an isomorphism

$$\{(\beta_1, ..., \beta_k)E :_E A\}/(\beta_1, ..., \beta_k)E$$
$$\approx \{(\gamma_k, \beta_1, ..., \beta_{k-1})E :_E A\}/(\gamma_k, \beta_1, ..., \beta_{k-1})E. \quad (5.1.6)$$

The arguments which led to (5.1.5) and (5.1.6) can now be repeated with $\gamma_k, \alpha_1, ..., \alpha_{k-1}$ and $\gamma_k, \beta_1, ..., \beta_{k-1}$ replacing $\alpha_1, \alpha_2, ..., \alpha_k$ and $\beta_1, \beta_2, ..., \beta_k$. It is found that there exists an element $\gamma_{k-1} \in A$ such that (i) $\gamma_{k-1}, \gamma_k, \alpha_1, ..., \alpha_{k-2}$ and $\gamma_{k-1}, \gamma_k, \beta_1, ..., \beta_{k-2}$ are R-sequences on E and (ii) there is an isomorphism

$$\{(\gamma_k, \alpha_1, ..., \alpha_{k-1})E :_E A\}/(\gamma_k, \alpha_1, ..., \alpha_{k-1})E$$
$$\approx \{(\gamma_{k-1}, \gamma_k, \alpha_1, ..., \alpha_{k-2})E :_E A\}/(\gamma_{k-1}, \gamma_k, \alpha_1, ..., \alpha_{k-2})E \quad (5.1.7)$$

with a similar isomorphism involving $\beta_1, \beta_2, ..., \beta_{k-1}$. Observe that, by combining the new isomorphisms with (5.1.5) and (5.1.6) we obtain

$$\{(\alpha_1, ..., \alpha_k)E :_E A\}/(\alpha_1, ..., \alpha_k)E$$
$$\approx \{(\gamma_{k-1}, \gamma_k, \alpha_1, ..., \alpha_{k-2})E :_E A\}/(\gamma_{k-1}, \gamma_k, \alpha_1, ..., \alpha_{k-2})E$$

and

$$\{(\beta_1, ..., \beta_k)E :_E A\}/(\beta_1, ..., \beta_k)E$$
$$\approx \{(\gamma_{k-1}, \gamma_k, \beta_1 ..., \beta_{k-2})E :_E A\}/(\gamma_{k-1}, \gamma_k, \beta_1, ..., \beta_{k-2})E.$$

It is now clear how the argument continues. Eventually we obtain an R-sequence $\gamma_1, \gamma_2, ..., \gamma_k$ on E such that

$$\{(\gamma_1, ..., \gamma_k)E :_E A\}/(\gamma_1, ..., \gamma_k)E$$

is isomorphic to both

$$\{(\alpha_1, ..., \alpha_k)E :_E A\}/(\alpha_1, ..., \alpha_k)E$$

and

$$\{(\beta_1, ..., \beta_k)E :_E A\}/(\beta_1, ..., \beta_k)E.$$

This completes the proof.

Definition. *If $\alpha_1, \alpha_2, ..., \alpha_n$ is an R-sequence on an R-module E and if the α_i all belong to an ideal A, then $\alpha_1, \alpha_2, ..., \alpha_n$ will be said to be an 'R-sequence on E in A'.*

As an application of the last theorem we now prove

Theorem 2. *Let E be a Noetherian R-module and A an ideal of R such that $AE \neq E$. Further let $\alpha_1, \alpha_2, ..., \alpha_m$ and $\beta_1, \beta_2, ..., \beta_n$ be R-sequences on E in A. If now $m < n$, then it is possible to find $\alpha_{m+1}, \alpha_{m+2}, ..., \alpha_n$ so that $\alpha_1, ..., \alpha_m, \alpha_{m+1}, ..., \alpha_n$ is an R-sequence on E in A.*

Proof. Since $\alpha_1, \alpha_2, ..., \alpha_m$ and $\beta_1, \beta_2, ..., \beta_m$ are both R-sequences on E and since they have the same number m of terms, the last theorem shows that there is an isomorphism

$$\{(\alpha_1, ..., \alpha_m)E :_E A\}/(\alpha_1, ..., \alpha_m)E \approx \{(\beta_1, ..., \beta_m)E :_E A\}/(\beta_1, ..., \beta_m)E$$

of R-modules. Now $m < n$ and

$$(\beta_1, ..., \beta_m)E :_E \beta_{m+1} = (\beta_1, ..., \beta_m)E;$$

consequently

$$(\beta_1, ..., \beta_m)E :_E A = (\beta_1, ..., \beta_m)E.$$

It follows, from the isomorphism, that

$$\{(\alpha_1, ..., \alpha_m)E :_E A\}/(\alpha_1, ..., \alpha_m)E$$

is a null module and this in turn implies that

$$(\alpha_1, ..., \alpha_m)E :_E A = (\alpha_1, ..., \alpha_m)E.$$

Accordingly A is not contained by any of the prime ideals belonging to the submodule $(\alpha_1, ..., \alpha_m)E$ of E. Hence, by Proposition 5 of section (2.3), there exists $\alpha_{m+1} \in A$ which is not in any of these prime ideals and which therefore satisfies

$$(\alpha_1, ..., \alpha_m)E :_E \alpha_{m+1} = (\alpha_1, ..., \alpha_m)E.$$

But $AE \neq E$; consequently $(\alpha_1, ..., \alpha_m, \alpha_{m+1})E \neq E$. Thus

$$\alpha_1, ..., \alpha_m, \alpha_{m+1}$$

is an R-sequence on E in A. If now $m+1 < n$, then we can repeat the argument. The theorem follows.

Theorem 3. *Let E be a Noetherian R-module and A an ideal of R such that $AE \neq E$. Then there is a largest integer k ($k \geqslant 0$) such that there exists an R-sequence on E in A having k terms. Furthermore, if $\alpha_1, \alpha_2, ..., \alpha_m$ is an arbitrary R-sequence on E in A, then it is possible to find elements $\alpha_{m+1}, \alpha_{m+2}, ..., \alpha_k$ so that $\alpha_1, ..., \alpha_m, \alpha_{m+1}, ..., \alpha_k$ is an R-sequence on E in A having the maximum number k of terms.*

Proof. In view of Theorem 2, it is only necessary to prove the first assertion. To this end we shall assume that for every $n > 0$ there exists at least one R-sequence on E in A having n terms. From this we shall derive a contradiction.

Let α_1 be an R-sequence on E in A. Since there exists an R-sequence on E in A with two terms, Theorem 2 shows that we can find α_2 so that α_1, α_2 is an R-sequence on E in A. Next, because there exists an R-sequence on E in A with three terms, it follows (again by Theorem 2) that we can find α_3 so that $\alpha_1, \alpha_2, \alpha_3$ is an R-sequence on E in A. And so on. In this way we obtain an infinite sequence $\alpha_1, \alpha_2, \alpha_3, \ldots$ such that, for each n, $\alpha_1, \alpha_2, \ldots, \alpha_n$ is an R-sequence on E in A.

Since E is a Noetherian R-module, the sequence

$$(\alpha_1)E \subseteq (\alpha_1, \alpha_2)E \subseteq (\alpha_1, \alpha_2, \alpha_3)E \subseteq \ldots$$

of submodules of E must terminate. We can therefore find an integer m so that $(\alpha_1, \ldots, \alpha_m)E = (\alpha_1, \ldots, \alpha_m, \alpha_{m+1})E$. Then

$$(\alpha_1, \ldots, \alpha_m)E = (\alpha_1, \ldots, \alpha_m)E :_E \alpha_{m+1}$$
$$= (\alpha_1, \ldots, \alpha_m, \alpha_{m+1})E :_E \alpha_{m+1}$$
$$= E.$$

However all the α_i belong to A and $AE \neq E$. This yields the required contradiction and completes the proof.

Definition. *Let E be a Noetherian R-module and A an ideal of R such that $AE \neq E$. The number of terms in a maximal R-sequence on E in A will be called the 'grade of A on E' and denoted by* gr $(A; E)$.

Here we have used the expression *maximal R-sequence on E in A* to mean an R-sequence on E in A having the largest possible number of terms. Theorem 3 ensures that gr $(A; E)$ is defined and finite under the conditions stated. It should be noted that, so far, we have no estimate of how large gr $(A; E)$ may be. We shall, however, return to this point later.

By taking $E = R$ we arrive at the following subsidiary

Definition. *Let A be a proper ideal of a Noetherian ring R. Then by the 'grade of A' is meant the grade of A on R when R is considered as a module with respect to itself. The grade of A is denoted by* gr (A). *Thus* gr $(A) = $ gr $(A; R)$.

Assuming that E is a Noetherian R-module and $AE \neq E$, we observe that gr $(A; E) = 0$ means that each element of A is a zero-divisor on E and hence is contained in some prime ideal belonging to the zero submodule of E. Accordingly, by Proposition 5 of section (2.3), A must be wholly contained by one of these prime ideals. *Hence*

$$\text{gr}\,(A; E) = 0 \quad \text{if and only if} \quad 0 :_E A \neq 0.$$

Another point to note is that if $\alpha_1, \alpha_2, \ldots, \alpha_s$ is an R-sequence on E in A, then

$$(\alpha_1, \ldots, \alpha_s)E :_E A = (\alpha_1, \ldots, \alpha_s)E$$

if $s < $ gr $(A; E)$ whereas

$$(\alpha_1, \ldots, \alpha_s)E :_E A \neq (\alpha_1, \ldots, \alpha_s)E$$

if $s = $ gr $(A; E)$. This is clear because in the former case the sequence can be extended and in the latter it can not.

Proposition 1. *Let E be a Noetherian R-module and A an ideal of R satisfying $AE \neq E$. If now B is an ideal of R and $B \subseteq \operatorname{Rad} A$, then $BE \neq E$ and $\operatorname{gr}(B;E) \leqslant \operatorname{gr}(A;E)$. Hence if $\operatorname{Rad} A = \operatorname{Rad} B$, then $\operatorname{gr}(B;E) = \operatorname{gr}(A;E)$.*

Proof. First we assume that $BE = E$ and drive a contradiction. Since E is a Noetherian R-module, it is finitely generated. Let

$$E = Re_1 + Re_2 + \ldots + Re_p.$$

Then for each $i \, (1 \leqslant i \leqslant p)$ we have a relation of the form

$$e_i = b_{i1}e_1 + b_{i2}e_2 + \ldots + b_{ip}e_p,$$

where $b_{ij} \in B$. Let B_0 be the ideal generated by the b_{ij} so that $B_0 \subseteq B$ and $E = B_0 E$. Since $B_0 \subseteq \operatorname{Rad} A$ and B_0 is finitely generated, it follows† that there is an integer m such that $B_0^m \subseteq A$. However from $E = B_0 E$ follows $E = B_0^m E$. Consequently $E = AE$ which is the required contradiction.

It has been shown that $BE \neq E$. Thus $\operatorname{gr}(B;E)$ is defined. Let $b_1, b_2, \ldots, b_s$ be a maximal R-sequence on E in B. Then $s = \operatorname{gr}(B;E)$. Choose n so that $b_1^n \in A$ and put $\beta_1 = b_1^n$. Since $0 :_E b_1 = 0$, it follows that $0 :_E \beta_1 = 0$. Hence β_1 is an R-sequence on E in $A \cap B$ and, by Theorem 3, it can be continued to a maximal R-sequence $\beta_1, b_2', \ldots, b_s'$ on E in B. Choose k so that $b_2'^k \in A$ and put $\beta_2 = b_2'^k$. Then, because $(\beta_1)E :_E b_2' = (\beta_1)E$, we have $(\beta_1)E :_E \beta_2 = (\beta_1)E$. Thus β_1, β_2 is an R-sequence on E in $A \cap B$ and it can be continued to a maximal R-sequence $\beta_1, \beta_2, b_3'' \ldots, b_s''$ on E in B. Proceeding in this way we finally obtain an R-sequence $\beta_1, \beta_2, \ldots, \beta_s$ on E in $A \cap B$. It follows that

$$\operatorname{gr}(A;E) \geqslant s = \operatorname{gr}(B;E).$$

The final assertion of the proposition needs no comment.

Proposition 2. *Let E be a Noetherian R-module and A, B ideals of R such that $AE \neq E$ and $BE \neq E$. Then*

$$\operatorname{gr}(AB;E) = \operatorname{gr}(A \cap B;E) = \min\{\operatorname{gr}(A;E), \operatorname{gr}(B;E)\}.$$

Proof. It is clear that

$$\operatorname{gr}(AB;E) \leqslant \operatorname{gr}(A \cap B;E) \leqslant \min\{\operatorname{gr}(A;E), \operatorname{gr}(B;E)\}.$$

Put $s = \operatorname{gr}(A;E)$ and $t = \operatorname{gr}(B;E)$. Let $\alpha_1, \alpha_2, \ldots, \alpha_s$ be an R-sequence on E in A and $\beta_1, \beta_2, \ldots, \beta_t$ an R-sequence on E in B. Without loss of generality we may assume that $s \leqslant t$.

Put $\gamma_1 = \alpha_1 \beta_1$. Then $\gamma_1 \in AB$ and, since $0 :_E \alpha_1 = 0$ and $0 :_E \beta_1 = 0$, we have $0 :_E \gamma_1 = 0$. Thus γ_1 is an R-sequence on E in AB. This can be

† See Proposition 8 of section (2.3).

extended to an R-sequence $\gamma_1, \alpha_2', ..., \alpha_s'$ on E in A. Similarly there exists an R-sequence $\gamma_1, \beta_2', ..., \beta_t'$ on E in B. Now set $\gamma_2 = \alpha_2'\beta_2'$. Then $\gamma_2 \in AB$ and we have $(\gamma_1)E:_E\gamma_2 = (\gamma_1)E$ because $(\gamma_1)E:_E\alpha_2' = (\gamma_1)E$ and $(\gamma_1)E:_E\beta_2' = (\gamma_1)E$. Thus γ_1, γ_2 is an R-sequence on E in AB. Further there exist sequences $\gamma_1, \gamma_2, \alpha_3'', ..., \alpha_s''$ and $\gamma_1, \gamma_2, \beta_3'', ..., \beta_t''$ which are R-sequences on E in A and B respectively. After s steps, this procedure produces an R-sequence $\gamma_1, \gamma_2, ..., \gamma_s$ on E in AB. Accordingly

$$\mathrm{gr}\,(AB;E) \geqslant s = \min\{\mathrm{gr}\,(A;E), \mathrm{gr}\,(B;E)\}$$

and the proof is complete.

There are certain situations in which the property of being an R-sequence on a module is not affected if the order of the elements in the sequence is changed. In this connection we recall that the Jacobson radical of a ring is defined as the intersection of all its maximal ideals. It will be shown presently that, for Noetherian modules, the order of the elements in an R-sequence is not important provided that the elements all belong to the Jacobson radical. In preparation for this we establish the following

Lemma 3. *Let E be a Noetherian R-module and α, β an R-sequence on E. If now α belongs to the Jacobson radical of R, then β, α is also an R-sequence on E.*

Proof. By Lemma 1, it is enough to show that $0:_E\beta = 0$. Assume that $\beta e = 0$ where $e \in E$. We contend that $e \in (\alpha^m)E$ for $m \geqslant 0$. Indeed this is trivial for $m = 0$. Also, if $e \in (\alpha^s)E$, where $s \geqslant 0$, then $e = \alpha^s e'$ for some $e' \in E$ and $\beta \alpha^s e' = 0$. But α is not a zero-divisor on E. Consequently $\beta e' = 0$ whence *a fortiori* $\beta e' \in \alpha E$. However $\alpha E:_E\beta = \alpha E$. We may therefore conclude that $e' \in \alpha E$, say $e' = \alpha e''$. Thus $e = \alpha^s e' = \alpha^{s+1}e''$ whence $e \in (\alpha^{s+1})E$. The assertion that $e \in (\alpha^m)E$ for all m follows by induction.

It has now been shown that

$$e \in \bigcap_{m=1}^{\infty} (\alpha^m)\,E = \bigcap_{m=1}^{\infty} (\alpha)^m\,E.$$

Hence $e = 0$ by Theorem 19 of section (4.6). Accordingly $0:_E\beta = 0$ and the proof is complete.

Theorem 4. *Let E be a Noetherian R-module and $\alpha_1, \alpha_2, ..., \alpha_s$ an R-sequence on E each of whose elements is contained in the Jacobson radical of R. If now $\{i_1, i_2, ..., i_s\}$ is an arbitrary rearrangement of $\{1, 2, ..., s\}$, then $\alpha_{i_1}, \alpha_{i_2}, ..., \alpha_{i_s}$ is also an R-sequence on E.*

Proof. It is enough to show that any two adjacent terms in the sequence $\alpha_1, \alpha_2, ..., \alpha_s$ may be interchanged without disturbing the property of being an R-sequence on E. Let us show that

$$\alpha_1, ..., \alpha_{i-1}, \alpha_{i+1}, \alpha_i, \alpha_{i+2}, ..., \alpha_s$$

is an R-sequence on E. Clearly all we have to do is to show that $\alpha_1, ..., \alpha_{i-1}, \alpha_{i+1}, \alpha_i$ is an R-sequence on E. By Lemma 1, this will follow if we prove that

$$(\alpha_1, ..., \alpha_{i-1})E :_E \alpha_{i+1} = (\alpha_1, ..., \alpha_{i-1})E.$$

Accordingly to establish the theorem it is sufficient to show that α_{i+1} is not a zero-divisor on $E' = E/(\alpha_1, ..., \alpha_{i-1})E$.

Now E' is a Noetherian R-module and α_i, α_{i+1} is an R-sequence on E'. Since α_i belongs to the Jacobson radical of R, Lemma 3 shows that α_{i+1}, α_i is an R-sequence on E'. In particular, α_{i+1} is not a zero-divisor on E'. This completes the proof.

Theorem 4 suggests that simplifications are likely to occur when we are concerned with ideals and elements that are contained in the Jacobson radical. The next result supports this idea.

Theorem 5. *Let $E \neq 0$ be a Noetherian R-module. Let B be an ideal and γ an element of R and suppose that both are contained in the Jacobson radical of R. Then $\mathrm{gr}\,((B, \gamma); E) \leqslant \mathrm{gr}\,(B; E) + 1$.*

The hypotheses ensure that $(B, \gamma) E \neq E$. For otherwise we could conclude, by Theorem 19 of section (4.6), that

$$E = \bigcap_{n=1}^{\infty} (B, \gamma)^n E = 0$$

and this is not the case. It therefore follows that the two grades (which occur in the statement of the theorem) are properly defined.

As a step towards proving this result we first establish

Lemma 4. *Let E, B and γ be as in the statement of Theorem 5. Assume also that $\mathrm{gr}\,((B, \gamma); E) \geqslant 1$. Then there exists $b \in B$ such that $\gamma + b$ is not a zero-divisor on E.*

Proof. Since $\mathrm{gr}\,((B, \gamma); E) \geqslant 1$, the ideal (B, γ) contains an element x which is not a zero-divisor on E. Let $x = r\gamma + b_0$, where $r \in R$ and $b_0 \in B$.

It will be shown that at least one of the elements

$$\gamma + b_0^\nu \quad (\nu = 1, 2, 3, ...)$$

is not a zero-divisor on E and this will establish the lemma. *Assume the contrary* and let $P_1, P_2, ..., P_s$ be the prime ideals which belong to the zero submodule of E. Then an element of R is a zero-divisor on E if and only if it is contained by one of $P_1, P_2, ..., P_s$.

Since $\gamma + b_0^\nu$ is a zero-divisor for every value of ν, it is possible to find positive integers m, n such that $m < n$ and both $\gamma + b_0^m$ and $\gamma + b_0^n$ belong to the same prime ideal P, where P occurs among $P_1, P_2, ..., P_s$. Then $b_0^m(1 - b_0^{n-m}) \in P$. Now $b_0 \in B$ and B is contained in the Jacobson radical. It follows that $1 - b_0^{n-m}$ is not contained in any maximal ideal and therefore it is a unit. Multiplying $b_0^m(1 - b_0^{n-m})$ by the inverse of $1 - b_0^{n-m}$, we see that $b_0^m \in P$ and therefore $b_0 \in P$. It follows that γ also belongs to P. Accordingly $x = r\gamma + b_0$ belongs to P and this is the required contradiction because x is not a zero-divisor on E.

Corollary. *Let E, B and γ satisfy the hypotheses of Theorem 5 and suppose, in addition, that every element of B is a zero-divisor on E. Then $\mathrm{gr}((B, \gamma); E) \leqslant 1$.*

Proof. Assume that $\mathrm{gr}((B, \gamma); E) \geqslant 1$. Then, by the lemma, there exists $b \in B$ such that $\gamma + b$ is not a zero-divisor on E. Let β be an arbitrary element of B. Then $\gamma + b, \beta$ is not an R-sequence on E. (For assume the contrary. Then, because $\gamma + b, \beta$ are contained in the Jacobson radical, it follows that $\beta, \gamma + b$ is also an R-sequence on E. However this is impossible since, by hypothesis, β is a zero-divisor on E.) Accordingly $(\gamma + b)E :_E \beta \neq (\gamma + b)E$ and therefore β is contained in one of the prime ideals $P_1', P_2', ..., P_m'$ which belong to the submodule $(\gamma + b)E$ of E. But β was an arbitrary element of B. It therefore follows, by Proposition 5 of section (2.3), that B itself is contained by one of $P_1', P_2', ..., P_m'$. Suppose for definiteness that $B \subseteq P_1'$. Then, since P_1' contains† the ideal $(\gamma + b)E : E$, we see that P_1' also contains $\gamma + b$. Thus

$$(B, \gamma) = (B, \gamma + b) \subseteq P_1'.$$

However $(\gamma + b)E :_E P_1' \neq (\gamma + b)E$ and therefore $\gamma + b$ is a maximal R-sequence on E contained in P_1'. Accordingly $\mathrm{gr}(P_1'; E) = 1$. Finally

$$\mathrm{gr}((B, \gamma); E) \leqslant \mathrm{gr}(P_1'; E) = 1$$

and now the proof is complete.

Proof of Theorem 5. Let E, B and γ be as in the statement of the theorem. Put $\mathrm{gr}(B; E) = s$ and $\mathrm{gr}((B, \gamma); E) = t$. Then there exists an R-sequence $\alpha_1, \alpha_2, ..., \alpha_s$ on E and contained in B. Also, by Theorem 3, this can be extended to an R-sequence $\alpha_1, ..., \alpha_s, \alpha_{s+1}, ..., \alpha_t$ on E in (B, γ).

† See the remarks in the paragraph following (4.4.5).

Write $\bar{E} = E/(\alpha_1, ..., \alpha_s)E$. Since $\alpha_1, \alpha_2, ..., \alpha_s$ is a maximal R-sequence on E in B, every element of B is a zero-divisor on $\bar{E}$. The corollary to Lemma 4 therefore shows that $\mathrm{gr}\,((B, \gamma); \bar{E}) \leqslant 1$. However $\alpha_{s+1}, \alpha_{s+2}, ..., \alpha_t$ is an R-sequence on $\bar{E}$ and all these elements are contained in (B, γ). Consequently $t - s \leqslant 1$ and the proof is complete.

Suppose that E is a Noetherian R-module and B an ideal of R such that $BE \neq E$. By Theorem 3, $\mathrm{gr}\,(B; E)$ is finite. We shall strengthen this result by showing that

$$\mathrm{gr}\,(B; E) \leqslant \mathrm{rank}\,[(\mathrm{Ann}_R E, B)/\mathrm{Ann}_R E],$$

where $\mathrm{Ann}_R E$ denotes (as usual) the annihilating ideal of E, and $(\mathrm{Ann}_R E, B)$ is used as an abbreviation for $(\mathrm{Ann}_R E) + B$. Observe that, by Theorem 2 of section (4.2), the ring $R/\mathrm{Ann}_R E$ is Noetherian and therefore each of its proper ideals has finite rank.

It is convenient to begin by establishing

Proposition 3. *Let E be a Noetherian R-module and $\alpha_1, \alpha_2, ..., \alpha_p$ elements of R. Then*

$$\mathrm{Rad}\,[(\mathrm{Ann}_R E, \alpha_1, ..., \alpha_p)] = \mathrm{Rad}\,[\mathrm{Ann}_R(E/(\alpha_1, ..., \alpha_p)E)]$$

and the prime ideals which contain $(\mathrm{Ann}_R E, \alpha_1, ..., \alpha_p)$ are the same as the prime ideals which contain $\mathrm{Ann}_R(E/(\alpha_1, ..., \alpha_p)E)$. If

$$(\alpha_1, ..., \alpha_p)E \neq E,$$

then $\quad \mathrm{rank}\,[\mathrm{Ann}_R(E/(\alpha_1, ..., \alpha_p)E)/\mathrm{Ann}_R E] \leqslant p$.

Proof. Put $A = \mathrm{Ann}_R E$ and $B = \mathrm{Ann}_R(E/(\alpha_1, ..., \alpha_p)E)$. Then $(A, \alpha_1, ..., \alpha_p) \subseteq B$ and

$$BE \subseteq (\alpha_1, ..., \alpha_p)E = (A, \alpha_1, ..., \alpha_p)E.$$

Since E is finitely generated, it follows, from Proposition 13 of section (2.7), that $B^n \subseteq (A, \alpha_1, ..., \alpha_p)$ for some positive integer n. The relations

$$B^n \subseteq (A, \alpha_1, ..., \alpha_p) \subseteq B$$

now show that $\mathrm{Rad}\,B = \mathrm{Rad}\,(A, \alpha_1, ..., \alpha_p)$ and also that a prime ideal contains B if and only if it contains $(A, \alpha_1, ..., \alpha_p)$.

Let us assume that $(\alpha_1, ..., \alpha_p)E \neq E$. Since

$$(A, \alpha_1, ..., \alpha_p)E = (\alpha_1, ..., \alpha_p)E,$$

it follows that $(A, \alpha_1, ..., \alpha_p) \neq R$.

Consider B/A and $(A, \alpha_1, ..., \alpha_p)/A$. Any prime ideal of R/A which contains the former will contain the latter and vice versa. Hence both are proper ideals and

$$\mathrm{rank}\,(B/A) = \mathrm{rank}\,((A, \alpha_1, ..., \alpha_p)/A).$$

Finally $(A, \alpha_1, ..., \alpha_p)/A$ is an ideal which can be generated by p elements namely the natural images of $\alpha_1, \alpha_2, ..., \alpha_p$. Since R/A is a Noetherian ring, we conclude, from Theorem 22 of section (4.8), that $\text{rank}\,((A, \alpha_1, ..., \alpha_p)/A) \leqslant p$. The proposition now follows.

Proposition 4. *Let E be a Noetherian R-module and $\alpha_1, \alpha_2, ..., \alpha_p$ an R-sequence on E. Then*

$$\text{rank}\,[(\text{Ann}_R E, \alpha_1, ..., \alpha_p)/\text{Ann}_R E]$$
$$= \text{rank}\,[\text{Ann}_R(E/(\alpha_1, ..., \alpha_p)E)/\text{Ann}_R E] = p.$$

Furthermore, if C is an ideal of R such that $CE \neq E$, then

$$(\text{Ann}_R E, C) \neq R \quad and \quad \text{gr}\,(C;E) \leqslant \text{rank}\,[(\text{Ann}_R E, C)/\text{Ann}_R E].$$

Proof. As in the proof of Proposition 3, put $A = \text{Ann}_R E$ and

$$B = \text{Ann}_R(E/(\alpha_1, ..., \alpha_p)E).$$

Since $(A, \alpha_1, ..., \alpha_p)E = (\alpha_1, ..., \alpha_p)E$ and $(\alpha_1, ..., \alpha_p)E \neq E$, it follows that $(A, \alpha_1, ..., \alpha_p) \neq R$.

Now suppose that $0 \leqslant i < p$. Then α_{i+1} is not a zero-divisor on $E/(\alpha_1, ..., \alpha_i)E$ and therefore α_{i+1} is not contained in any minimal prime ideal of its zero submodule. Moreover, by Theorem 14 of section (2.9), these minimal prime ideals are the same as the minimal prime ideals of $\text{Ann}_R(E/(\alpha_1, ..., \alpha_i)E) = B_i$ say. Now, by Proposition 3, the prime ideals which contain B_i are the same as the prime ideals containing $(A, \alpha_1, ..., \alpha_i)$. Accordingly α_{i+1} is not contained in any minimal prime ideal of the latter.

Let $\bar{\alpha}_j$ denote the natural image of α_j in the Noetherian ring $\bar{R} = R/A$. The above remarks show that $\bar{\alpha}_{i+1}$ is not contained in any minimal prime ideal of $(\bar{\alpha}_1, \bar{\alpha}_2, ..., \bar{\alpha}_i)$ and therefore

$$\text{rank}\,(\bar{\alpha}_1, \bar{\alpha}_2, ..., \bar{\alpha}_{i+1}) > \text{rank}\,(\bar{\alpha}_1, \bar{\alpha}_2, ..., \bar{\alpha}_i)$$

for $0 \leqslant i \leqslant p-1$. Thus $\text{rank}\,(\bar{\alpha}_1, \bar{\alpha}_2, ..., \bar{\alpha}_p) \geqslant p$ and, since the opposite inequality holds by Theorem 22 of section (4.8), it follows that

$$\text{rank}\,[(A, \alpha_1, ..., \alpha_p)/A] = \text{rank}\,(\bar{\alpha}_1, \bar{\alpha}_2, ..., \bar{\alpha}_p) = p.$$

Furthermore, as we saw in the proof of Proposition 3,

$$\text{rank}\,[(A, \alpha_1, ..., \alpha_p)/A] = \text{rank}\,[B/A].$$

Thus the first part of the proposition is established.

Now suppose that $\alpha_1, \alpha_2, ..., \alpha_p$ is a maximal R-sequence on E contained in C so that $\text{gr}\,(C;E) = p$. Since $(A, C)E = CE$ and $CE \neq E$, it follows that $(A, C) \neq R$. Finally

$$\text{gr}\,(C;E) = p = \text{rank}\,[(A, \alpha_1, ..., \alpha_p)/A] \leqslant \text{rank}\,[(A, C)/A],$$

because $(A, \alpha_1, ..., \alpha_p) \subseteq (A, C)$. This completes the proof.

Corollary 1. *Let R be a Noetherian ring and $\alpha_1, \alpha_2, \ldots, \alpha_p$ an R-sequence. Then* rank $(\alpha_1, \alpha_2, \ldots, \alpha_p) = p$.

Corollary 2. *Let R be a Noetherian ring and C a proper ideal of R. Then* gr $(C) \leqslant$ rank C.

Both corollaries are obtained by taking $E = R$ in the proposition just proved.

5.2 The theory of grade for semi-local rings

The theory of grade can be considerably extended if we restrict our attention to semi-local rings. The additional results that can be obtained in this way form the subject of the present section.

Let R be a semi-local ring and $M_1, M_2, \ldots, M_h$ its maximal ideals. Put

$$J = M_1 \cap M_2 \cap \ldots \cap M_h \tag{5.2.1}$$

so that J is the Jacobson radical. For brevity we shall refer to J as the *radical* of R.

We shall now make a number of simple observations in order to draw attention to certain basic facts which we shall use tacitly in what follows. By Proposition 1 Cor. of section (4.1), every finitely generated R-module is a Noetherian R-module. Let E be such a module and B an ideal contained in the radical J. Then, by Theorem 19 of section (4.6),

$$\bigcap_{n=1}^{\infty} B^n E = 0. \tag{5.2.2}$$

Hence if $E \neq 0$, then $BE \neq E$. Note that as a special case of (5.2.2)

$$\bigcap_{n=1}^{\infty} J^n = (0). \tag{5.2.3}$$

Next, if elements $\alpha_1, \alpha_2, \ldots, \alpha_p$ *belonging to the radical of R* form an R-sequence on E, then they will continue to do so if their order is changed in any way. This follows from Theorem 4. Observe, also, that if K is a submodule of the Noetherian R-module E, then the prime ideals belonging to the submodule K are the same as those that belong to the zero submodule of E/K. This follows from Theorem 20 Cor. of section (2.10).

Let $\sigma: R \to \bar{R}$ be an epimorphism of the semi-local ring R on to a non-null ring $\bar{R}$. Then, as we saw in section (4.9), $\bar{R}$ is also a semi-local ring. We can suppose that the maximal ideals $M_1, M_2, \ldots, M_h$ of R are so numbered that $M_1, M_2, \ldots, M_l$ contain the kernel of σ whereas

$$M_{l+1}, M_{l+2}, \ldots, M_h$$

do not. On this understanding $\sigma(M_1), \ldots, \sigma(M_t)$ are the maximal ideals of $\bar{R}$ and $\sigma(M_j) = \bar{R}$ for $t+1 \leqslant j \leqslant h$. Now $M_i + M_j = R$ if $i \neq j$. Hence, by Proposition 4 of section (4.2),

$$J = M_1 \cap M_2 \cap \ldots \cap M_t \cap \ldots \cap M_h = M_1 M_2 \ldots M_t \ldots M_h.$$

Accordingly

$$\sigma(J) = \sigma(M_1)\,\sigma(M_2) \ldots \sigma(M_t) \ldots \sigma(M_h)$$
$$= \sigma(M_1)\,\sigma(M_2) \ldots \sigma(M_t)$$

and this is the radical of $\bar{R}$. *Thus the epimorphism σ maps the radical of R on to the radical of $\bar{R}$.*

At this stage, it becomes convenient to introduce two new definitions. The first of these is of general interest because it enables us to apply the concept of *dimension* to modules. The second definition plays only a minor role and it will disappear from the discussion after it has served its immediate purpose.

Definition. *Let $E \neq 0$ be an R-module. Then by the 'dimension' of E we shall understand the dimension of the ring $R/\mathrm{Ann}_R E$. The dimension of the module E will be denoted by $\mathrm{Dim}\, E$.*

Thus $\mathrm{Dim}\, E = \mathrm{Dim}\,(R/\mathrm{Ann}_R E)$ or, equally well, $\mathrm{Dim}\, E$ is equal to the dimension $\dim\,(\mathrm{Ann}_R E)$ of the ideal of elements which annihilate E. Note that the definition does not require that R be a semi-local ring or even that it be Noetherian. Another point to note is that on taking $E = R$ we find that there is no conflict with the notion of dimension as previously applied to rings. Should E happen to be a Noetherian R-module and $P_1, P_2, \ldots, P_t$ the prime ideals belonging to its zero submodule, then

$$\mathrm{Dim}\, E = \max_{1 \leqslant i \leqslant t} \dim P_i. \tag{5.2.4}$$

This is because† the minimal prime ideals belonging to the zero submodule of E are the same as the minimal prime ideals of $\mathrm{Ann}_R E$.

Now suppose that R is a semi-local ring. Let $E \neq 0$ be a finitely generated R-module and denote by $P_1, P_2, \ldots, P_t$ the prime ideals belonging to its zero submodule. We put

$$s(E) = \max_{1 \leqslant i \leqslant t} \dim P_i - \min_{1 \leqslant j \leqslant t} \dim P_j \tag{5.2.5}$$

and call $s(E)$ the *span* of the R-module E. The fact that R is a semi-local ring ensures that the dimensions occurring in (5.2.5) are finite. Note that $s(E) \geqslant 0$ and

$$\min_{1 \leqslant j \leqslant t} \dim P_j = \mathrm{Dim}\, E - s(E). \tag{5.2.6}$$

† See Theorem 14 of section (2.9).

Also $s(E) = 0$ when and only when the dimensions of the prime ideals belonging to the zero submodule of E are all equal, in which case the common value of the dimensions is $\mathrm{Dim}\,E$.

Theorem 6. *Let R be a semi-local ring, $E \neq 0$ a finitely generated R-module, and γ an element which belongs to the radical J of R and is not a zero-divisor on E. Then $\mathrm{Dim}\,(E/\gamma E) = \mathrm{Dim}\,E - 1$ and*

$$s(E/\gamma E) \geqslant s(E).$$

Proof. Put $\mathrm{Dim}\,E = r$, $s(E) = p$, $A = \mathrm{Ann}_R E$ and $\bar{R} = R/A$. Further let $\bar{\gamma}$ denote the image of γ under the natural mapping $R \to \bar{R}$. By Proposition 3, the prime ideals which contain $\mathrm{Ann}_R(E/\gamma E)$ are the same as those which contain (A, γ). Hence

$$\mathrm{Dim}\,(E/\gamma E) = \dim(A, \gamma) = \mathrm{Dim}\,R/(A, \gamma) = \mathrm{Dim}\,\bar{R}/(\bar{\gamma})$$

because $R/(A, \gamma)$ is ring-isomorphic to $\bar{R}/(\bar{\gamma})$.

By hypothesis, γ is not a zero-divisor on E. It follows that γ is not contained in any minimal prime ideal belonging to zero submodule of E and therefore not contained in any minimal prime ideal of A. This shows that $\dim(A, \gamma) < \dim A = r$ and therefore

$$\mathrm{Dim}\,\bar{R}/(\bar{\gamma}) = \dim(A, \gamma) < r = \mathrm{Dim}\,\bar{R}.$$

Now $\bar{R}$ is a semi-local ring and $\bar{\gamma}$ belongs to each of its maximal ideals. Consequently, by Proposition 19 of section (4.9),

$$\mathrm{Dim}\,\bar{R}/(\bar{\gamma}) \geqslant \mathrm{Dim}\,\bar{R} - 1.$$

Accordingly $\mathrm{Dim}\,(E/\gamma E) = \mathrm{Dim}\,\bar{R}/(\bar{\gamma}) = r - 1$

and the first assertion is proved.

Since $s(E) = p$, it follows, from (5.2.6), that there exists a prime ideal P which belongs to the zero submodule of E and is such that $\dim P = r - p$. Now $0:_E P \neq 0$. Consequently there exists $y \in E$ such that $y \neq 0$ and $Py = 0$. By (5.2.2) and the fact that γ belongs to the radical of R, we have

$$\bigcap_{n=1}^{\infty} \gamma^n E = 0.$$

Accordingly there exists an integer k such that $y \in \gamma^k E$ and $y \notin \gamma^{k+1}E$. Let $y = \gamma^k z$, where $z \in E$. Then $z \notin \gamma E$ but $\gamma^k Pz = Py = 0$. However γ is not a zero-divisor on E so this implies that $Pz = 0$. It follows that

$$\gamma E :_E P \neq \gamma E$$

because the left-hand side contains z whereas the right-hand side does not. This shows that $P \subseteq P'$, where P' is one of the prime ideals belonging to the submodule γE of E. Furthermore P' must strictly contain P. (For if $P = P'$, then $\gamma \in (\gamma E : E) \subseteq P' = P$ which is impossible because γ is not a zero-divisor on E.) Accordingly

$$\dim P' < \dim P = r - p$$

and therefore, since P' belongs to the zero submodule of $E/\gamma E$,

$$s(E/\gamma E) \geqslant \mathrm{Dim}\,(E/\gamma E) - \dim P' \geqslant (r-1) - (r-p-1) = p.$$

This completes the proof.

Corollary. *Let R be a semi-local ring with radical J, $E \neq 0$ a finitely generated R-module, and $\alpha_1, \alpha_2, \ldots, \alpha_m\ (m \geqslant 0)$ an R-sequence on E in J. Then*

$$\mathrm{Dim}\,[E/(\alpha_1, \ldots, \alpha_m)E] = \mathrm{Dim}\,E - m$$

and

$$s[E/(\alpha_1, \ldots, \alpha_m)E] \leqslant \mathrm{Dim}\,E - \mathrm{gr}\,(J;E).$$

Remark. It is important to note that the case $m = 0$ is included in the corollary. Thus the proof will show that

$$s(E) \leqslant \mathrm{Dim}\,E - \mathrm{gr}\,(J;E). \tag{5.2.7}$$

This in turn will imply, because of (5.2.6), that

$$\mathrm{gr}\,(J;E) \leqslant \min_{1 \leqslant j \leqslant t} \dim P_j \leqslant \mathrm{Dim}\,E, \tag{5.2.8}$$

where $P_1, P_2, \ldots, P_t$ denote the prime ideals belonging to the zero submodule of E.

Proof. We begin with the assertion that

$$\mathrm{Dim}\,[E/(\alpha_1, \ldots, \alpha_m)E] = \mathrm{Dim}\,E - m.$$

This is trivial if $m = 0$. Now suppose that $m > 0$ and put

$$E_i = E/(\alpha_1, \ldots, \alpha_i)E.$$

Then, for $0 \leqslant i \leqslant m-1$, α_{i+1} is not a zero-divisor on E_i and therefore, by Theorem 6,

$$\mathrm{Dim}\,(E_i/\alpha_{i+1}E_i) = \mathrm{Dim}\,E_i - 1.$$

However $\alpha_{i+1}E_i = (\alpha_1, \ldots, \alpha_i, \alpha_{i+1})E/(\alpha_1, \ldots, \alpha_i)E$ and therefore

$$E_i/\alpha_{i+1}E_i$$

is isomorphic to E_{i+1}. Accordingly $\mathrm{Dim}\,E_{i+1} = \mathrm{Dim}\,E_i - 1$ and the first assertion of the corollary follows.

We now turn our attention to the statement that

$$s[E/(\alpha_1, \ldots, \alpha_m)E] \leqslant \mathrm{Dim}\,E - \mathrm{gr}\,(J;E).$$

If $\alpha_1, \alpha_2, \ldots, \alpha_m$ happens to be a maximal R-sequence on E in J, then $\mathrm{gr}\,(J;E) = m$ and every element of J is a zero-divisor on $E/(\alpha_1, \ldots, \alpha_m)E$. It follows that J is contained in one of the prime ideals belonging to the zero submodule of $E/(\alpha_1, ..., \alpha_m)E$. But J is an intersection of maximal ideals. This shows that one of the maximal ideals of R belongs to the zero submodule of $E/(\alpha_1, ..., \alpha_m)E$. Therefore, by the definition of the span of a module,

$$s[E/(\alpha_1, ..., \alpha_m)E] = \text{Dim}\,[E/(\alpha_1, ..., \alpha_m)E]$$
$$= \text{Dim}\,E - m$$
$$= \text{Dim}\,E - \text{gr}\,(J;E).$$

Finally, suppose that $\alpha_1, \alpha_2, ..., \alpha_m$ is not a maximal R-sequence on E in J. Then Theorem 3 shows that we can extend it in such a way that it becomes one. Let $\alpha_1, ..., \alpha_m, \alpha_{m+1}, ..., \alpha_t$ be a maximal R-sequence on E in J and put $E_i = E/(\alpha_1, ..., \alpha_i)E$ as before. By Theorem 6 and the fact that $E_i/\alpha_{i+1}E_i$ is isomorphic to E_{i+1}, we see that $s(E_{i+1}) \geqslant s(E_i)$ and therefore that $s(E_t) \geqslant s(E_m)$. However, the remarks of the last paragraph show that $s(E_t) = \text{Dim}\,E - \text{gr}\,(J;E)$, so all is proved.

The following lemma collects together some simple results which will be required in the proofs of the main theorems.

Lemma 5. *Let R be a semi-local ring, $E \neq 0$ a finitely generated R-module, and $\alpha_1, \alpha_2, ..., \alpha_m$ elements contained in the radical J of R. Assume that* $$\text{Dim}\,[E/(\alpha_1, ..., \alpha_m)E] = \text{Dim}\,E - m.$$
Then $$\text{Dim}\,[E/(\alpha_1, ..., \alpha_i)E] = \text{Dim}\,E - i$$
for $i = 0, 1, ..., m$. Moreover, if it is the case that for each value of i satisfying $0 \leqslant i < m$ the prime ideals belonging to the zero submodule of $E/(\alpha_1, ..., \alpha_i)E$ all have dimension equal to $\text{Dim}\,E - i$, then $\alpha_1, \alpha_2, ..., \alpha_m$ is an R-sequence on E.

Proof. Put $A = \text{Ann}_R E$ and $\bar{R} = R/A$. By Proposition 3, the prime ideals which contain $\text{Ann}_R(E/(\alpha_1, ..., \alpha_m)E)$ are the same as those containing $(A, \alpha_1, ..., \alpha_m)$. It follows that

$$\dim\,(A, \alpha_1, ..., \alpha_m) = \text{Dim}\,[E/(\alpha_1, ..., \alpha_m)E]$$
$$= \text{Dim}\,E - m$$
$$= \text{Dim}\,\bar{R} - m.$$

Let $\bar{\alpha}_j$ denote the natural image of α_j in $\bar{R}$. Then $\bar{R}$ is a semi-local ring whose radical contains the elements $\bar{\alpha}_1, \bar{\alpha}_2, ..., \bar{\alpha}_m$. Passing to the ring $\bar{R}/(\bar{\alpha}_1, ..., \bar{\alpha}_m)$, we see that

$$\text{Dim}\,\bar{R}/(\bar{\alpha}_1, ..., \bar{\alpha}_m) = \dim\,(\bar{\alpha}_1, \bar{\alpha}_2, ..., \bar{\alpha}_m)$$
$$= \dim\,(A, \alpha_1, ..., \alpha_m)$$
$$= \text{Dim}\,\bar{R} - m.$$

Hence, by Proposition 19 of section (4.9), $\bar{\alpha}_1, \bar{\alpha}_2, ..., \bar{\alpha}_m$ is a subset of a system of parameters of $\bar{R}$. Suppose now that $0 \leqslant i \leqslant m$. Then $\bar{\alpha}_1, \bar{\alpha}_2, ..., \bar{\alpha}_i$ is also a subset of a system of parameters of $\bar{R}$ and therefore by the same proposition, $\operatorname{Dim} \bar{R}/(\bar{\alpha}_1, ..., \bar{\alpha}_i) = \operatorname{Dim} \bar{R} - i$. Accordingly

$$\begin{aligned} \dim(A, \alpha_1, ..., \alpha_i) &= \dim(\bar{\alpha}_1, \bar{\alpha}_2, ..., \bar{\alpha}_i) \\ &= \operatorname{Dim} \bar{R}/(\bar{\alpha}_1, ..., \bar{\alpha}_i) \\ &= \operatorname{Dim} \bar{R} - i \\ &= \operatorname{Dim} E - i. \end{aligned}$$

However, by Proposition 3, $(A, \alpha_1, ..., \alpha_i)$ and the annihilating ideal of $E/(\alpha_1, ..., \alpha_i)E$ have the same dimension. This yields

$$\operatorname{Dim}[E/(\alpha_1, ..., \alpha_i)E] = \operatorname{Dim} E - i$$

and establishes the first part of the lemma.

We now assume that (for each i satisfying $0 \leqslant i < m$) the prime ideals belonging to the zero submodule of $E/(\alpha_1, ..., \alpha_i)E = E_i$ (say) all have dimension equal to $\operatorname{Dim} E - i$. Then α_{i+1} cannot be contained by any prime ideal belonging to the zero submodule of E_i. For assume that one of these prime ideals, P say, were to contain α_{i+1}. We should then have $\dim P = \operatorname{Dim} E - i$ and also $(A, \alpha_1, ..., \alpha_i, \alpha_{i+1}) \subseteq P$ because P must contain any ideal, such as $(A, \alpha_1, ..., \alpha_i)$, which annihilates E_i. Accordingly $\dim(A, \alpha_1, ..., \alpha_i, \alpha_{i+1}) \geqslant \dim P = \operatorname{Dim} E - i$ and now we have a contradiction because we have already seen that

$$\dim(A, \alpha_1, ..., \alpha_i, \alpha_{i+1}) = \operatorname{Dim} E - i - 1.$$

It follows that α_{i+1} is not a zero-divisor on E_i. As this holds for $0 \leqslant i < m$, we see that $\alpha_1, \alpha_2, ..., \alpha_m$ is an R-sequence on E. The proof is now complete.

Theorem 7. *Let R be a semi-local ring with radical J and $E \neq 0$ a finitely generated R-module. Then the following two statements are equivalent:*

(a) $\operatorname{gr}(J; E) = \operatorname{Dim} E$;

(b) *whenever $\alpha_1, \alpha_2, ..., \alpha_m$ ($m \geqslant 0$) belong to J and*
$$\operatorname{Dim}[E/(\alpha_1, ..., \alpha_m)E] = \operatorname{Dim} E - m,$$

then all the prime ideals belonging to the zero submodule of $E/(\alpha_1, ..., \alpha_m)E$ have the same dimension.

Proof. First assume that (a) is true. Let $\alpha_1, \alpha_2, ..., \alpha_m$ belong to J and be such that $\operatorname{Dim}[E/(\alpha_1, ..., \alpha_m)E] = \operatorname{Dim} E - m.$

Put $E_i = E/(\alpha_1, ..., \alpha_i)E$. We have to show that $s(E_m) = 0$. This will be accomplished by proving, using induction on i, that $s(E_i) = 0$ for $i = 0, 1, ..., m$.

The case $i = 0$ is trivial because, by (5.2.7),

$$0 \leqslant s(E) \leqslant \operatorname{Dim} E - \operatorname{gr}(J;E).$$

We shall therefore suppose that $0 < i \leqslant m$ and also that $s(E_j) = 0$ for $0 \leqslant j < i$. By Lemma 5, we have

$$\operatorname{Dim}[E/(\alpha_1, ..., \alpha_j)E] = \operatorname{Dim} E - j$$

for $0 \leqslant j \leqslant m$. It follows (using the inductive hypothesis) that if $0 \leqslant j < i$, then all the prime ideals belonging to the zero submodule of $E/(\alpha_1, ..., \alpha_j)E$ must have dimension equal to $\operatorname{Dim} E - j$. Hence, again using Lemma 5, $\alpha_1, \alpha_2, ..., \alpha_i$ is an R-sequence on E. Accordingly, by the corollary to Theorem 6,

$$0 \leqslant s(E_i) \leqslant \operatorname{Dim} E - \operatorname{gr}(J;E).$$

Thus $s(E_i) = 0$ and we conclude that (a) implies (b).

Now assume that (b) is true. Put

$$A = \operatorname{Ann}_R E, \quad \bar{R} = R/A \quad \text{and} \quad \operatorname{Dim} E = p.$$

Then $\bar{R}$ is a p-dimensional semi-local ring. Consider the natural epimorphism $R \to \bar{R}$. This maps the radical of R on to the radical of $\bar{R}$. We can therefore find elements $\alpha_1, \alpha_2, ..., \alpha_p$ in J so that their images $\bar{\alpha}_1, \bar{\alpha}_2, ..., \bar{\alpha}_p$ form a system of parameters of $\bar{R}$. Then

$$\operatorname{Dim}[E/(\alpha_1, ..., \alpha_p)E] = \dim(A, \alpha_1, ..., \alpha_p)$$
$$= \dim(\bar{\alpha}_1, \bar{\alpha}_2, ..., \bar{\alpha}_p)$$
$$= 0$$
$$= \operatorname{Dim} E - p.$$

Hence, by Lemma 5, $\operatorname{Dim}[E/(\alpha_1, ..., \alpha_i)E] = \operatorname{Dim} E - i$ for $0 \leqslant i \leqslant p$. Furthermore, all the prime ideals belonging to the zero submodule of $E/(\alpha_1, ..., \alpha_i)E$ have dimension equal to $\operatorname{Dim} E - i$ because we are assuming (b) to be true. Accordingly, by the second part of Lemma 5, $\alpha_1, \alpha_2, ..., \alpha_p$ is an R-sequence on E and therefore

$$\operatorname{gr}(J;E) \geqslant p = \operatorname{Dim} E.$$

However the opposite inequality holds by virtue of (5.2.8) and so the proof is complete.

Proposition 5. *Let R be a semi-local ring with radical J and $E \neq 0$ a finitely generated R-module satisfying $\operatorname{gr}(J;E) = \operatorname{Dim} E$. If now $\alpha_1, \alpha_2, ..., \alpha_m$ belong to J and*

$$\operatorname{Dim}[E/(\alpha_1, ..., \alpha_m)E] = \operatorname{Dim} E - m,$$

then $\alpha_1, \alpha_2, ..., \alpha_m$ is an R-sequence on E.

Proof. The first part of Lemma 5 shows that

$$\mathrm{Dim}\,[E/(\alpha_1, ..., \alpha_i)E] = \mathrm{Dim}\,E - i$$

for $0 \leqslant i \leqslant m$. Next, it follows (from Theorem 7) that the prime ideals belonging to the zero submodule of $E/(\alpha_1, ..., \alpha_i)E$ all have dimension equal to $\mathrm{Dim}\,E - i$. Hence, by the second part of Lemma 5, $\alpha_1, \alpha_2, ..., \alpha_m$ is an R-sequence on E. This establishes the proposition.

Theorem 8. *Let R be a semi-local ring with radical J and $E \neq 0$ a finitely generated R-module satisfying $\mathrm{gr}\,(J;E) = \mathrm{Dim}\,E$. If now P is any prime ideal containing the annihilator $\mathrm{Ann}_R E$ of E, then*

$$\mathrm{gr}\,(P;E) = \mathrm{rank}\,(P/\mathrm{Ann}_R E)$$

and $\qquad \mathrm{Dim}\,E = \mathrm{rank}\,(P/\mathrm{Ann}_R E) + \mathrm{dim}\,P.$

Proof. We see, from Proposition 13 of section (2.7), that $PE \neq E$. Hence, by Proposition 4, $\mathrm{gr}\,(P;E) \leqslant \mathrm{rank}\,(P/\mathrm{Ann}_R E)$. Also, for trivial reasons,

$$\mathrm{rank}\,(P/\mathrm{Ann}_R E) + \mathrm{dim}\,P = \mathrm{rank}\,(P/\mathrm{Ann}_R E) + \mathrm{dim}\,(P/\mathrm{Ann}_R E)$$
$$\leqslant \mathrm{Dim}\,(R/\mathrm{Ann}_R E)$$
$$= \mathrm{Dim}\,E.$$

It is therefore sufficient to show that

$$\mathrm{gr}\,(P;E) + \mathrm{dim}\,P \geqslant \mathrm{Dim}\,E.$$

To this end let $\alpha_1, \alpha_2, ..., \alpha_k$ be a maximal R-sequence on E in $P \cap J$. By the corollary to Theorem 6,

$$\mathrm{Dim}\,[E/(\alpha_1, ..., \alpha_k)E] = \mathrm{Dim}\,E - k$$

and, by Theorem 7, each of the prime ideals belonging to the zero submodule of $E/(\alpha_1, ..., \alpha_k)E$ has dimension equal to $\mathrm{Dim}\,E - k$. Furthermore, one of these prime ideals, P' say, must contain $P \cap J$ for otherwise there would exist an R-sequence on E in $P \cap J$ which was longer than $\alpha_1, \alpha_2, ..., \alpha_k$. Thus $\mathrm{dim}\,P' = \mathrm{Dim}\,E - k$ and either $P \subseteq P'$ or $J \subseteq P'$.

If $P \subseteq P'$, then every element of P is a zero-divisor on $E/(\alpha_1, ..., \alpha_k)E$ and therefore $\alpha_1, \alpha_2, ..., \alpha_k$ is a maximal R-sequence on E in P. Thus $\mathrm{gr}\,(P;E) = k$ and therefore

$$\mathrm{gr}\,(P;E) + \mathrm{dim}\,P \geqslant k + \mathrm{dim}\,P' = \mathrm{Dim}\,E.$$

On the other hand, if $J \subseteq P'$, then P' is a maximal ideal of R and so $\mathrm{Dim}\,E - k = \mathrm{dim}\,P' = 0$. Accordingly

$$\mathrm{gr}\,(P;E) + \mathrm{dim}\,P \geqslant \mathrm{gr}\,(P;E) \geqslant k = \mathrm{Dim}\,E$$

as required.

Theorem 9. *Let R be a semi-local ring with radical J and $E \neq 0$ a finitely generated R-module. Then the following three statements are equivalent*:

    (a) $\operatorname{gr}(J;E) = \operatorname{Dim} E = \operatorname{Dim} R$;

    (b) *every system of parameters is an R-sequence on E*;

    (c) *there is at least one system of parameters that is an R-sequence on E*.

Remarks. By (5.2.8), we have $\operatorname{gr}(J;E) \leqslant \operatorname{Dim} E \leqslant \operatorname{Dim} R$. Thus (a) holds if and only if $\operatorname{gr}(J;E) = \operatorname{Dim} R$. Again if $\operatorname{Dim} R = 0$, then the empty set is the only system of parameters. This is to be regarded as forming an R-sequence on every non-zero R-module.

Proof. We shall give a cyclic demonstration. First suppose that (a) is true and let $\alpha_1, \alpha_2, \ldots, \alpha_d$ be a system of parameters of R. Then $d = \operatorname{Dim} R$ and, since $E/(\alpha_1, \ldots, \alpha_d)E$ is a non-zero module annihilated by $(\alpha_1, \alpha_2, \ldots, \alpha_d)$,

$$\operatorname{Dim}[E/(\alpha_1, \ldots, \alpha_d)E] = 0 = \operatorname{Dim} E - d.$$

That $\alpha_1, \alpha_2, \ldots, \alpha_d$ is an R-sequence on E now follows from Proposition 5. Thus (a) implies (b). Obviously (b) implies (c).

Now assume that $\alpha_1, \alpha_2, \ldots, \alpha_d$ is a system of parameters which is also an R-sequence on E. Then

$$\operatorname{gr}(J;E) \geqslant d = \operatorname{Dim} R$$

and the equations

$$\operatorname{gr}(J;E) = \operatorname{Dim} E = \operatorname{Dim} R$$

follow by virtue of the remarks made immediately after the statement of the theorem. The proof is now complete.

5.3 Semi-regular rings

In this section the theory of grade will be applied to the study of an important class of Noetherian rings. This class is conveniently described with the aid of two preliminary definitions.

Let A be a proper ideal in a non-null Noetherian ring R.

Definition. *The ideal A will be said to be of the 'fundamental class' if it can be generated by r elements, where $r = \operatorname{rank} A$.*

By Theorem 22 of section (4.8), A cannot be generated by fewer than $\operatorname{rank} A$ elements; hence an ideal of the fundamental class is in an extreme situation in respect of rank. Again, according to the convention whereby the empty set is regarded as generating the zero ideal, the zero ideal is necessarily of the fundamental class. Indeed it is the only ideal of rank zero to have this property.

For the guidance of the reader, it needs to be stated that what are here called ideals of the fundamental class are more usually described as ideals of the *principal class*. However the former terminology has been preferred because an ideal of the principal class might be confused with a *principal ideal* by which we mean, of course, an ideal that can be generated by a single element.

Definition. *The ideal A is said to be 'unmixed in respect of rank' if all the prime ideals belonging to A have the same rank.*

Clearly if A is unmixed in respect of rank, then all the prime ideals belonging to A have the same rank as A itself. Also A has no embedded prime ideals.†

We come now to the central concept of this section.

Definition. *A non-null Noetherian ring will be called 'semi-regular' if every proper ideal of the fundamental class is unmixed in respect of rank.*

Semi-regular rings are also known as *Macaulay–Cohen* rings.

Theorem 10. *Let R be a local ring of dimension $d \geqslant 0$. Then the following three statements are equivalent*:

(a) *R is semi-regular*;

(b) *there exists at least one system of parameters which is an R-sequence*;

(c) *every system of parameters is an R-sequence.*

Remark. It should be noted that if $d = 0$, then R is semi-regular because, in that case, there exists only one prime ideal. See also the remarks following Theorem 9.

Proof. Let M be the maximal ideal of R so that rank $M = d$. We begin by assuming that (a) is true.

By Proposition 15 of section (4.8), there exist elements $\alpha_1, \alpha_2, \ldots, \alpha_d$ belonging to M such that rank $(\alpha_1, \alpha_2, \ldots, \alpha_i) = i$ for $0 \leqslant i \leqslant d$. Accordingly, $(\alpha_1, \alpha_2, \ldots, \alpha_i)$ is of the fundamental class and therefore, since R is semi-regular, every prime ideal belonging to it has rank equal to i. If therefore $i < d$, then α_{i+1} does not belong to any of these prime ideals for otherwise $(\alpha_1, \alpha_2, \ldots, \alpha_i, \alpha_{i+1})$ would only have rank i. Accordingly

$$(\alpha_1, \alpha_2, \ldots, \alpha_i) : \alpha_{i+1} = (\alpha_1, \alpha_2, \ldots, \alpha_i)$$

† See section (2.9) for the explanation of this term.

which shows that $\alpha_1, \alpha_2, ..., \alpha_d$ is an R-sequence. Moreover $(\alpha_1, \alpha_2, ..., \alpha_d)$ has rank equal to d and therefore $\alpha_1, \alpha_2, ..., \alpha_d$ is a system of parameters. It has thus been proved that (a) implies (b).

Now suppose that (b) is true. On taking $E = R$ in Theorem 9 we find that every system of parameters is an R-sequence. Thus (b) implies (c).

Finally assume that (c) holds. Then, again using Theorem 9, $\text{gr}\,(M) = \text{Dim}\,R$ and so it follows, from Theorem 8, that

$$\dim P + \text{rank}\,P = \text{Dim}\,R$$

for every prime ideal P. Let $\beta_1, \beta_2, ..., \beta_r$ be elements of R which generate an ideal of rank r. By considering the prime ideals which belong to $(\beta_1, \beta_2, ..., \beta_r)$, we see at once that $\dim\,(\beta_1, \beta_2, ..., \beta_r) = d - r$. Thus the dimension of the R-module $R/(\beta_1, \beta_2, ..., \beta_r)$ is $\text{Dim}\,R - r$. We can therefore apply Theorem 7 (with $E = R$) to deduce that all the prime ideals belonging to $(\beta_1, \beta_2, ..., \beta_r)$ have dimension $d - r$. But this means that all these prime ideals have rank r. Accordingly $(\beta_1, \beta_2, ..., \beta_r)$ is unmixed in respect of rank and therefore R has been shown to be semi-regular.

Corollary. *Let R be a local ring which is semi-regular. Then*

$$\dim P + \text{rank}\,P = \text{Dim}\,R$$

for every prime ideal P.

In order to see this it is sufficient to remark that, in the course of the proof just given, it was shown that the relation

$$\dim P + \text{rank}\,P = \text{Dim}\,R$$

is a consequence of condition (c) in the statement of the theorem.

Lemma 6. *Let R be a semi-regular ring and P one of its prime ideals. Then R_P is a semi-regular local ring.*

Proof. Let rank $P = d$. By Proposition 15 of section (4.8), there exist elements $\alpha_1, \alpha_2, ..., \alpha_d$ in P such that $\text{rank}\,(\alpha_1, \alpha_2, ..., \alpha_i) = i$ for $0 \leqslant i \leqslant d$. Since R is semi-regular, all the prime ideals belonging to $(\alpha_1, \alpha_2, ..., \alpha_i)$ have rank equal to i. It follows that

$$(\alpha_1, \alpha_2, ..., \alpha_i) : \alpha_{i+1} = (\alpha_1, \alpha_2, ..., \alpha_i)$$

provided $0 \leqslant i < d$. We note, before proceeding, that P is a minimal prime ideal of $(\alpha_1, \alpha_2, ..., \alpha_d)$.

Let $\chi : R \to R_P$ be the canonical mapping. By Proposition 3 of section (3.2) and by (3.5.6),

$$(\chi(\alpha_1), \ldots, \chi(\alpha_i)) : \chi(\alpha_{i+1}) = (\chi(\alpha_1), \ldots, \chi(\alpha_i)),$$

for $0 \leqslant i < d$. It follows that $\chi(\alpha_1), \chi(\alpha_2), \ldots, \chi(\alpha_d)$ is both an R_P-sequence and a system of parameters in R_P. That R_P is a semi-regular local ring is now a consequence of Theorem 10. This completes the proof.

Theorem 11. *Let R be a non-null Noetherian ring. Then R is semi-regular if and only if R_M is semi-regular for every maximal ideal M.*

Proof. We shall assume that R_M is semi-regular for every maximal ideal M and deduce from this that R itself is semi-regular. The converse follows from Lemma 6.

Let $\alpha_1, \alpha_2, \ldots, \alpha_r$ be r elements which generate an ideal of rank r. Then every *minimal* prime ideal of $(\alpha_1, \alpha_2, \ldots, \alpha_r)$ has rank r. Suppose now that P is an arbitrary prime ideal belonging to $(\alpha_1, \alpha_2, \ldots, \alpha_r)$. The theorem will follow if we can show that the rank of P is equal to r.

There exists a maximal ideal M such that $(\alpha_1, \alpha_2, \ldots, \alpha_r) \subseteq P \subseteq M$. On passing to the ring R_M of fractions, we find, using Theorem 11 of section (3.7), that $(\alpha_1, \alpha_2, \ldots, \alpha_r)$ extends to an ideal of the fundamental class whose rank is r and which has P_M as one of its prime ideals. Since R_M is semi-regular, it follows that rank $P_M = r$. However

$$\text{rank } P_M = \text{rank } P.$$

Consequently rank $P = r$ and the theorem is proved.

The next result provides a generalization of Lemma 6.

Theorem 12. *Let R be a semi-regular ring and S a non-empty multiplicatively closed subset of R not containing the zero element. Then R_S is also a semi-regular ring.*

Proof. We begin by noting that R_S is a non-null Noetherian ring. Let M be a maximal ideal of R_S. Then $M = P_S$, where P is some prime ideal of R which does not meet S. By Theorem 11, it will suffice to show that $(R_S)_M$, that is $(R_S)_{P_S}$, is semi-regular. However, by Proposition 19 of section (3.8), $(R_S)_{P_S}$ is ring-isomorphic to R_P and we know that R_P is semi-regular by virtue of Lemma 6. This completes the proof.

Theorem 13. *Let R be a semi-regular ring and P', P prime ideals of R satisfying $P' \subseteq P$. Then*

$$\text{rank } P = \text{rank } (P/P') + \text{rank } P'.$$

If we localize at P, then the required result follows immediately from Lemma 6 and the corollary to Theorem 10.

Theorem 14. *Let R be a semi-regular ring and A an ideal of the fundamental class. Then R/A is also a semi-regular ring.*

Proof. Let $r = \operatorname{rank} A$. Then, since A belongs to the fundamental class, there exist elements $\alpha_1, \alpha_2, \ldots, \alpha_r$ which generate A. Put $\bar{R} = R/A$ and assume that $\beta_1, \beta_2, \ldots, \beta_s$ belong to R and are such that

$$\operatorname{rank}(\bar{\beta}_1, \bar{\beta}_2, \ldots, \bar{\beta}_s) = s.$$

Here $\bar{\beta}_i$ denotes the natural image of β_i in $\bar{R}$.

The prime ideals belonging to the $\bar{R}$-ideal $(\bar{\beta}_1, \bar{\beta}_2, \ldots, \bar{\beta}_s)$ are just the ideals Π/A, where Π is a typical prime ideal belonging to

$$(\alpha_1, \ldots, \alpha_r, \beta_1, \ldots, \beta_s).$$

To establish the theorem we must show that $\operatorname{rank}(\Pi/A) = s$ in every case.

Let P be a *minimal* prime ideal of $(\alpha_1, \ldots, \alpha_r, \beta_1, \ldots, \beta_s)$. Then P/A is a minimal prime ideal of $(\bar{\beta}_1, \ldots, \bar{\beta}_s)$. Accordingly $\operatorname{rank}(P/A) \leqslant s$ and, since $\operatorname{rank}(\bar{\beta}_1, \ldots, \bar{\beta}_s) = s$, we have in fact $\operatorname{rank}(P/A) = s$. It follows that there exists a minimal prime ideal P' of A such that $A \subseteq P' \subseteq P$ and $\operatorname{rank}(P/P') = s$. However all the prime ideals belonging to A have rank equal to r and so, in particular, $\operatorname{rank} P' = r$. Hence, by Theorem 13,

$$\operatorname{rank} P = \operatorname{rank} P' + \operatorname{rank}(P/P') = r + s.$$

This proves that

$$\operatorname{rank}(\alpha_1, \ldots, \alpha_r, \beta_1, \ldots, \beta_s) = r + s$$

and shows that the R-ideal $(\alpha_1, \ldots, \alpha_r, \beta_1, \ldots, \beta_s)$ is of the fundamental class.

Now assume that Π is an arbitrary prime ideal belonging to

$$(\alpha_1, \ldots, \alpha_r, \beta_1, \ldots, \beta_s).$$

Since R is semi-regular and the ideal is of the fundamental class, we may conclude that $\operatorname{rank} \Pi = r + s$. Next we can choose a minimal prime ideal P^* of A such that $P^* \subseteq \Pi$ and

$$\operatorname{rank}(\Pi/P^*) = \operatorname{rank}(\Pi/A).$$

Then, using Theorem 13 and the fact that $\operatorname{rank} P^* = r$, we obtain

$$\operatorname{rank}(\Pi/A) = \operatorname{rank} \Pi - \operatorname{rank} P^*$$
$$= r + s - r$$
$$= s.$$

This completes the proof

Our remaining discussion of semi-regular rings will be concerned with semi-regular polynomial rings. For our investigations, we shall need to use certain properties of polynomial rings which have not yet been established. Because of this we shall take the opportunity to digress a little in order to fill in some of the gaps in our treatment of polynomials.

5.4 General properties of polynomial rings

As usual, R denotes a commutative ring with an identity element and $R[X_1, X_2, ..., X_n]$ the ring of polynomials in $X_1, X_2, ..., X_n$ with coefficients in R. Accordingly a typical element f of $R[X_1, X_2, ..., X_n]$ has a *unique* representation

$$f = \Sigma r_{\mu_1 \mu_2 ... \mu_n} X_1^{\mu_1} X_2^{\mu_2} ... X_n^{\mu_n}, \qquad (5.4.1)$$

where the $r_{\mu_1 \mu_2 ... \mu_n}$ belong to R and only a finite number of them are different from zero.

Let A be an ideal of the ring R. *In order that the polynomial f, of (5.4.1), should belong to the extension $AR[X_1, X_2, ..., X_n]$ of A it is both necessary and sufficient that all the coefficients $r_{\mu_1 \mu_2 ... \mu_n}$ belong to A.* One immediate consequence of this observation is the relation

$$AR[X_1, X_2, ..., X_n] \cap R = A. \qquad (5.4.2)$$

Suppose now that $\{A_i\}_{i \in I}$ is a family of ideals of R. Then the same observation shows that

$$(\bigcap_{i \in I} A_i) R[X_1, X_2, ..., X_n] = \bigcap_{i \in I} (A_i R[X_1, X_2, ..., X_n]). \qquad (5.4.3)$$

As before, let A be an ideal of R. If f is the polynomial described in (5.4.1) and $c \in R$, then cf belongs to $AR[X_1, X_2, ..., X_n]$ if and only if $cr_{\mu_1 \mu_2 ... \mu_n} \in A$; that is if and only if $r_{\mu_1 \mu_2 ... \mu_n} \in (A:c)$ for each coefficient $r_{\mu_1 \mu_2 ... \mu_n}$. Accordingly

$$AR[X_1, X_2, ..., X_n] : c = (A:c) R[X_1, X_2, ..., X_n]. \qquad (5.4.4)$$

Let C also be an R-ideal. Then

$$A:C = \bigcap_{c \in C} (A:c)$$

and, using (5.4.4), we find that

$$AR[X_1, ..., X_n] : CR[X_1, ..., X_n] = \bigcap_{c \in C} (AR[X_1, ..., X_n] : c)$$
$$= \bigcap_{c \in C} ((A:c) R[X_1, ..., X_n]).$$

It therefore follows, from (5.4.3), that

$$AR[X_1, ..., X_n]:CR[X_1, ..., X_n] = (\bigcap_{c \in C} (A:c)) R[X_1, ..., X_n]$$

which may be rewritten as

$$AR[X_1, ..., X_n]:CR[X_1, ..., X_n] = (A:C) R[X_1, ..., X_n]. \quad (5.4.5)$$

Still assuming that A is an R-ideal, let $\phi:R \to R/A$ be the natural mapping. One obtains a ring-epimorphism

$$R[X_1, X_2, ..., X_n] \to (R/A)[X_1, X_2, ..., X_n]$$

by operating with ϕ on the coefficients of each polynomial in

$$R[X_1, X_2, ..., X_n].$$

The kernel of this epimorphism is clearly $AR[X_1, X_2, ..., X_n]$. Consequently there is induced a ring-isomorphism

$$R[X_1, ..., X_n]/AR[X_1, ..., X_n] \approx (R/A)[X_1, ..., X_n]. \quad (5.4.6)$$

This isomorphism is frequently used to identify the two rings which occur in (5.4.6).

Another useful identification arises in the following way. Let S be a non-empty multiplicatively closed subset of R. Then it is also a multiplicatively closed subset of $R[X_1, X_2, ..., X_n]$ and therefore we may form the ring $R[X_1, X_2, ..., X_n]_S$ of fractions. Let

$$f = \Sigma r_{\nu_1 \nu_2 ... \nu_n} X_1^{\nu_1} X_2^{\nu_2} ... X_n^{\nu_n}$$

be a typical element of $R[X_1, X_2, ..., X_n]$. One now obtains a ring-isomorphism

$$R[X_1, X_2, ..., X_n]_S \approx R_S[X_1, X_2, ..., X_n] \quad (5.4.7)$$

by means of the mapping

$$\left[\frac{f}{s}\right] \to \Sigma \left[\frac{r_{\nu_1 \nu_2 ... \nu_n}}{s}\right] X_1^{\nu_1} X_2^{\nu_2} ... X_n^{\nu_n}.$$

Here, of course, s denotes an element of S.

Let D be the monoid consisting of all sequences $(\nu_1, \nu_2, ..., \nu_n)$ of n non-negative integers, it being understood that addition of such sequences is defined by

$$(\nu_1, \nu_2, ..., \nu_n) + (\nu_1', \nu_2', ..., \nu_n') = (\nu_1 + \nu_1', \nu_2 + \nu_2', ..., \nu_n + \nu_n').$$

The monoid D is torsionless. Furthermore $R[X_1, X_2, ..., X_n]$ may be regarded as a D-graded ring† in which the homogeneous elements of degree $(\nu_1, \nu_2, ..., \nu_n)$ are those having the form $rX_1^{\nu_1} X_2^{\nu_2} ... X_n^{\nu_n}$, where r belongs to R.

---

† See section (2.11) for the definition of a graded ring.

Let A be an ideal of R. Then $AR[X_1, X_2, ..., X_n]$ is a homogeneous ideal and therefore, by Proposition 32 of section (2.13),

$$\text{Rad}\,(AR[X_1, ..., X_n])$$

is also homogeneous, *We now contend that*

$$(\text{Rad}\,A)\,R[X_1, X_2, ..., X_n] = \text{Rad}\,(AR[X_1, X_2, ..., X_n]). \quad (5.4.8)$$

Indeed it is clear that the left-hand side is contained in the right-hand side. To prove the opposite inclusion, it is enough to show that if a homogeneous element $rX_1^{\nu_1}X_2^{\nu_2}\ldots X_n^{\nu_n}$ belongs to

$$\text{Rad}\,(AR[X_1, ..., X_n]),$$

then $r \in \text{Rad}\,A$. However, if

$$(rX_1^{\nu_1}X_2^{\nu_2}\ldots X_n^{\nu_n})^m \in AR[X_1, X_2, ..., X_n],$$

then $r^m \in A$ and therefore $r \in \text{Rad}\,A$ as required.

Proposition 6. *Let P be a prime ideal of R. Then $PR[X_1, X_2, ..., X_n]$ is a prime ideal of the polynomial ring $R[X_1, X_2, ..., X_n]$.*

Proof. First we note that $PR[X_1, X_2, ..., X_n]$ is a proper homogeneous ideal of $R[X_1, X_2, ..., X_n]$. Hence, by Lemma 13 of section (2.13), it is sufficient to check that the characteristic property of a prime ideal holds in the case of *homogeneous* elements. However this is obvious.

Corollary. *If R is an integral domain, then the polynomial ring*

$$R[X_1, X_2, ..., X_n]$$

is also an integral domain.

This is the special case of Proposition 6 in which P is the zero ideal.

Our next result concerns an important property of zero-divisors in a polynomial ring.

Proposition 7. *Let $f(X_1, X_2, ..., X_n)$ be a zero-divisor in the polynomial ring $R[X_1, X_2, ..., X_n]$. Then there exists c in R such that $c \neq 0$ and $cf(X_1, X_2, ..., X_n) = 0$.*

Proof. It has already been observed that $R[X_1, X_2, ..., X_n]$ is graded by the monoid consisting of all sequences of n non-negative integers. This monoid is torsionless. It can therefore be given a total order which is compatible with its monoid structure.† Let

$$f(X_1, ..., X_n) = \alpha X_1^{\mu_1}X_2^{\mu_2}\ldots X_n^{\mu_n} + \beta X_1^{\nu_1}X_2^{\nu_2}\ldots X_n^{\nu_n} + \ldots +$$
$$\omega X_1^{\sigma_1}X_2^{\sigma_2}\ldots X_n^{\sigma_n},$$

---

† See Theorem 22 of section (2.12). However, it may be noted that the familiar 'lexicographical order' has the required property. (See Exercise 15 of Chapter 2.)

where $\alpha, \beta, \ldots, \omega$ are in R and

$$(\mu_1, \mu_2, \ldots, \mu_n) > (\nu_1, \nu_2, \ldots, \nu_n) > \ldots > (\sigma_1, \sigma_2, \ldots, \sigma_n).$$

By hypothesis, there exists $g(X_1, X_2, \ldots, X_n) \neq 0$ such that

$$f(X_1, \ldots, X_n)\, g(X_1, \ldots, X_n) = 0.$$

We choose $g(X_1, X_2, \ldots, X_n)$ so that it satisfies these conditions and, in addition, has the smallest possible number of non-zero terms. Let

$$g(X_1, X_2, \ldots, X_n) = cX_1^{m_1} X_2^{m_2} \ldots X_n^{m_n} + \ldots,$$

where $c \in R$, $c \neq 0$, and $X_1^{m_1} X_2^{m_2} \ldots X_n^{m_n}$ is the highest power-product actually occurring in $g(X_1, X_2, \ldots, X_n)$.

Since $f(X_1, \ldots, X_n)\, g(X_1, \ldots, X_n) = 0$ we see, by considering the term of degree $(\mu_1 + m_1, \mu_2 + m_2, \ldots, \mu_n + m_n)$, that $\alpha c = 0$. Thus $\alpha g(X_1, X_2, \ldots, X_n)$ has fewer non-zero terms than $g(X_1, X_2, \ldots, X_n)$ and

$$f(X_1, X_2, \ldots, X_n)\, \alpha g(X_1, X_2, \ldots, X_n) = 0.$$

Hence, by the choice of $g(X_1, X_2, \ldots, X_n)$, we must have

$$\alpha g(X_1, X_2, \ldots, X_n) = 0.$$

Next we observe that

$$[f(X_1, X_2, \ldots, X_n) - \alpha X_1^{\mu_1} X_2^{\mu_2} \ldots X_n^{\mu_n}]\, g(X_1, X_2, \ldots, X_n) = 0,$$

that is

$$(\beta X_1^{\nu_1} X_2^{\nu_2} \ldots X_n^{\nu_n} + \ldots + \omega X_1^{\sigma_1} X_2^{\sigma_2} \ldots X_n^{\sigma_n})\, g(X_1, X_2, \ldots, X_n) = 0.$$

This time consideration of the term of degree

$$(\nu_1 + m_1, \nu_2 + m_2, \ldots, \nu_n + m_n)$$

shows that $\beta c = 0$. Thus $\beta g(X_1, X_2, \ldots, X_n)$ has fewer non-zero terms than $g(X_1, X_2, \ldots, X_n)$ and

$$f(X_1, X_2, \ldots, X_n)\, \beta g(X_1, X_2, \ldots, X_n) = 0.$$

Consequently, by the choice of $g(X_1, X_2, \ldots, X_n)$, we have

$$\beta g(X_1, X_2, \ldots, X_n) = 0.$$

Proceeding in this way we show, in succession, that

$$\alpha c = 0, \quad \beta c = 0, \ldots, \omega c = 0.$$

Accordingly $cf(X_1, X_2, \ldots, X_n) = 0$ and the proof is complete.

We shall now prove that a primary ideal of the ring R remains primary when it is extended to $R[X_1, X_2, \ldots, X_n]$.

Proposition 8. *Let Q be a P-primary ideal of the ring R. Then*

$$QR[X_1, X_2, \ldots, X_n]$$

is a primary ideal of the polynomial ring $R[X_1, X_2, \ldots, X_n]$ and the prime ideal to which it belongs is $PR[X_1, X_2, \ldots, X_n]$.

Proof. It is sufficient to show that $QR[X_1, X_2, ..., X_n]$ is a primary ideal, for then the remaining assertion will follow immediately from (5.4.8). Now $QR[X_1, X_2, ..., X_n]$ is a proper homogeneous ideal. Hence, by Lemma 14 of section (2.13), in order to verify that it has the characteristic property of a primary ideal we may confine our attention to homogeneous elements. However in that case the verification is trivial.

Proposition 9. *Let A be an ideal of R and suppose that it has a primary decomposition*

$$A = Q_1 \cap Q_2 \cap ... \cap Q_s \qquad (5.4.9)$$

in R. Then

$$AR[X_1, ..., X_n]$$
$$= Q_1 R[X_1, ..., X_n] \cap Q_2 R[X_1, ..., X_n] \cap ... \cap Q_s R[X_1, ..., X_n] \quad (5.4.10)$$

is a primary decomposition of $AR[X_1, X_2, ..., X_n]$ in $R[X_1, X_2, ..., X_n]$. Furthermore, if (5.4.9) is a normal decomposition, then so too is (5.4.10). It follows that if $P_i (1 \leqslant i \leqslant t)$ are the prime ideals belonging to A, then the $P_i R[X_1, X_2, ..., X_n]$, where $1 \leqslant i \leqslant t$, are the prime ideals belonging to $AR[X_1, X_2, ..., X_n]$.

Proof. Suppose that the primary ideal Q_i belongs to the prime ideal P_i. Then, by Proposition 8, $Q_i R[X_1, X_2, ..., X_n]$ is also primary and its radical is $P_i R[X_1, X_2, ..., X_n]$. Next, the relation (5.4.10) follows from (5.4.9) by virtue of (5.4.3). These two remarks establish the first assertion.

Now assume that (5.4.9) is a normal decomposition. It immediately follows, from (5.4.2), that the prime ideals $P_i R[X_1, X_2, ..., X_n]$ are distinct. Furthermore we cannot have

$$Q_i R[X_1, X_2, ..., X_n] \supseteq \bigcap_{j \neq i} Q_j R[X_1, X_2, ..., X_n]$$

because, on taking intersections with R, this would yield

$$Q_i \supseteq \bigcap_{j \neq i} Q_j$$

which is contrary to the assumption that (5.4.9) is a normal decomposition. The proposition follows.

Proposition 10. *Let R be a Noetherian ring and P a prime ideal of R. Then $PR[X_1, X_2, ..., X_n]$ is a prime ideal of the polynomial ring*

$$R[X_1, X_2, ..., X_n]$$

and it has the same rank as P.

Proof. We first note that, by Theorem 1 Cor. of section (4.1),

$$R[X_1, X_2, ..., X_n]$$

is a Noetherian ring and, by Proposition 6, $PR[X_1, X_2, ..., X_n]$ is a prime ideal. Let rank $P = r$. Then there exists a strictly decreasing sequence
$$P \supset P_1 \supset P_2 \supset ... \supset P_r$$

of $r+1$ prime ideals of the ring R. It follows, from Proposition 6 and (5.4.2), that

$$PR[X_1, ..., X_n] \supset P_1 R[X_1, ..., X_n] \supset ... \supset P_r R[X_1, ..., X_n]$$

is also a strictly decreasing sequence of prime ideals. Thus

$$\text{rank}\,(PR[X_1, ..., X_n]) \geqslant r.$$

Next, by Proposition 15 of section (4.8), there exist r elements $\alpha_1, \alpha_2, ..., \alpha_r$ such that P is a minimal prime ideal of $(\alpha_1, \alpha_2, ..., \alpha_r)$. Hence, by Proposition 9, $PR[X_1, ..., X_n]$ is a minimal prime ideal of $(\alpha_1, \alpha_2, ..., \alpha_r) R[X_1, ..., X_n]$. It therefore follows, from Theorem 22 of section (4.8), that
$$\text{rank}\,(PR[X_1, X_2, ..., X_n]) \leqslant r.$$

This completes the proof of the theorem.

Suppose that R is a ring (commutative and possessing an identity element) and $X_1, X_2, ..., X_n$ are interdeterminates. For $1 \leqslant i \leqslant n$ put $R_i = R[X_1, X_2, ..., X_i]$ and let $R_0 = R$. Then, as we noted in section (4.2),
$$R_{i+1} = R_i[X_{i+1}].$$

In other terms, $R[X_1, X_2, ..., X_{i+1}]$ may be regarded as the ring of polynomials in X_{i+1} having as coefficients polynomials in $X_1, X_2, ..., X_i$. This useful observation enables us, on occasion, to reduce a problem concerning polynomials in several variables to the case where there is only a single variable. In view of this, we shall take the opportunity to make some elementary observations concerning polynomials in one variable. The variable itself will be denoted by X.

Let $f(X) = a_0 + a_1 X + ... + a_m X^m$ belonging to $R[X]$. The highest power of X which has a non-zero coefficient is called the *degree* of $f(X)$. This degree will be denoted by $\partial^\circ f(X)$ or simply $\partial^\circ f$. Thus if $a_m \neq 0$, then $\partial^\circ f = m$. In this case a_m is called the *leading coefficient* and $a_m X^m$ the *leading term* of $f(X)$. Clearly if $f(X)$ and $g(X)$ both belong to $R[X]$, then
$$\partial^\circ (f+g) \leqslant \max\,(\partial^\circ f, \partial^\circ g) \tag{5.4.11}$$

and indeed there is equality if it happens that $\partial^\circ f \neq \partial^\circ g$.

Let $\partial^\circ f(X) = m$ and $\partial^\circ g(X) = n$. Then

$$f(X) = a_0 + a_1 X + a_2 X^2 + \ldots + a_m X^m$$

where $a_m \neq 0$, and

$$g(X) = b_0 + b_1 X + b_2 X^2 + \ldots + b_n X^n$$

where $b_n \neq 0$. Now

$$f(X)g(X) = (a_0 b_0) + (a_0 b_1 + a_1 b_0) X + \ldots + (a_m b_n) X^{m+n}.$$

Of course it may happen that $a_m b_n = 0$, but in any case $\partial^\circ(fg) \leqslant m+n$, that is

$$\partial^\circ(fg) \leqslant \partial^\circ f + \partial^\circ g. \tag{5.4.12}$$

If however $a_m b_n \neq 0$, then we obtain

$$\partial^\circ(fg) = \partial^\circ f + \partial^\circ g, \tag{5.4.13}$$

in which case we shall say that *the degree formula holds for $f(X)$ and $g(X)$*. It is important to note that the degree formula holds in the following situations:

(1) *either $f(X)$ or $g(X)$ has a leading coefficient which is not a zero-divisor*;

(2) *either $f(X)$ or $g(X)$ has a leading coefficient which is a unit*;

(3) *the ring R, of coefficients, is an integral domain.*

There is one respect in which the above discussion is incomplete. Since the null or zero polynomial has no non-zero coefficients, its degree is not defined. However it is convenient to assign to the null polynomial the conventional degree *minus infinity*. If this is done, then (5.4.11), (5.4.12) and, under appropriate conditions, (5.4.13) remain valid even if the null polynomial is present, provided, of course, that certain natural conventions concerning the use of infinite quantities are observed. Here the situation is exactly the same as in elementary algebra so we shall not elaborate this point.

Lemma 7. *Let $g(X)$ be a non-null polynomial (with coefficients in R) whose leading coefficient is a unit. If now $f(X)$ belongs to $R[X]$, then $f(X)$ has a unique representation in the form*

$$f(X) = q(X)g(X) + r(X),$$

where $q(X)$, $r(X)$ belong to $R[X]$ and $\partial^\circ r(X) < \partial^\circ g(X)$.

Proof. We begin by showing that such a representation is always possible. *Assume the contrary.* Then there will exist at least one polynomial which cannot be expressed in the desired form. From among such polynomials we select one, say $\phi(x)$, whose degree is as small as possible. Let $\partial^\circ \phi = n$ and $\partial^\circ g = m$. Then $n \geqslant m$ for otherwise

$$\phi(X) = 0g(X) + \phi(X)$$

would be a representation of the type in question.

Let $g(X) = a_0 X^m + \ldots + a_m$ and $\phi(X) = b_0 X^n + \ldots + b_n$. By hypothesis, a_0 is a unit. We can therefore form

$$\phi_1(X) = \phi(X) - a_0^{-1} b_0 X^{n-m} g(X)$$

which is a polynomial having a smaller degree than $\phi(X)$. Accordingly there exist $q_1(X)$ and $r_1(X)$ in $R[X]$ such that $\partial^\circ r_1(X) < \partial^\circ g(X)$ and

$$\phi_1(X) = q_1(X) g(X) + r_1(X).$$

But then $\quad \phi(X) = \{q_1(X) + a_0^{-1} b_0 X^{n-m}\} g(X) + r_1(X)$

and now we have a contradiction because this is a representation of $\phi(X)$ of the required type.

It has thus been shown that any element of $R[X]$ can be represented in the manner described and so it only needs to be proved that the representation is unique. Suppose therefore that

$$q(X) g(X) + r(X) = q'(X) g(X) + r'(X),$$

where $q(X)$, $q'(X)$, $r(X)$, $r'(X)$ are all in $R[X]$ and $\partial^\circ r(X) < \partial^\circ g(X)$, $\partial^\circ r'(X) < \partial^\circ g(X)$. We wish to show that both $q(X) = q'(X)$ and $r(X) = r'(X)$. However, once the first of these is proved, the second will follow immediately.

Let us assume that $q(X) \neq q'(X)$. Then $q(X) - q'(X)$ is not the null polynomial and

$$r'(X) - r(X) = \{q(X) - q'(X)\} g(X).$$

Now the leading coefficient of $g(X)$ is a unit; hence the degree formula may be applied to the right-hand side. This yields

$$\partial^\circ \{r'(X) - r(X)\} = \partial^\circ \{q(X) - q'(X)\} + \partial^\circ g(X)$$
$$\geqslant \partial^\circ g(X)$$

because, since $q(X) - q'(X)$ is not the null polynomial, its degree is not negative. However

$$\partial^\circ \{r'(X) - r(X)\} \leqslant \max \{\partial^\circ r'(X), \partial^\circ r(X)\} < \partial^\circ g(X).$$

This gives a contradiction and establishes the lemma.

The next result is both familiar and very important. It forms a natural companion to Theorem 7 of section (4.2). A proof has been included for the sake of completeness.

Theorem 15. *Let F be a field. Then each ideal of the polynomial ring $F[X]$ can be generated by a single element.*

Proof. Let A be an ideal of $F[X]$. Since the zero ideal is always singly generated, we shall assume that $A \neq (0)$. In what follows we suppose that $\phi(X)$ is a non-zero polynomial belonging to A and, subject to this condition, it is assumed to have the smallest possible degree. It will be proved that $A \subseteq (\phi)$. Since the opposite inclusion is obvious, this will establish the theorem.

Let $f(X) \in A$. By Lemma 7, we can express $f(X)$ in the form

$$f(X) = q(X)\,\phi(X) + r(X),$$

where $q(X)$ and $r(X)$ belong to $F[X]$ and $\partial^{\circ}r(X) < \partial^{\circ}\phi(X)$. But then $r(X) \in A$, whence, by the choice of $\phi(X)$, we see that $r(X) = 0$. Thus $f(X)$ belongs to (ϕ) and, as already explained, the theorem follows.

Corollary. *Let F be a field. Then $F[X]$ is not a field but every non-zero prime ideal is a maximal ideal.*

Proof. It is clear that X is not a unit in $F[X]$. Consequently $F[X]$ is not a field. Now suppose that $P_1 \subset P_2$ (strict inclusion), where P_1, P_2 are prime ideals. Then, by the theorem, there exist polynomials ϕ_1, ϕ_2 such that $P_1 = (\phi_1)$ and $P_2 = (\phi_2)$. Since ϕ_1 belongs to P_2, we have $\phi_1 = \phi_2\psi$ where $\psi \in F[X]$. But then $\phi_2\psi \in P_1$ and $\phi_2 \notin P_1$. Accordingly $\psi \in P_1$, say $\psi = \phi_1\psi'$ where $\psi' \in F[X]$. Thus $\phi_1 = \phi_1\phi_2\psi'$. *We now contend that* $\phi_1 = 0$. For otherwise $1 = \phi_2\psi'$ whence, because the degree formula holds in $F[X]$, ϕ_2 is a non-zero constant and hence a unit. However this is impossible because $P_2 = (\phi_2)$ is a proper ideal. It follows that ϕ_1 is the null polynomial and therefore $P_1 = (0)$. The corollary is thus established.

Theorem 16. *Let R be a Noetherian ring and $X_1, X_2, \ldots, X_n$ indeterminates. Then*
$$\operatorname{Dim} R[X_1, X_3, \ldots, X_n] = \operatorname{Dim} R + n.$$

Proof. It is enough to prove the theorem for the case $n = 1$. We shall therefore assume that we have this situation and, in order to simplify the notation, we shall denote the single variable by X.

Let P be a prime ideal of R. Then, by Proposition 10,

$$\operatorname{rank}(PR[X]) = \operatorname{rank} P.$$

It follows that $\operatorname{Dim} R[X] \geqslant \operatorname{Dim} R$ which establishes the theorem in the case where $\operatorname{Dim} R = \infty$. From now on we shall suppose that $\operatorname{Dim} R < \infty$.

There exists a maximal ideal M of R which satisfies rank $M = \mathrm{Dim}\,R$ and which is therefore such that

$$\mathrm{rank}\,(MR[X]) = \mathrm{Dim}\,R.$$

Now $R/M = F$ (say) is a field and, by (5.4.6), $R[X]/MR[X]$ is ring-isomorphic to $F[X]$. However, by the corollary to Theorem 15, $F[X]$ is not a field and therefore $MR[X]$ is not a maximal ideal of $R[X]$. It follows that

$$\mathrm{Dim}\,R[X] \geqslant \mathrm{Dim}\,R + 1.$$

Next, let Π be a maximal ideal of $R[X]$. To complete the proof it will suffice to show that rank Π does not exceed $\mathrm{Dim}\,R + 1$. Put $P = \Pi \cap R$. Then P is a prime ideal of R and its complement S, in R, is a non-empty multiplicatively closed set.

We now identify $R[X]_S$ with $R_S[X]$ by means of (5.4.7). Then Π_S is a maximal ideal of $R_S[X]$, R_S is a local ring, and $\Pi_S \cap R_S$ is the maximal ideal of R_S. Moreover Π_S and Π have the same rank while

$$\mathrm{Dim}\,R_S = \mathrm{rank}\,P \leqslant \mathrm{Dim}\,R.$$

Consequently it is enough to show that

$$\mathrm{rank}\,\Pi_S \leqslant \mathrm{Dim}\,R_S + 1.$$

This conclusion may be restated in the following form: *when proving that* rank Π *is at most* $\mathrm{Dim}\,R + 1$ *we may assume that* (i) R *is a local ring and* (ii) Π *contracts to the maximal ideal of* R. In what follows it will be supposed that we have this situation.

Let M be the maximal ideal of R, so that $\Pi \cap R = M$, and let $\alpha_1, \alpha_2, \ldots, \alpha_d$ be a system of parameters of R. We know that the rings $R[X]/MR[X]$ and $(R/M)[X]$ may be identified and, by Theorem 15, each ideal of $(R/M)[X]$, in particular the ideal $\Pi/MR[X]$, is singly generated. Thus there exists $f \in R[X]$ such that

$$MR[X] + fR[X] = \Pi.$$

Let Π' be an arbitrary prime ideal of $R[X]$ containing $\alpha_1, \alpha_2, \ldots, \alpha_d, f$. Then $\Pi' \cap R$ is a prime R-ideal containing the system of parameters $\alpha_1, \alpha_2, \ldots, \alpha_d$. It follows that $\Pi' \cap R = M$ and therefore Π' contains

$$MR[X] + fR[X] = \Pi.$$

This shows that Π must be a minimal prime ideal of $(\alpha_1, \alpha_2, \ldots, \alpha_d, f)$. Hence, by Theorem 22 of section (4.8),

$$\mathrm{rank}\,\Pi \leqslant d + 1 = \mathrm{Dim}\,R + 1.$$

The proof of the theorem is now complete.

5.5 Semi-regular polynomial rings

In this, the final section of this chapter, we return to the study of semi-regular rings which was begun in section (5.3).

Theorem 17. *Let R be a semi-regular ring. Then the polynomial ring $R[X_1, X_2, ..., X_n]$ is also semi-regular.*

Proof. It is clearly sufficient to prove the theorem for the case in which there is only one variable. In what follows we shall assume that we have this situation and we shall denote the single variable by X. Obviously $R[X]$ is a non-null Noetherian ring.

Let Π be a maximal ideal of $R[X]$. If we can show that the ring of fractions of $R[X]$ with respect to Π is semi-regular, then the required result will follow by virtue of Theorem 11.

Put $P = \Pi \cap R$ and denote the complement of P in R by S. Then P is a prime R-ideal and S is a (non-empty) multiplicatively closed subset of $R[X]$. Furthermore, by Proposition 19 of section (3.8), $R[X]_\Pi$ and the ring of fractions of $R[X]_S$ with respect to Π_S are isomorphic. Consequently it is sufficient to prove that the latter is semi-regular. We now use (5.4.7) in order to identify $R[X]_S$ with $R_S[X]$. On this understanding, Π_S is a maximal ideal of $R_S[X]$, R_S is a semi-regular local ring,† and Π_S contracts, in R_S, to the maximal ideal of that ring.

The above remarks show that, for the purpose of proving that $R[X]_\Pi$ is a semi-regular ring, we may now add the assumptions that R is a (semi-regular) local ring and that the maximal ideal Π of $R[X]$ contracts to the maximal ideal M (say) of R. These assumptions are made in what follows.

Let $\alpha_1, \alpha_2, ..., \alpha_d$ be a system of parameters of R. By Theorem 10,

$$(\alpha_1, \alpha_2, ..., \alpha_i)R : \alpha_{i+1} = (\alpha_1, \alpha_2, ..., \alpha_i)R$$

for $0 \leqslant i < d$. Hence, by (5.4.4),

$$(\alpha_1, \alpha_2, ..., \alpha_i)R[X] : \alpha_{i+1} = (\alpha_1, \alpha_2, ..., \alpha_i)R[X]$$

so that $\alpha_1, \alpha_2, ..., \alpha_d$ is also an $R[X]$-sequence. Note that, since M is the only prime ideal belonging to $(\alpha_1, \alpha_2, ..., \alpha_d)R$, $MR[X]$ is the only prime ideal belonging to $(\alpha_1, \alpha_2, ..., \alpha_d)R[X]$. Here we have made use of Proposition 9.

Next, $\Pi/MR[X]$ is a maximal ideal of the ring $R[X]/MR[X]$ and this ring can be identified with $(R/M)[X]$ by virtue of (5.4.6). Moreover

† See Theorem 12.

R/M is a field; consequently $\Pi/MR[X]$ is a non-zero principal ideal.†
Thus there exists $f \in R[X]$ such that $f \notin MR[X]$ and

$$MR[X] + fR[X] = \Pi.$$

This shows us that‡ Π is a minimal prime ideal of $(\alpha_1, ..., \alpha_d, f)$, while
from $f \notin MR[X]$ we conclude that

$$(\alpha_1, \alpha_2, ..., \alpha_d)R[X]{:}f = (\alpha_1, \alpha_2, ..., \alpha_d)R[X].$$

Accordingly $\alpha_1, \alpha_2, ..., \alpha_d, f$ is an $R[X]$-sequence and rank $\Pi = d + 1$.

We are now ready to pass to the ring $R[X]_\Pi$. Let $\alpha_1^*, \alpha_2^*, ..., \alpha_d^*, f^*$
be the images of $\alpha_1, \alpha_2, ..., \alpha_d, f$ under the canonical mapping

$$R[X] \to R[X]_\Pi.$$

We observe, in the first place, that $R[X]_\Pi$ is a local ring of dimension
$d + 1$. Next, by Proposition 3 of section (3.2), the extension of

$$(\alpha_1, ..., \alpha_d, f)R[X] \quad \text{in} \quad R[X]_\Pi$$

is generated by $\alpha_1^*, ..., \alpha_d^*, f^*$. In addition, Theorem 11 of section (3.7)
shows that the only prime ideal belonging to this extension is the
maximal ideal of $R[X]_\Pi$. Thus $\alpha_1^*, ..., \alpha_d^*, f^*$ is a system of parameters
in $R[X]_\Pi$.

Finally, by (3.5.6), we have

$$(\alpha_1^*, ..., \alpha_i^*)\,R[X]_\Pi{:}\alpha_{i+1}^* = (\alpha_1^*, ..., \alpha_i^*)\,R[X]_\Pi,$$

for $0 \leqslant i < d$, and

$$(\alpha_1^*, ..., \alpha_d^*)\,R[X]_\Pi{:}f^* = (\alpha_1^*, ..., \alpha_d^*)\,R[X]_\Pi.$$

Accordingly the system of parameters $\alpha_1^*, ..., \alpha_d^*, f^*$ is also an $R[X]_\Pi$-
sequence. It now follows, from Theorem 10, that $R[X]_\Pi$ is a semi-
regular local ring and with this the proof is complete.

Theorem 18. *Let F be a field. Then the polynomial ring $F[X_1, X_2, ..., X_n]$
is semi-regular and its dimension is n.*

Proof. Since a field is obviously zero-dimensional and semi-regular,
Theorem 18 is an immediate consequence of Theorems 16 and 17.

Exercises on Chapter 5

In these exercises R always denotes a commutative ring with an identity
element.

1. R is an integral domain, α, β is an R-sequence, and X is an indeterminate.
Prove that $\alpha X + \beta$ generates a prime ideal in $R[X]$.

† See Theorem 15 and its corollary.
‡ A similar situation was encountered in the proof of Theorem 16. The argument
has therefore not been repeated at length.

2. R is a Noetherian ring and A a proper ideal of R. Show that if $\operatorname{gr}(A) = n$ and A can be generated by n elements, then A can be generated by an R-sequence of length n.

3. R is a local ring and A, B non-zero proper ideals such that $AB = (0)$. Prove that $\operatorname{gr}(A + B) \leqslant 1$.

4. Let $E \neq 0$ be a Noetherian R-module, α an element of R which is not a zero-divisor on E, and P a prime ideal belonging to the zero submodule of E. If now P^* is a prime ideal containing (P, α), show that there exists a prime ideal P' which belongs to the zero submodule of $E/\alpha E$ and satisfies $(P, \alpha) \subseteq P' \subseteq P^*$.

5. Let R be a semi-local ring with radical J, $E \neq 0$ a finitely generated R-module satisfying $\operatorname{gr}(J; E) = \operatorname{Dim} E$, and P a prime ideal containing $\operatorname{Ann}_R E$. If now $\alpha_1, \alpha_2, \ldots, \alpha_k$ is a maximal R-sequence on E in $P \cap J$, prove that P belongs to the zero submodule of $E/(\alpha_1, \alpha_2, \ldots, \alpha_k)E$.

6. A semi-local ring R admits a system of parameters which forms an R-sequence. Show that every system of parameters forms an R-sequence and that R is semi-regular.

7. R is a d-dimensional $(d \geqslant 1)$ semi-local ring with the property that whenever $x_1, x_2, \ldots, x_d$ is a system of parameters no maximal ideal belongs to $(x_1, x_2, \ldots, x_{d-1})$. Show that every system of parameters forms an R-sequence.

8. Show that the direct sum of a finite number of semi-regular rings is semi-regular.

9. Let R be a ring and let $I = (a_1, a_2, \ldots, a_k)$, where $a_1, a_2, \ldots, a_k$ $(k \geqslant 1)$ is an R-sequence. Show that if $F(X_1, X_2, \ldots, X_k)$ is a form of degree s with coefficients in R and $F(a_1, a_2, \ldots, a_k) \in I^{s+1}$, then all the coefficients of F are in I. Show also that if $I : x = I$, then $I^n : x = I^n$ for every positive integer n.

10. Let R be a ring and let $I = (a_1, a_2, \ldots, a_k)$, where $a_1, a_2, \ldots, a_k$ is an R-sequence. Show that if $1 \leqslant j \leqslant k$, then $I^n : a_j = I^{n-1}$ for all $n \geqslant 1$.

11. R is a Noetherian ring and $I = (a_1, a_2, \ldots, a_k)$, where $a_1, a_2, \ldots, a_k$ $(k \geqslant 1)$ belong to the Jacobson radical of R. Show that the following two statements are equivalent:

(i) $a_1, a_2, \ldots, a_k$ is an R-sequence;

(ii) if $F(X_1, X_2, \ldots, X_k)$ is a form (with coefficients in R) such that

$$F(a_1, a_2, \ldots, a_k) = 0,$$

then all the coefficients of F are in I.

12. R is a semi-regular ring and A is an R-ideal of the fundamental class. Prove that the residue class ring R/A^m is semi-regular for every positive integer m.

13. R is a semi-regular ring and $\alpha_1, \alpha_2, \ldots, \alpha_m$ are m elements belonging to it which generate an ideal of rank m. Show that elements $\lambda_1, \lambda_2, \ldots, \lambda_m$ of R satisfy $\lambda_1 \alpha_1 + \lambda_2 \alpha_2 + \ldots + \lambda_m \alpha_m = 0$ if and only if they can be expressed in the form

$$\lambda_i = \sum_{j=1}^{m} d_{ij} \alpha_j \quad (1 \leqslant i \leqslant m),$$

where the d_{ij} belong to R and are such that $d_{ii} = 0$ $(1 \leqslant i \leqslant m)$ and $d_{ij} = -d_{ji}$ $(1 \leqslant i, j \leqslant m)$.

14. Let R be a semi-regular ring and α, β an R-sequence. Prove that $R[\beta/\alpha]$ is also semi-regular, where β/α denotes the element of the full ring of fractions of R which has β as numerator and α as denominator.

6

HILBERT RINGS AND THE ZEROS THEOREM

General remarks. In this chapter we continue the study of polynomial rings. Historically, the ideas presented here developed out of a single result, due to D. Hilbert, known as the *Zeros Theorem*. An analysis of this result drew attention to a certain special class of rings whose members are now known as *Hilbert rings*. The abstract theory of Hilbert rings is presented in the first section, the remaining ones being devoted to applications.

Throughout this chapter, the rings considered are assumed to be commutative and it is supposed that each possesses an identity element.

6.1 Hilbert rings

A commutative ring R is called a *Hilbert ring* if each prime ideal is the intersection of all the maximal ideals that contain it. A null ring is regarded as a Hilbert ring because, in this case, there is no prime ideal which fails to have the property described in the definition. Obviously any field is a Hilbert ring and, more generally, so is any ring in which every prime ideal is a maximal ideal. Also, it is easy to see that the ring formed by the integers provides another example. By contrast, a local ring is a Hilbert ring when and only when it is zero-dimensional.

Proposition 1. *Let R be a Hilbert ring and $\phi: R \to R'$ an epimorphism of R on to a ring R'. Then R' is also a Hilbert ring.*

Proof. We know† that there is a natural one-one correspondence between the prime ideals of R' and those prime R-ideals that contain the kernel of ϕ. Let P' be a prime ideal of R' and P the corresponding prime ideal of R. Then the maximal R'-ideals containing P' correspond to the maximal R-ideals containing P. Now the intersection of the latter is P because R is a Hilbert ring. It follows that the maximal R'-ideals containing P' have P' as their intersection. This completes the proof.

† See Proposition 2 of section (2.2).

Corollary. *If R is a Hilbert ring, then so is R/A for every ideal A.*

The next result provides an alternative characterization of Hilbert rings.

Proposition 2. *Let R be a ring (commutative and possessing an identity element). Then the following two statements are equivalent:*

    (a) R is a Hilbert ring;

    (b) for every ideal A, the intersection† of all the maximal ideals containing A is the radical, Rad A, of A.

Proof. Assume that R is a Hilbert ring. Let A be an ideal and denote by B the intersection of all the maximal ideals containing A. Since every maximal ideal which contains A also contains Rad A, it follows that Rad A ⊆ B. Now let P be a minimal prime ideal of A. Then B is certainly contained in the intersection of all the maximal ideals that contain P. But this intersection is P itself, because R is a Hilbert ring. Thus B ⊆ P and therefore B is contained in the intersection of all the minimal prime ideals of A. Accordingly, by Theorem 4 of section (2.3), B ⊆ Rad A. This proves that (a) implies (b). Obviously (b) implies (a) because a prime ideal is its own radical.

We shall now establish certain lemmas with the object of showing that, if R is a Hilbert ring and X an indeterminate, then R[X] is also a Hilbert ring. It will be convenient to refer to an integral domain which happens also to be a Hilbert ring, as a *Hilbert domain*. Since the zero ideal of an integral domain is a prime ideal, the zero ideal of a Hilbert domain R is the intersection of all the maximal ideals, that is to say it is the Jacobson radical of R. In fact certain basic results concerning Hilbert domains hold for any integral domain whose Jacobson radical is the zero ideal. We shall therefore, when the time comes, present them in the more general form.

Lemma 1. *Let F be a field and X an indeterminate. Then F[X] is a Hilbert domain.*

Proof. It is clear that F[X] is an integral domain. Indeed this is a special case of Proposition 6 Cor. of section (5.4). Next, by the corollary to Theorem 15 of section (5.4), every non-zero prime ideal of F[X] is a maximal ideal. Hence it is sufficient to show that if A is the intersection of *all* the maximal ideals of F[X], then A is the zero ideal.

---

† We adopt the usual convention relating to the intersection of an empty set of ideals.

By Theorem 15 of section (5.4), A is a principal ideal, say $A = (\phi)$. Then $1 + \phi$ is not contained in any maximal ideal and therefore it is a unit. It follows that $1 + \phi$ is a constant and so ϕ itself is a constant. Since (ϕ) is a proper ideal and F is a field, this constant must be zero. The proof is now complete.

Let R be a ring, A an ideal of R, and X an indeterminate. The natural mapping $R \to R/A$, when applied to the coefficients of the polynomials in $R[X]$, induces a ring-epimorphism of $R[X]$ on to $(R/A)[X]$. We shall call this the *natural epimorphism of $R[X]$ on to $(R/A)[X]$*. Note that its kernel is $AR[X]$.

Lemma 2. *Let R be an integral domain whose Jacobson radical is the zero ideal and let X be an indeterminate. Then the intersection of all the maximal ideals of $R[X]$ is the zero ideal.*

Proof. Let $\phi(X) = c_0 X^n + c_1 X^{n-1} + \ldots + c_n$ $(c_0 \neq 0)$ be a non-zero element of $R[X]$. We shall show that there exists a maximal ideal of $R[X]$ which does not contain ϕ and this will complete the proof. Note that, since the Jacobson radical of R is the ideal (0), there exists a maximal ideal of M, of R, such that $c_0 \notin M$.

Consider the natural ring-epimorphism $R[X] \to (R/M)[X]$. The image $\overline{\phi}(X)$ of $\phi(X)$ is not zero and, since R/M is a field, $(R/M)[X]$ is a Hilbert domain by virtue of Lemma 1. Accordingly there is a maximal ideal of $(R/M)[X]$ which does not contain $\overline{\phi}$. Let the $R[X]$-ideal Π be the inverse image of this maximal ideal. Then Π is a maximal ideal of $R[X]$ and it does not contain ϕ. The proof is now complete.

Suppose that R is an integral domain and $\Pi \neq (0)$ a prime ideal of the polynomial ring $R[X]$. Choose $f(X) \in \Pi$ so that (i) $f(X) \neq 0$ and (ii) the degree of $f(X)$ is as small as possible. Let

$$f(X) = aX^s + bX^{s-1} + \ldots + d,$$

where $a \neq 0$. We denote by K the quotient field of R and by Ω the subring of K consisting of all those elements which can be expressed in the form r/a^m, where $r \in R$ and $m \geqslant 0$ is an integer. Suppose now that $h(X)$ is an element of $R[X]$. Since $h(X)$ and $f(X)$ both belong to $\Omega[X]$ and the leading coefficient of $f(X)$ is a unit in Ω, it follows, by Lemma 7 of section (5.4), that there exist $q^*(X)$ and $r^*(X)$, in $\Omega[X]$, such that

$$h(X) = q^*(X)f(X) + r^*(X) \tag{6.1.1}$$

and $\partial^0 r^*(X) < \partial^0 f(X)$. Let us now multiply the equation (6.1.1) by a^n, where n is a sufficiently large positive integer. This yields

$$a^n h(X) = q(X)f(X) + r(X),$$

where $q(X), r(X)$ belong to $R[X]$ and $\partial^0 r(X) < \partial^0 f(X)$. In particular, we see that if $h(X) \in \Pi$, then $a^n h(X) = q(X) f(X)$.

Proposition 3. *Let R be an integral domain whose Jacobson radical is the zero ideal, and let Π be a prime ideal of the polynomial ring $R[X]$. If now $\Pi \cap R = (0)$, then Π is the intersection of all the maximal ideals of $R[X]$ that contain it.*

Proof. We may suppose that $\Pi \neq (0)$ for otherwise the desired result follows from Lemma 2. We choose a non-zero polynomial $f(X)$ belonging to Π and having the smallest possible degree. Let $f(X) = aX^s + bX^{s-1} + \ldots + d$, where $a \neq 0$. Then, since $\Pi \cap R = (0)$, we must have $s \geqslant 1$. In what follows, the quotient field of R is denoted by K.

Now assume that $g(X) \in R[X]$ and $g(X) \notin \Pi$. If we can show that there exists a maximal ideal of $R[X]$ which contains Π but does not contain $g(X)$, then the proof will be complete.

By Theorem 15 of section (5.4), every ideal of $K[X]$ is a principal ideal. Let
$$(g, f) K[X] = (\phi') K[X].$$
Then there exist $\lambda'(X), \mu'(X), \sigma'(X), \tau'(X)$, all in $K[X]$, such that $g\lambda' + f\mu' = \phi'$, $g = \sigma'\phi'$ and $f = \tau'\phi'$. Now each element of K can be written as a fraction in which the numerator and denominator belong to R. In particular, the coefficients of $\lambda'(X), \mu'(X), \sigma'(X), \tau'(X)$ and $\phi'(X)$ can all be represented in this way. It follows that there exists $\gamma \in R$ such that $\gamma \neq 0$ and, if we put $\gamma\lambda'(X) = \lambda(X)$, $\gamma\mu'(X) = \mu(X)$ and so on, then $\lambda(X), \mu(X), \sigma(X), \tau(X)$ and $\phi(X)$ all belong to $R[X]$. We now have
$$g(X)\lambda(X) + f(X)\mu(X) = \phi(X) \tag{6.1.2}$$
together with $\gamma^2 g(X) = \sigma(X)\phi(X)$, $\gamma^2 f(X) = \tau(X)\phi(X)$. Since $\gamma \notin \Pi$ and $g(X) \notin \Pi$, we see that $\phi(X) \notin \Pi$. On the other hand, $\tau(X)\phi(X) \in \Pi$ because $f(X) \in \Pi$. Consequently $\tau(X)$ belongs to Π. But $\tau(X)$ cannot have a smaller degree than $f(X)$. Hence from $\gamma^2 f(X) = \tau(X)\phi(X)$ we conclude that $\phi(X)$ is a non-zero constant; say $\phi(X) = \omega$, where $\omega \in R$ and $\omega \neq 0$. Note that (6.1.2) can be rewritten as
$$g(X)\lambda(X) + f(X)\mu(X) = \omega. \tag{6.1.3}$$

By hypothesis, the intersection of all the maximal ideals of R is the zero ideal. Accordingly there exists a maximal ideal M of R such that $a\omega \notin M$.

We contend that $\Pi + MR[X] \neq R[X]$. For otherwise there would exist $l(X) \in \Pi$ and $m(X) \in MR[X]$ such that $l(X) + m(X) = 1$. But, if

n is a sufficiently large positive integer,† $a^n l(X) = f(X) q(X)$ for some $q(X)$ in $R[X]$. Thus $a^n = f(X) q(X) + a^n m(X)$. Denote by $\bar{q}(X)$ the sum of those terms in $q(X)$ whose coefficients do *not* belong to M. Then

$$a^n = f(X) \bar{q}(X) + \bar{m}(X) \qquad (6.1.4)$$

for some $\bar{m}(X)$ in $MR[X]$. Now $\bar{q}(X) \neq 0$ because $a^n \notin M$. Let $\bar{q}(X)$ be of degree p and let α be its leading coefficient. Then $\alpha \notin M$. Also, since $s \geqslant 1$, the terms of degree $s + p$ in (6.1.4) show that $a\alpha \in M$. But this is impossible because neither a nor α is in M. Thus $\Pi + MR[X] \neq R[X]$ as was stated. Hence $\Pi + MR[X]$ is contained in a maximal ideal Π^* (say).

Since $M \subseteq \Pi^* \cap R$ and M is a maximal ideal of R, we have

$$M = \Pi^* \cap R.$$

Finally, $g(X)$ cannot be contained in Π^* for otherwise, by (6.1.3), ω would belong to $\Pi^* \cap R = M$ and this is not so because $a\omega \notin M$. The proof is therefore complete.

The following general observation is used in the proof of our first theorem on Hilbert rings and it will be needed again later. Let R be any commutative ring (with an identity element), Π an ideal of $R[X]$, and A an ideal of R contained in $\Pi \cap R$. The natural ring-epimorphism $\phi : R[X] \to (R/A)[X]$ has $AR[X]$ as its kernel and $AR[X]$ is contained in Π. Accordingly $\Pi = \phi^{-1}(\Pi')$, where Π' is a certain ideal of $(R/A)[X]$. Let $r \in R$. Then $\phi(r) \in \Pi' \cap (R/A)$ if and only if $r \in \Pi \cap R$. Thus $\Pi \cap R$ is the inverse image of $\Pi' \cap (R/A)$ with respect to the natural mapping $R \to R/A$.

Theorem 1. *Let R be a Hilbert ring. Then the polynomial ring*

$$R[X_1, X_2, \ldots, X_n]$$

is also a Hilbert ring.

Proof. It is enough to prove the theorem for the case in which there is only one indeterminate. We shall therefore assume that we have this situation. The indeterminate will be denoted by X.

Let Π be a prime ideal of $R[X]$. We must show that Π is the intersection of all the maximal ideals that contain it. Put $P = \Pi \cap R$ and let $\phi : R[X] \to (R/P)[X]$ be the natural ring-epimorphism. Then Π is the inverse image of an ideal Π' of $(R/P)[X]$ and, since Π is prime, so too is Π'. Clearly it will suffice to show that Π' is the intersection of all the maximal ideals containing it.

---

† See the remarks in the paragraph preceding the statement of Proposition 3.

Since $P = \Pi \cap R$, it is a prime ideal of R. Consequently, by Proposition 1, R/P is a Hilbert domain and therefore its Jacobson radical is the zero ideal. Now the inverse image of $\Pi' \cap (R/P)$, with respect to the natural mapping $R \to R/P$, is $\Pi \cap R = P$. It follows that

$$\Pi' \cap (R/P) = (0).$$

Finally, Proposition 3 shows that Π' is the intersection of all the maximal ideals of $(R/P)[X]$ that contain it. Thus the proof is complete.

Corollary 1. *If F is a field, then the polynomial ring $F[X_1, X_2, ..., X_n]$ is a Hilbert ring.*

This is now obvious because, as previously observed, every field is a Hilbert ring.

Corollary 2. *Let R be a Hilbert ring and $\alpha_1, \alpha_2, ..., \alpha_n$ elements from a (commutative) extension ring of R. Then $R[\alpha_1, \alpha_2, ..., \alpha_n]$ is a Hilbert ring.*

Proof. There exists a ring-epimorphism of the polynomial ring

$$R[X_1, X_2, ..., X_n] \quad \text{on to} \quad R[\alpha_1, \alpha_2, ..., \alpha_n]$$

in which X_i is mapped into α_i and each element of R is left fixed. Since $R[X_1, X_2, ..., X_n]$ is a Hilbert ring, the desired result follows from Proposition 1.

Proposition 4. *Let R be an integral domain whose Jacobson radical is its zero ideal, and let X be an indeterminate. If now there exists a maximal ideal Π of $R[X]$ such that $\Pi \cap R = (0)$, then R is a field.*

Proof. Since X does not have an inverse in $R[X]$, $R[X]$ is not a field. It follows that $\Pi \neq (0)$. We can therefore choose $f(X) \in \Pi$ so that $f(X) \neq 0$ and the degree of $f(X)$ is as small as possible. Let

$$f(X) = aX^s + bX^{s-1} + cX^{s-2} + ... + d,$$

where $a \neq 0$. Then, because $\Pi \cap R = (0)$, we have $s \geqslant 1$. Now the intersection of all the maximal ideals of R is the zero ideal. Hence there is a maximal ideal M of R such that $a \notin M$. If it can be shown that $M = (0)$, then the proof will be complete. We shall, in fact, assume that $M \neq (0)$ and derive a contradiction.

Choose $\alpha \in M$ so that $\alpha \neq 0$. Then $\alpha \notin \Pi$ and so, since Π is a maximal ideal of $R[X]$,
$$\alpha R[X] + \Pi = R[X].$$

Accordingly $\alpha g(X) + h(X) = 1$ for suitable polynomials $g(X)$ and $h(X)$, where $h(X) \in \Pi$. Now if n is a sufficiently large positive integer,† then $a^{n}h(X) = f(X)q_{1}(X)$ and $a^{n}g(X) = f(X)q_{2}(X) + r(X)$, where $q_{1}(X)$, $q_{2}(X), r(X)$ belong to $R[X]$ and $\partial^{0}r(X) < \partial^{0}f(X)$. Hence

$$a^{n} = \alpha r(X) + f(X)\,(\alpha q_{2}(X) + q_{1}(X)).$$

This shows that $a^{n} - \alpha r(X)$ belongs to Π and, by construction, it has a smaller degree than $f(X)$. We must therefore have $a^{n} - \alpha r(X) = 0$ whence, by considering the constant term, $a^{n} \in (\alpha) \subseteq M$. This, however, gives a contradiction because $a \notin M$. The proposition follows.

Theorem 2. *Let R be a Hilbert ring and Π a maximal ideal of the polynomial ring $R[X_{1}, X_{2}, ..., X_{n}]$. Then $\Pi \cap R$ is a maximal ideal of R.*

Proof. We first consider the case in which $n = 1$. To simplify the notation, we shall write X in place of X_{1}.

Put $P = \Pi \cap R$, where Π is a maximal ideal of $R[X]$. The natural mapping $R[X] \rightarrow (R/P)[X]$ maps Π on to a maximal ideal Π' of $(R/P)[X]$ and, since $\Pi' \cap (R/P)$ corresponds to $\Pi \cap R = P$ under the natural mapping $R \rightarrow R/P$, we see that $\Pi' \cap (R/P) = (0)$. Now R/P is a Hilbert domain and therefore its Jacobson radical is the zero ideal. Consequently, by Proposition 4, R/P is a field. Thus P is a maximal ideal of R and the proof is complete in this case.

We next turn our attention to the situation in which there may be more than one variable. Put $R' = R[X_{1}, X_{2}, ..., X_{n-1}]$. Then, by Theorem 1, R' is a Hilbert ring and Π is a maximal ideal of $R'[X_{n}]$. Hence, by the result just proved for polynomial rings in one variable, $\Pi \cap R'$ is a maximal ideal of R', i.e. $\Pi \cap R[X_{1}, X_{2}, ..., X_{n-1}]$ is a maximal ideal of $R[X_{1}, X_{2}. ..., X_{n-1}]$.

The argument may now be repeated with $R[X_{1}, X_{2}, ..., X_{n-1}]$ and $\Pi \cap R[X_{1}, X_{2}, ..., X_{n-1}]$ replacing $R[X_{1}, X_{2}, ..., X_{n}]$ and Π. This shows that $\Pi \cap R[X_{1}, X_{2}, ..., X_{n-2}]$ is a maximal ideal of $R[X_{1}, X_{2}, ..., X_{n-2}]$. Further applications of the same idea eventually yield the desired result.

Theorem 2 can be used to extend our knowledge of the properties of polynomials with coefficients in a field. This will be done in the next section, but first we prepare the way by establishing two lemmas.

Lemma 3. *Let R be a Hilbert ring and Π a maximal ideal of the polynomial ring $R[X]$. Put $P = R \cap \Pi$. Then there exists $f(X) \in R[X]$ such*

---

† See the remarks in the paragraph immediately preceding Proposition 3.

that $f(X) \notin PR[X]$, $(P,f)R[X] = \Pi$, *and the leading coefficient of* $f(X)$ *is the identity element of* R.

Proof. The image of Π under the natural mapping $R[X] \to (R/P)[X]$ is a maximal ideal Π' (say) of the latter ring. By Theorem 2, P is a maximal ideal of R and therefore R/P is a field. It follows, from Theorem 15 of section (5.4) and its corollary, that Π' can be generated by a non-zero polynomial, say by $f'(X)$. Since R/P is a field, we can arrange that the leading coefficient of $f'(X)$ is the identity element of R/P. We can now find $f(X) \in R[X]$ such that $f(X)$ has 1_R as its leading coefficient and $f(X)$ is mapped into $f'(X)$ by the natural mapping $R[X] \to (R/P)[X]$. Since $f'(X) \neq 0$, we have $f(X) \notin PR[X]$. Since $(P,f)R[X]$ is mapped on to $(f')R[X] = \Pi'$, the relation $(P,f)R[X] = \Pi$ holds as well. Thus the lemma is established.

Lemma 4. *Let R be a Hilbert ring and Π a maximal ideal of the polynomial ring $R[X_1, X_2, \ldots, X_n]$ in $n \, (n \geqslant 1)$ variables. Then, for each value of $i \, (1 \leqslant i \leqslant n)$, there exists a non-zero element in $R[X_i] \cap \Pi$ whose leading coefficient is the identity element of R.*

Put differently, the lemma states that, for each value of i, Π contains a polynomial of the form

$$X_i^{m_i} + a_1 X_i^{m_i - 1} + \ldots + a_{m_i},$$

where a_1, a_2 etc. belong to R. Note that, since Π is a proper ideal, the degrees of these polynomials are strictly positive.

Proof. Put $R_i = R[X_i]$. Then R_i is a Hilbert ring and

$$R[X_1, \ldots, X_i, \ldots, X_n]$$

is the polynomial ring in $X_1, \ldots, X_{i-1}, X_{i+1}, \ldots, X_n$ with coefficients in R_i. Hence, by Theorem 2, $\Pi \cap R_i$ is a maximal ideal of R_i, i.e. $\Pi \cap R[X_i]$ is a maximal ideal of $R[X_i]$. The desired result now follows by applying Lemma 3 to R, $R[X_i]$ and $\Pi \cap R[X_i]$.

6.2 Polynomials with coefficients in a field

We have already observed that every field is a Hilbert ring. The results of the last section can therefore be applied to $F[X_1, X_2, \ldots, X_n]$, where F is an arbitrary field and $X_1, X_2, \ldots, X_n$ are indeterminates.

Theorem 3. *Let F be a field and Π any maximal ideal of the polynomial ring $F[X_1, X_2, \ldots, X_n]$. Then rank $\Pi = n$ and Π can be generated by n elements.*

Proof. We use induction on n. If $n = 1$, then the statements are true by virtue of Theorem 15 of section (5.4) and its corollary. From now on we suppose that $n > 1$ and assume that both assertions hold for all smaller values of the inductive variable.

Put $R = F[X_1, X_2, ..., X_{n-1}]$ and $P = \Pi \cap R$. Then Π is a maximal ideal of $R[X_n]$ and, by Theorem 1, R is a Hilbert ring. Hence, by Theorem 2, P is a maximal ideal of R. The induction hypothesis now shows that rank $P = n - 1$ and $P = (f_1, f_2, ..., f_{n-1})$, where $f_1, f_2, ..., f_{n-1}$ are suitable elements of $F[X_1, X_2, ..., X_{n-1}]$. Note that, since R is a Noetherian ring, it follows, by Proposition 10 of section (5.4), that

$$\operatorname{rank}(PF[X_1, ..., X_{n-1}, X_n]) = \operatorname{rank} P = n - 1.$$

We now apply Lemma 3. This shows that there exists

$$f \in F[X_1, ..., X_{n-1}, X_n]$$

such that $f \notin PR[X_1, ..., X_{n-1}, X_n]$ and

$$\Pi = PF[X_1, ..., X_{n-1}, X_n] + (f) F[X_1, ..., X_{n-1}, X_n].$$

Thus $\Pi = (f_1, ..., f_{n-1}, f)$ and hence, by Theorem 22 of section (4.8), rank $\Pi \leqslant n$. On the other hand

$$\operatorname{rank} \Pi > \operatorname{rank}(PF[X_1, ..., X_{n-1}, X_n]) = n - 1$$

because $f \notin PF[X_1, ..., X_{n-1}, X_n]$. The theorem follows.

The next theorem is stated for polynomials with coefficients in an Artin ring rather than in a field. This is because the extra generality presents no problem and will be useful later.

Theorem 4. *Let A be a non-null Artin ring and Π^* a maximal ideal of the polynomial ring $A[X_1, X_2, ..., X_n]$. Then rank $\Pi^* = n$. Furthermore, if Π is an arbitrary prime ideal of $A[X_1, X_2, ..., X_n]$, then*

$$\dim \Pi + \operatorname{rank} \Pi = n.$$

Proof. There exists a minimal prime ideal Π_0 of $A[X_1, X_2, ..., X_n]$ such that $\Pi^* \supseteq \Pi_0$ and rank $(\Pi^*/\Pi_0) = \operatorname{rank} \Pi^*$. Put $P_0 = A \cap \Pi_0$. Then P_0 is a prime ideal of the Artin ring A and hence, by Theorem 8 of section (4.2), a maximal ideal of that ring. It follows that A/P_0 is a field. Next, by Proposition 6 of section (5.4), $P_0 A[X_1, X_2, ..., X_n]$ is a prime ideal of the polynomial ring and so, since it is contained in the minimal prime ideal Π_0, we have $\Pi_0 = P_0 A[X_1, X_2, ..., X_n]$.

The natural mapping of

$$A[X_1, X_2, ..., X_n] \quad \text{on to} \quad (A/P_0)[X_1, X_2, ..., X_n]$$

maps Π^* on to a maximal ideal Π' (say) and has Π_0 as its kernel. Thus we see that the prime ideals between Π^* and Π_0 are in one-one correspondence with the prime ideals of $(A/P_0)[X_1, X_2, ..., X_n]$ that are contained in Π'. From this it follows that

$$\operatorname{rank} \Pi^* = \operatorname{rank}(\Pi^*/\Pi_0) = \operatorname{rank} \Pi'.$$

But $\operatorname{rank} \Pi' = n$ by Theorem 3 and so the first assertion is established.

For the second part of the theorem, we choose the maximal ideal Π^* so that $\Pi \subseteq \Pi^*$ and $\operatorname{rank}(\Pi^*/\Pi) = \dim \Pi$. Obviously A is a semi-regular ring because it is Noetherian and all its prime ideals are of zero rank. Accordingly, by Theorem 17 of section (5.5),

$$A[X_1, X_2, ..., X_n]$$

is also a semi-regular ring. We can therefore make use of Theorem 13 of section (5.3) to deduce that

$$\dim \Pi + \operatorname{rank} \Pi = \operatorname{rank}(\Pi^*/\Pi) + \operatorname{rank} \Pi = \operatorname{rank} \Pi^* = n.$$

The proof is now complete.

Corollary. *Let A be a non-null Artin ring and suppose that Π_1, Π_2 are prime ideals of the polynomial ring $A[X_1, X_2, ..., X_n]$ such that* (i) $\Pi_1 \subset \Pi_2$ *(strict inclusion) and* (ii) *there is no prime ideal strictly between Π_1 and Π_2. Then $\operatorname{rank} \Pi_2 = \operatorname{rank} \Pi_1 + 1$ and $\dim \Pi_1 = \dim \Pi_2 + 1$.*

Proof. Clearly we need only establish the first assertion. By hypothesis, $\operatorname{rank}(\Pi_2/\Pi_1) = 1$. Now, as we have already seen, $A[X_1, X_2, ..., X_n]$ is a semi-regular ring. Hence, by Theorem 13 of section (5.3),

$$\operatorname{rank} \Pi_2 = \operatorname{rank}(\Pi_2/\Pi_1) + \operatorname{rank} \Pi_1 = \operatorname{rank} \Pi_1 + 1$$

as required.

Let F be a field and α an element belonging to some extension ring of F. Then α is said to be *algebraic with respect to F* if it satisfies a relation of the form

$$a_0 \alpha^n + a_1 \alpha^{n-1} + a_2 \alpha^{n-2} + ... + a_n = 0,$$

where $a_0, a_1, ..., a_n$ belong to F and $a_0 \neq 0$. Since F is a field, the relation may be rewritten as

$$\alpha^n + a_0^{-1} a_1 \alpha^{n-1} + a_0^{-1} a_2 \alpha^{n-2} + ... + a_0^{-1} a_n = 0.$$

Thus α *is algebraic with respect to F if and only if it is integral with respect to F.* However, the former expression is the one customarily used and so it will be employed here in spite of the fact that it is super-fluous.

Suppose now that $\alpha_1, \alpha_2, ..., \alpha_n$ all belong to some (commutative) extension ring of the field F. Then there is a unique epimorphism of the polynomial ring $F[X_1, X_2, ..., X_n]$ on to the ring $F[\alpha_1, \alpha_2, ..., \alpha_n]$ in which X_i is mapped into α_i and the individual elements of F are left fixed. If $f(X_1, X_2, ..., X_n)$ belongs to $F[X_1, X_2, ..., X_n]$, then, of course, its image under the homomorphism is denoted by $f(\alpha_1, \alpha_2, ..., \alpha_n)$.

Theorem 5. *Let F be a field and $\alpha_1, \alpha_2, ..., \alpha_n$ elements of some extension ring. Then $F[\alpha_1, \alpha_2, ..., \alpha_n]$ is an Artin ring if and only if each α_i is algebraic with respect to F.*

Proof. First suppose that $F[\alpha_1, \alpha_2, ..., \alpha_n]$ is an Artin ring. Then, by Theorem 8 of section (4.2), it has only a finite number of prime ideals and each of these is a maximal ideal. In what follows, the prime ideals of $F[\alpha_1, \alpha_2, ..., \alpha_n]$ are denoted by $\Pi'_1, \Pi'_2, ..., \Pi'_s$.

Let Π_ν be the inverse image of Π'_ν with respect to the ring-epimorphism $F[X_1, X_2, ..., X_n] \to F[\alpha_1, \alpha_2, ..., \alpha_n]$ described above. Then Π_ν is a maximal ideal of $F[X_1, X_2, ..., X_n]$ and therefore, by Lemma 4, there exists $f_\nu(X_1)$ belonging to $F[X_1] \cap \Pi_\nu$ such that $f_\nu(X_1) \neq 0$. Put $\phi(X_1) = f_1(X_1)f_2(X_1) ... f_s(X_1)$. Then $\phi(X_1) \neq 0$ and

$$\phi(\alpha_1) \in \Pi'_1 \Pi'_2 ... \Pi'_s.$$

Further, by Lemma 4 Cor. of section (4.2), there exists an integer m such that $(\Pi'_1 \Pi'_2 ... \Pi'_s)^m = (0)$. Accordingly $\phi^m(\alpha_1) = 0$ and therefore α_1 is algebraic with respect to F. Similarly $\alpha_2, \alpha_3, ..., \alpha_n$ are all algebraic with respect to F.

To prove the converse, assume that each α_i is algebraic and therefore integral with respect to F. Then, by Lemma 7 Cor. of section (2.5), $F[\alpha_1, \alpha_2, ..., \alpha_n]$ is finitely generated as an F-module. That

$$F[\alpha_1, \alpha_2, ..., \alpha_n]$$

is an Artin ring now follows from Theorem 5 of section (4.2).

6.3 The Zeros Theorem

Let F be a field and $a_1, a_2, ..., a_n$ elements of F. Then, for any non-negative integer ν, $X_i^\nu - a_i^\nu$ belongs to the ideal

$$(X_1 - a_1, X_2 - a_2, ..., X_n - a_n)$$

of the polynomial ring $F[X_1, X_2, ..., X_n]$. It follows that if

$$f(X_1, X_2, ..., X_n)$$

belongs to $F[X_1, X_2, ..., X_n]$, then

$$f(X_1, X_2, ..., X_n) \equiv f(a_1, a_2, ..., a_n) \qquad (6.3.1)$$

modulo $(X_1 - a_1, X_2 - a_2, ..., X_n - a_n)$.

Consider the ring-epimorphism $F[X_1, X_2, ..., X_n] \to F$ in which

$$f(X_1, X_2, ..., X_n) \to f(a_1, a_2, ..., a_n).$$

It is clear, from (6.3.1), that the kernel of the mapping is

$$(X_1 - a_1, X_2 - a_2, ..., X_n - a_n)$$

and, since $F[X_1, X_2, ..., X_n]$ is mapped on to a *field*, the kernel is a maximal ideal. Thus *if $a_1, a_2, ..., a_n$ belong to F, then*

$$(X_1 - a_1, X_2 - a_2, ..., X_n - a_n)$$

is a maximal ideal of $F[X_1, X_2, ..., X_n]$. It will now be shown that for a certain kind of field, every maximal ideal of $F[X_1, X_2, ..., X_n]$ has this form.

The field F is said to be *algebraically closed* if each non-zero element $f(X)$, of $F[X]$, can be represented in the form

$$f(X) = c(X - \alpha_1)(X - \alpha_2) ... (X - \alpha_p),$$

where $c, \alpha_1, \alpha_2, ..., \alpha_p$ are elements of F. For example, the complex numbers form an algebraically closed field.

Theorem 6. *If the field F is algebraically closed and Π is a maximal ideal of $F[X_1, X_2, ..., X_n]$, then there exist elements $a_1, a_2, ..., a_n$ in F such that $\Pi = (X_1 - a_1, X_2 - a_2, ..., X_n - a_n)$.*

Proof. By Lemma 4 there exists $f(X_1)$ belonging to $F[X_1] \cap \Pi$ such that $f(X_1)$ is not zero and has the identity element as its leading coefficient. Since F is algebraically closed, $f(X_1)$ can be expressed in the form

$$f(X_1) = (X_1 - \alpha_1)(X_1 - \alpha_2) ... (X_1 - \alpha_p),$$

where $\alpha_1, \alpha_2, ..., \alpha_p$ belong to F. It follows that one of the factors $X_1 - \alpha_i$ must belong to Π. Thus we see that there exists $a_1 \in F$ such that $X_1 - a_1$ belongs to Π. In exactly the same way it can be shown that, for each value of j $(1 \leqslant j \leqslant n)$, there is an element $a_j \in F$ for which $X_j - a_j$ is in Π. It follows that

$$(X_1 - a_1, X_2 - a_2, ..., X_n - a_n) \subseteq \Pi.$$

But $(X_1 - a_1, X_2 - a_2, ..., X_n - a_n)$ is a maximal ideal. Hence

$$(X_1 - a_1, X_2 - a_2, ..., X_n - a_n) = \Pi$$

and the theorem is proved.

The next result is the celebrated *Zeros Theorem.*

Theorem 7. *Let F be a field and F^* an algebraically closed extension field of F. Further, let $\phi(X_1, X_2, ..., X_n)$ and $f_i(X_1, X_2, ..., X_n)$, where*

$1 \leqslant i \leqslant p$, *belong to the polynomial ring* $F[X_1, X_2, ..., X_n]$. *Then the following statements are equivalent:*

(a) ϕ *belongs to the radical of the* $F[X_1, X_2, ..., X_n]$*-ideal* $(f_1, f_2, ..., f_p)$;

(b) $\phi(\alpha_1, \alpha_2, ..., \alpha_n) = 0$ *whenever* $\alpha_1, \alpha_2, ..., \alpha_n$ *belong to* F^* *and* $f_i(\alpha_1, \alpha_2, ..., \alpha_n) = 0$ *for* $1 \leqslant i \leqslant p$.

Proof. First assume that (a) is true. Then there exists a positive integer m and polynomials $g_i(X_1, X_2, ..., X_n)$, where $1 \leqslant i \leqslant p$ and the coefficients of $g_i(X_1, X_2, ..., X_n)$ are in F, such that

$$\phi^m(X_1, X_2, ..., X_n) = \sum_{i=1}^{p} g_i(X_1, X_2, ..., X_n) f_i(X_1, X_2, ..., X_n).$$

Hence, if $\alpha_1, \alpha_2, ..., \alpha_n$ belong to F^* and $f_i(\alpha_1, \alpha_2, ..., \alpha_n) = 0$ for $1 \leqslant i \leqslant p$, then $\phi^m(\alpha_1, \alpha_2, ..., \alpha_n) = 0$. Accordingly $\phi(\alpha_1, \alpha_2, ..., \alpha_n) = 0$ and we have shown that (a) implies (b).

Now assume that (b) is true and let Π^* be an arbitrary maximal ideal† of $F^*[X_1, X_2, ..., X_n]$ containing $(f_1, f_2, ..., f_p) F^*[X_1, ..., X_n]$. By Theorem 6, there exist $\alpha_1, \alpha_2, ..., \alpha_n$ in F^* such that

$$\Pi^* = (X_1 - \alpha_1, X_2 - \alpha_2, ..., X_n - \alpha_n).$$

Then, since f_i belongs to Π^*, we have $f_i(\alpha_1, \alpha_2, ..., \alpha_n) = 0$. It follows that $\phi(\alpha_1, \alpha_2, ..., \alpha_n) = 0$ and therefore $\phi(X_1, X_2, ..., X_n)$ belongs to Π^*. Thus ϕ is contained in every maximal ideal of $F^*[X_1, X_2, ..., X_n]$ that contains $(f_1, f_2, ..., f_p) F^*[X_1, X_2, ..., X_n]$. It is therefore contained in the intersection of all such maximal ideals. But $F^*[X_1, X_2, ..., X_n]$ is a Hilbert ring; consequently, by Proposition 2, ϕ belongs to the radical of $(f_1, f_2, ..., f_p) F^*[X_1, ..., X_n]$. Accordingly there exists a positive integer m such that

$$\phi^m(X_1, X_2, ..., X_n) \in (f_1, f_2, ..., f_p) F^*[X_1, ..., X_n] \cap F[X_1, ..., X_n].$$

If therefore it can be shown that

$$(f_1, f_2, ..., f_p) F^*[X_1, ..., X_n] \cap F[X_1, ..., X_n]$$
$$= (f_1, f_2, ..., f_p) F[X_1, ..., X_n],$$

then the proof will be complete. However, this is a special case of the following general proposition on polynomial rings and field extensions.

Proposition 5. *Let* F^* *be an arbitrary extension field of a field* F, *and let* A *be an ideal of the polynomial ring* $F[X_1, X_2, ..., X_n]$. *Then*

$$AF^*[X_1, ..., X_n] \cap F[X_1, ..., X_n] = A.$$

† Note that the argument is not affected if the set of such maximal ideals is empty.

Proof. Since $F[X_1, X_2, ..., X_n]$ is a Noetherian ring, the ideal A is finitely generated. Let $A = (f_1, f_2, ..., f_p)$. We shall now assume that $\phi(X_1, X_2, ..., X_n)$ belongs to

$$AF^*[X_1, ..., X_n] \cap F[X_1, ..., X_n]$$

and deduce that ϕ belongs to A. This is clearly sufficient.

There exist $g_i^*(X_1, X_2, ..., X_n)$, where $1 \leqslant i \leqslant p$, in

$$F^*[X_1, X_2, ..., X_n]$$

and such that

$$\phi(X_1, ..., X_n) = \sum_{i=1}^{p} g_i^*(X_1, ..., X_n) f_i(X_1, ..., X_n).$$

Consider the F-space which is generated by the identity element and the non-zero coefficients of the polynomials $g_i^*(X_1, X_2, ..., X_n)$. This space is finitely generated. Hence, by Theorem 21 of section (1.13), it possesses a finite base $\omega_1, \omega_2, ..., \omega_m$ (say). Each coefficient of $g_i^*(X_1, X_2, ..., X_n)$ can be represented as a linear combination (with coefficients in F) of $\omega_1, \omega_2, ..., \omega_m$. Consequently we have relations

$$g_i^*(X_1, X_2, ..., X_n) = \sum_{\mu=1}^{m} \omega_\mu g_{i\mu}(X_1, X_2, ..., X_n),$$

where $g_{i\mu}(X_1, X_2, ..., X_n)$ belongs to $F[X_1, X_2, ..., X_n]$. We also have a relation in the form of

$$1_F = \sum_{\mu=1}^{m} \omega_\mu e_\mu, \tag{6.3.2}$$

where $e_1, e_2, ..., e_m$ are in F, for the identity element. Accordingly

$$\sum_{\mu=1}^{m} \omega_\mu e_\mu \phi(X_1, ..., X_n) = \sum_{\mu=1}^{m} \omega_\mu \left[\sum_{i=1}^{p} g_{i\mu}(X_1, ..., X_n) f_i(X_1, ..., X_n) \right]. \tag{6.3.3}$$

Let us now take an arbitrary power-product $X_1^{\nu_1} X_2^{\nu_2} ... X_n^{\nu_n}$ and denote by a_μ and b_μ its coefficients in $e_\mu \phi(X_1, ..., X_n)$ and

$$\sum_{i=1}^{p} g_{i\mu}(X_1, ..., X_n) f_i(X_1, ..., X_n)$$

respectively. Then, comparing the coefficients of $X_1^{\nu_1} X_2^{\nu_2} ... X_n^{\nu_n}$ on the two sides of (6.3.3), we see that

$$\sum_{\mu=1}^{m} \omega_\mu a_\mu = \sum_{\mu=1}^{m} \omega_\mu b_\mu,$$

whence $a_\mu = b_\mu$ because a_μ and b_μ are in F and $\omega_1, \omega_2, ..., \omega_m$ are linearly independent with respect to this field. It follows that

$$e_\mu \phi(X_1, ..., X_n) = \sum_{i=1}^{p} g_{i\mu}(X_1, ..., X_n) f_i(X_1, ..., X_n)$$

and therefore $e_\mu \phi(X_1, \ldots, X_n)$ belongs to A for $\mu = 1, 2, \ldots, m$. However, by (6.3.2), at least one e_μ is not zero. It follows that

$$\phi(X_1, X_2, \ldots, X_n)$$

itself belongs to A and now the proof is complete.

Exercises on Chapter 6

In these exercises R denotes a commutative ring with an identity element.

1. The ring R has the property that every non-maximal prime ideal is the intersection of a set of prime ideals which strictly contain it. Prove that R is a Hilbert ring.

2. Given that the polynomial ring $R[X_1, X_2, \ldots, X_n]$ is a Hilbert ring, prove that R is a Hilbert ring.

3. The ring R is such that every maximal ideal in the polynomial ring $R[X]$ contracts to a maximal ideal of R. Prove that R is a Hilbert ring.

4. The ring R' is an integral extension of the ring R. Prove that if one is a Hilbert ring so is the other.

5. Show that if Z denotes the ring of integers, then every maximal ideal of the polynomial ring $Z[X_1, X_2, \ldots, X_n]$ has rank equal to $n + 1$ and can be generated by $n + 1$ elements.

6. R is an integral domain whose Jacobson radical is its zero ideal, and $\alpha_1, \alpha_2, \ldots, \alpha_n$ belong to some commutative extension ring of R. Show that if $R[\alpha_1, \alpha_2, \ldots, \alpha_n]$ is an integral domain, then it also has the property that its Jacobson radical is its zero ideal.

7. R is an integral domain whose Jacobson radical is its zero ideal, and $\alpha_1, \alpha_2, \ldots, \alpha_n$ belong to some commutative extension ring of R. Show that if $R[\alpha_1, \alpha_2, \ldots, \alpha_n]$ is a field, then so is R.

7

MULTIPLICITY THEORY

General remarks. This chapter is devoted to the development of a general theory of algebraic multiplicities. This body of results arises directly out of the concept of length which was first encountered in Chapter 1. Since the notion of length does not require the ring of operators to be commutative, it is natural to avoid introducing commutativity into the discussion until a point is reached where its use seems unavoidable. In fact one can go a surprisingly long way before this happens, but first it is necessary to adapt some of the general theorems of commutative algebra to meet the needs of the present situation. The results that require to be generalized are the Artin-Rees Theorem, the Intersection Theorem, and the theory of the Jacobson radical. These are reviewed and re-established in suitably modified forms in section (7.2).

We continue to use the convention that R always denotes a ring with an identity element. Throughout sections (7.8) to (7.10) inclusive it is also assumed that R is commutative, but in none of the remaining sections of the chapter is this condition imposed. Considerable use will be made of some of the results obtained in Chapter 4. Now although Chapter 4 is primarily concerned with the theory of commutative rings, it will be recalled† that *all* the results of section (4.1) are valid in the non-commutative case.

7.1 Preliminary considerations

Let R be a ring with an identity element. In what follows, the term R-module will always mean a *left* R-module. If an R-module E satisfies the maximal condition for submodules, then E will be called a *Noetherian* R-module. This is in conformity with the definition already introduced in connection with commutative rings.

There are certain other terms and certain constructions, already employed in the commutative case, which must be adapted to our present purposes. Let E be an R-module and B a left ideal. Denote by

† See the 'General Remarks' at the beginning of Chapter 4.

BE the submodule of E which consists of all elements that can be expressed in the form $b_1 e_1 + b_2 e_2 + \ldots + b_s e_s$, where $b_i \in B$ and $e_i \in E$. If now A is also a left ideal, then, as is easily verified, $(AB)E = A(BE)$. We may therefore omit the brackets and write ABE for the module in question. Note that it makes sense to speak of the mth power A^m of A. More generally, if $A_1, A_2, \ldots, A_m$ are left ideals, then the extended product $A_1 A_2 \ldots A_m$ is well defined and is also a left ideal.

Definition. *An element γ of R is called a 'central element' if $\gamma r = r\gamma$ for every r in R.*

For example, both 0_R and 1_R are central elements. A simple verification shows that the central elements of R form a ring. This ring, which is called the *centre* of R, has the same identity element as R itself.

Let G be a set of central elements of R. Then the left ideal generated by G coincides with the right ideal generated by the same set. Hence the totality of all elements which can be expressed in the form $r_1 \gamma_1 + r_2 \gamma_2 + \ldots + r_s \gamma_s$, where $r_i \in R$ and $\gamma_i \in G$, is a two-sided ideal. We may therefore speak of the ideal generated by a set of central elements without specifying whether it is a left ideal, a right ideal, or a two-sided ideal that we have in mind.

Definition. *An ideal which can be generated by a set of central elements will be called a 'central ideal'.*

The above remarks show that central ideals are two-sided. Let G and G' be sets of central elements and A and A' the central ideals which they generate. Then $A + A'$ is generated by $G \cup G'$ whereas AA' is generated by the set of all products $\gamma\gamma'$, where γ belongs to G and γ' to G'. Hence the sum and product of two central ideals are again central ideals. Further the set of products $\gamma\gamma'$ coincides with the corresponding set obtained by interchanging the roles of G and G'. Hence $AA' = A'A$ and therefore, in a product of central ideals, the order of the factors may be changed in any way and this will not alter the value of the product.

Suppose now that A is a given central ideal. Then A is a two-sided ideal. Let us assume that, as a two-sided ideal, it is finitely generated. This means that there exists a finite set $u_1, u_2, \ldots, u_s$ of elements of R such that A is the smallest two-sided ideal which contains them. In these circumstances each u_i can be expressed in the form

$$u_i = r_{i1} \gamma_{i1} + r_{i2} \gamma_{i2} + \ldots + r_{in_i} \gamma_{in_i} \quad (1 \leqslant i \leqslant s),$$

where γ_{ij} denotes a central element contained in A and r_{ij} belongs to R. It is clear that the ideal generated by the various γ_{ij} is A itself. Hence *A can be generated by a finite number of central elements*. In this way we see that the expression *finitely generated central ideal* can be used without essential ambiguity.

In the case of a finite set $\gamma_1, \gamma_2, ..., \gamma_p$ of central elements, we may sometimes use $(\gamma_1, \gamma_2, ..., \gamma_p)$ to denote the central ideal which they generate. If now $\gamma_1', \gamma_2', ..., \gamma_q'$ are also central elements, then

$$(\gamma_1, \gamma_2, ..., \gamma_p) + (\gamma_1', \gamma_2', ..., \gamma_q') = (\gamma_1, ..., \gamma_p, \gamma_1', ..., \gamma_q')$$

and $\quad (\gamma_1, \gamma_2, ..., \gamma_p)(\gamma_1', \gamma_2', ..., \gamma_q') = (\gamma_1 \gamma_1', ..., \gamma_i \gamma_j', ..., \gamma_p \gamma_q').$

It follows, from the latter relation, that if $A = (\gamma_1, \gamma_2, ..., \gamma_p)$, then A^m is generated by the collection of products $\gamma_1^{n_1} \gamma_2^{n_2} ... \gamma_p^{n_p}$, where $n_1, n_2, ..., n_p$ are non-negative integers satisfying $n_1 + n_2 + ... + n_p = m$.

Let E be an R-module and $\gamma_1, \gamma_2, ..., \gamma_p$ central elements. Then $(\gamma_1, \gamma_2, ..., \gamma_p)E$ consists of all elements which can be expressed in the form $\gamma_1 e_1 + \gamma_2 e_2 + ... + \gamma_p e_p$, where $e_1, e_2, ..., e_p$ belong to E. For this reason we often use $\gamma_1 E + \gamma_2 E + ... + \gamma_s E$ as an alternative to

$$(\gamma_1, \gamma_2, ..., \gamma_p)E.$$

In particular γE is used in place of $(\gamma)E$.

Suppose that we have an *epimorphism* of a ring R on to a ring R'. It is clear that every central element of R is carried into a central element of R'. For example, if A is a two-sided ideal of R and γ belongs to the centre of R, then the natural image of γ in the residue ring R/A is a central element of that ring.

We complete this section by noting some simple relations connecting central elements, submodules and factor modules. Let $K \subseteq L$ be submodules of an R-module E and let γ, γ' be central elements. The mapping $E \to E$ in which $e \to \gamma e$ is an endomorphism of R-modules. The inverse image of K with respect to this mapping is denoted by $K :_E \gamma$ so that an element e, of E, belongs to $K :_E \gamma$ if and only if $\gamma e \in K$. (More generally, if A is a subset of R, then $K :_E A$ denotes the set of all elements $e \in E$ such that $ae \in K$ for every $a \in A$. This is a submodule of E if A happens to be a two-sided ideal.) Evidently

$$(K :_E \gamma) :_E \gamma' = K :_E \gamma \gamma' \tag{7.1.1}$$

and, as is easily verified,

$$\gamma E \cap K = \gamma(K :_E \gamma). \tag{7.1.2}$$

Now consider the natural mapping $\phi: E \to E/K$. Since $K + \gamma E$ is mapped on to $\gamma(E/K)$, we have an isomorphism

$$E/(K + \gamma E) \approx (E/K)/\gamma(E/K). \qquad (7.1.3)$$

Also if e belongs to E, then $\phi(e)$ belongs to $(L/K):_{E/K}\gamma$ if and only if e is in $L:_E\gamma$. It follows that

$$(L:_E\gamma)/K = (L/K):_{E/K}\gamma. \qquad (7.1.4)$$

A useful special case of this relation is obtained by putting $L = K$. This yields

$$(K:_E\gamma)/K = 0:_{E/K}\gamma. \qquad (7.1.5)$$

7.2 Key theorems on central ideals

In this section we shall establish certain basic results involving central ideals and Noetherian modules. The first of these is a generalization of the Artin–Rees Theorem which, the reader will recall, was first encountered as Theorem 20 of section (4.7).

Theorem 1. *Let K be a submodule of a Noetherian R-module E and let A be a central ideal. Then there exists an integer $q \geqslant 0$ such that*

$$A^nE \cap K = A^{n-q}(A^qE \cap K)$$

for all $n \geqslant q$.

Proof. Let A_0 be an ideal of R which can be generated by a *finite* number of central elements contained in A. Then $A_0 \subseteq A$ and A_0E is a submodule of E. Since E is a Noetherian R-module, we can choose A_0 so that A_0E is a maximal member of the family of all submodules which can be obtained in this way. Now suppose that γ is an arbitrary central element contained in A. Then

$$\gamma E \subseteq (A_0 + R\gamma)E = A_0E$$

and therefore $AE = A_0E$. It follows that

$$A^2E = AA_0E = A_0AE = A_0^2E$$

whence $\qquad\qquad A^3E = AA_0^2E = A_0^2AE = A_0^3E.$

Proceeding in this way we find that $A^nE = A_0^nE$ for all $n \geqslant 0$.

Let us assume that there exists an integer $q \geqslant 0$ such that

$$A_0^nE \cap K = A_0^{n-q}(A_0^qE \cap K)$$

for all $n \geqslant q$. Then, when $n \geqslant q$,

$$A^nE \cap K = A_0^nE \cap K = A_0^{n-q}(A_0^qE \cap K) \subseteq A^{n-q}(A^qE \cap K)$$

which implies that $A^nE \cap K = A^{n-q}(A^qE \cap K)$. Thus for the purposes of the proof we may assume that A is *finitely generated*.

From here on the proof is virtually the same as that of Theorem 20 of section (4.7). Indeed the particular argument presented there was chosen because it can be adapted to the present conditions. The reader is therefore advised to read through this argument again in the light of the following remarks. First, the treatment in Chapter 4 uses the Basis Theorem as established as Theorem 1 of section (4.1). This, it will be recalled, is one of the results of Chapter 4 that holds for non-commutative as well as for commutative rings. Next, although the first paragraph of the original proof was devoted to showing that we could assume that A was finitely generated, the argument used on that occasion is not applicable in the present instance. This particular paragraph should therefore be replaced by the discussion given above. Finally, in selecting a finite set $\gamma_1, \gamma_2, \ldots, \gamma_s$ of elements to generate A, it is necessary to arrange that each γ_i is a central element of R. Provided these points are kept in mind, the reader should not experience any difficulty in establishing the theorem.

The next result corresponds to the Intersection Theorem† of Krull. When we encountered this result before, the proof given made use of the theory of primary decompositions and this is not available to us in our present circumstances. However, we can get around this difficulty with the aid of Theorem 1.

Theorem 2. *Let E be a Noetherian R-module and A a central ideal. Then an element e, of E, belongs to $\bigcap_{n=1}^{\infty} A^n E$ if and only if $e = \alpha e$ for some element α belonging to A.*

Proof. If $e = \alpha e$, where $\alpha \in A$, then $e = \alpha^n e$ for every n and therefore e belongs to the intersection. Now suppose that e is in $\bigcap_{n=1}^{\infty} A^n E$. By Theorem 1, there exists an integer $q \geqslant 0$ such that

$$A^n E \cap Re = A^{n-q}(A^q E \cap Re)$$

for all $n \geqslant q$. Taking $n = q+1$ we find that $Re = A(Re)$. Consequently $e \in A(Re)$ and therefore $e = \alpha e$ for some α in A. This completes the proof.

We shall also need a result which can be regarded as an adaptation of part of Theorem 19 of section (4.6). The difficulty here is that, for non-commutative rings, the concept of the Jacobson radical presents certain problems not encountered in the commutative case. We shall, in fact, define the Jacobson radical of a general ring R as the intersection of all its maximal left ideals. The reader will probably feel that

† See Theorem 18 of section (4.6).

this should be called the left Jacobson radical and that, by analogy, there should be a second Jacobson radical to be distinguished by the adjective 'right'. However, as we shall see very shortly, the intersection of all the maximal left ideals always coincides with the intersection of all the maximal right ideals. For this reason, the proposed distinction becomes unnecessary very quickly.

Before proceeding, let us note that *every proper left ideal of R is contained in a maximal left ideal.* Indeed if I is a left ideal different from R itself, then the set Ω of proper left ideals containing I forms a non-empty inductive system when its members are partially ordered by the inclusion relation. Zorn's lemma then shows that there exists a maximal member, L say, of Ω. Clearly L is a maximal left ideal of R and I is contained in L.

Proposition 1. *If x belongs to the Jacobson radical J, of R, then $1+x$ has a two-sided inverse.*

Proof. We have to show that there exists an element ρ, of R, such that $\rho(1+x) = (1+x)\rho = 1$. Clearly $1+x$ is not contained in any maximal left ideal so the same is true for $R(1+x)$. It follows that $R(1+x) = R$ and therefore $\rho(1+x) = 1$ for a suitable element ρ of R. Next

$$\rho = 1-\rho x = 1+x' \text{ say.}$$

Here $x' = -\rho x$ belongs to J because J is certainly a left ideal. Repeating the argument we now see that there exists an element ρ' such that $\rho'(1+x') = 1$, i.e. $\rho'\rho = 1$. Finally

$$\rho' = \rho'(\rho(1+x)) = (\rho'\rho)(1+x) = 1+x$$

and therefore $(1+x)\rho = \rho'\rho = 1$. This completes the proof.

Lemma 1. *Let $x \in J$, where J is the Jacobson radical of R, and let r be an arbitrary element of the ring. Then $xr \in J$.*

Proof. Let L be a maximal left ideal. The lemma will follow if we show that xr belongs to L. Assume the contrary. Then $Rxr + L$ is a left ideal which strictly contains L. Hence $Rxr + L = R$ and therefore

$$\rho x r + \lambda = 1$$

for suitable elements $\rho \in R$ and $\lambda \in L$. Put $y = \rho x$. Then $y \in J$ and $1-yr = \lambda$. Now

$$(1-ry)r = r(1-yr) = r\lambda$$

and, by Proposition 1, $1 - ry$ has a two-sided inverse. If we multiply the two sides of the equation $(1 - ry)r = r\lambda$ on the left by this inverse, then it is seen that $r \in L$. Consequently $1 = \lambda + yr$ belongs to L and now we have a contradiction because L is a proper ideal.

Theorem 3. *The intersection of all the maximal left ideals of R coincides with the intersection of all its maximal right ideals.*

Proof. We shall show that the intersection, J say, of all the maximal left ideals is contained in the intersection of all the maximal right ideals. The theorem will then follow by symmetry.

Let x belong to J and let Q be an arbitrary maximal right ideal. It is now sufficient to show that x belongs to Q. Assume that $x \notin Q$. Then $Q + xR = R$ and therefore $1 = q + xr$ for suitable elements $q \in Q$ and $r \in R$. By Lemma 1, $xr \in J$ and so, by Proposition 1, $1 - xr = q$ has a two-sided inverse q^{-1} say. But then $qq^{-1} = 1$ belongs to the right ideal Q and now we have the desired contradiction because Q is a proper ideal. This completes the proof.

Theorem 4. *Let E be a Noetherian R-module and A a central ideal contained in the Jacobson radical J of R. Then $\bigcap\limits_{n=1}^{\infty} A^n E = 0$.*

Proof. Let e belong to $\bigcap A^n E$. Then, by Theorem 2, we have $e = \alpha e$ for some element α which belongs to A and therefore also to J. Thus $(1 - \alpha)e = 0$ and, by Proposition 1, $1 - \alpha$ has a two-sided inverse. If we multiply the relation $(1 - \alpha)e = 0$ by this inverse, then we see that $e = 0$. This completes the proof.

7.3 Multiplicity systems

Let E be an R-module and let $\gamma_1, \gamma_2, ..., \gamma_s$ $(s \geqslant 0)$ be central elements of R. What we shall do is define, under certain conditions, the multiplicity of the system $\gamma_1, \gamma_2, ..., \gamma_s$ on (or in relation to) the module E. In order to have a concise way of referring to these conditions, we make the following definition.

Definition. *The central elements $\gamma_1, \gamma_2, ..., \gamma_s$ $(s \geqslant 0)$ will be said to form a 'multiplicity system' on E if the R-module $E/(\gamma_1 E + \gamma_2 E + ... + \gamma_s E)$ has finite length. When $s = 0$ this condition is to be understood as meaning that $L_R(E)$ is finite.*

Before we proceed to enumerate the basic properties of multiplicity systems, we shall establish an elementary inequality concerning lengths of modules which will prove useful on many occasions. The inequality is due to D. J. Wright.

Proposition 2. *Let E be an R-module and $\gamma_1, \gamma_2, ..., \gamma_s$ central elements of R. Then*

$$L_R\{E/(\gamma_1^{n_1} E + \gamma_2^{n_2} E + ... + \gamma_s^{n_s} E)\}$$
$$\leqslant n_1 n_2 ... n_s L_R\{E/(\gamma_1 E + \gamma_2 E + ... + \gamma_s E)\}$$

for arbitrary positive integers $n_1, n_2, ..., n_s$.

Remark. In this result it is not necessary for the lengths involved to be finite.

Proof. It is clearly sufficient to show that

$$L_R\{E/(\gamma_1^{n_1} E + \gamma_2 E + ... + \gamma_s E)\} \leqslant n_1 L_R\{E/(\gamma_1 E + \gamma_2 E + ... + \gamma_s E)\}$$

for once this has been established the more general result will follow by repeated applications of the special case. Put

$$E' = E/(\gamma_2 E + ... + \gamma_s E).$$

Then, by (7.1.3), we have an isomorphism

$$E'/\gamma_1^{n_1} E' \approx E/(\gamma_1^{n_1} E + \gamma_2 E + ... + \gamma_s E).$$

Accordingly

$$L_R\{E'/\gamma_1^{n_1} E'\} = L_R\{E/(\gamma_1^{n_1} E + \gamma_2 E + ... + \gamma_s E)\}$$

and so what we have to prove may be written as

$$L_R\{E'/\gamma_1^{n_1} E'\} \leqslant n_1 L_R\{E'/\gamma_1 E'\}.$$

This conclusion may equally well be described by saying that we need only consider the case in which $s = 1$.

Put $\gamma = \gamma_1$ and $n = n_1$. We have to show that $L_R(E/\gamma^n E)$ does not exceed $n L_R(E/\gamma E)$. But

$$L_R(E/\gamma^n E) = \sum_{i=1}^{n} L_R(\gamma^{i-1} E/\gamma^i E),$$

so it is only necessary to prove that $L_R(\gamma^{i-1} E/\gamma^i E) \leqslant L_R(E/\gamma E)$. Next, multiplication by γ^{i-1} produces an epimorphism $E \to \gamma^{i-1} E$. The inverse image of $\gamma^i E$ with respect to this mapping is $(0 :_E \gamma^{i-1}) + \gamma E$. Thus we have an isomorphism

$$\gamma^{i-1} E/\gamma^i E \approx E/(0 :_E \gamma^{i-1} + \gamma E)$$

and therefore

$$L_R\{\gamma^{i-1} E/\gamma^i E\} = L_R\{E/(0 :_E \gamma^{i-1} + \gamma E)\} \leqslant L_R\{E/\gamma E\}.$$

This completes the proof.

Let $\gamma_1, \gamma_2, ..., \gamma_s$ be a multiplicity system on E. We make a number of simple observations.

(a) *The elements* $\gamma_1, \gamma_2, ..., \gamma_s$ *remain a multiplicity system on E if we alter their order in any manner. They also continue to form a multiplicity system on E if we delete any γ_i for which $\gamma_i E = 0$.*

(b) *For arbitrary positive integers* $n_1, n_2, ..., n_s$, *the central elements* $\gamma_1^{n_1}, \gamma_2^{n_2}, ..., \gamma_s^{n_s}$ *also form a multiplicity system on E.*

To see this we have only to observe that, by Proposition 2,

$$L_R\{E/(\gamma_1^{n_1} E + ... + \gamma_s^{n_s} E)\}$$
$$\leqslant n_1 n_2 ... n_s L_R\{E/(\gamma_1 E + \gamma_2 E + ... + \gamma_s E)\}.$$

Since the latter expression is finite, this establishes (b).

(c) *If E' is a factor module of E, then* $\gamma_1, \gamma_2, ..., \gamma_s$ *is a multiplicity system on E'.*

We can prove this as follows. Let $E' = E/K$. Then

$$\gamma_1 E' + \gamma_2 E' + ... + \gamma_s E'$$

is just $(K + \gamma_1 E + ... + \gamma_s E)/K$ and so we have an isomorphism

$$E'/(\gamma_1 E' + \gamma_2 E' + ... + \gamma_s E') \approx E/(K + \gamma_1 E + ... + \gamma_s E).$$

It follows that

$$L_R\{E'/(\gamma_1 E' + ... + \gamma_s E')\} = L_R\{E/(K + \gamma_1 E + ... + \gamma_s E)\}$$
$$\leqslant L_R\{E/(\gamma_1 E + ... + \gamma_s E)\}$$

and this is finite. Thus (c) has been demonstrated.

It is not true, without adding an extra condition, that $\gamma_1, \gamma_2, ..., \gamma_s$ is necessarily a multiplicity system on every submodule of E. However we do have the following result.

Lemma 2. *Let* $0 \to E' \to E \to E'' \to 0$ *be an exact sequence of R-modules and let* $\gamma_1, \gamma_2, ..., \gamma_s$ *be central elements. Then*

$$L_R\{E/(\gamma_1 E + ... + \gamma_s E)\}$$
$$\leqslant L_R\{E'/(\gamma_1 E' + ... + \gamma_s E')\} + L_R\{E''/(\gamma_1 E'' + ... + \gamma_s E'')\}.$$

Hence if $\gamma_1, \gamma_2, ..., \gamma_s$ *is a multiplicity system on both E' and E'', then it is also a multiplicity system on E.*

Proof. Without loss of generality we may suppose that E' is a submodule of E and $E'' = E/E'$. Then

$$\gamma_1 E'' + \gamma_2 E'' + ... + \gamma_s E'' = (E' + \gamma_1 E + ... + \gamma_s E)/E'$$

and therefore

$$L_R\{E''/(\gamma_1 E'' + ... + \gamma_s E'')\} = L_R\{E/(E' + \gamma_1 E + ... + \gamma_s E)\},$$

because the modules occurring in this equation are isomorphic. Accordingly

$$L_R\{E/(\gamma_1 E + \ldots + \gamma_s E)\}$$
$$= L_R\{E/(E' + \gamma_1 E + \ldots + \gamma_s E)\}$$
$$+ L_R\{(E' + \gamma_1 E + \ldots + \gamma_s E)/(\gamma_1 E + \ldots + \gamma_s E)\}$$
$$= L_R\{E''/(\gamma_1 E'' + \ldots + \gamma_s E'')\} + L_R\{E'/(E' \cap (\gamma_1 E + \ldots + \gamma_s E))\}.$$

Here we have made use of the isomorphism

$$(E' + \gamma_1 E + \ldots + \gamma_s E)/(\gamma_1 E + \ldots + \gamma_s E) \approx E'/(E' \cap (\gamma_1 E + \ldots + \gamma_s E))$$

which holds by virtue of Theorem 3 of section (1.6). Again

$$\gamma_1 E' + \gamma_2 E' + \ldots + \gamma_s E' \subseteq E' \cap (\gamma_1 E + \ldots + \gamma_s E) \subseteq E'$$

and so

$$L_R\{E'/(E' \cap (\gamma_1 E + \ldots + \gamma_s E))\} \leqslant L_R\{E'/(\gamma_1 E' + \ldots + \gamma_s E')\}.$$

The lemma follows.

It has already been remarked that if $\gamma_1, \gamma_2, \ldots, \gamma_s$ is a multiplicity system on an R-module E, then there may exist submodules of E for which $\gamma_1, \gamma_2, \ldots, \gamma_s$ is not a multiplicity system. However if confine our attention to .Noetherian modules, this awkward possibility cannot arise as is shown by our next result.

Proposition 3. *Let $0 \to E' \to E \to E'' \to 0$ be an exact sequence of Noetherian R-modules and let $\gamma_1, \gamma_2, \ldots, \gamma_s$ be central elements. Then $\gamma_1, \gamma_2, \ldots, \gamma_s$ is a multiplicity system on E if and only if it is a multiplicity system on both E' and E''.*

Proof. We shall assume that $\gamma_1, \gamma_2, \ldots, \gamma_s$ is a multiplicity system on E and deduce that it is also a multiplicity system on E'. This will suffice because all the other assertions contained in the statement of the proposition have already been established. Without loss of generality we may assume that E' is a submodule of E.

Put $A = \gamma_1 R + \gamma_2 R + \ldots + \gamma_s R$. Then, by Theorem 1, there exists a non-negative integer q such that

$$A^{q+1} E \cap E' = A(A^q E \cap E') \subseteq AE'.$$

Accordingly

$$L_R\{E'/(\gamma_1 E' + \ldots + \gamma_s E')\} = L_R\{E'/AE'\}$$
$$\leqslant L_R\{E'/(A^{q+1} E \cap E')\}$$
$$= L_R\{(A^{q+1} E + E')/A^{q+1} E\}$$
$$\leqslant L_R\{E/A^{q+1} E\}$$

because, by Theorem 3 of section (1.6), the modules $E'/(A^{q+1}E \cap E')$ and $(A^{q+1}E + E')/A^{q+1}E$ are isomorphic. Now

$$\gamma_1^{q+1}E + \gamma_2^{q+1}E + \ldots + \gamma_s^{q+1}E$$

is contained in $A^{q+1}E$. Consequently

$$L_R\{E'/(\gamma_1 E' + \ldots + \gamma_s E')\} \leqslant L_R\{E/(\gamma_1^{q+1}E + \ldots + \gamma_s^{q+1}E)\}$$
$$\leqslant (q+1)^s L_R\{E/(\gamma_1 E + \ldots + \gamma_s E)\} < \infty$$

by virtue of Proposition 2. This shows that $\gamma_1, \gamma_2, \ldots, \gamma_s$ is a multiplicity system on E' and completes the proof.

We conclude this section with a remark concerning the behaviour of multiplicity systems in relation to a reduction of the ring of operators. Let $\gamma_1, \gamma_2, \ldots, \gamma_s$ be a multiplicity system on an R-module E and let I be a two-sided ideal such that $IE = 0$. Put $\bar{R} = R/I$. Then, by Proposition 24 of section (1.14), E has a well defined structure as an $\bar{R}$-module. Denote the natural image of γ_i in $\bar{R}$ by $\bar{\gamma}_i$. Then $\bar{\gamma}_i$ is a central element of $\bar{R}$ and

$$E/(\gamma_1 E + \gamma_2 E + \ldots + \gamma_s E) = E/(\bar{\gamma}_1 E + \bar{\gamma}_2 E + \ldots + \bar{\gamma}_s E)$$

because $\gamma_i E = \bar{\gamma}_i E$. What we have here is both an R-module and an $\bar{R}$-module and, moreover, its R-submodules and its $\bar{R}$-submodules coincide. Accordingly

$$L_{\bar{R}}\{E/(\bar{\gamma}_1 E + \ldots + \bar{\gamma}_s E)\} = L_R\{E/(\gamma_1 E + \ldots + \gamma_s E)\} < \infty.$$

Hence $\bar{\gamma}_1, \bar{\gamma}_2, \ldots, \bar{\gamma}_s$ *is a multiplicity system on E when E is considered as an $\bar{R}$-module.*

7.4 The multiplicity symbol

Let E be a Noetherian R-module and $\gamma_1, \gamma_2, \ldots, \gamma_s$ a multiplicity system on E. We shall now define the multiplicity of $\gamma_1, \gamma_2, \ldots, \gamma_s$ on (or with respect to) E. This multiplicity will turn out to be a non-negative integer and we shall use the symbol $e_R(\gamma_1, \gamma_2, \ldots, \gamma_s | E)$ to denote it. The definition of the multiplicity symbol uses induction on s.

First suppose that $s = 0$. In this case the empty set is a multiplicity system on E and therefore, by our convention, $L_R(E)$ is finite. We may therefore put

$$e_R(\cdot | E) = L_R(E). \tag{7.4.1}$$

Now assume that $s \geqslant 1$ and that the multiplicity symbol has been defined for Noetherian modules and multiplicity systems with only $s-1$ elements. By Lemma 1 of section (4.1), both the factor module $E/\gamma_1 E$ and the submodule $0:_E \gamma_1$ are Noetherian. Next, by Proposition 3, since $\gamma_1, \gamma_2, \ldots, \gamma_s$ is a multiplicity system on E, the γ_i also form a multiplicity system on the two modules derived from it. But γ_1

annihilates $E/\gamma_1 E$. Hence if we remove γ_1 the remaining elements will still form a multiplicity system on this module. Thus $\gamma_2, \gamma_3, ..., \gamma_s$ is a multiplicity system on $E/\gamma_1 E$ and similar considerations show that $\gamma_2, \gamma_3, ..., \gamma_s$ is a multiplicity system on $0:_E \gamma_1$ as well. Accordingly, by virtue of our assumptions, $e_R(\gamma_2, \gamma_3, ..., \gamma_s | E/\gamma_1 E)$ and

$$e_R(\gamma_2, \gamma_3, ..., \gamma_s | 0:_E \gamma_1)$$

are both defined and so we may put

$$e_R(\gamma_1, \gamma_2, ..., \gamma_s | E)$$
$$= e_R(\gamma_2, \gamma_3, ..., \gamma_s | E/\gamma_1 E) - e_R(\gamma_2, \gamma_3, ..., \gamma_s | 0:_E \gamma_1). \quad (7.4.2)$$

The general multiplicity symbol is now fully determined by means of (7.4.1) and (7.4.2).

It is clear that $e_R(\gamma_1, ..., \gamma_s | E)$ is an integer though we have no grounds, as yet, for saying that it is non-negative. It is obvious, too, that the value of the multiplicity symbol does not change if we replace E by any module which is isomorphic to it. Further $e_R(\gamma_1, ..., \gamma_s | E)$ has the value zero whenever E is a null module.

We note one other elementary property. Suppose that E is a Noetherian R-module and $\gamma_1, \gamma_2, ..., \gamma_s$ is a multiplicity system on E. Suppose also that I is a two-sided ideal and that $IE = 0$. Put $\bar{R} = R/I$ and let $\bar{\gamma}_i$ denote the natural image of γ_i in $\bar{R}$. Then E is an $\bar{R}$-module and as such it is Noetherian because the $\bar{R}$-submodules of E are just the same as its R-submodules. Moreover, as we saw at the end of the last section, $\bar{\gamma}_1, \bar{\gamma}_2, ..., \bar{\gamma}_s$ is a multiplicity system on the $\bar{R}$-module E. Accordingly both $\gamma_1, \gamma_2, ..., \gamma_s$ and $\bar{\gamma}_1, \bar{\gamma}_2, ..., \bar{\gamma}_s$ have a multiplicity on E. We claim that

$$e_R(\gamma_1, \gamma_2, ..., \gamma_s | E) = e_{\bar{R}}(\bar{\gamma}_1, \bar{\gamma}_2, ..., \bar{\gamma}_s | E). \quad (7.4.3)$$

For if $s = 0$, then there is no problem because (7.4.3) simply asserts that $L_R(E)$ and $L_{\bar{R}}(E)$ are equal. Now suppose that $s > 0$ and that the result in question has been proved in the case of multiplicity systems with $s - 1$ elements. Then, since

$$E/\gamma_1 E = E/\bar{\gamma}_1 E \quad \text{and} \quad 0:_E \gamma_1 = 0:_E \bar{\gamma}_1,$$

it follows that

$$\begin{aligned} e_R(\gamma_1, \gamma_2, ..., \gamma_s | E) &= e_R(\gamma_2, ..., \gamma_s | E/\gamma_1 E) - e_R(\gamma_2, ..., \gamma_s | 0:_E \gamma_1) \\ &= e_{\bar{R}}(\bar{\gamma}_2, ..., \bar{\gamma}_s | E/\gamma_1 E) - e_{\bar{R}}(\bar{\gamma}_2, ..., \bar{\gamma}_s | 0:_E \gamma_1) \\ &= e_{\bar{R}}(\bar{\gamma}_2, ..., \bar{\gamma}_s | E/\bar{\gamma}_1 E) - e_{\bar{R}}(\bar{\gamma}_2, ..., \bar{\gamma}_s | 0:_E \bar{\gamma}_1) \\ &= e_{\bar{R}}(\bar{\gamma}_1, \bar{\gamma}_2, ..., \bar{\gamma}_s | E). \end{aligned}$$

Thus (7.4.3) is established.

The multiplicity symbol has an important additive property. The following lemma will be useful in establishing this fact.

Lemma 3. *Let* $0 \to E' \to E \to E'' \to 0$ *be an exact sequence of R-modules and let* γ *be a central element. Then an exact sequence of the form*

$$0 \to 0:_{E'}\gamma \xrightarrow{\phi'} 0:_{E}\gamma \xrightarrow{\psi'} 0:_{E''}\gamma \xrightarrow{f} E'/\gamma E' \xrightarrow{\phi^*} E/\gamma E \xrightarrow{\psi^*} E''/\gamma E'' \to 0$$

(7.4.4)

can be constructed.

Proof. Without loss of generality we may suppose that E' is a submodule of E and $E'' = E/E'$. In what follows $\phi : E' \to E$ denotes the inclusion mapping and $\psi : E \to E''$ the natural mapping of E on to the factor module $E/E' = E''$. Then $\phi(0:_{E'}\gamma)$ is contained in $0:_{E}\gamma$ and $\phi(\gamma E') \subseteq \gamma E$. Thus, by restriction, ϕ gives rise to a mapping ϕ' of $0:_{E'}\gamma$ into $0:_{E}\gamma$ and it induces a mapping $\phi^* : E'/\gamma E' \to E/\gamma E$. Likewise $\psi(0:_{E}\gamma)$ is contained in $0:_{E''}\gamma$ and $\psi(\gamma E) \subseteq \gamma E''$. Thus from ψ we obtain mappings $0:_{E}\gamma \to 0:_{E''}\gamma$ and $E/\gamma E \to E''/\gamma E''$ which we denote by ψ' and ψ^* respectively.

We next define the mapping f. Let e'' belong to $0:_{E''}\gamma$. It is possible to choose $e \in E$ so that $\psi(e) = e''$ and then, because $\gamma e'' = 0$, we have $\psi(\gamma e) = 0$ and therefore $\gamma e \in E'$. Now put

$$f(e'') = \text{image of } \gamma e \text{ in } E'/\gamma E' \tag{7.4.5}$$

and observe that if $e_1 \in E$ also has the property that $\psi(e_1) = e''$, then $e - e_1$ belongs to E' and therefore $\gamma e - \gamma e_1$ belongs to $\gamma E'$. Thus γe and γe_1 have the same image in $E'/\gamma E'$ which shows that f is well defined by (7.4.5). An easy verification shows that f is R-linear.

Now that the mappings in (7.4.4) have all been defined, it is necessary to check that the sequence is exact. Most of this is very simple. We shall verify that $\mathrm{Ker}\, f = \mathrm{Im}\, \psi'$, $\mathrm{Im}\, f = \mathrm{Ker}\, \phi^*$ and leave the remaining details to the reader.

By (7.4.5), $f(e'') = 0$ if and only if $\gamma e \in \gamma E'$ that is to say $\gamma(e - e') = 0$ for some e' in E'. Thus $e'' \in \mathrm{Ker}\, f$ when and only when there exists $\xi \in E$ such that $\gamma \xi = 0$ and $\psi(\xi) = e''$. In other terms, e'' is in $\mathrm{Ker}\, f$ when and only when it is the image, under ψ, of an element of $0:_{E}\gamma$. This proves that $\mathrm{Ker}\, f = \mathrm{Im}\, \psi'$.

From (7.4.5) we see that $\phi^* f(e'') = 0$. Accordingly $\mathrm{Im}\, f \subseteq \mathrm{Ker}\, \phi^*$. Now assume that $\eta \in \mathrm{Ker}\, \phi^*$ and let η be the image of e' in $E'/\gamma E'$. Then $\phi^*(\eta)$ is the image of e' in $E/\gamma E$ whence, since $\phi^*(\eta) = 0$, $e' = \gamma e$ for some e in E. Put $e'' = \psi(e)$. Then $\gamma e'' = \psi(\gamma e) = \psi(e') = 0$ which shows that e'' belongs to $0:_{E''}\gamma$. Further, by (7.4.5), $f(e'')$ is the image of

$\gamma e = e'$ in $E'/\gamma E'$, that is to say $f(e'') = \eta$. Thus $\eta \in \mathrm{Im} f$ and therefore $\mathrm{Ker}\, \phi^* \subseteq \mathrm{Im} f$. The lemma follows.

The next result concerns the additive property of the multiplicity symbol to which reference has already been made.

Theorem 5. *Let* $0 \to E' \to E \to E'' \to 0$ *be an exact sequence of Noetherian R-modules and suppose that* $\gamma_1, \gamma_2, ..., \gamma_s$ *is a multiplicity system on each term. Then*

$$e_R(\gamma_1, ..., \gamma_s | E) = e_R(\gamma_1, ..., \gamma_s | E') + e_R(\gamma_1, ..., \gamma_s | E'').$$

It is convenient to prove Theorem 5 and the following corollary together.

Corollary 1. *Let* $0 \to E_p \to ... \to E_1 \to E_0 \to 0$ *be an exact sequence of Noetherian R-modules and suppose that* $\gamma_1, \gamma_2, ..., \gamma_s$ *is a multiplicity system on each term. Then*

$$\sum_{i=0}^{p} (-1)^i e_R(\gamma_1, ..., \gamma_s | E_i) = 0.$$

Proof. The theorem and the corollary will be proved simultaneously using induction on s. When $s = 0$ there is no problem because everything needed is provided by Theorem 20 of section (1.12). We now assume that $s \geqslant 1$ and that both the theorem and the corollary are known to be true when we have to do with multiplicity systems containing only $s-1$ elements.

Turning now to Theorem 5, we observe that, by Lemma 3, we have an exact sequence

$$0 \to 0:_{E'}\gamma_1 \to 0:_{E}\gamma_1 \to 0:_{E''}\gamma_1 \to E'/\gamma_1 E' \to E/\gamma_1 E \to E''/\gamma_1 E'' \to 0.$$

In this sequence every term is a Noetherian R-module and admits $\gamma_2, \gamma_3, ..., \gamma_s$ as a multiplicity system. Consequently we are in a position to apply Cor. 1. This yields

$$e_R(\gamma_2, ..., \gamma_s | E/\gamma_1 E) - e_R(\gamma_2, ..., \gamma_s | 0:_E \gamma_1)$$
$$= e_R(\gamma_2, ..., \gamma_s | E'/\gamma_1 E') - e_R(\gamma_2, ..., \gamma_s | 0:_{E'} \gamma_1)$$
$$+ e_R(\gamma_2, ..., \gamma_s | E''/\gamma_1 E'') - e_R(\gamma_2, ..., \gamma_s | 0:_{E''} \gamma_1).$$

But this can be written as

$$e_R(\gamma_1, ..., \gamma_s | E) = e_R(\gamma_1, ..., \gamma_s | E') + e_R(\gamma_1, ..., \gamma_s | E'')$$

in view of (7.4.2). Accordingly Theorem 5 is established for the value of s under consideration. To derive the corollary for this value we simply adapt the argument used for the corresponding result in the case of lengths.†

† See Theorem 20 of section (1.12).

Corollary 2. *Let E_1, E_2 be Noetherian submodules of an R-module E and let $\gamma_1, \gamma_2, ..., \gamma_s$ be a multiplicity system on both E_1 and E_2. Then $\gamma_1, \gamma_2, ..., \gamma_s$ is a multiplicity system on the Noetherian modules*

$$E_1 + E_2, E_1 \cap E_2,$$

and

$$e_R(\gamma_1, ..., \gamma_s | E_1 + E_2) + e_R(\gamma_1, ..., \gamma_s | E_1 \cap E_2)$$
$$= e_R(\gamma_1, ..., \gamma_s | E_1) + e_R(\gamma_1, ..., \gamma_s | E_2). \quad (7.4.6)$$

Proof. The facts that $E_1 + E_2$ and $E_1 \cap E_2$ are Noetherian R-modules follow from Proposition 1 and Lemma 1 both of section (4.1). Next, by Lemma 1 Cor. 1 of section (4.1), $E_1 \oplus E_2$ is a Noetherian R-module. Since we have an exact sequence

$$0 \to E_1 \to E_1 \oplus E_2 \to E_2 \to 0,$$

Proposition 3 shows that $\gamma_1, \gamma_2, ..., \gamma_s$ is a multiplicity system on $E_1 \oplus E_2$. Again $E_1 + E_2$ is isomorphic to a factor module of $E_1 \oplus E_2$ and $E_1 \cap E_2$ is a submodule of E_1. Hence, again by virtue of Proposition 3, $\gamma_1, \gamma_2, ..., \gamma_s$ is a multiplicity system on $E_1 + E_2$ and $E_1 \cap E_2$. This shows that all the multiplicities occurring in (7.4.6) are properly defined.

Consider the exact sequence

$$0 \to E_1 \to E_1 + E_2 \to (E_1 + E_2)/E_1 \to 0.$$

By Theorem 3 of section (1.6), we have an isomorphism

$$(E_1 + E_2)/E_1 \approx E_2/(E_1 \cap E_2)$$

so that the sequence may be rewritten as

$$0 \to E_1 \to E_1 + E_2 \to E_2/(E_1 \cap E_2) \to 0. \quad (7.4.7)$$

We also have an exact sequence

$$0 \to E_1 \cap E_2 \to E_2 \to E_2/(E_1 \cap E_2) \to 0. \quad (7.4.8)$$

The desired result now follows by applying Theorem 5 to (7.4.7) and (7.4.8).

Corollary 3. *Let E_1 and E_2 be submodules of an R-module E such that E/E_1 and E/E_2 are Noetherian. If now $\gamma_1, \gamma_2, ..., \gamma_s$ is a multiplicity system on E/E_1 and E/E_2, then it is also a multiplicity system on the Noetherian modules $E/(E_1 \cap E_2)$ and $E/(E_1 + E_2)$. Furthermore*

$$e_R(\gamma_1, ..., \gamma_s | E/(E_1 \cap E_2)) + e_R(\gamma_1, ..., \gamma_s | E/(E_1 + E_2))$$
$$= e_R(\gamma_1, ..., \gamma_s | E/E_1) + E_R(\gamma_1, ..., \gamma_s | E/E_2). \quad (7.4.9)$$

Proof. The module $E/(E_1 + E_2)$ is Noetherian and admits $\gamma_1, \gamma_2, ..., \gamma_s$ as a multiplicity system because it is isomorphic to a factor module of E/E_1. On the other hand $E/(E_1 \cap E_2)$ is Noetherian by Proposition 2 of section (4.1). Indeed the proof of this result shows that $E/(E_1 \cap E_2)$ is isomorphic to a submodule of $(E/E_1) \oplus (E/E_2)$ and the direct sum is known to be Noetherian by virtue of Lemma 1 Cor. 1 of section (4.1). Next, by applying Proposition 3 to the exact sequence

$$0 \to E/E_1 \to (E/E_1) \oplus (E/E_2) \to E/E_2 \to 0,$$

we see that $\gamma_1, \gamma_2, ..., \gamma_s$ is a multiplicity system on $(E/E_1) \oplus (E/E_2)$ and hence on $E/(E_1 \cap E_2)$. Thus all the multiplicities which occur in (7.4.9) are well defined.

The module $E_1/(E_1 \cap E_2)$ is isomorphic to $(E_1 + E_2)/E_2$ so we can rewrite the exact sequence

$$0 \to E_1/(E_1 \cap E_2) \to E/(E_1 \cap E_2) \to E/E_1 \to 0$$

as $\qquad 0 \to (E_1 + E_2)/E_2 \to E/(E_1 \cap E_2) \to E/E_1 \to 0. \qquad (7.4.10)$

We also have an exact sequence

$$0 \to (E_1 + E_2)/E_2 \to E/E_2 \to E/(E_1 + E_2) \to 0. \qquad (7.4.11)$$

The corollary now follows by applying Theorem 5 to the sequences (7.4.10) and (7.4.11).

Our next main result will show that the value of $e_R(\gamma_1, \gamma_2, ..., \gamma_s | E)$ is independent of the order in which the γ_i occur. The proof involves a fair amount of direct computation which it is convenient to concentrate in a separate lemma.

Lemma 4. *Let E be a Noetherian R-module and $\gamma_1, \gamma_2, ..., \gamma_s$ $(s \geqslant 2)$ a multiplicity system on E. Then*

$$e_R(\gamma_1, \gamma_2, \gamma_3, ..., \gamma_s | E) = e_R(\gamma_2, \gamma_1, \gamma_3, ..., \gamma_s | E).$$

Proof. We simply apply (7.4.2) twice and examine the result to see whether it is symmetrical in γ_1 and γ_2. However the expressions which occur become a little complicated. For this reason we introduce some auxiliary notation which is as follows. When K is a Noetherian R-module with the property that $\gamma_3, \gamma_4, ..., \gamma_s$ is a multiplicity system on it, we write $\qquad [K] = e_R(\gamma_3, ..., \gamma_s | K).$

Observe that if $L \subseteq M$ are submodules of a Noetherian R-module N and $\gamma_3, \gamma_4, ..., \gamma_s$ is a multiplicity system on N/L, then $\gamma_3, \gamma_4, ..., \gamma_s$ is a multiplicity system on each of M/L and N/M, and we have

$$[N/L] = [N/M] + [M/L]. \qquad (7.4.12)$$

This is clear if we apply Proposition 3 and Theorem 5 to the exact sequence

$$0 \to M/L \to N/L \to N/M \to 0.$$

We are now ready to start computing. By (7.4.2),

$$e_R(\gamma_1, \gamma_2, \ldots, \gamma_s | E) = e_R(\gamma_2, \gamma_3, \ldots, \gamma_s | E/\gamma_1 E)$$
$$- e_R(\gamma_2, \gamma_3, \ldots, \gamma_s | 0:_E \gamma_1).$$

Let us now use the same relation to express each term on the right-hand side as the difference of two multiplicities. This gives

$$e_R(\gamma_1, \gamma_2, \ldots, \gamma_s | E) = [1] - [2] - [3] + [4],$$

where

$$[1] = [(E/\gamma_1 E)/\gamma_2(E/\gamma_1 E)],$$
$$[2] = [0:_{E/\gamma_1 E} \gamma_2],$$
$$[3] = [(0:_E \gamma_1)/\gamma_2(0:_E \gamma_1)],$$
$$[4] = [0:_{(0:_E \gamma_1)} \gamma_2].$$

Now, by (7.1.3), $(E/\gamma_1 E)/\gamma_2(E/\gamma_1 E)$ is isomorphic to $E/(\gamma_1 E + \gamma_2 E)$. On the other hand it is obvious that

$$0:_{(0:_E \gamma_1)} \gamma_2 = (0:_E \gamma_1) \cap (0:_E \gamma_2).$$

Accordingly

$$[1] = [E/(\gamma_1 E + \gamma_2 E)]$$

and

$$[4] = [(0:_E \gamma_1) \cap (0:_E \gamma_2)]$$

both of which involve γ_1 and γ_2 symmetrically. Thus we have now only to concern ourselves with $[2] + [3]$.

By (7.1.5), we have

$$0:_{E/\gamma_1 E} \gamma_2 = (\gamma_1 E:_E \gamma_2)/\gamma_1 E$$

and it is evident that $\gamma_1 E \subseteq \gamma_1 E + (0:_E \gamma_2) \subseteq \gamma_1 E:_E \gamma_2$. Hence, by (7.4.12),

$$[2] = [(\gamma_1 E + (0:_E \gamma_2))/\gamma_1 E] + [(\gamma_1 E:_E \gamma_2)/(\gamma_1 E + (0:_E \gamma_2))]$$
$$= [(0:_E \gamma_2)/\gamma_1 E \cap (0:_E \gamma_2)] + [(\gamma_1 E:_E \gamma_2)/(\gamma_1 E + (0:_E \gamma_2))]$$

because by Theorem 3 of section (1.6), $(\gamma_1 E + (0:_E \gamma_2))/\gamma_1 E$ is isomorphic to $(0:_E \gamma_2)/(\gamma_1 E \cap (0:_E \gamma_2))$. Next, by (7.1.2) and (7.1.1),

$$\gamma_1 E \cap (0:_E \gamma_2) = \gamma_1(0:_E \gamma_1 \gamma_2).$$

Thus

$$[2] = [5] + [6],$$

where

$$[5] = [(0:_E \gamma_2)/\gamma_1(0:_E \gamma_1 \gamma_2)]$$

and

$$[6] = [(\gamma_1 E:_E \gamma_2)/(\gamma_1 E + (0:_E \gamma_2))].$$

Again, multiplication by γ_2 produces a homomorphism of $\gamma_1 E :_E \gamma_2$ on to $\gamma_2(\gamma_1 E :_E \gamma_2)$ which, by (7.1.2), is the same as $\gamma_1 E \cap \gamma_2 E$. Thus we have a homomorphism

$$(\gamma_1 E :_E \gamma_2) \to \gamma_1 E \cap \gamma_2 E$$

whose kernel is $0 :_E \gamma_2$. Now $\gamma_1 E + (0 :_E \gamma_2)$ contains this kernel and is mapped on to $\gamma_1 \gamma_2 E$. It follows, by Theorem 1 of section (1.6), that we have an isomorphism

$$(\gamma_1 E :_E \gamma_2)/(\gamma_1 E + (0 :_E \gamma_2)) \approx (\gamma_1 E \cap \gamma_2 E)/\gamma_1 \gamma_2 E.$$

Accordingly $[6] = [(\gamma_1 E \cap \gamma_2 E)/\gamma_1 \gamma_2 E]$

and this, too, is symmetrical in γ_1 and γ_2.

It remains for us to examine the sum of [3] and [5]. Evidently

$$\gamma_2(0 :_E \gamma_1) \subseteq \gamma_2(0 :_E \gamma_1 \gamma_2) \subseteq 0 :_E \gamma_1.$$

Hence, by (7.4.12),

$$[3] + [5] = [7] + \{8\},$$

where $[7] = [\gamma_2(0 :_E \gamma_1 \gamma_2)/\gamma_2(0 :_E \gamma_1)]$

and $\{8\} = [(0 :_E \gamma_1)/\gamma_2(0 :_E \gamma_1 \gamma_2)] + [(0 :_E \gamma_2)/\gamma_1(0 :_E \gamma_1 \gamma_2)].$

The latter of these has the desired symmetry. Also multiplication by γ_2 induces an epimorphism

$$(0 :_E \gamma_1 \gamma_2) \to \gamma_2(0 :_E \gamma_1 \gamma_2)$$

which is such that the inverse image of $\gamma_2(0 :_E \gamma_1)$ is $(0 :_E \gamma_1) + (0 :_E \gamma_2)$. Accordingly we have an isomorphism

$$(0 :_E \gamma_1 \gamma_2)/((0 :_E \gamma_1) + (0 :_E \gamma_2)) \approx \gamma_2(0 :_E \gamma_1 \gamma_2)/\gamma_2(0 :_E \gamma_1),$$

and therefore $[7] = [(0 :_E \gamma_1 \gamma_1)/((0 :_E \gamma_1) + (0 :_E \gamma_2))]$

which is another symmetrical expression. Collecting terms we find that $e_R(\gamma_1, \gamma_2, \ldots, \gamma_s | E)$ is equal to

$$[1] + [4] - [6] - [7] - \{8\}$$

and here each term has the property that it remains unaltered when γ_1 and γ_2 are interchanged. The lemma follows.

Proposition 4. *Let E be a Noetherian R-module and $\gamma_1, \gamma_2, \ldots, \gamma_s$ a multiplicity system on E. If now $\{i_1, i_2, \ldots, i_s\}$ is a permutation of $\{1, 2, \ldots, s\}$, then*

$$e_R(\gamma_1, \gamma_2, \ldots, \gamma_s | E) = e_R(\gamma_{i_1}, \gamma_{i_2}, \ldots, \gamma_{i_s} | E).$$

Remark. We shall refer to this property of the multiplicity symbol as the *Exchange Property*.

Proof. It is sufficient to show that the value of

$$e_R(\gamma_1, ..., \gamma_m, \gamma_{m+1}, ..., \gamma_s | E)$$

is unaltered if we exchange γ_m and γ_{m+1}. Now by $m-1$ applications of (7.4.2) we obtain an expression for $e_R(\gamma_1, ..., \gamma_s | E)$ as a finite sum as follows:

$$e_R(\gamma_1, ..., \gamma_m, \gamma_{m+1}, ..., \gamma_s | E) = \sum_\nu \epsilon_\nu e_R(\gamma_m, \gamma_{m+1}, ..., \gamma_s | E_\nu).$$

Here each ϵ_ν is equal to ± 1 and each E_ν is a Noetherian R-module which admits $\gamma_m, \gamma_{m+1}, ..., \gamma_s$ as a multiplicity system. Furthermore, the numbers ϵ_ν and the modules E_ν are determined solely by E and $\gamma_1, ..., \gamma_{m-1}$, that is to say they are quite independent of

$$\gamma_m, \gamma_{m+1}, ..., \gamma_s.$$

Hence if we interchange γ_m and γ_{m+1} we obtain

$$e_R(\gamma_1, ..., \gamma_{m+1}, \gamma_m, ..., \gamma_s | E) = \sum_\nu \epsilon_\nu e_R(\gamma_{m+1}, \gamma_m, ..., \gamma_s | E_\nu)$$

where ϵ_ν and E_ν have the same meanings as before. But, by Lemma 4,

$$e_R(\gamma_m, \gamma_{m+1}, ..., \gamma_s | E_\nu) = e_R(\gamma_{m+1}, \gamma_m, ..., \gamma_s | E_\nu).$$

It follows that

$$e_R(\gamma_1, ..., \gamma_m, \gamma_{m+1}, ..., \gamma_s | E) = e_R(\gamma_1, ..., \gamma_{m+1}, \gamma_m, ..., \gamma_s | E)$$

and with this the proposition is established.

Proposition 5. *Let E be a Noetherian R-module and $\gamma_1, \gamma_2, ..., \gamma_s$ a multiplicity system on E. Assume that for some particular value of i we have $\gamma_i^m E = 0$, where m is a positive integer. Then*

$$e_R(\gamma_1, \gamma_2, ..., \gamma_s | E) = 0.$$

Proof. The exchange property of the multiplicity symbol shows that we may assume that $i = 1$. Having made this assumption we proceed to prove the proposition by induction on m.

First suppose that $m = 1$. Then $\gamma_1 E = 0$ and therefore $E/\gamma_1 E = E$ and $0 :_E \gamma_1 = E$. Accordingly, by (7.4.2),

$$e_R(\gamma_1, \gamma_2, ..., \gamma_s | E) = e_R(\gamma_2, ..., \gamma_s | E) - e_R(\gamma_2, ..., \gamma_s | E) = 0.$$

Now assume that $m > 1$ and that the desired result has already been established for smaller values of the inductive variable. By Theorem 5 and the exact sequence

$$0 \to \gamma_1 E \to E \to E/\gamma_1 E \to 0,$$

we see that

$$e_R(\gamma_1, \gamma_2, ..., \gamma_s | E) = e_R(\gamma_1, \gamma_2, ..., \gamma_s | \gamma_1 E) + e_R(\gamma_1, \gamma_2, ..., \gamma_s | E/\gamma_1 E).$$

However $\gamma_1^{m-1}(\gamma_1 E) = 0$ and $\gamma_1(E/\gamma_1 E) = 0$. Hence, by the inductive hypothesis, both $e_R(\gamma_1, ..., \gamma_s | \gamma_1 E)$ and $e_R(\gamma_1, ..., \gamma_s | 0 :_E \gamma_1)$ are zero. The proposition follows.

The following lemma is useful because it helps us to exploit Proposition 5.

Lemma 5. *Let E be a Noetherian R-module and γ a central element. Put $F_m = E/(0 :_E \gamma^m)$. Then $0 :_{F_m} \gamma = 0$ provided that m is sufficiently large.*

Proof. Since E is a Noetherian R-module, the ascending sequence

$$0 :_E \gamma \subseteq 0 :_E \gamma^2 \subseteq 0 :_E \gamma^3 \subseteq \ ...$$

of submodules of E must terminate. Suppose now that m is large enough to ensure that $0 :_E \gamma^m = 0 :_E \gamma^{m+1}$. Then, by (7.1.5) and (7.1.1),

$$0 :_{F_m} \gamma = ((0 :_E \gamma^m) :_E \gamma)/(0 :_E \gamma^m)$$

$$= (0 :_E \gamma^{m+1})/(0 :_E \gamma^m)$$

$$= (0 :_E \gamma^m)/(0 :_E \gamma^m)$$

$$= 0$$

as required.

Theorem 6. *Let E be a Noetherian R-module and $\gamma_1, \gamma_2, ..., \gamma_s$ a multiplicity system on E. Then*

$$0 \leqslant e_R(\gamma_1, ..., \gamma_s | E) \leqslant L_R\{E/(\gamma_1 E + ... + \gamma_s E)\}. \qquad (7.4.13)$$

Proof. We first show, using induction on s, that $e_R(\gamma_1, ..., \gamma_s | E)$ is non-negative. If $s = 0$, then this is obvious. We therefore assume that $s \geqslant 1$ and also that the non-negative character of multiplicities has been established in the case of multiplicity systems having only $s - 1$ members.

Put $F = E/(0 :_E \gamma_1^m)$, where m is chosen large enough to ensure that $0 :_F \gamma_1 = 0$. This is possible by Lemma 5. Next, applying Theorem 5 to the exact sequence
$$0 \to 0 :_E \gamma_1^m \to E \to F \to 0$$

and making use of Proposition 5, we see that

$$e_R(\gamma_1, \gamma_2, ..., \gamma_s | E) = e_R(\gamma_1, \gamma_2, ..., \gamma_s | F)$$

$$= e_R(\gamma_2, ..., \gamma_s | F/\gamma_1 F)$$

because $0 :_F \gamma_1 = 0$. However, by the inductive hypothesis,

$$e_R(\gamma_2, ..., \gamma_s | F/\gamma_1 F)$$

is non-negative and so the same is true of $e_R(\gamma_1, \gamma_2, ..., \gamma_s | E)$.

It remains for us to establish the second inequality in (7.4.13). If $s \geqslant 1$ we have

$$e_R(\gamma_1, \gamma_2, ..., \gamma_s|E) = e_R(\gamma_2, ..., \gamma_s|E/\gamma_1 E) - e_R(\gamma_2, ..., \gamma_s|0:_E\gamma_1)$$

whence $\quad e_R(\gamma_1, \gamma_2, ..., \gamma_s|E) \leqslant e_R(\gamma_2, ..., \gamma_s|E/\gamma_1 E) \qquad (7.4.14)$

because $e_R(\gamma_2, ..., \gamma_s|0:_E\gamma_1)$ is non-negative. In this inequality let us replace $\gamma_1, \gamma_2, ..., \gamma_s$ and E by $\gamma_2, \gamma_3, ..., \gamma_s$ and $E/\gamma_1 E = E'$ (say). This shows that

$$e_R(\gamma_2, \gamma_3, ..., \gamma_s|E/\gamma_1 E) \leqslant e_R(\gamma_3, ..., \gamma_s|E'/\gamma_2 E').$$

But, by (7.1.3), $E'/\gamma_2 E'$ is isomorphic to $E/(\gamma_1 E + \gamma_2 E)$. Hence

$$e_R(\gamma_1, \gamma_2, ..., \gamma_s|E) \leqslant e_R(\gamma_3, ..., \gamma_s|E/(\gamma_1 E + \gamma_2 E)).$$

Proceeding in this way we eventually arrive at the inequality

$$e_R(\gamma_1, \gamma_2, ..., \gamma_s|E) \leqslant e_R(\cdot |E/(\gamma_1 E + ... + \gamma_s E)).$$

In view of (7.4.1) this completes the proof.

Corollary. *If* $\gamma_1, \gamma_2, ..., \gamma_s$ *is a multiplicity system on a Noetherian R-module E and* $\gamma_1 E + \gamma_2 E + ... + \gamma_s E = E$, *then*

$$e_R(\gamma_1, ..., \gamma_s|E) = 0.$$

Theorem 7. *Let E be a Noetherian R-module and let* $\gamma_1, ..., \gamma_i, ..., \gamma_s$ *and* $\gamma_1, ..., \gamma_i', ..., \gamma_s$ *be multiplicity systems on E. Then* $\gamma_1, ..., \gamma_i\gamma_i', ..., \gamma_s$ *is also a multiplicity system on E and*

$$e_R(\gamma_1, ..., \gamma_i\gamma_i', ..., \gamma_s|E)$$
$$= e_R(\gamma_1, ..., \gamma_i, ..., \gamma_s|E) + e_R(\gamma_1, ..., \gamma_i', ..., \gamma_s|E). \quad (7.4.15)$$

Proof. Put $F = \gamma_1 E + ... + \gamma_i E + ... + \gamma_s E$. Then F is a submodule of E and therefore, by Proposition 3, $\gamma_1, ..., \gamma_i', ..., \gamma_s$ is a multiplicity system on F. Next

$$L_R\{E/(\gamma_1 F + ... + \gamma_i' F + ... + \gamma_s F)\}$$
$$= L_R\{E/F\} + L_R\{F/(\gamma_1 F + ... + \gamma_i' F + ... + \gamma_s F)\}$$

and, in this equation, the terms on the right-hand side are finite. Again

$$\gamma_1 F + ... + \gamma_i' F + ... + \gamma_s F \subseteq \gamma_1 E + ... + \gamma_i\gamma_i' E + ... + \gamma_s E \subseteq E$$

and so we see that

$$L_R\{E/(\gamma_1 E + ... + \gamma_i\gamma_i' E + ... + \gamma_s E)\} < \infty.$$

Accordingly $\gamma_1, ..., \gamma_i\gamma_i', ..., \gamma_s$ is a multiplicity system on E.

We now prove (7.4.15) and begin by observing that, in view of Proposition 4, we may suppose that $i = s$. But, by repeated applications of (7.4.2),

$$e_R(\gamma_1, ..., \gamma_{s-1}, \gamma_s|E) = \sum_\nu \epsilon_\nu e_R(\gamma_s|E_\nu).$$

Here the right-hand side is a finite sum in which each ϵ_ν has the value ± 1 and the E_ν are certain Noetherian R-modules on which γ_s is, by itself, a multiplicity system. Further the numbers ϵ_ν and the modules E_ν depend only on E and the elements $\gamma_1, \gamma_2, \ldots, \gamma_{s-1}$. It follows that

$$e_R(\gamma_1, \ldots, \gamma_{s-1}, \gamma_s'|E) = \sum_\nu \epsilon_\nu e_R(\gamma_s'|E_\nu)$$

and
$$e_R(\gamma_1, \ldots, \gamma_{s-1}, \gamma_s \gamma_s'|E) = \sum_\nu \epsilon_\nu e_R(\gamma_s \gamma_s'|E_\nu)$$

with the same ϵ_ν and E_ν as before. We see from this that it is only necessary to prove the theorem in the case $s = 1$. In what follows we assume we have this situation and, to simplify the notation, write γ and γ' in place of γ_s and γ_s'.

By definition, we have

$$e_R(\gamma|E) = L_R(E/\gamma E) - L_R(0:_E \gamma)$$

with similar expressions for $e_R(\gamma'|E)$ and $e_R(\gamma\gamma'|E)$. Next multiplication by γ produces an epimorphism $E \rightarrow \gamma E$ with the property that the inverse image of $\gamma\gamma' E$ is $(0:_E \gamma) + \gamma' E$. Accordingly we have an isomorphism
$$\gamma E/\gamma\gamma' E \approx E/((0:_E \gamma) + \gamma' E)$$
whence
$$L_R(\gamma E/\gamma\gamma' E) = L_R(E/\gamma' E) - L_R((0:_E \gamma + \gamma' E)/\gamma' E).$$

However $(0:_E \gamma + \gamma' E)/\gamma' E$ is isomorphic to

$$(0:_E \gamma)/(\gamma' E \cap (0:_E \gamma)) = (0:_E \gamma)/\gamma'(0:_E \gamma\gamma')$$
and so
$$L_R(\gamma E/\gamma\gamma' E) = L_R(E/\gamma' E) - L_R(0:_E \gamma) + L_R(\gamma'(0:_E \gamma\gamma'))$$

which yields

$$L_R(E/\gamma\gamma' E) = L_R(E/\gamma E) - L_R(0:_E \gamma) + L_R(E/\gamma' E) + L_R(\gamma'(0:_E \gamma\gamma'))$$

on adding $L_R(E/\gamma E)$ to both sides. To complete the proof it will now suffice to show that

$$L_R(\gamma'(0:_E \gamma\gamma')) = L_R(0:_E \gamma\gamma') - L_R(0:_E \gamma').$$

However this is clear because $\gamma'(0:_E \gamma\gamma')$ is isomorphic to

$$(0:_E \gamma\gamma')/(0:_E \gamma')$$

as may be seen from the epimorphism

$$(0:_E \gamma\gamma') \rightarrow \gamma'(0:_E \gamma\gamma')$$

produced by means of multiplication by γ'.

Corollary 1. *Let E be a Noetherian R-module, $\gamma_1, \gamma_2, ..., \gamma_s$ a multiplicity system on E, and $n_1, n_2, ..., n_s$ positive integers. Then*

$$\gamma_1^{n_1}, \gamma_2^{n_2}, ..., \gamma_s^{n_s}$$

is also a multiplicity system on E and

$$e_R(\gamma_1^{n_1}, \gamma_2^{n_2}, ..., \gamma_s^{n_s}|E) = n_1 n_2 ... n_s e_R(\gamma_1, \gamma_2, ..., \gamma_s|E).$$

This follows by repeated applications of the theorem.

Corollary 2. *Let E be a Noetherian R-module and $\gamma_1, \gamma_2, ..., \gamma_s$ a multiplicity system on E. Then*

$$0 \leqslant e_R(\gamma_1, ..., \gamma_s|E) \leqslant \frac{L_R\{E/(\gamma_1^{n_1}E + ... + \gamma_s^{n_s}E)\}}{n_1 n_2 ... n_s}$$

for arbitrary positive integers $n_1, n_2, ..., n_s$.

Proof. Since $n_1 n_2 ... n_s e_R(\gamma_1, \gamma_2, ..., \gamma_s|E)$ is equal to

$$e_R(\gamma_1^{n_1}, \gamma_2^{n_2}, ..., \gamma_s^{n_s}|E),$$

the desired result follows from Theorem 6.

Corollary 3. *Let E be a Noetherian R-module and $\gamma_1, \gamma_2, ..., \gamma_s$ a multiplicity system on E. Suppose that*

$$\gamma_i^m E \subseteq \gamma_1 E + ... + \gamma_{i-1}E + \gamma_{i+1}E + ... + \gamma_s E,$$

where m is a positive integer. Then $e_R(\gamma_1, ..., \gamma_s|E) = 0$.

Remark. It may be noted that this result considerably strengthens Proposition 5.

Proof. By Proposition 4, we may suppose that $i = 1$. If now $n > m$, then
$$\gamma_1^n E + \gamma_2 E + ... + \gamma_s E = \gamma_2 E + ... + \gamma_s E$$
and so, by Cor. 2,

$$0 \leqslant n e_R(\gamma_1, \gamma_2, ..., \gamma_s|E) \leqslant L_R\{E/(\gamma_2 E + ... + \gamma_s E)\} < \infty.$$

The desired result follows by dividing through by n and then letting n tend to infinity.

It is of considerably interest to know under what conditions the multiplicity $e_R(\gamma_1, ..., \gamma_s|E)$ and the length $L_R\{E/(\gamma_1 E + ... + \gamma_s E)\}$ are equal. We first give a sufficient condition for this to happen.

Theorem 8. *Let E be a Noetherian R-module and $\gamma_1, \gamma_2, ..., \gamma_s$ a multiplicity system on E. If now*

$$(\gamma_1 E + \gamma_2 E + ... + \gamma_i E):_E \gamma_{i+1} = \gamma_1 E + \gamma_2 E + ... + \gamma_i E$$

for $0 \leqslant i \leqslant s-1$, then

$$e_R(\gamma_1, ..., \gamma_s|E) = L_R\{E/(\gamma_1 E + ... + \gamma_s E)\}.$$

Proof. Put $E_i = E/(\gamma_1 E + \ldots + \gamma_i E)$. Then, by our hypothesis and (7.1.5), we have $0:_{E_i}\gamma_{i+1} = 0$ for $0 \leqslant i \leqslant s-1$. Also, by (7.1.3),

$$E_i/\gamma_{i+1}E_i \quad \text{and} \quad E_{i+1}$$

are isomorphic. Accordingly

$$e_R(\gamma_1, \gamma_2, \ldots, \gamma_s | E) = e_R(\gamma_2, \gamma_3, \ldots, \gamma_s | E_1)$$

$$= e_R(\gamma_3, \ldots, \gamma_s | E_2)$$

$$\cdots\cdots\cdots\cdots\cdots\cdots\cdots\cdots$$

$$= e_R(\cdot | E_s)$$

$$= L_R\{E/(\gamma_1 E + \ldots + \gamma_s E)\}$$

as required.

Our next result provides an important partial converse to Theorem 8.

Theorem 9. *Let E be a Noetherian R-module and $\gamma_1, \gamma_2, \ldots, \gamma_s$ a multiplicity system on E whose elements are contained in the Jacobson radical of R. Then the following two statements are equivalent:*

    (a) $e_R(\gamma_1, \gamma_2, \ldots, \gamma_s | E) = L_R\{E/(\gamma_1 E + \ldots + \gamma_s E)\}$;

    (b) *for $0 \leqslant i \leqslant s-1$, we have*

$$(\gamma_1 E + \ldots + \gamma_i E):_E \gamma_{i+1} = \gamma_1 E + \ldots + \gamma_i E.$$

Proof. By Theorem 8, we need only prove that (a) implies (b). For this we use induction on s.

When $s = 1$ we have

$$e_R(\gamma_1 | E) = L_R(E/\gamma_1 E) - L_R(0:_E\gamma_1)$$

and so, since we are assuming (a) to be true, $L_R(0:_E\gamma_1) = 0$. Accordingly $0:_E\gamma_1 = 0$ which is all we need to prove in this case.

From here on we assume that $s > 1$ and also that the assertion '(a) implies (b)' has been established in all situations where the multiplicity system has only $s-1$ elements. Let $n_1, n_2, \ldots, n_s$ be positive integers. Then, by Theorem 6, Proposition 2, and Theorem 7 Cor. 1,

$$e_R(\gamma_1^{n_1}, \ldots, \gamma_s^{n_s} | E) \leqslant L_R\{E/(\gamma_1^{n_1} E + \ldots + \gamma_s^{n_s} E)\}$$

$$\leqslant n_1 n_2 \ldots n_s L_R\{E/(\gamma_1 E + \ldots + \gamma_s E)\}$$

$$= n_1 n_2 \ldots n_s e_R(\gamma_1, \ldots, \gamma_s | E)$$

$$= e_R(\gamma_1^{n_1}, \ldots, \gamma_s^{n_s} | E).$$

Thus $\quad\quad e_R(\gamma_1^{n_1}, \ldots, \gamma_s^{n_s} | E) = L_R\{E/(\gamma_1^{n_1} E + \ldots + \gamma_s^{n_s} E)\}$ (7.4.16)

for arbitrary positive integers $n_1, n_2, \ldots, n_s$.

Put $K = E/(0:_E\gamma_1)$. It then follows from the exact sequence

$$0 \to 0:_E\gamma_1 \to E \to K \to 0$$

and Proposition 5, that

$$e_R(\gamma_1^{n_1}, ..., \gamma_s^{n_s}|E) = e_R(\gamma_1^{n_1}, ..., \gamma_s^{n_s}|K).$$

Hence, by (7.4.16) and Theorem 6,

$$L_R\{E/(\gamma_1^{n_1}E + ... + \gamma_s^{n_s}E)\} = e_R(\gamma_1^{n_1}, ..., \gamma_s^{n_s}|K)$$
$$\leqslant L_R\{K/(\gamma_1^{n_1}K + ... + \gamma_s^{n_s}K)\}$$
$$= L_R\{E/((0:_E\gamma_1) + \gamma_1^{n_1}E + ... + \gamma_s^{n_s}E)\}$$

because the modules

$$K/(\gamma_1^{n_1}K + ... + \gamma_s^{n_s}K) \quad \text{and} \quad E/((0:_E\gamma_1) + \gamma_1^{n_1}E + ... + \gamma_s^{n_s}E)$$

are isomorphic. But, since

$$\gamma_1^{n_1}E + ... + \gamma_s^{n_s}E \subseteq (0:_E\gamma_1) + \gamma_1^{n_1}E + ... + \gamma_s^{n_s}E \subseteq E,$$

it now follows that

$$\gamma_1^{n_1}E + ... + \gamma_s^{n_s}E = (0:_E\gamma_1) + \gamma_1^{n_1}E + ... + \gamma_s^{n_s}E$$

and therefore $\quad (0:_E\gamma_1) \subseteq \gamma_1^{n_1}E + \gamma_2^{n_2}E + ... + \gamma_s^{n_s}E$

for arbitrary positive integers $n_1, n_2, ..., n_s$.

Put $A = \gamma_1 R + \gamma_2 R + ... + \gamma_s R$. Then for each positive integer n we have
$$(0:_E\gamma_1) \subseteq \gamma_1^n E + \gamma_2^n E + ... + \gamma_s^n E \subseteq A^n E.$$

However, by Theorem 4, the intersection of all the

$$A^n E \quad (n = 1, 2, 3, ...)$$

is the zero submodule of E because A is a central ideal contained in the Jacobson radical of R. Accordingly $0:_E\gamma_1 = 0$.

Put $\bar{E} = E/\gamma_1 E$. Since $0:_E\gamma_1 = 0$ and we have an isomorphism

$$E/(\gamma_1 E + \gamma_2 E + ... + \gamma_s E) \approx \bar{E}/(\gamma_2\bar{E} + ... + \gamma_s\bar{E}),$$

it follows that

$$e_R(\gamma_2, ..., \gamma_s|\bar{E}) = e_R(\gamma_1, \gamma_2, ..., \gamma_s|E)$$
$$= L_R\{E/(\gamma_1 E + \gamma_2 E + ... + \gamma_s E)\}$$
$$= L_R\{\bar{E}/(\gamma_2\bar{E} + ... + \gamma_s\bar{E})\}.$$

It is now possible to apply the inductive hypothesis. This shows that

$$(\gamma_2\bar{E} + ... + \gamma_i\bar{E}):_E\gamma_{i+1} = \gamma_2\bar{E} + ... + \gamma_i\bar{E}$$

for $1 \leqslant i < s$. But $\gamma_2\bar{E} + ... + \gamma_i\bar{E} = (\gamma_1 E + \gamma_2 E + ... + \gamma_i E)/\gamma_1 E$. Accordingly, by (7.1.4),

$$(\gamma_1 E + \gamma_2 E + ... + \gamma_i E):_E\gamma_{i+1} = \gamma_1 E + \gamma_2 E + ... + \gamma_i E$$

not only for $i = 0$ but also for $1 \leqslant i < s$. This completes the proof.

7.5 The limit formula of Lech

Historically the modern theory of multiplicities began with the
study of the asymptotic behaviour of powers of ideals. This part of
the theory centres round two expressions for $e_R(\gamma_1, ..., \gamma_s | E)$ as a
limit. One of these limit formulae is due to P. Samuel and the other to
C. Lech. From our point of view it is easier to begin with the latter.

Consider first the case in which we have a Noetherian R-module E
and a central element γ which, by itself, forms a multiplicity system
on E. Then, because

$$0:_E \gamma \subseteq 0:_E \gamma^2 \subseteq 0_E \gamma^3 \subseteq \cdots$$

is an ascending sequence of submodules of E, there exists an integer m
such that $0:_E \gamma^n = 0:_E \gamma^m$ for all $n \geqslant m$. Accordingly when $n \geqslant m$,

$$e_R(\gamma^n | E) = L_R(E/\gamma^n E) - L_R(0:_E \gamma^n)$$
$$= L_R(E/\gamma^n E) - L_R(0:_E \gamma^m).$$

Now, by Theorem 7 Cor., $e_R(\gamma^n | E) = n e_R(\gamma | E)$. It follows that

$$L_R(E/\gamma^n E) = n e_R(\gamma | E) + C$$

for all $n \geqslant m$, where C is independent of n. In particular we see that

$$\operatorname*{Lim}_{n \to \infty} \frac{L_R(E/\gamma^n E)}{n} = e_R(\gamma | E). \tag{7.5.1}$$

This is the simplest case of Lech's formula. The general result is contained in the next theorem.

Theorem 10 *Let E be a Noetherian R-module and $\gamma_1, \gamma_2, ..., \gamma_s$ a multiplicity system on E. Then*

$$\operatorname*{Lim}_{\min(n_i) \to \infty} \frac{L_R\{E/(\gamma_1^{n_1} E + ... + \gamma_s^{n_s} E)\}}{n_1 n_2 ... n_s} = e_R(\gamma_1, ..., \gamma_s | E). \tag{7.5.2}$$

Proof. We use induction on s and begin by observing that the case $s = 1$
has already been dealt with in (7.5.1). It will therefore be supposed that
$s > 1$ and that the formula corresponding to (7.5.2) has been established for the case of multiplicity systems having $s - 1$ members. The
object of the first part of the proof is to show that, without loss of
generality, we may impose the extra condition that $0:_E \gamma_1$ is the zero
submodule of E. Once this has been done the argument proceeds
rapidly to its conclusion.

Put $F = E/(0:_E \gamma_1^p)$, where p is chosen sufficiently large to ensure
that $0:_F \gamma_1 = 0$. This is possible by Lemma 5. It follows, from the exact
sequence

$$0 \to 0:_E \gamma_1^p \to E \to F \to 0$$

and Proposition 5, that

$$e_R(\gamma_1, \gamma_2, ..., \gamma_s|E) = e_R(\gamma_1, \gamma_2, ..., \gamma_s|F). \tag{7.5.3}$$

Next, from the isomorphism

$$F/(\gamma_1^{n_1}F + ... + \gamma_s^{n_s}F) \approx E/((0:_E\gamma_1^p) + \gamma_1^{n_1}E + ... + \gamma_s^{n_s}E),$$

we deduce that

$$0 \leqslant L_R\{E/(\gamma_1^{n_1}E + ... + \gamma_s^{n_s}E)\} - L_R\{F/(\gamma_1^{n_1}F + ... + \gamma_s^{n_s}F)\}$$
$$= L_R\{((0:_E\gamma_1^p) + \gamma_1^{n_1}E + ... + \gamma_s^{n_s}E)/(\gamma_1^{n_1}E + ... + \gamma_s^{n_s}E)\}$$
$$= L_R\{(0:_E\gamma_1^p)/((0:_E\gamma_1^p) \cap (\gamma_1^{n_1}E + ... + \gamma_s^{n_s}E))\}$$

by one of our standard isomorphism theorems. Again

$$(0:_E\gamma_1^p) \cap (\gamma_1^{n_1}E + ... + \gamma_s^{n_s}E) \supseteq \gamma_2^{n_2}(0:_E\gamma_1^p) + ... + \gamma_s^{n_s}(0:_E\gamma_1^p)$$

and so, by Proposition 2,

$$L_R\{(0:_E\gamma_1^p)/((0:_E\gamma_1^p) \cap (\gamma_1^{n_1}E + ... + \gamma_s^{n_s}E))\}$$
$$\leqslant n_2 n_3 ... n_s L_R\{(0:_E\gamma_1^p)/(\gamma_2(0:_E\gamma_1^p) + ... + \gamma_s(0:_E\gamma_1^p))\}$$
$$= n_2 n_3 ... n_s C$$

say. Now C *is finite*. To see this it is sufficient to show that $\gamma_2, \gamma_3, ..., \gamma_s$ is a multiplicity system on $0:_E\gamma_1^p$. However this is clear because $\gamma_1^p, \gamma_2, ..., \gamma_s$ is certainly a multiplicity system on the module in question and $\gamma_1^p(0:_E\gamma_1^p) = 0$.

These various observations show that we have inequalities

$$0 \leqslant \frac{L_R\{E/(\gamma_1^{n_1}E + ... + \gamma_s^{n_s}E)\} - L_R\{F/(\gamma_1^{n_1}F + ... + \gamma_s^{n_s}F)\}}{n_1 n_2 ... n_s} \leqslant \frac{C}{n_1}$$

and so, by (7.5.3), it is enough to show that

$$\underset{\min(n_i) \to \infty}{\text{Lim}} \frac{L_R\{F/(\gamma_1^{n_1}F + ... + \gamma_s^{n_s}F)\}}{n_1 n_2 ... n_s} = e_R(\gamma_1, ..., \gamma_s|F).$$

In view of this it is permissible to assume, for the remainder of the proof, that the additional condition $0:_E\gamma_1 = 0$ is satisfied.

Put $\bar{E} = E/\gamma_1 E$. Then, by Theorem 7 Cor. 2 and Proposition 2,

$$0 \leqslant n_1 n_2 ... n_s e_R(\gamma_1, \gamma_2, ..., \gamma_s|E) \leqslant L_R\{E/(\gamma_1^{n_1}E + ... + \gamma_s^{n_s}E)\}$$
$$\leqslant n_1 L_R\{E/(\gamma_1 E + \gamma_2^{n_2}E + ... + \gamma_s^{n_s}E)\}$$
$$= n_1 L_R\{\bar{E}/(\gamma_2^{n_2}\bar{E} + ... + \gamma_s^{n_s}\bar{E})\}$$

because $E/(\gamma_1 E + \gamma_2^{n_2}E + ... + \gamma_s^{n_s}E)$ and $\bar{E}/(\gamma_2^{n_2}\bar{E} + ... + \gamma_s^{n_s}\bar{E})$ are isomorphic. Accordingly

$$\frac{L_R\{E/(\gamma_1^{n_1}E + ... + \gamma_s^{n_s}E)\}}{n_1 n_2 ... n_s}$$

lies between $e_R(\gamma_1, \ldots, \gamma_s | E)$ and

$$\frac{L_R\{\bar{E}/(\gamma_2^{n_2}\bar{E} + \ldots + \gamma_s^{n_s}\bar{E})\}}{n_2 n_3 \ldots n_s}.$$

However our inductive hypothesis allows us to conclude that this latter expression tends to $e_R(\gamma_2, \ldots, \gamma_s | \bar{E})$ and this is equal to

$$e_R(\gamma_1, \gamma_2, \ldots, \gamma_s | E)$$

because we now have $0:_E \gamma_1 = 0$. The theorem follows.

7.6 Hilbert functions

It was stated earlier that two expressions for $e_R(\gamma_1, \ldots, \gamma_s | E)$ as a limit had played important roles in the historical development of our subject. One of these limit formulae has now been derived. The second has close associations with the arithmetical theory of graded modules. This is a fascinating topic in its own right quite apart from its applications to multiplicity theory. In this section we shall establish some of the basic facts of this theory in order to apply them later.

The notion of a *graded ring* was introduced in section (2.11). On that occasion we had agreed to restrict our attention to commutative rings but, as the reader will easily verify, the extension of the *definition* to the non-commutative case presents no difficulty. Also when one has a graded (but not necessarily commutative) ring, it is possible to define the notion of a *graded module* over the ring exactly as before.

The graded rings which concern us here are polynomial rings. To be quite explicit, let R be a non-null ring and let us form the polynomial ring $R[X_1, X_2, \ldots, X_s]$ in the indeterminates $X_1, X_2, \ldots, X_s$, where $s \geqslant 0$. It is understood that the X_i are to commute with one another so that, with a self explanatory notation,

$$(rX_1^{\mu_1} X_2^{\mu_2} \ldots X_s^{\mu_s})(\rho X_1^{\nu_1} X_2^{\nu_2} \ldots X_s^{\nu_s}) = (r\rho) X_1^{\mu_1+\nu_1} X_2^{\mu_2+\nu_2} \ldots X_s^{\mu_s+\nu_s}.$$

Thus $X_1, X_2, \ldots, X_s$ all belong to the *centre* of $R[X_1, X_2, \ldots, X_s]$. We shall regard $R[X_1, X_2, \ldots, X_s]$ as being graded, in the usual way, by the non-negative integers. Accordingly if $m \geqslant 0$ is an integer, then a polynomial is *homogeneous of degree* m if it can be expressed in the form

$$\sum_{\mu_1 + \ldots + \mu_s = m} r_{\mu_1 \mu_2 \ldots \mu_s} X_1^{\mu_1} X_2^{\mu_2} \ldots X_s^{\mu_s}.$$

It will often be convenient to use $R[X]$ as an abbreviation for

$$R[X_1, X_2, \ldots, X_s].$$

When $s = 0$, $R[X]$ is just R itself and the grading on R is the trivial one, i.e. all non-zero elements are homogeneous of degree zero.

Suppose next that M is a graded $R[X]$-module. Then, for each $n \geqslant 0$, the homogeneous elements of M of degree n form a subgroup M_n of the additive group of M and we have

$$M = M_0 \oplus M_1 \oplus \dots \oplus M_n \oplus \dots.$$

Furthermore if an element of M_n is multiplied by a homogeneous polynomial of degree m, then the result belongs to M_{m+n}. Now the homogeneous polynomials of degree zero form a subring of $R[X]$ which we naturally identify with R. On this understanding, each M_n is an R-module. Put

$$H(n, M) = L_R(M_n). \tag{7.6.1}$$

Then $H(n, M)$ is a function of n for $n = 0, 1, 2, \dots$ and its value is always a non-negative integer or 'plus infinity'.

Definition. *The arithmetical function $H(n, M)$ is called the 'Hilbert function' of the graded $R[X]$-module M.*

From the point of view of multiplicity theory, it is not $H(n, M)$ that interests us most but the related function

$$H^*(n, M) = H(0, M) + H(1, M) + \dots + H(n, M). \tag{7.6.2}$$

We shall refer to $H^*(n, M)$ as the *cumulative Hilbert function of M*.

Before beginning the study of these functions, we recall certain fundamental notions belonging to the theory of graded modules.[†] To this end let M be a graded $R[X]$-module and N an $R[X]$-submodule of M. It will be remembered that N is called a *homogeneous submodule* if it can be generated by homogeneous elements. (An equivalent definition is to say that whenever u belongs to N then all the homogeneous components of u also belong to N.) Suppose that we have this situation. Then the grading on M induces a grading on N, and M/N can also be regarded as a graded $R[X]$-module in a natural way. These gradings are such that in the usual exact sequence

$$0 \to N \to M \to M/N \to 0 \tag{7.6.3}$$

the mappings preserve degrees, that is to say a homogeneous element is always mapped into a homogeneous element of the *same degree*. It follows that for each integer $n \geqslant 0$, (7.6.3) gives rise to an exact sequence

$$0 \to N_n \to M_n \to (M/N)_n \to 0$$

† These matters are fully discussed in section (2.11).

of R-modules. Accordingly

$$L_R(M_n) = L_R(N_n) + L_R((M/N)_n),$$

that is to say

$$H(n, M) = H(n, N) + H(n, M/N). \qquad (7.6.4)$$

More generally, if $0 \to N \to M \to K \to 0$ is an exact sequence of graded $R[X]$-modules in which the mappings preserve degrees, then

$$H(n, M) = H(n, N) + H(n, K) \qquad (7.6.5)$$

for all $n \geqslant 0$ and therefore

$$H^*(n, M) = H^*(n, N) + H^*(n, K) \qquad (7.6.6)$$

for all $n \geqslant 0$ as well.

It will be realized that the Hilbert function of a graded module is only likely to be interesting if the values which it takes are finite. In fact the situation which is most rich in useful consequences is that in which the module is both finitely generated and has a Hilbert function which is never infinite. Accordingly we make the following

Definition. *A graded $R[X]$-module M will be called a 'Hilbert $R[X]$-module' if M is finitely generated and $H(n, M) < \infty$ for all $n \geqslant 0$.*

Let M be a Hilbert $R[X]$-module and suppose that u is an element of M which is homogeneous of degree p. Then $Ru \subseteq M_p$ and therefore

$$L_R(Ru) \leqslant L_R(M_p) = H(p, M) < \infty.$$

Let I be the kernel of the R-epimorphism $R \to Ru$ in which $r \to ru$. Then R/I is isomorphic to Ru and therefore $L_R(R/I) < \infty$. In particular we see that R/I is a Noetherian R-module. It follows, by Theorem 1 of section (4.1), that† the $R[X]$-module $(R/I)[X]$, which consists of all polynomials in $X_1, X_2, ..., X_s$ with coefficients in the module R/I, is also Noetherian.

If $\phi(X)$ belongs to $R[X]$, let $\overline{\phi}(X)$ denote the element of $(R/I)[X]$ obtained by applying the natural mapping $R \to R/I$ to the coefficients of $\phi(X)$. It is easy to check that there exists a well defined $R[X]$-epimorphism $(R/I)[X] \to R[X]u$ in which $\overline{\phi}(X) \to \phi(X)u$. Since $(R/I)[X]$ is a Noetherian $R[X]$-module, it follows that $R[X]u$ *is a Noetherian $R[X]$-module* as well. Put $U = R[X]u$. Then

$$U = U_p \oplus U_{p+1} \oplus U_{p+2} \oplus ...,$$

where $U_p = Ru$, and

$$X_1 U + X_2 U + ... + X_s U = U_{p+1} \oplus U_{p+2} \oplus ....$$

---

† The reader will remember that the theorem quoted holds for non-commutative as well as for commutative rings.

This shows that

$$L_R\{U/(X_1 U + ... + X_s U)\} = L_R\{U_p\} = L_R\{Ru\} < \infty$$

and so $X_1, X_2, ..., X_s$ is a multiplicity system on $R[X]u$.

We are now ready to prove

Proposition 6. *Let M be a Hilbert $R[X]$-module. Then M is a Noetherian $R[X]$-module and $X_1, X_2, ..., X_s$ is a multiplicity system on M.*

Proof. Since M is a Hilbert $R[X]$-module, it is finitely generated. Let

$$M = R[X]u_1 + R[X]u_2 + ... + R[X]u_q$$

then it is clear that we can arrange for each u_i to be homogeneous. From the remarks made above, it now follows that, for $1 \leqslant i \leqslant q$, $R[X]u_i$ is a Noetherian $R[X]$-module. Accordingly M is a Noetherian $R[X]$-module by Proposition 1 of section (4.1). We also know, from the previous discussion, that $X_1, X_2, ..., X_s$ is a multiplicity system on each of the submodules $R[X]u_1, R[X]u_2, ..., R[X]u_q$. It therefore follows, from Theorem 5 Cor. 2, that $X_1, X_2, ..., X_s$ is a multiplicity system on M. This establishes the proposition.

Corollary. *Let N be a homogeneous submodule of a Hilbert $R[X]$-module M. Then N and M/N, when endowed with the usual gradings, are also Hilbert $R[X]$-modules.*

Proof. The proposition shows that M is a Noetherian $R[X]$-module. Consequently both N and M/N are finitely generated. Next, since $H(n, M) < \infty$, it follows, from (7.6.4), that $H(n, N)$ and $H(n, M/N)$ are also finite for all n. This completes the proof.

The following lemma tells us all about the Hilbert function of a Hilbert module in the case where the number of variables is zero. This provides us with a starting point for certain arguments which use induction on the number of variables.

Lemma 6. *Let M be a Hilbert $R[X_1, ..., X_s]$-module and suppose that $s = 0$. Then $H(n, M) = 0$ for all large values of n. Furthermore $L_R(M)$ is finite and $H^*(n, M) = L_R(M)$ for all large values of n.*

Proof. Since M is finitely generated and $s = 0$, there exist homogeneous elements $u_1, u_2, ..., u_q$ such that $M = Ru_1 + Ru_2 + ... + Ru_q$. Let the degree of u_i be n_i. Then $Ru_i \subseteq M_{n_i}$ and $L_R(Ru_i) \leqslant H(n_i, M)$ which is finite. It follows that $L_R(M) < \infty$. But

$$M = M_0 \oplus M_1 \oplus M_2 \oplus ...$$

this being a direct sum of R-modules. Consequently, since M has finite length, $L_R(M_n) = 0$ when n is sufficiently large, say when $n > p$. Accordingly $M_n = 0$ when $n > p$ and so

$$M = M_0 \oplus M_1 \oplus \ldots \oplus M_p.$$

It follows that

$$\begin{aligned} H^*(n, M) &= L_R(M_0) + L_R(M_1) + \ldots + L_R(M_n) \\ &= L_R(M_0) + L_R(M_1) + \ldots + L_R(M_p) \\ &= L_R(M) \end{aligned}$$

provided that $n > p$. This completes the proof.

Theorem 11. *Let M be a Hilbert $R[X_1, X_2, \ldots, X_s]$-module. Then for all large values of n, $H(n, M)$ is given by a relation of the form*

$$H(n, M) = \sum_{\nu=0}^{s-1} c_\nu \binom{n+\nu}{\nu}, \tag{7.6.7}$$

where $c_0, c_1, \ldots, c_{s-1}$ are integers which are independent of n.

Remarks. This is the key result concerning Hilbert functions. Before proceeding to the proof we make a few general observations. First $\binom{n+\nu}{\nu}$ denotes the usual binomial coefficient. It is given explicitly by

$$\binom{n+\nu}{\nu} = \frac{(n+\nu)(n+\nu-1)\ldots(n+1)}{1.2.3\ldots\nu}$$

so, for a *fixed* value of ν ($\nu \geqslant 0$), $\binom{n+\nu}{\nu}$ is a polynomial in n of degree ν whose leading term is $n^\nu/\nu!$. The theorem therefore shows that, for large values of n,

$$H(n, M) = c_{s-1} \frac{n^{s-1}}{(s-1)!} + \ldots$$

where, in this equation, the terms indicated by $+\ldots$ constitute a polynomial in n whose degree is smaller than $s-1$.

The next point to which attention should be drawn is that the representation (7.6.7) is necessarily *unique*. To see this we first observe that if $f(t)$ is a polynomial (with complex coefficients) in an indeterminate t and $f(n) = 0$ whenever n is a sufficiently large positive integer, then $f(t)$ is a null polynomial. This is because a non-null polynomial has at most a finite number of roots. Now suppose that

$$c_0', c_1', \ldots, c_q' \quad \text{and} \quad c_0'', c_1'', \ldots, c_q''$$

are two sequences and each consisting of $q+1$ complex numbers, and suppose further that

$$\sum_{\nu=0}^{q} c_\nu' \binom{n+\nu}{\nu} = \sum_{\nu=0}^{q} c_\nu'' \binom{n+\nu}{\nu} \tag{7.6.8}$$

for all sufficiently large positive integers n. We can then rewrite each side as a polynomial in n. However, if we do this, then the remarks just made show that the two polynomials so obtained will have identical coefficients. This in turn implies that $c_i' = c_i''$ for $i = 0, 1, 2, \ldots, q$. Thus (7.6.8) can only hold for all large n if the sequences $c_0', c_1', \ldots, c_q'$ and $c_0'', c_1'', \ldots, c_q''$ are identical. This shows that, provided a representation of the form (7.6.7) is possible, it must be unique.

For the proof of Theorem 11, we require two well-known properties of binomial coefficients. The first of these is the relation

$$\binom{n+\nu}{\nu} - \binom{n+\nu-1}{\nu-1} = \binom{n-1+\nu}{\nu} \qquad (7.6.9)$$

which is valid provided that $n \geqslant 1$ and $\nu \geqslant 1$. The second is

$$\sum_{k=0}^{n} \binom{k+\nu}{\nu} = \binom{n+\nu+1}{\nu+1}. \qquad (7.6.10)$$

This holds for $n \geqslant 0$ and $\nu \geqslant 0$ and may be established by applying the binomial theorem to the identity

$$\frac{t}{1-t} + \frac{t}{(1-t)^2} + \cdots + \frac{t}{(1-t)^{n+1}} \equiv \frac{1}{(1-t)^{n+1}} - 1.$$

Proof of Theorem 11. The argument uses induction on the number s of indeterminates and we begin by observing that Lemma 6 shows that the theorem is true when $s = 0$ provided that we make the natural convention concerning empty sums. Accordingly from now on it will be assumed that $s \geqslant 1$ and that the result in question has been established for Hilbert modules over polynomials rings in $s-1$ indeterminates.

Consider the exact sequence

$$0 \to 0\!:_M X_s \to M \to M \to M/X_s M \to 0, \qquad (7.6.11)$$

where the mapping $M \to M$ consists in multiplying the elements of M by X_s. It is clear that $0\!:_M X_s$ and $X_s M$ are homogeneous submodules of M and therefore $0\!:_M X_s$ and $M/X_s M$ have natural gradings derived from the given grading on M. On this understanding both $(0\!:_M X_s) \to M$ and $M \to M/X_s M$ preserve degrees when applied to homogeneous elements. By contrast, the mapping $M \to M$ raises the degree of a homogeneous element by unity because we must multiply by X_s. Hence for each $n \geqslant 0$ we have an exact sequence

$$0 \to (0\!:_M X_s)_{n-1} \to M_{n-1} \to M_n \to (M/X_s M)_n \to 0$$

of R-modules provided that we interpret $(0:_M X_s)_{n-1}$ and M_{n-1} as being null modules when $n = 0$. Further all these R-modules have finite length. Consequently

$$L_R\{(M/X_s M)_n\} - L_R\{(0:_M X_s)_{n-1}\} = H(n, M) - H(n-1, M) \quad (7.6.12)$$

on the understanding that $H(n-1, M)$ is taken to be zero when $n = 0$.

By Proposition 6 Cor., Both $M/X_s M$ and $0:_M X_s$ are Hilbert $R[X_1, X_2, ..., X_s]$-modules. However both are annihilated by X_s. It follows that if we regard them as $R[X_1, X_2, ..., X_{s-1}]$-modules and keep the gradings unchanged, then they will both be Hilbert modules over the smaller ring which, we note, is a polynomial ring in only $s-1$ variables. Accordingly, by our assumptions, we have

$$L_R\{(M/X_s M)_n\} = \sum_{\nu=0}^{s-2} a_\nu \binom{n+\nu}{\nu} \quad (n \text{ large})$$

and
$$L_R\{(0:_M X_s)_n\} = \sum_{\nu=0}^{s-2} b_\nu \binom{n+\nu}{\nu} \quad (n \text{ large}),$$

where the a_ν and b_ν are integers which are independent of n. Next, using (7.6.9), we obtain

$$L_R\{(0:_M X_s)_{n-1}\} = b_0 + \sum_{\nu=1}^{s-2} b_\nu \left[\binom{n+\nu}{\nu} - \binom{n+\nu-1}{\nu-1} \right]$$

for large values of n. If we now substitute for $L_R\{(M/X_s M)_n\}$ and $L_R\{(0:_M X_s)_{n-1}\}$ in (7.6.12), we find that

$$H(n, M) - H(n-1, M) = \sum_{\nu=0}^{s-2} d_\nu \binom{n+\nu}{\nu} \quad (n \text{ large})$$

where $d_0, d_1, ..., d_{s-2}$ are certain integers which do not depend on n. Accordingly we may write, for all values of k ($k \geqslant 0$),

$$H(k, M) - H(k-1, M) = \sum_{\nu=0}^{s-2} d_\nu \binom{k+\nu}{\nu} + w_k, \quad (7.6.13)$$

where the w_k ($k = 0, 1, 2, ...$) are integers which are zero from some point onwards, say $w_k = 0$ when $k \geqslant p$. Finally suppose that $n \geqslant p$ and sum (7.6.13) over the range $k = 0, 1, 2, ..., n$. This yields

$$\begin{aligned} H(n, M) &= \sum_{\nu=0}^{s-2} d_\nu \sum_{k=0}^{n} \binom{k+\nu}{\nu} + \sum_{k=0}^{\infty} w_k \\ &= \sum_{\nu=0}^{s-2} d_\nu \binom{n+\nu+1}{\nu+1} + \sum_{k=0}^{\infty} w_k \end{aligned}$$

by (7.6.10). The proof is now complete.

Let us review this result. It tells us that if M is a Hilbert $R[X_1, ..., X_s]$-module, then there exists unique integers

$$h_0(M), h_1(M), ..., h_{s-1}(M)$$

such that
$$H(n, M) = \sum_{\nu=0}^{s-1} h_\nu(M) \binom{n+\nu}{\nu} \qquad (7.6.14)$$

for all large values of n. We shall call the integers $h_\nu(M)$ the *Hilbert coefficients* of M. Next (7.6.14) shows that

$$H(k, M) = \sum_{\nu=0}^{s-1} h_\nu(M) \binom{k+\nu}{\nu} + \mu_k \qquad (7.6.15)$$

for all $k \geqslant 0$, where $\mu_0, \mu_1, \mu_2, ...$ is a sequence of integers with the property that $\mu_k = 0$ once k is sufficiently large. Suppose that $\mu_k = 0$ when $k > q$. If now $n > q$ and we sum (7.6.15) for $k = 0, 1, ..., n$, then we find that

$$H^*(n, M) = \sum_{\nu=0}^{s-1} h_\nu(M) \sum_{k=0}^{n} \binom{k+\nu}{\nu} + \sum_{k=0}^{\infty} \mu_k$$

$$= \sum_{k=0}^{\infty} \mu_k + \sum_{\nu=0}^{s-1} h_\nu(M) \binom{n+\nu+1}{\nu+1}$$

by virtue of (7.6.10). It follows that there exist unique integers $h_0^*(M), h_1^*(M), ..., h_s^*(M)$ such that

$$H^*(n, M) = \sum_{\nu=0}^{s} h_\nu^*(M) \binom{n+\nu}{\nu} \qquad (7.6.16)$$

provided that n is sufficiently large. Furthermore

$$h_0(M), h_1(M), ..., h_{s-1}(M)$$

are the same as $h_1^*(M), h_2^*(M), ..., h_s^*(M)$. In particular, when $s \geqslant 1$ we have
$$h_s^*(M) = h_{s-1}(M). \qquad (7.6.17)$$

Let us assume that $s \geqslant 1$. Then (7.6.12) shows that

$$H(m, M/X_s M) - H(m-1, 0:_M X_s) = H(m, M) - H(m-1, M)$$

this being true for all $m \geqslant 0$. If therefore we sum over the range $m = 0, 1, ..., n$, we find that

$$H^*(n, M/X_s M) - H^*(n-1, 0:_M X_s) = H(n, M) \qquad (7.6.18)$$

for every n. But when n is large each of these functions is expressible as a polynomial in n. In fact for all large n

$$H^*(n, M/X_s M) = h_{s-1}^*(M/X_s M) \frac{n^{s-1}}{(s-1)!} + ..., \qquad (7.6.19)$$

$$H^*(n, 0:_M X_s) = h^*_{s-1}(0:_M X_s) \frac{n^{s-1}}{(s-1)!} + \ldots \qquad (7.6.20)$$

and
$$H(n, M) = h_{s-1}(M) \frac{n^{s-1}}{(s-1)!} + \ldots, \qquad (7.6.21)$$

where in each case $+\ldots$ denotes a polynomial in n whose degree is smaller than $s-1$. From (7.6.20) we obtain

$$H^*(n-1, 0:_M X_s) = h^*_{s-1}(0:_M X_s) \frac{n^{s-1}}{(s-1)!} + \ldots \qquad (7.6.22)$$

and (7.6.21) can be rewritten as

$$H(n, M) = h^*_s(M) \frac{n^{s-1}}{(s-1)!} + \ldots \qquad (7.6.23)$$

because $h_{s-1}(M) = h^*_s(M)$. Let us use (7.6.19), (7.6.22) and (7.6.23) to substitute in (7.6.18). Then, comparing terms of degree $s-1$, we obtain

$$h^*_s(M) = h^*_{s-1}(M/X_s M) - h^*_{s-1}(0:_M X_s). \qquad (7.6.24)$$

This holds provided that $s \geqslant 1$.

Theorem 12. *Let M be a Hilbert $R[X_1, \ldots, X_s]$-module Then*
$$h^*_s(M) = e_{R[X]}(X_1, \ldots, X_s | M).$$

Proof. We use induction on s. If $s = 0$, then $R[X] = R$ and
$$e_{R[X]}(X_1, \ldots, X_s | M)$$

is just $L_R(M)$. On the other hand, Lemma 6 shows that $h^*_s(M)$ is also equal to $L_R(M)$ in this particular case.

From now on we assume that $s \geqslant 1$ and also that the result in question has been proved when only $s-1$ indeterminates are involved. By Proposition 6, $e_{R[X]}(X_1, \ldots, X_s | M)$ is defined. Now

$$e_{R[X]}(X_1, \ldots, X_{s-1}, X_s | M)$$
$$= e_{R[X]}(X_1, \ldots, X_{s-1} | M/X_s M) - e_{R[X]}(X_1, \ldots, X_{s-1} | 0:_M X_s).$$

Further, the residue class ring $R[X]/(X_s)$ is naturally isomorphic to the polynomial ring $R[X_1, \ldots, X_{s-1}] = \Lambda$ (say) and the isomorphism is such that, for $1 \leqslant i \leqslant s-1$, the image of X_i in $R[X]/(X_s)$ corresponds to X_i considered as an element of Λ. In addition, the ideal (X_s) annihilates $M/X_s M$. We can therefore apply (7.4.3) which shows that

$$e_{R[X]}(X_1, \ldots, X_{s-1} | M/X_s M) = e_\Lambda(X_1, \ldots, X_{s-1} | M/X_s M).$$

But $M/X_s M$ is a Hilbert module with respect to $R[X_1, \ldots, X_{s-1}] = \Lambda$. Accordingly, by the inductive assumption,

$$e_\Lambda(X_1, \ldots, X_{s-1} | M/X_s M) = h^*_{s-1}(M/X_s M)$$

and therefore
$$e_{R[X]}(X_1, \ldots, X_{s-1} | M/X_s M) = h_{s-1}^*(M/X_s M).$$
Similarly we can show that
$$e_{R[X]}(X_1, \ldots, X_{s-1} | 0 :_M X_s) = h_{s-1}^*(0 :_M X_s).$$
Thus
$$e_{R[X]}(X_1, \ldots, X_s | M) = h_{s-1}^*(M/X_s M) - h_{s-1}^*(0 :_M X_s)$$
$$= h_s^*(M)$$
by (7.6.24). The proof is now complete.

7.7 The limit formula of Samuel

The results of the last section will now be applied to extend our knowledge of the properties of general multiplicities. To this end let E be a Noetherian R-module and $\gamma_1, \gamma_2, \ldots, \gamma_s$ a multiplicity system on E. We put
$$A = \gamma_1 R + \gamma_2 R + \ldots + \gamma_s R,$$
so that A is a finitely generated central ideal, and we use $X_1, X_2, \ldots, X_s$ to denote indeterminates. Note that there is one indeterminate for each element of the multiplicity system.

If $n \geqslant 0$ is an integer, then $\gamma_1^n E + \gamma_2^n E + \ldots + \gamma_s^n E$ is contained in $A^n E$. Accordingly $L_R(E/A^n E) < \infty$ for all n. Put
$$M = (E/AE) \oplus (AE/A^2E) \oplus (A^2E/A^3E) \oplus \ldots. \qquad (7.7.1)$$
We claim that M can be given the structure of a graded $R[X_1, \ldots, X_s]$-module in which the group of homogeneous elements of degree n is just $A^n E/A^{n+1}E$. To explain this structure let an element η of $A^n E/A^{n+1}E$ be represented by the element y of $A^n E$ and let
$$\phi(X) = \sum_{\mu_1 + \ldots + \mu_s = m} r_{\mu_1 \mu_2 \ldots \mu_s} X_1^{\mu_1} X_2^{\mu_2} \ldots X_s^{\mu_s}$$
be a homogeneous element of $R[X_1, \ldots, X_s]$ of degree m. Then $\phi(X)\eta$ is that element of $A^{m+n}E/A^{m+n+1}E$ which is represented by
$$\sum_{\mu_1 + \ldots + \mu_s = m} r_{\mu_1 \mu_2 \ldots \mu_s} \gamma_1^{\mu_1} \gamma_2^{\mu_2} \ldots \gamma_s^{\mu_s} y.$$
We leave the reader to check that, in this way, M can be made into a graded $R[X]$-module with the prescribed homogeneous elements.

Since E is a Noetherian R-module, it is finitely generated. Let $E = Re_1 + Re_2 + \ldots + Re_q$ and denote by $\bar{e}_i$ the natural image of e_i in E/AE so that $\bar{e}_i$ is a homogeneous element of M of degree zero. Then
$$M = R[X]\bar{e}_1 + R[X]\bar{e}_2 + \ldots + R[X]\bar{e}_q,$$

where, as usual, $R[X]$ is used as an abbreviation $R[X_1, ..., X_s]$. Now

$$L_R(M_n) = L_R(A^n E / A^{n+1} E) \qquad (7.7.2)$$

which is finite. Accordingly *when M is regarded as a graded $R[X]$-module, in the manner described above, it becomes a Hilbert $R[X]$-module.* Moreover, by (7.7.2),

$$L_R(E/A^n E) = L_R(M_0) + L_R(M_1) + ... + L_R(M_{n-1}),$$

that is to say $\qquad L_R(E/A^n E) = H^*(n-1, M). \qquad (7.7.3)$

Hence, by (7.6.16), once n has become sufficiently large, $L_R(E/A^n E)$ is equal to a polynomial in n of degree not exceeding s. Indeed (7.6.16) shows that, for large n,

$$L_R(E/A^n E) = h_s^*(M) \frac{n^s}{s!} + ..., \qquad (7.7.4)$$

where this time $+ ...$ denotes a polynomial in n whose degree is smaller than s. Put

$$e(s, A, E) = \operatorname*{Lim}_{n \to \infty} \frac{L_R(E/A^n E)}{n^s/s!} = h_s^*(M). \qquad (7.7.5)$$

The important fact, to be proved shortly, is that $e(s, A, E)$ is actually equal to $e_R(\gamma_1, ..., \gamma_s | E)$. Note that when $s = 0$ we obtain

$$e(0, (0), E) = L_R(E) \qquad (7.7.6)$$

directly from the definition so the assertion is certainly true in this case.

Now suppose that $s \geqslant 1$ and put $\bar{E} = E/\gamma_1 E$, $\bar{A} = \gamma_2 R + ... + \gamma_s R$. Then $\bar{E}$ is a Noetherian R-module and it admits $\gamma_2, \gamma_3, ..., \gamma_s$ as a multiplicity system. Also, because γ_1 annihilates $\bar{E}$, $(\bar{A})^n \bar{E} = A^n \bar{E}$ and therefore $\bar{E}/(\bar{A})^n \bar{E}$ is isomorphic to $E/(\gamma_1 E + A^n E)$. Accordingly

$$L_R\{\bar{E}/(\bar{A})^n \bar{E}\} = L_R\{E/A^n E\} - L_R\{(\gamma_1 E + A^n E)/A^n E\}$$
$$= L_R\{E/A^n E\} - L_R\{\gamma_1 E/(\gamma_1 E \cap A^n E)\}$$
$$= L_R\{E/A^n E\} - L_R\{\gamma_1 E/\gamma_1(A^n E:_E \gamma_1)\}$$

by (7.1.2). However, the epimorphism $E \to \gamma_1 E$ produced by multiplication by γ_1 is such that the inverse image of $\gamma_1(A^n E:_E \gamma_1)$ is just $A^n E:_E \gamma_1$. Thus $\gamma_1 E/\gamma_1(A^n E:_E \gamma_1)$ is isomorphic to $E/(A^n E:_E \gamma_1)$ and so

$$L_R\{\bar{E}/(\bar{A})^n \bar{E}\} = L_R\{E/A^n E\} - L_R\{E/(A^n E:_E \gamma_1)\}$$
$$\geqslant L_R\{E/A^n E\} - L_R\{E/A^{n-1} E\}$$

because $A^{n-1} E \subseteq (A^n E:_E \gamma_1)$. Next, by (7.7.4) and (7.7.5),

$$L_R\{E/A^n E\} = e(s, A, E) \frac{n^s}{s!} + ... \qquad (7.7.7)$$

for large n, where $+\ldots$ has its usual significance. It follows that, when n is large,

$$L_R\{E/A^nE\} - L_R\{E/A^{n-1}E\} = e(s, A, E)\frac{n^{s-1}}{(s-1)!} + \ldots$$

and therefore

$$L_R\{\bar{E}/(\bar{A})^n\bar{E}\} \geqslant e(s, A, E)\frac{n^{s-1}}{(s-1)!} + \ldots.$$

Let us divide through by $n^{s-1}/(s-1)!$ and afterwards let n tend to infinity. This yields $\quad e(s-1, \bar{A}, \bar{E}) \geqslant e(s, A, E)$.

If it should happen that $s \geqslant 2$, then the above argument can be repeated. In this way we obtain

$$e(s-2, \bar{\bar{A}}, \bar{\bar{E}}) \geqslant e(s-1, \bar{A}, \bar{E}),$$

where $\bar{\bar{A}} = \gamma_3 R + \ldots + \gamma_s R$ and $\bar{\bar{E}} = \bar{E}/\gamma_2\bar{E}$. But, by (7.1.3), $\bar{\bar{E}}$ is isomorphic to $E/(\gamma_1 E + \gamma_2 E)$ and so we have established that

$$e(s, A, E) \leqslant e(s-1, \bar{A}, E/\gamma_1 E) \leqslant e(s-2, \bar{\bar{A}}, E/(\gamma_1 E + \gamma_2 E)).$$

It is now clear how the argument proceeds. In fact we eventually obtain $\quad e(s, A, E) \leqslant e(0, (0), E/(\gamma_1 E + \ldots + \gamma_s E))$

$$= L_R(E/(\gamma_1 E + \ldots + \gamma_s E)),$$

by (7.7.6), or, changing the notation,

$$\operatorname*{Lim}_{n\to\infty} \frac{L_R(E/A^nE)}{n^s/s!} \leqslant L_R(E/(\gamma_1 E + \ldots + \gamma_s E)).$$

In this relation we now replace $\gamma_1, \gamma_2, \ldots, \gamma_s$ by $\gamma_1^p, \gamma_2^p, \ldots, \gamma_s^p$, where p is an arbitrary positive integer, and observe, at the same time, that

$$(\gamma_1^p R + \ldots + \gamma_s^p R)^n E \subseteq (\gamma_1 R + \ldots + \gamma_s R)^{np} E.$$

This device shows that

$$\operatorname*{Lim}_{n\to\infty} \frac{L_R\{E/(\gamma_1 R + \ldots + \gamma_s R)^{np} E\}}{(np)^s/s!} \leqslant \operatorname*{Lim}_{n\to\infty} \frac{L_R\{E/(\gamma_1^p R + \ldots + \gamma_s^p R)^n E\}}{(np)^s/s!}$$

$$\leqslant \frac{L_R\{E/(\gamma_1^p E + \ldots + \gamma_s^p E)\}}{p^s}.$$

But the first of these limits is $e(s, A, E)$. Hence

$$e(s, A, E) \leqslant \frac{L_R\{E/(\gamma_1^p E + \ldots + \gamma_s^p E)\}}{p^s}.$$

Now let p tend to infinity and apply Theorem 10. This yields

$$e(s, A, E) \leqslant e_R(\gamma_1, \ldots, \gamma_s | E). \tag{7.7.8}$$

It remains for us to establish the opposite inequality. To this end let $n_1, n_2, \ldots, n_s$ be positive integers and put

$$U = \bigoplus_{n=0}^{\infty} \{(A^n E \cap (A^{n+1}E + \gamma_1^{n_1}E + \ldots + \gamma_s^{n_s}E))/A^{n+1}E\}.$$

It is easy to check that U is a homogeneous $R[X]$-submodule of M (see (7.7.1)) and that

$$X_1^{n_1} M + X_2^{n_2} M + \ldots + X_s^{n_s} M \subseteq U.$$

Accordingly

$$L_R\{M/(X_1^{n_1} M + \ldots + X_s^{n_s} M)\} \geqslant L_R\{M/U\}. \qquad (7.7.9)$$

Put $F = \gamma_1^{n_1}E + \gamma_2^{n_2}E + \ldots + \gamma_s^{n_s}E$. Then we have an isomorphism

$$M/U \approx \bigoplus_{n=0}^{\infty} \{A^n E/(A^n E \cap (A^{n+1}E + F))\}$$

of R-modules. But

$$A^n E/(A^n E \cap (A^{n+1}E + F)) \approx (A^n E + A^{n+1}E + F)/(A^{n+1}E + F)$$

$$\approx (A^n E + F)/(A^{n+1}E + F)$$

and so
$$M/U \approx \bigoplus_{n=0}^{\infty} \{(A^n E + F)/(A^{n+1}E + F)\}.$$

However, when $n \geqslant n_1 + n_2 + \ldots + n_s$ we have $A^n E \subseteq F$ and therefore $A^n E + F = F$. It follows that

$$L_R\{M/U\} = \sum_{n \geqslant 0} L_R\{(A^n E + F)/(A^{n+1}E + F)\} = L_R\{E/F\},$$

and so, by (7.7.9),

$$L_R\{E/(\gamma_1^{n_1}E + \ldots + \gamma_s^{n_s}E)\} \leqslant L_R\{M/(X_1^{n_1} M + \ldots + X_s^{n_s} M)\}.$$

The next step is to show that

$$L_R\{M/(X_1^{n_1} M + \ldots + X_s^{n_s} M)\} = L_{R[X]}\{M/(X_1^{n_1} M + \ldots + X_s^{n_s} M)\}.$$

$$(7.7.10)$$

Let us leave the proof of this for the moment and discuss its consequences. Since we now have

$$\frac{L_R\{E/(\gamma_1^{n_1}E + \ldots + \gamma_s^{n_s}E)\}}{n_1 n_2 \ldots n_s} \leqslant \frac{L_{R[X]}\{M/(X_1^{n_1} M + \ldots + X_s^{n_s} M)\}}{n_1 n_2 \ldots n_s}$$

it follows, from Theorem 10, that

$$e_R(\gamma_1, \ldots, \gamma_s | E) \leqslant e_{R[X]}(X_1, \ldots, X_s | M).$$

But, by Theorem 12 and (7.7.5),

$$e_{R[X]}(X_1, \ldots, X_s | M) = h_s^*(M) = e(s, A, E).$$

Thus $e_R(\gamma_1, \ldots, \gamma_s | E) \leqslant e(s, A, E)$ which, in view of (7.7.8), leads us to

Theorem 13. *Let E be a Noetherian R-module and $\gamma_1, \gamma_2, \ldots, \gamma_s$ a multiplicity system on E. If now $A = \gamma_1 R + \gamma_2 R + \ldots + \gamma_s R$, then*

$$\lim_{n \to \infty} \frac{L_R\{E/A^n E\}}{n^s/s!} = e_R(\gamma_1, \gamma_2, \ldots, \gamma_s | E). \qquad (7.7.11)$$

Furthermore

$$e_R(\gamma_1, \gamma_2, \ldots, \gamma_s | E) = e_{R[X]}(X_1, \ldots, X_s | M),$$

where M is the $R[X_1, \ldots, X_s]$-module given by (7.7.1).

Remark. The expression for $e_R(\gamma_1, \ldots, \gamma_s | E)$ provided by (7.7.11) is the limit formula of P. Samuel.

Proof. After the previous discussion it is only necessary to prove (7.7.10). To this end let I be the $R[X]$-ideal generated by $X_1, X_2, \ldots, X_s$ and let n be an integer satisfying $n \geqslant n_1 + n_2 + \ldots + n_s$. Then

$$I^n M \subseteq X_1^{n_1} M + \ldots + X_s^{n_s} M.$$

Now

$$L_R\{M/(X_1^{n_1} M + \ldots + X_s^{n_s} M)\} \geqslant L_{R[X]}\{M/(X_1^{n_1} M + \ldots + X_s^{n_s} M)\}$$

and

$$L_R\{(X_1^{n_1} M + \ldots + X_s^{n_s} M)/I^n M\} \geqslant L_{R[X]}\{(X_1^{n_1} M + \ldots + X_s^{n_s} M)/I^n M\}$$

because every $R[X]$-module is also an R-module. Hence the desired result will follow if we can show that

$$L_R\{M/I^n M\} = L_{R[X]}\{M/I^n M\} < \infty.$$

But

$$L_R\{M/I^n M\} = \sum_{\nu=0}^{n-1} L_R\{I^\nu M/I^{\nu+1} M\},$$

$$L_{R[X]}\{M/I^n M\} = \sum_{\nu=0}^{n-1} L_{R[X]}\{I^\nu M/I^{\nu+1} M\},$$

and each of $X_1, X_2, \ldots, X_s$ annihilates $I^\nu M/I^{\nu+1} M$. Thus the $R[X]$-submodules of $I^\nu M/I^{\nu+1} M$ are the same as its R-submodules and so

$$L_R\{I^\nu M/I^{\nu+1} M\} = L_{R[X]}\{I^\nu M/I^{\nu+1} M\}.$$

Accordingly

$$L_R\{M/I^n M\} = L_{R[X]}\{M/I^n M\}.$$

Moreover the latter expression is finite because, by Proposition 6, $X_1, X_2, \ldots, X_s$ is a multiplicity system on M. The proof is now complete.

As an application of Samuel's limit formula we shall prove

Theorem 14. *Let E be a Noetherian R-module and $\gamma_1, \gamma_2, ..., \gamma_s$ and $\gamma_1', \gamma_2', ..., \gamma_s'$ multiplicity systems on E. Suppose that*

$$\gamma_1 E + \gamma_2 E + ... + \gamma_s E \subseteq \gamma_1' E + \gamma_2' E + ... + \gamma_s' E.$$

Then $\qquad\qquad e_R(\gamma_1, ..., \gamma_s | K) \geqslant e_R(\gamma_1', ..., \gamma_s' | K),$

where K is any submodule or factor module of E.

Proof. We may suppose that $s > 0$. Put

$$\gamma_1 R + ... + \gamma_s R = A \quad \text{and} \quad \gamma_1' R + ... + \gamma_s' R = A'.$$

Then we are given that $AE \subseteq A'E$. Hence

$$A^2 E \subseteq AA'E = A'AE \subseteq A'^2 E$$

and, in general, $A^n E \subseteq A'^n E$. Accordingly

$$L_R\{E/A^n E\} \geqslant L_R\{E/A'^n E\}$$

and so $\qquad\qquad e_R(\gamma_1, ..., \gamma_s | E) \geqslant e_R(\gamma_1', ..., \gamma_s' | E)$

by virtue of (7.7.11).

Now let K be a factor module of E. Then from

$$\gamma_1 E + ... + \gamma_s E \subseteq \gamma_1' E + ... + \gamma_s' E$$

follows $\gamma_1 K + ... + \gamma_s K \subseteq \gamma_1' K + ... + \gamma_s' K$, and so we obtain

$$e_R(\gamma_1, ..., \gamma_s | K) \geqslant e_R(\gamma_1', ..., \gamma_s' | K)$$

for reasons entirely similar to those already given.

Finally, suppose that K is a submodule of E. Then, by Theorem 1, there exists an integer $q \geqslant 0$ such that $A'(A'^q E \cap K) = A'^{q+1} E \cap K$. Now $\gamma_1'^q(K/(A'^q E \cap K)) = 0$. Also $\gamma_1^q(K/(A'^q E \cap K)) = 0$ because

$$\gamma_1^q K \subseteq A^q E \cap K \subseteq A'^q E \cap K.$$

It therefore follows, from Proposition 5, that

$$e_R(\gamma_1', ..., \gamma_s' | K) = e_R(\gamma_1', ..., \gamma_s' | A'^q E \cap K) \qquad (7.7.12)$$

and $\qquad e_R(\gamma_1, ..., \gamma_s | K) = e_R(\gamma_1, ..., \gamma_s | A'^q E \cap K). \qquad (7.7.13)$

Next

$$A(A'^q E \cap K) \subseteq AA'^q E \cap K = A'^q AE \cap K \subseteq A'^{q+1} E \cap K$$
$$= A'(A'^q E \cap K)$$

and therefore

$$\gamma_1(A'^q E \cap K) + ... + \gamma_s(A'^q E \cap K) \subseteq \gamma_1'(A'^q E \cap K) + ... + \gamma_s'(A'^q E \cap K).$$

Accordingly

$$e_R(\gamma_1, ..., \gamma_s | A'^q E \cap K) \geqslant e_R(\gamma_1', ..., \gamma_s' | A'^q E \cap K)$$

by the case already considered. Finally

$$e_R(\gamma_1, ..., \gamma_s | K) \geqslant e_R(\gamma_1', ..., \gamma_s' | K)$$

by (7.7.12) and (7.7.13). This completes the proof.

Corollary 1. *Let E be a Noetherian R-module and $\gamma_1, \gamma_2, ..., \gamma_s$ and $\gamma_1', \gamma_2', ..., \gamma_s'$ two multiplicity systems on E. If now*

$$\gamma_1 E + ... + \gamma_s E = \gamma_1' E + ... + \gamma_s' E,$$

then $\quad\quad e_R(\gamma_1, ..., \gamma_s | K) = e_R(\gamma_1', ..., \gamma_s' | K)$

where K is any submodule or factor module of E.

Corollary 2. *Let $\gamma_1, \gamma_2, ..., \gamma_s$ and $\gamma_1', \gamma_2', ..., \gamma_s'$ be central elements of R such that $\gamma_1 R + ... + \gamma_s R = \gamma_1' R + ... + \gamma_s' R$. Then for every Noetherian R-module E which admits both $\gamma_1, \gamma_2, ..., \gamma_s$ and $\gamma_1', \gamma_2', ..., \gamma_s'$ as multiplicity systems we have*

$$e_R(\gamma_1, ..., \gamma_s | E) = e_R(\gamma_1', ..., \gamma_s' | E).$$

7.8 Localization and extension

As is to be expected, the theory of multiplicities and the theory of Hilbert functions can be developed in more detail when one has to do with *commutative* rings. This will be exemplified by the results of the remaining sections of this chapter. Accordingly *from now on we shall assume that R is a commutative ring with an identity element.* Of course, in this situation every element of R is a central element.

Let E be a Noetherian R-module and $\gamma_1, \gamma_2, ..., \gamma_s$ a multiplicity system on E. Then $E/(\gamma_1, ..., \gamma_s)E$ is finitely generated and moreover has finite length. It follows, from Theorem 2 Cor. of section (4.2), that R/I is an Artin ring, where $I = \mathrm{Ann}_R\{E/(\gamma_1, ..., \gamma_s)E\}$. Accordingly there are only a finite number of prime ideals which contain I and each of these is a maximal ideal. However, by Proposition 3 of section (5.1), the prime ideals which contain I are the same as those which contain $(\mathrm{Ann}_R E, \gamma_1, ..., \gamma_s)$. Hence there are only a finite number of prime ideals which contain $(\mathrm{Ann}_R E, \gamma_1, ..., \gamma_s)$ and each is a maximal ideal. Indeed $R/(\mathrm{Ann}_R E, \gamma_1, ..., \gamma_s)$ *is also an Artin ring.* (To see this note that $R/(\mathrm{Ann}_R E, \gamma_1, ..., \gamma_s)$ is a Noetherian ring because it is a homomorphic image of $R/\mathrm{Ann}_R E$ which is Noetherian by Theorem 2 of section (4.2). The Artinian character of $R/(\mathrm{Ann}_R E, \gamma_1, ..., \gamma_s)$ then follows by applying Proposition 7 Cor. of section (4.4) to its zero ideal.) Now put $\bar{R} = R/\mathrm{Ann}_R E$ and let $\bar{\gamma}_i$ denote the natural image of γ_i

in $\bar{R}$. Then $\bar{R}/(\bar{\gamma}_1, ..., \bar{\gamma}_s)$ is an Artin ring because it is isomorphic to $R/(\mathrm{Ann}_R E, \gamma_1, ..., \gamma_s)$ and therefore $L_{\bar{R}}\{\bar{R}/(\bar{\gamma}_1, ..., \bar{\gamma}_s)\} < \infty$. Accordingly $\bar{R} = R/\mathrm{Ann}_R E$ *is a Noetherian ring and* $\bar{\gamma}_1, \bar{\gamma}_2, ..., \bar{\gamma}_s$ *is a multiplicity system on* $\bar{R}$ *considered as a module with respect to itself. Hence*

$$e_{\bar{R}}(\bar{\gamma}_1, ..., \bar{\gamma}_s|\bar{R})$$

is defined. The reader should pay particular attention to the statements in italics because they will find several applications in the sequel.

As before, let E be a Noetherian R-module and $\gamma_1, \gamma_2, ..., \gamma_s$ a multiplicity system on E. We shall use M to denote a typical maximal ideal of R and $\phi_M \colon R \to R_M$ will denote the canonical ring-homomorphism of R into its localization at M. Since localization functors are exact, the exact sequence

$$0 \to (\gamma_1, ..., \gamma_s)E \to E \to E/(\gamma_1, ..., \gamma_s)E \to 0$$

shows that we have an R_M-isomorphism

$$(E/(\gamma_1, ..., \gamma_s)E)_M \approx E_M/((\gamma_1, ..., \gamma_s)E)_M.$$

But, by Proposition 4 of section (3.2) and Proposition 3 of the same section,

$$((\gamma_1, ..., \gamma_s)E)_M = (\gamma_1, ..., \gamma_s)_M E_M = (\phi_M \gamma_1, ..., \phi_M \gamma_s)E_M$$

and so the R_M-isomorphism takes the form

$$(E/(\gamma_1, ..., \gamma_s)E)_M \approx E_M/(\phi_M \gamma_1, ..., \phi_M \gamma_s)E_M. \qquad (7.8.1)$$

Accordingly, by Theorem 4 of section (3.2),

$$L_{R_M}\{E_M/(\phi_M \gamma_1, ..., \phi_M \gamma_s)E_M\} \leqslant L_R\{E/(\gamma_1, ..., \gamma_s)E\} < \infty.$$

We can therefore sum up as follows: *if E is a Noetherian R-module and $\gamma_1, \gamma_2, ..., \gamma_s$ a multiplicity system on E then, for every maximal ideal M, E_M is a Noetherian R_M-module*† *and* $\phi_M \gamma_1, \phi_M \gamma_2, ..., \phi_M \gamma_s$ *a multiplicity system on* E_M.

The above remarks show that, when the conditions are as stated, $e_{R_M}(\phi_M \gamma_1, ..., \phi_M \gamma_s|E_M)$ is defined for every maximal ideal M. Now if, $\mathrm{Ann}_R E \nsubseteq M$, then $E_M = 0$ and the multiplicity is zero. Again if

$$(\gamma_1, ..., \gamma_s) \nsubseteq M,$$

then $(\phi_M \gamma_1, ..., \phi_M \gamma_s) = R_M$ and therefore $E_M = (\phi_M \gamma_1, ..., \phi_M \gamma_s)E_M$. Thus the multiplicity is again zero, this time by virtue of Theorem 6 Cor. There remains the case in which $(\mathrm{Ann}_R E, \gamma_1, ..., \gamma_s) \subseteq M$ and, as we have already observed, there are only finitely many M's which have this property. This shows that *there are at most a finite number of*

† See Theorem 3 of section (3.2).

maximal ideals M for which $e_{R_M}(\phi_M\gamma_1, ..., \phi_M\gamma_s|E_M) \neq 0$ and these all occur among the maximal ideals which contain $(\operatorname{Ann}_R E, \gamma_1, ..., \gamma_s)$.

The next result will be referred to as the *Localization Principle* for multiplicities.

Theorem 15. *Let R be a commutative ring, E a Noetherian R-module, and $\gamma_1, \gamma_2, ..., \gamma_s$ a multiplicity system on E. Then*

$$e_R(\gamma_1, ..., \gamma_s|E) = \sum_M e_{R_M}(\phi_M\gamma_1, ..., \phi_M\gamma_s|E_M),$$

where M denotes a typical maximal ideal of R and ϕ_M the canonical mapping of R into R_M.

We shall not give a separate proof of this result because it is a special case of the *Extension Formula* which will be established presently.

Suppose that the (commutative) ring S is an integral extension of R and let us use Π to denote a typical maximal ideal of S. By Theorem 11 of section (2.6), $\Pi \cap R$ is a maximal ideal of R. Now S/Π is an R-module that is annihilated by $\Pi \cap R$. It can therefore be regarded as a vector space over the field $R/(\Pi \cap R)$. Naturally we use $[S/\Pi : R/(\Pi \cap R)]$ to denote the dimension of this vector space. Finally we observe that any S-module is automatically an R-module. After these preliminaries we come to the *Extension Formula* for multiplicities.

Theorem 16. *Let the commutative ring S be an integral extension of the commutative ring R, let E be an S-module, and let $\gamma_1, \gamma_2, ..., \gamma_s$ be elements of R. Assume that* (i) *E is Noetherian both as an R-module and as an S-module, and* (ii) *$\gamma_1, \gamma_2, ..., \gamma_s$ is a multiplicity system on E both when it is considered as an R-module and when it is considered as an S-module. If now Π denotes a typical maximal ideal of S, then*

$$e_R(\gamma_1, ..., \gamma_s|E) = \sum_\Pi e_{S_\Pi}(\phi_\Pi\gamma_1, ..., \phi_\Pi\gamma_s|E_\Pi) [S/\Pi : R/(\Pi \cap R)],$$

where ϕ_Π denotes the canonical mapping of S into S_Π.

Remarks. There are a number of points that need to be noted. First of all if Π is such that $e_{S_\Pi}(\phi_\Pi\gamma_1, ..., \phi_\Pi\gamma_s|E_\Pi)$ is zero whereas

$$[S/\Pi : R/(\Pi \cap R)]$$

is infinite, then their product is to be regarded as having the value zero. Secondly, one obtains Theorem 15 from Theorem 16 by allowing S to coincide with R. Finally, it is possible to replace conditions (i) and (ii) by the following: *when E is considered as an R-module it is Noetherian* *and admits* $\gamma_1, \gamma_2, ..., \gamma_s$ *as a multiplicity system.* For this secures that E is Noetherian as an S-module and, since

$$L_S\{E/(\gamma_1 E + ... + \gamma_s E)\} \leqslant L_R\{E/(\gamma_1 E + ... + \gamma_s E)\},$$

the requirement that $\gamma_1, \gamma_2, ..., \gamma_s$ be a multiplicity system on the S-module E is satisfied as well.

Proof. We use induction on s and begin by observing that when $s = 0$ the result in question reduces to Theorem 13 of section (3.9).

Now assume that $s \geqslant 1$ and that the theorem has already been proved in the case of multiplicity systems having only $s - 1$ elements. It is clear that $E/\gamma_1 E$ and $0 :_E \gamma_1$ do not depend on whether we regard R or S as the ring of operators. Again, we have an isomorphism $(E/\gamma_1 E)_\Pi \approx E_\Pi/(\phi_\Pi \gamma_1) E_\Pi$ of R_Π-modules and, by (3.5.5), we also have $(0 :_E \gamma_1)_\Pi = 0 :_{E_\Pi} \phi_\Pi \gamma_1$. Accordingly, by the inductive hypothesis,

$$e_R(\gamma_2, ..., \gamma_s | E/\gamma_1 E)$$
$$= \sum_\Pi e_{S_\Pi}(\phi_\Pi \gamma_2, ..., \phi_\Pi \gamma_s | E_\Pi/(\phi_\Pi \gamma_1) E_\Pi) [S/\Pi : R/(\Pi \cap R)]$$

and

$$e_R(\gamma_2, ..., \gamma_s | 0 :_E \gamma_1)$$
$$= \sum_\Pi e_{S_\Pi}(\phi_\Pi \gamma_2, ..., \phi_\Pi \gamma_s | 0 :_{E_\Pi} \phi_\Pi \gamma_1) [S/\Pi : R/(\Pi \cap R)].$$

The desired result now follows by subtraction.

The next result is useful, apart from its intrinsic interest, because it helps us to apply the localization principle.

Proposition 7. *Let R be a local ring and $\gamma_1, \gamma_2, ..., \gamma_s$ non-units forming a multiplicity system on R. Then $s \geqslant \mathrm{Dim}\, R$. Moreover $s = \mathrm{Dim}\, R$ if and only if $e_R(\gamma_1, ..., \gamma_s | R) > 0$.*

Proof. Put $d = \mathrm{Dim}\, R$ and let M denote the maximal ideal of R. Since $L_R\{R/(\gamma_1, ..., \gamma_s)\} < \infty$, it follows, from Proposition 7 Cor. of section (4.4), that every prime ideal containing $(\gamma_1, ..., \gamma_s)$ is a maximal ideal. But M is the only maximal ideal of R. Thus $(\gamma_1, ..., \gamma_s)$ is M-primary and therefore $s \geqslant d$ by Theorem 23 of section (4.9). For the remainder of the discussion we separate the cases $s = d$ and $s > d$.

Case (i) $s = d$. For this situation $\gamma_1, \gamma_2, ..., \gamma_s$ is a system of parameters and we must show that $e_R(\gamma_1, ..., \gamma_s | R)$ is strictly positive. This will be done by induction on s. When $s = 0$ the assertion is trivial. We shall therefore suppose that $s \geqslant 1$ and make the obvious induction hypothesis.

Since $\operatorname{Dim} R = s$, there exists a prime ideal P such that

$$\operatorname{Dim}(R/P) = s.$$

Now

$$e_R(\gamma_1, ..., \gamma_s|R) \geqslant e_R(\gamma_1, ..., \gamma_s|R/P) = e_{R/P}(\phi\gamma_1, ..., \phi\gamma_s|R/P)$$

by (7.4.3), where $\phi: R \to R/P$ is the natural mapping. It is therefore sufficient to show that $e_{R/P}(\phi\gamma_1, ..., \phi\gamma_s|R/P)$ is strictly positive. This implies that, for the remainder of the proof, we may assume that R is an integral domain. However, in this situation, γ_1 is not a zero-divisor of the ring R and so, again using (7.4.3), we have

$$e_R(\gamma_1, ..., \gamma_s|R) = e_R(\gamma_2, ..., \gamma_s|R/(\gamma_1)) = e_{R/(\gamma_1)}(\bar{\gamma}_2, ..., \bar{\gamma}_s|R/(\gamma_1)),$$

where $\bar{\gamma}_i$ denotes the natural image of γ_i in $R/(\gamma_i)$. But, by Proposition 20 of section (4.9), $R/(\gamma_1)$ is an $(s-1)$-dimensional local ring and moreover $\bar{\gamma}_2, \bar{\gamma}_3, ..., \bar{\gamma}_s$ are non-units of this ring. Accordingly, by the inductive hypothesis, $e_{R/(\gamma_1)}(\bar{\gamma}_2, ..., \bar{\gamma}_s|R/(\gamma_1))$ is strictly positive and the desired result follows.

Case (ii) $s > d$. This time it must be shown that $e_R(\gamma_1, ..., \gamma_s|R) = 0$. We do this by means of an induction with respect to d. If $d = 0$, then γ_1 is nilpotent and so $\gamma_1^m R = 0$ for a suitable integer m. Accordingly $e_R(\gamma_1, ..., \gamma_s|R) = 0$ by Proposition 5.

From now on we assume that $d \geqslant 1$ and suppose that the desired result has been proved for local rings of smaller dimension. If γ_1 is nilpotent, then $e_R(\gamma_1, ..., \gamma_s|R) = 0$ by Proposition 5. We shall therefore suppose that γ_1 is not nilpotent. Choose an integer p so that γ_1 is not a zero-divisor on $R^* = R/(0:\gamma_1^p)$. This is possible by Lemma 5. Furthermore, because γ_1 is not nilpotent, R^* is a local ring with

$$\operatorname{Dim} R^* \leqslant \operatorname{Dim} R < s.$$

Now, by Proposition 5, $e_R(\gamma_1, ..., \gamma_s|0:\gamma_1^p) = 0$. Consequently

$$e_R(\gamma_1, \gamma_2, ..., \gamma_s|R) = e_R(\gamma_1, \gamma_2, ..., \gamma_s|R^*)$$
$$= e_{R^*}(\gamma_1^*, \gamma_2^*, ..., \gamma_s^*|R^*),$$

where γ_i^* is the natural image of γ_i in R^*. But $\gamma_1^*, \gamma_2^*, ..., \gamma_s^*$ are non-units in R^* and γ_1^* is not a zero-divisor. This shows that for the remainder of the proof we may assume that γ_1 is not a zero-divisor of R. But then

$$e_R(\gamma_1, \gamma_2, ..., \gamma_s|R) = e_R(\gamma_2, ..., \gamma_s|R/(\gamma_1))$$
$$= e_{\bar{R}}(\bar{\gamma}_2, ..., \bar{\gamma}_s|\bar{R}),$$

where $\bar{R} = R/(\gamma_1)$ and, as before, $\bar{\gamma}_i$ denotes the natural image of γ_i in $\bar{R}$. However, $\bar{R}$ is a local ring, the $\bar{\gamma}_i$ are non-units, and

$$\operatorname{Dim} \bar{R} = \operatorname{Dim} R - 1 < s - 1.$$

Accordingly, by the inductive hypothesis, $e_{\bar{R}}(\bar{\gamma}_2, ..., \bar{\gamma}_s | \bar{R}) = 0$. This completes the proof.

We shall use this result to derive a useful criterion for a multiplicity to be zero.

Proposition 8. *Let R be a commutative ring, E a Noetherian R-module, and $\gamma_1, \gamma_2, ..., \gamma_s$ $(s \geqslant 0)$ a multiplicity system on E. Suppose that for every maximal ideal M containing $(\mathrm{Ann}_R E, \gamma_1, ..., \gamma_s)$ we have*

$$\mathrm{rank}\,(M/\mathrm{Ann}_R E) < s. \ Then \ e_R(\gamma_1, ..., \gamma_s | E) = 0.$$

Proof. Put $\bar{R} = R/\mathrm{Ann}_R E$ and let $\bar{\gamma}_i$ denote the natural image of γ_i in $\bar{R}$. Then, as we have had occasion to remark earlier, $\bar{R}$ is a Noetherian ring and $\bar{\gamma}_1, \bar{\gamma}_2, ..., \bar{\gamma}_s$ is a multiplicity system on $\bar{R}$. Furthermore, in the present instance, rank $\bar{M} < s$ for every maximal ideal $\bar{M}$ containing $(\bar{\gamma}_1, \bar{\gamma}_2, ..., \bar{\gamma}_s)$. By (7.4.3),

$$e_R(\gamma_1, ..., \gamma_s | E) = e_{\bar{R}}(\bar{\gamma}_1, ..., \bar{\gamma}_s | E).$$

Consequently the proposition will follow if we show that

$$e_{\bar{R}}(\bar{\gamma}_1, ..., \bar{\gamma}_s | E) = 0.$$

Now, as an $\bar{R}$-module, E is finitely generated. Suppose that it can be generated by m elements. Then Proposition 23 Cor. 2 of section (1.13) shows that E is a homomorphic image of the direct sum

$$\bar{R} \oplus \bar{R} \oplus ... \oplus \bar{R} \ (m \ \mathrm{terms}).$$

Consequently it will suffice to establish that $e_{\bar{R}}(\gamma_1, ..., \gamma_s | \bar{R}) = 0$. The same conclusion may be expressed as follows: *for the purpose of the proof we may assume that R is a Noetherian ring and $E = R$.*

Let us impose these extra conditions. Then, by Theorem 15 and with the same notation,

$$e_R(\gamma_1, ..., \gamma_s | R) = \sum_M e_{R_M}(\phi_M \gamma_1, ..., \phi_M \gamma_s | R_M).$$

Moreover in this sum we need only consider maximal ideals M which contain $(\gamma_1, \gamma_2, ..., \gamma_s)$. For such an M, the elements

$$\phi_M \gamma_1, \phi_M \gamma_2, ..., \phi_M \gamma_s$$

are non-units of R_M and Dim $R_M = \mathrm{rank}\, M < s$. Accordingly, by Proposition 7, $e_{R_M}(\phi_M \gamma_1, ..., \phi_M \gamma_s | R_M) = 0$. The proposition follows.

Theorem 17. *Let R be a commutative ring and E a Noetherian R-module. Furthermore let $\gamma_1, \gamma_2, ..., \gamma_s$ and $\omega_1, \omega_2, ..., \omega_s$ be two multiplicity systems* *on E such that each of the ideals* $(\gamma_1, \gamma_2, ..., \gamma_s)$ *and* $(\omega_1, \omega_2, ..., \omega_s)$ *contains a power of the other. If now*

$$e_R(\gamma_1, \gamma_2, ..., \gamma_s | E) = L_R\{E/(\gamma_1, \gamma_2, ..., \gamma_s)E\},$$

then $\qquad e_R(\omega_1, \omega_2, ..., \omega_s | E) = L_R\{E/(\omega_1, \omega_2, ..., \omega_s)E\}$

as well.

Proof. This result will be established with the aid of the theory of grade. We therefore begin by reducing the theorem to a special case where this theory is directly applicable.

Put $R^* = R/\text{Ann}_R E$ and let γ_i^* respectively ω_j^* be the natural image of γ_i respectively ω_j in R^*. Then

$$L_R\{E/(\gamma_1, ..., \gamma_s)E\} = L_{R^*}\{E/(\gamma_1^*, ..., \gamma_s^*)E\}$$

and we also have

$$e_R(\gamma_1, ..., \gamma_s | E) = e_{R^*}(\gamma_1^*, ..., \gamma_s^* | E)$$

by (7.4.3). Of course there are entirely similar equations involving $L_R\{E/(\omega_1, ..., \omega_s)E\}$ and $e_R(\omega_1, ..., \omega_s | E)$. Now R^* is a Noetherian ring, each of the ideals $(\gamma_1^*, \gamma_2^*, ..., \gamma_s^*)$ and $(\omega_1^*, \omega_2^*, ..., \omega_s^*)$ contains a power of the other, and $\text{Ann}_{R^*} E = (0)$. These observations justify us in assuming, for the remainder of the proof, that R is a Noetherian ring and $\text{Ann}_R E = (0)$.

Let M denote a typical maximal ideal of R and $\phi_M : R \to R_M$ the canonical mapping. By Theorem 15,

$$e_R(\gamma_1, ..., \gamma_s | E) = \sum_M e_{R_M}(\phi_M \gamma_1, ..., \phi_M \gamma_s | E_M)$$

and there is a similar expression for $e_R(\omega_1, ..., \omega_s | E)$. Next, as has already been observed, we have an R_M-isomorphism

$$(E/(\gamma_1, ..., \gamma_s)E)_M \approx E_M/(\phi_M \gamma_1, ..., \phi_M \gamma_s)E_M.$$

Hence, by Theorem 12 of section (3.9),

$$L_R\{E/(\gamma_1, ..., \gamma_s)E\} = \sum_M L_{R_M}\{E_M/(\phi_M \gamma_1, ..., \phi_M \gamma_s)E_M\}.$$

Naturally there is a similar expression for $L_R\{E/(\omega_1, ..., \omega_s)E\}$.

Our hypotheses imply that

$$\sum_M e_{R_M}(\phi_M \gamma_1, ..., \phi_M \gamma_s | E_M) = \sum_M L_{R_M}\{E_M/(\phi_M \gamma_1, ..., \phi_M \gamma_s)E_M\}.$$

However, by (7·4·13), we have

$$e_{R_M}(\phi_M \gamma_1, ..., \phi_M \gamma_s | E_M) \leqslant L_{R_M}\{E_M/(\phi_M \gamma_1, ..., \phi_M \gamma_s)E_M\}.$$

It follows that

$$e_{R_M}(\phi_M \gamma_1, ..., \phi_M \gamma_s | E_M) = L_{R_M}\{E_M/(\phi_M \gamma_1, ..., \phi_M \gamma_s)E_M\}$$

for every maximal ideal M. Note that (i) R_M is a local ring, (ii) each of the ideals $(\phi_M \gamma_1, ..., \phi_M \gamma_s)$ and $(\phi_M \omega_1, ..., \phi_M \omega_s)$ contains a power of the other, and (iii) $\mathrm{Ann}_{R_M} E_M = (0)$ because, by Proposition 8 of section (3.5), $(\mathrm{Ann}_R E)_M = \mathrm{Ann}_{R_M} E_M$. Further, if we can deduce from these facts that $e_{R_M}(\phi_M \omega_1, ..., \phi_M \omega_s | E_M)$ and $L_{R_M}\{E_M/(\phi_M \omega_1, ..., \phi_M \omega_s)E_M\}$ are equal, then the theorem will follow.

These remarks show that we may assume, from here on, that R is a local ring with maximal ideal M and that $\mathrm{Ann}_R E = (0)$. It is now necessary to distinguish between two possibilities.

If $(\gamma_1, \gamma_2, ..., \gamma_s) = R$, then also $(\omega_1, \omega_2, ..., \omega_s) = R$ and we have

$$e_R(\omega_1, ..., \omega_s | E) = L_R\{E/(\omega_1, ..., \omega_s)E\} = 0$$

by (7.4.13). Thus there remains only the case in which each γ_i and ω_j is a non-unit. Before proceeding, observe that in this situation

$$(\gamma_1, ..., \gamma_s) E \neq E;$$

for otherwise $\qquad E = \bigcap_{n=1}^{\infty} (\gamma_1, ..., \gamma_s)^n E = 0$

by Theorem 19 of section (4.6), because $\gamma_1, \gamma_2, ..., \gamma_s$ belong to the Jacobson radical of R. This, however, contradicts the assumption that $\mathrm{Ann}_R E = (0)$.

We now apply Theorem 9 which shows that $\gamma_1, \gamma_2, ..., \gamma_s$ is an R-sequence on E. Accordingly, by Proposition 4 of section (5.1),

$$\mathrm{rank}\, M \geqslant \mathrm{rank}\, (\gamma_1, ..., \gamma_s) = s.$$

On the other hand, M is a minimal prime ideal of $(\gamma_1, ..., \gamma_s)$ because $R/(\mathrm{Ann}_R E, \gamma_1, ..., \gamma_s)$ is an Artin ring and $\mathrm{Ann}_R E = (0)$. Accordingly $\mathrm{rank}\, M \leqslant s$ and $(\gamma_1, \gamma_2, ..., \gamma_s)$ is M-primary. Thus $\mathrm{Dim}\, R = s$ and $\gamma_1, \gamma_2, ..., \gamma_s$ is a system of parameters. But then $\omega_1, \omega_2, ..., \omega_s$ is also a system of parameters. Now we have shown that $\gamma_1, \gamma_2, ..., \gamma_s$ is an R-sequence on E. It therefore follows, by Theorem 9 of section (5.2), that $\omega_1, \omega_2, ..., \omega_s$ is also an R-sequence on E. Finally, by Theorem 8, $e_R(\omega_1, ..., \omega_s | E)$ is equal to $L_R\{E/(\omega_1, ..., \omega_s)E\}$ and with this the proof is complete.

7.9 The associative law for multiplicities

In this section R denotes a (commutative) *Noetherian* ring and we shall be concerned with a sequence $\gamma_1, \gamma_2, ..., \gamma_s$, of elements of R, forming a multiplicity system on R itself. Note that this secures that

$$e_R(\gamma_1, ..., \gamma_s | E)$$

is defined for every finitely generated R-module E. (Indeed such a module E is necessarily Noetherian and $\gamma_1, \gamma_2, ..., \gamma_s$ is a multiplicity system on E because E is a homomorphic image of a direct sum $R \oplus R \oplus ... \oplus R$ with a finite number of terms.) Next, from the fact that $L_R\{R/(\gamma_1, ..., \gamma_s)\}$ is finite it follows that $R/(\gamma_1, ..., \gamma_s)$ is an Artin ring. Accordingly any prime ideal which contains $(\gamma_1, \gamma_2, ..., \gamma_s)$ is a maximal ideal of R and a minimal prime ideal of $(\gamma_1, \gamma_2, ..., \gamma_s)$. Hence, by Theorem 22 of section (4.8), if M is such a prime ideal, then rank $M \leqslant s$.

Proposition 9. *Let R be a commutative Noetherian ring, $\gamma_1, \gamma_2, ..., \gamma_s$ a multiplicity system on R, and $E \neq 0$ a finitely generated R-module. If now rank $(\operatorname{Ann}_R E) > 0$, then $e_R(\gamma_1, ..., \gamma_s|E) = 0$. In particular, if A is a proper ideal such that rank $A > 0$, then $e_R(\gamma_1, ..., \gamma_s|R/A) = 0$.*

Proof. If M is a maximal ideal containing $(\operatorname{Ann}_R E, \gamma_1, ..., \gamma_s)$, then it is a maximal ideal containing $(\gamma_1, \gamma_2, ..., \gamma_s)$ and therefore

$$\text{rank } M \leqslant s$$

by the above remarks. But then rank $(M/\operatorname{Ann}_R E) < s$ because rank $(\operatorname{Ann}_R E) > 0$. That $e_R(\gamma_1, ..., \gamma_s|E) = 0$ now follows by Proposition 8.

The next result has a great deal of intrinsic interest quite apart from its use in the theory of multiplicities.

Proposition 10. *Let R be a commutative Noetherian ring, E a finitely generated R-module, and K a proper submodule of E. Then there exists a finite sequence $e_1, e_2, ..., e_m$ of elements of E such that if*

$$E_i = K + Re_1 + Re_2 + ... + Re_i \quad and \quad E_0 = K,$$

then (i) $E_m = E$, *and* (ii) $E_{i-1}:e_i = P_i$, *where P_i is a prime ideal, for $i = 1, 2, ..., m$. It follows that there is an isomorphism $R/P_i \approx E_i/E_{i-1}$ of R-modules for $i = 1, 2, ..., m$.*

Proof. Denote by Ω the set of all ideals of the form $K:e$, where $e \in E$, $e \notin K$. Since R is a Noetherian ring, we can choose $e_1 \in E$, $e_1 \notin K$, so that $K:e_1 = P_1$ (say) is maximal in Ω. *We contend that P_1 is a prime ideal.* For let a, b be elements of R such that $ab \in P_1$ but $b \notin P_1$. Then $be_1 \notin K$. However $K:be_1$ contains $K:e_1$. Consequently, by the choice of e_1, $K:be_1 = K:e_1 = P_1$. However $abe_1 \in K$. Consequently a belongs to $K:be_1 = P_1$. This establishes our contention that P_1 is prime.

Next $K + Re_1 = E_1$ strictly contains $K = E_0$. If E_1 coincides with E there is no need to continue the sequence. If not, then repeating the argument, with E_1 replacing K, we can show that there exists $e_2 \in E$ such that $E_1 : e_2 = P_2$, where P_2 is a prime ideal. Then

$$K + Re_1 + Re_2 = E_2$$

strictly contains E_1. If E_2 coincides with E, then we stop. If not, then we find $e_3 \in E$ so that $E_2 : e_3 = P_3$, where P_3 is a prime ideal. And so on. Eventually we must arrive at a module E_m such that $E_m = E$; for otherwise $E_1, E_2, E_3, \ldots$ would form an infinite strictly increasing sequence which is impossible because E is a Noetherian R-module.

Finally suppose that $1 \leqslant i \leqslant m$ and let $\bar{e}_i$ denote the natural image of e_i in E_i/E_{i-1}. The mapping $R \to E_i/E_{i-1}$ in which $r \to r\bar{e}_i$ is then an epimorphism with kernel P_i. Accordingly we have an isomorphism $R/P_i \approx E_i/E_{i-1}$ of R-modules and this completes the proof.

In the next section, when we are discussing refinements of the theory of Hilbert functions, we shall need Proposition 10 in a form adapted to graded rings and modules. The adaptation is contained in the following corollary.

Corollary. *Let R, E and K satisfy all the hypotheses of Proposition 10. Assume further that (a) R is a graded ring, (b) E is a graded module with respect to R, and (c) K is a homogeneous submodule of E. Finally let the monoid used to grade R and E be torsionless. In these circumstances the elements e_i of Proposition 10 may be chosen so that they are homogeneous. If this is done then the E_i ($i = 0, 1, \ldots, m$) will be homogeneous submodules of E and the P_i homogeneous prime ideals. Let R/P_i and E_i/E_{i-1} be endowed with the natural factor gradings. Then the isomorphism*

$$R/P_i \approx E_i/E_{i-1}$$

(described in the proof of Proposition 10) when applied to a homogeneous element of R/P_i will raise its degree by an amount equal to the degree of e_i.

Proof. The original arguments apply provided minor adjustments are made. This time we denote by Ω the set of all ideals of the form $K : e$, where $e \in E$, $e \notin K$, and e is homogeneous. Then Ω is not empty and all its members are homogeneous ideals. Since the ring R is Noetherian, we can choose a homogeneous element $e_1 \in E$, $e_1 \notin K$, so that $K : e_1 = P_1$ (say) is maximal in Ω. Once again P_1 is a prime ideal. Indeed to prove this it is enough, by Lemma 13 of section (2.13), to show that if a, b are *homogeneous* elements of R, ab belongs to P_1, and $b \notin P_1$, then $a \in P_1$. But with the extra condition of homogeneity, the argument proceeds as before. The remaining parts of the proof of Proposition 10 need only trivial modifications.

The next result is a special case of the Associative Law for multiplicities and provides a natural stepping-stone to the full theorem.

Proposition 11. *Let R be a commutative Noetherian ring and E a finitely generated R-module. Further, let $\gamma_1, \gamma_2, ..., \gamma_s$ be a multiplicity system on R itself. Then*

$$e_R(\gamma_1, ..., \gamma_s|E) = \sum_P L_{R_P}(E_P)\, e_{R/P}(\psi_P\gamma_1, ..., \psi_P\gamma_s|R/P),$$

where ψ_P denotes the natural mapping $R \to R/P$ and the summation is taken over all prime ideals of rank zero.

Remark. Since a prime has rank zero only when it is a minimal prime ideal of the zero ideal, the fact that R is a Noetherian ring ensures that the sum contains only a finite number of terms.

Proof. By Proposition 10 there exist submodules $E_0, E_1, ..., E_m$ of E and prime ideals $P_1, P_2, ..., P_m$ such that

$$0 = E_0 \subseteq E_1 \subseteq ... \subseteq E_m = E$$

and, for each $i\,(1 \leqslant i \leqslant m)$, there exists an isomorphism

$$E_i/E_{i-1} \approx R/P_i$$

of R-modules. Hence, by Theorem 5,

$$e_R(\gamma_1, ..., \gamma_s|E) = \sum_{i=1}^{m} e_R(\gamma_1, ..., \gamma_s|R/P_i).$$

Moreover Proposition 9 shows that, in this sum, we need only consider values of i for which P_i has rank zero.

Suppose now that P is a prime ideal whose rank is zero and let λ_P denote the number of times P occurs in the sequence $\{P_1, P_2, ..., P_m\}$. The above remarks then show that

$$e_R(\gamma_1, ..., \gamma_s|E) = \sum_P \lambda_P e_R(\gamma_1, ..., \gamma_s|R/P).$$

Next $\quad L_{R_P}(E_P) = \sum_{i=1}^{m} L_{R_P}\{(E_i/E_{i-1})_P\} = \sum_{i=1}^{m} L_{R_P}\{(R/P_i)_P\}.$

However $(R/P_i)_P$ is a zero R_P-module if $P \neq P_i$ and it is a simple R_P-module when $P = P_i$. It follows that $L_{R_P}(E_P) = \lambda_P$. Thus

$$e_R(\gamma_1, ..., \gamma_s|E) = \sum_P L_{R_P}(E_P)\, e_R(\gamma_1, ..., \gamma_s|R/P)$$
$$= \sum_P L_{R_P}(E_P)\, e_{R/P}(\psi_P\gamma_1, ..., \psi_P\gamma_s|R/P)$$

by (7.4.3). This completes the proof.

The next theorem is the general Associative Law.

Theorem 18. *Let R be a commutative Noetherian ring and E a finitely generated R-module. Furthermore let $\gamma_1, \gamma_2, ..., \gamma_s$ be a multiplicity system on R itself. If now i is any integer satisfying $0 \leqslant i \leqslant s$, then*

$$e_R(\gamma_1, ..., \gamma_s | E)$$
$$= \sum_P e_{R_P}(\phi_P \gamma_1, ..., \phi_P \gamma_i | E_P) \, e_{R/P}(\psi_P \gamma_{i+1}, ..., \psi_P \gamma_s | R/P).$$

Here P ranges over all the minimal prime ideals of $(\gamma_1, \gamma_2, ..., \gamma_i)$, $\phi_P : R \to R_P$ is the canonical mapping of R into R_P, and $\psi_P : R \to R/P$ is the natural mapping of R on to R/P.

Remark. Note that if P is a minimal prime ideal of $(\gamma_1, \gamma_2, ..., \gamma_i)$, then $\phi_P \gamma_1, \phi_P \gamma_2, ..., \phi_P \gamma_i$ is a multiplicity system on R_P considered as a module with respect to itself. This is because R_P is a local ring whose maximal ideal is the only prime ideal containing $\phi_P \gamma_1, \phi_P \gamma_2, ..., \phi_P \gamma_i$. The assertion therefore follows from Proposition 7 Cor. of section (4.4).

Proof. We use induction on s. However before we start the inductive argument it should be noted that the theorem certainly holds in the special case $i = 0$ because it then reduces to Proposition 11. In particular, the theorem is true when $s = 0$ since, in that situation, $i = 0$ as well.

It will now be assumed that $s \geqslant 1$ and that the theorem has been proved in the case of multiplicity systems having only $s - 1$ elements. It will also be supposed that $1 \leqslant i \leqslant s$ for, as has already been observed, the case $i = 0$ presents no problem.

Put $R^* = R/(\gamma_1)$ and let γ_i^* denote the natural image of γ_i in R^*. Then

$$e_R(\gamma_1, \gamma_2, ..., \gamma_s | E) = e_{R^*}(\gamma_2^*, ..., \gamma_s^* | E/\gamma_1 E) - e_{R^*}(\gamma_2^*, ..., \gamma_s^* | 0 :_E \gamma_1).$$

Next let P be a typical minimal prime ideal of $(\gamma_1, \gamma_2, ..., \gamma_i)$ and put $P^* = P/(\gamma_1)$. Then P^* is a typical minimal prime ideal of $(\gamma_2^*, ..., \gamma_i^*)$. Let $\phi_{P^*} : R^* \to R^*_{P^*}$ be the canonical mapping of R^* into $R^*_{P^*}$ and ψ_{P^*} the natural mapping of R^* on to R^*/P^*. Then, by the inductive hypothesis,

$$e_{R^*}(\gamma_2^*, ..., \gamma_s^* | E/\gamma_1 E)$$
$$= \sum_P e_{R^*_{P^*}}(\phi_{P^*} \gamma_2^*, ..., \phi_{P^*} \gamma_i^* | (E/\gamma_1 E)_{P^*}) \, e_{R^*/P^*}(\psi_{P^*} \gamma_{i+1}^*, ..., \psi_{P^*} \gamma_s^* | R^*/P^*)$$

and

$$e_{R^*}(\gamma_2^*, ..., \gamma_s^* | 0 :_E \gamma_1)$$
$$= \sum_P e_{R^*_{P^*}}(\phi_{P^*} \gamma_2^*, ..., \phi_{P^*} \gamma_i^* | (0 :_E \gamma_1)_{P^*}) \, e_{R^*/P^*}(\psi_{P^*} \gamma_{i+1}^*, ..., \psi_{P^*} \gamma_s^* | R^*/P^*).$$

Now the rings R^*/P^* and R/P are isomorphic under an isomorphism which makes $\psi_{P^*}\gamma_j^*$ correspond to $\psi_P\gamma_j$. Hence

$$e_{R^*/P^*}(\psi_{P^*}\gamma_{i+1}^*, ..., \psi_{P^*}\gamma_s^* \,|\, R^*/P^*)$$
$$= e_{R/P}(\psi_P\gamma_{i+1}, ..., \psi_P\gamma_s \,|\, R/P).$$

Next, by Proposition 9 of section (3.5), the rings $R_P/(\phi_P\gamma_1)R_P$ and $R_{P^*}^*$ may be identified in a natural manner. Furthermore, if we denote $R_P/(\phi_P\gamma_1)R_P$ by $\overline{R}_P$ and the image of $\phi_P\gamma_j$ in $\overline{R}_P$ by $\overline{\phi_P\gamma_j}$, then the identification of the rings leads to the identification of $\overline{\phi_P\gamma_j}$ and $\phi_{P^*}\gamma_j^*$. Again,† assuming that $\overline{R}_P$ and $R_{P^*}^*$ are identified, the $R_{P^*}^*$-module $(E/\gamma_1 E)_{P^*}$ is isomorphic to the $\overline{R}_P$-module $E_P/(\phi_P\gamma_1)E_P$ and the $R_{P^*}^*$-module $(0:_E\gamma_1)_{P^*}$ to the $\overline{R}_P$-module $0:_{E_P}\phi_P\gamma_1$. It follows that

$$e_{R^*{}_{P^*}}(\phi_{P^*}\gamma_2^*, ..., \phi_{P^*}\gamma_i^* \,|\, (E/\gamma_1 E)_{P^*})$$
$$= e_{\overline{R_P}}(\overline{\phi_P\gamma_2}, ..., \overline{\phi_P\gamma_i} \,|\, E_P/(\phi_P\gamma_1)E_P)$$
$$= e_{R_P}(\phi_P\gamma_2, ..., \phi_P\gamma_i \,|\, E_P/(\phi_P\gamma_1)E_P)$$

and $\quad e_{R^*{}_{P^*}}(\phi_{P^*}\gamma_2^*, ..., \phi_{P^*}\gamma_i^* \,|\, (0:_E\gamma_1)_{P^*})$
$$= e_{\overline{R_P}}(\overline{\phi_P\gamma_2}, ..., \overline{\phi_P\gamma_i} \,|\, (0:_{E_P}\phi_P\gamma_1))$$
$$= e_{R_P}(\phi_P\gamma_2, ..., \phi_P\gamma_i \,|\, (0:_{E_P}\phi_P\gamma_1)).$$

Thus the difference between

$$e_{R^*{}_{P^*}}(\phi_{P^*}\gamma_2^*, ..., \phi_{P^*}\gamma_i^* \,|\, (E/\gamma_1 E)_{P^*})$$
and $\quad e_{R^*{}_{P^*}}(\phi_{P^*}\gamma_2^*, ..., \phi_{P^*}\gamma_i^* \,|\, (0:_E\gamma_1)_{P^*})$
is simply $\quad e_{R_P}(\phi_P\gamma_1, \phi_P\gamma_2, ..., \phi_P\gamma_i \,|\, E_P).$

Hence

$$\sum_P e_{R_P}(\phi_P\gamma_1, ..., \phi_P\gamma_i \,|\, E_P)\, e_{R/P}(\psi_P\gamma_{i+1}, ..., \psi_P\gamma_s \,|\, R/P)$$
$$= e_{R^*}(\gamma_2^*, ..., \gamma_s^* \,|\, E/\gamma_1 E) - e_{R^*}(\gamma_2^*, ..., \gamma_s^* \,|\, 0:_E\gamma_1)$$
$$= e_R(\gamma_1, \gamma_2, ..., \gamma_s \,|\, E).$$

This completes the proof of the associative law.

7.10 Further consideration of Hilbert functions

In the present section we return to the consideration of Hilbert functions. This was begun in section (7.6) but was taken only to the point where it could be applied to the study of multiplicities. As before

---

† See Propositions 10 and 11 of section (3.5) for the justification of the assertions that follow.

we shall be concerned with a polynomial ring $R[X_1, X_2, ..., X_s] = R[X]$ (say) in s indeterminates and we shall grade the ring by means of the non-negative integers in the usual way. However we now add the condition that the coefficient ring R is to be *commutative*.

Let E be a graded $R[X]$-module. The R-module formed by the homogeneous elements of degree m $(m \geqslant 0)$ is, of course, denoted by E_m. It is convenient to make the convention that if k is a *negative* integer then by E_k we shall always mean the zero R-module. Put

$$H(n, E) = L_R(E_n).$$

Thus for $n \geqslant 0$ $H(n, E)$ is the Hilbert function of E and we have extended the definition by making the function zero for all negative values of the argument.

Suppose now that E is a Hilbert $R[X_1, X_2, ..., X_s]$-module. Then

$$E = R[X] e_1 + R[X] e_2 + ... + R[X] e_p,$$

where $e_1, e_2, ..., e_p$ are suitably chosen homogeneous elements of E. Let e_i be of degree m_i. Then

$$L_R(Re_i) \leqslant H(m_i, E) < \infty$$

whence $L_R(Re_1 + Re_2 + ... + Re_p) < \infty.$

Put $A = \mathrm{Ann}_R(Re_1 + ... + Re_p)$. Then, by Theorem 2 Cor. of section (4.2), R/A is an *Artin ring*. Furthermore, the $R[X]$-ideal $AR[X]$ annihilates E so we can regard E as a module with respect to the residue class ring $R[X]/AR[X]$. But† $R[X]/AR[X]$ may be identified with $(R/A)[X]$ and therefore E may be considered as a module with respect to the new polynomial ring $(R/A)[X]$. Indeed if we keep the original grading on E, then E is a graded $(R/A)[X]$-module *whose Hilbert function is exactly the same as before*. Since it is the Hilbert function we propose to investigate, we shall sacrifice no generality if we postulate that, *from now on, R will always denote a (commutative) Artin ring*.

Suppose next that E is any finitely generated, graded, $R[X]$-module. The fact that R is an Artin ring ensures that $R[X]$ is a Noetherian ring. Hence E is a Noetherian $R[X]$-module. It follows, by Theorem 12 of section (4.3), that E_m is a Noetherian and therefore finitely generated R-module. Accordingly

$$H(m, E) = L_R(E_m) < \infty.$$

This shows that (provided R is an Artin ring) *every finitely generated, graded, R[X]-module is a Hilbert R[X]-module*.

† See (5.4.6).

Lemma 7. *Let R be an Artin ring and P a homogeneous prime ideal of the polynomial ring $R[X_1, X_2, ..., X_s]$, where the latter is graded in the usual way by the non-negative integers. Then $\dim P = 0$ if and only if $(X_1, X_2, ..., X_s) \subseteq P$.*

Proof. Assume first that $(X_1, X_2, ..., X_s) \subseteq P$. Then consideration of the natural mapping $R[X] \to R[X]/P$ shows that $R[X]/P$ is ring-isomorphic to $R/(R \cap P)$. But, since $R \cap P$ is a prime ideal of an Artin ring, it is a maximal ideal of that ring. Thus $R/(R \cap P)$ is a field and therefore $R[X]/P$ is a field as well. This shows that P is a maximal ideal and therefore $\dim P = 0$.

Assume next that for some i $(1 \leqslant i \leqslant s)$ we have $X_i \notin P$. Then (X_i, P) is a homogeneous ideal and it is clear that it cannot contain the identity element of $R[X]$. If now P' is any minimal prime ideal of (X_i, P), then P' strictly contains P and therefore $\dim P > 0$. The lemma follows.

Lemma 8. *Let R be an Artin ring and $E \neq 0$ a finitely generated, graded, $R[X_1, X_2, ..., X_s]$-module. Then the Hilbert function $H(n, E)$ is zero for all large values of n if and only if $\operatorname{Dim} E = 0$.*

Remark. We recall that the dimension, $\operatorname{Dim} E$, of the module E is, by definition, the same as the dimension of the ring $R[X]/\operatorname{Ann}_{R[X]}E$. Thus $\operatorname{Dim} E$ is equal to the maximum value of $\dim P$, where P ranges over all the minimal prime ideals of $\operatorname{Ann}_{R[X]}E$ or (equivalently) over all the minimal prime ideals belonging to the zero submodule of E. Note that, by Theorem 24 of section (2.13), all these prime ideals are homogeneous. Again, by Theorem 4 of section (6.2), $\operatorname{Dim} E$ cannot exceed s.

Proof. First assume that $H(n, E) = 0$ for all $n \geqslant q$. Then there are no non-zero homogeneous elements of degree q or greater. Hence

$$(X_1, X_2, ..., X_s)^q E = 0.$$

Let P be a minimal prime ideal of $\operatorname{Ann}_{R[X]}E$. Then P is homogeneous and contains $(X_1, X_2, ..., X_s)$. Consequently $\dim P = 0$ by Lemma 7. This shows that $\operatorname{Dim} E = 0$.

Next suppose that $\operatorname{Dim} E = 0$. Then, by Lemma 7, every minimal prime ideal of $\operatorname{Ann}_{R[X]}E$, since it is homogeneous and of dimension zero, contains the ideal $(X_1, X_2, ..., X_s)$. Thus the radical of $\operatorname{Ann}_{R[X]}E$ contains $(X_1, X_2, ..., X_s)$ and therefore $(X_1, X_2, ..., X_s)^l E = 0$ for some positive integer l. Let $E = R[X]e_1 + R[X]e_2 + ... + R[X]e_p$, where e_i is homogeneous of degree m_i. If now

$$n \geqslant \max(l_1 + m_1, l_2 + m_2, ..., l_p + m_p),$$

then every homogeneous element of degree n is zero and therefore $H(n, E) = 0$. This establishes the lemma.

Theorem 19. *Let R be an Artin ring and $E \neq 0$ a finitely generated, graded module over the polynomial ring $R[X_1, X_2, ..., X_s]$. Then, when n is sufficiently large, the Hilbert function $H(n, E)$ is equal to a polynomial in n whose degree is equal to $\mathrm{Dim}\, E - 1$.*

Remark. For this result it is necessary to treat the null polynomial as though its degree were equal to -1. Although this is at variance with the convention introduced in section (5.4) in a quite different context, it will be found that this is not, in fact, a source of confusion.

Proof. Put $d = \mathrm{Dim}\, E$. We shall proceed by induction on d. Note that when $d = 0$ the theorem follows from Lemma 8.

From this point on it will be supposed that $d \geqslant 1$. (The reader may like to be warned that some of the observations which follow now will be used again in the proof of Theorem 20.) By Proposition 10 Cor., there exist homogeneous submodules $E_0, E_1, ..., E_q$ of E and homogeneous prime ideals $P_1, P_2, ..., P_q$ such that

$$E = E_q \supseteq E_{q-1} \supseteq ... \supseteq E_1 \supseteq E_0 = 0,$$

and $R[X]/P_i$ is isomorphic to E_i/E_{i-1} under an isomorphism which increases the degree of a homogeneous element of $R[X]/P_i$ by a fixed amount, k_i say. Accordingly

$$H(n - k_i, R[X]/P_i) = H(n, E_i/E_{i-1}).$$

But, by (7.6.5), $\qquad H(n, E) = \sum_{i=1}^{q} H(n, E_i/E_{i-1}).$

Consequently $\qquad H(n, E) = \sum_{i=1}^{q} H(n - k_i, R[X]/P_i). \qquad (7.10.1)$

Next $\mathrm{Ann}_{R[X]}E$ annihilates E_i/E_{i-1} which is isomorphic to $R[X]/P_i$. Hence $\mathrm{Ann}_{R[X]}E \subseteq P_i$. Suppose now that P is a prime ideal containing $\mathrm{Ann}_{R[X]}E$. Then, since $P_i(E_i/E_{i-1}) = 0$, we have $P_i E_i \subseteq E_{i-1}$ and therefore $P_1 P_2 ... P_q E = 0$. Thus $P_1 P_2 ... P_q \subseteq \mathrm{Ann}_{R[X]}E \subseteq P$ and so $P_j \subseteq P$ for some j. *It follows that the minimal prime ideals of $\mathrm{Ann}_{R[X]}E$ are precisely those prime ideals P_i $(1 \leqslant i \leqslant q)$ which do not strictly contain any other member of the sequence $P_1, P_2, ..., P_q$.* In particular, it follows that

$$d = \max(\dim P_1, \dim P_2, ..., \dim P_q). \qquad (7.10.2)$$

Let us now suppose that the theorem has already been established in the case of modules whose dimensions are smaller than d and let P be a d-dimensional homogeneous prime ideal. Since $d \geqslant 1$, Lemma 7 shows that we can find j such that $1 \leqslant j \leqslant s$ and $X_j \notin P$. Consider the exact sequence

$$0 \to R[X]/P \to R[X]/P \to R[X]/(P, X_j) \to 0,$$

where the mapping $R[X]/P \to R[X]/P$ consists in multiplication by X_j. If we now endow each module in the sequence with the natural factor grading, then $R[X]/P \to R[X]/P$ raises the degree of a homogeneous element by unity whereas $R[X]/P \to R[X]/(P, X_j)$ preserves degrees. It follows that, with a self-explanatory notation, we have exact sequences

$$0 \to (R[X]/P)_{n-1} \to (R[X]/P)_n \to (R[X]/(P, X_j))_n \to 0$$

whence

$$H(n, R[X]/P) - H(n-1, R[X]/P) = H(n, R[X]/(P, X_j)).$$

By Theorem 11, when n is large $H(n, R[X]/P)$ is equal to a polynomial in n of degree δ say. Since, by Lemma 8, $H(n, R[X]/P)$ does not vanish for all large n, we must have $\delta \geqslant 0$. It follows that, for large n, $H(n, R[X]/P) - H(n-1, R[X]/P)$ is equal to a polynomial in n of degree $\delta - 1$ provided we observe the convention that the degree of the null polynomial is -1. Next, since P is homogeneous, $R[X]/(P, X_j)$ is a non-null module and its dimension is certainly smaller than d. Hence, by the inductive hypothesis, $H(n, R[X]/(P, X_j))$ is, for large values of n, equal to a polynomial in n whose degree is

$$\mathrm{Dim}\,(R[X]/(P, X_j)) - 1.$$

Thus $\delta = \mathrm{Dim}\,(R[X]/(P, X_j))$. Now let P' be any minimal prime ideal of (P, X_j). Then, from Theorem 21 of section (4.8), it follows that $\mathrm{rank}\,(P'/P) = 1$ and therefore there is no prime ideal strictly between P and P'. Accordingly, by Theorem 4 Cor. of section (6.2),

$$\dim P' = \dim P - 1 = d - 1.$$

It follows that $\mathrm{Dim}\,(R[X]/(P, X_j)) = d - 1$ and therefore $\delta = d - 1$. Thus to sum up: *if P is a d-dimensional homogeneous prime ideal, then, for all large values of n, $H(n, R[X]/P)$ is equal to a polynomial in n of degree $d - 1$.* Observe that the coefficient of n^{d-1} in the polynomial in question must, since it is non-zero, be strictly positive. This is because $H(n, R[X]/P) \geqslant 0$ for all n.

We know, from (7.10.2), that $\text{Dim}\,(R[X]/P_i) \leqslant d$ for each i. Hence, for all large n, $H(n-k_i, R[X]/P_i)$ is equal to a polynomial in n of degree $\text{Dim}\,(R[X]/P_i) - 1$. As i varies some of the values of $\text{Dim}\,(R[X]/P_i) - 1$ may be less than $d-1$ but there will be at least one case where the value is $d-1$ itself. When this happens the coefficient of n^{d-1} in the polynomial associated with $H(n-k_i, R[X]/P_i)$ will be strictly positive. It now follows, from (7.10.1), that, when n is large, $H(n, E)$ is equal to a polynomial in n whose degree is $d-1$ and with this the theorem is established.

Let us re-examine this result. If E is a finitely generated, graded $R[X]$-module and $\text{Dim}\,E = d \geqslant 1$, then, by (7.6.14),

$$H(n, E) = \sum_{\nu=0}^{s-1} h_\nu(E) \binom{n+\nu}{\nu} \quad (n\,\text{large}),$$

where the $h_\nu(E)$ are the Hilbert coefficients of E. These we know to be integers. It now follows, from Theorem 19, that $h_\nu(E) = 0$ for $d \leqslant \nu \leqslant s-1$ and $h_{d-1}(E) > 0$.

Definition. *If* $\text{Dim}\,E = d\,(d \geqslant 1)$, *then the strictly positive integer* $h_{d-1}(E)$ *will be called the 'order' of* E *and it will be denoted by* $\text{Ord}\,E$.

Thus when n is large we have a representation for $H(n, E)$ of the form

$$H(n, E) = \text{Ord}\,E\,\frac{n^{d-1}}{(d-1)!} + \dots,$$

where $+\dots$ indicates terms of lower degree. Suppose now that

$$P_1, P_2, \dots, P_q$$

are the homogeneous prime ideals which occur in the proof of Theorem 19 and let I be the subset of $\{1, 2, \dots, q\}$ such that $\dim P_i = d$ precisely when $i \in I$. Then $\text{Dim}\,(R[X]/P_j) < d$ if $j \notin I$ and so, by (7.10.1),

$$\text{Ord}\,E = \sum_{i \in I} \text{Ord}\,(R[X]/P_i). \qquad (7.10.3)$$

Now let P be an arbitrary d-dimensional homogeneous prime ideal and let λ_P be the number of times it occurs in the sequence $\{P_1, P_2, \dots, P_q\}$. Then (7.10.3) may be written as

$$\text{Ord}\,E = \sum_P \lambda_P \text{Ord}\,(R[X]/P). \qquad (7.10.4)$$

Further

$$L_{R[X]_P}(E_P) = \sum_{i=1}^{q} L_{R[X]_P}\{(E_i/E_{i-1})_P\} = \sum_{i=1}^{q} L_{R[X]_P}\{(R[X]/P_i)_P\}.$$

Now, because $\dim P_i \leqslant d$, we must have either $P_i = P$ or $P_i \nsubseteq P$. If $P_i = P$, then $(R[X]/P_i)_P$ is an $R[X]_P$-module of unit length whereas if $P_i \nsubseteq P$, then $(R[X]/P_i)_P = 0$. This shows that $L_{R[X]_P}(E_P) = \lambda_P$ and, on substituting in (7.10.4), we obtain

Theorem 20. *Let R be an Artin ring and E a finitely generated, graded $R[X_1, X_2, ..., X_s]$-module satisfying $\operatorname{Dim} E \geqslant 1$. Then*

$$\operatorname{Ord} E = \sum_P L_{R[X]_P}(E_P) \operatorname{Ord}(R[X]/P),$$

where the summation is taken over all homogeneous prime ideals P such that $\operatorname{Dim}(R[X]/P) = \operatorname{Dim} E$.

Exercises on Chapter 7

In these exercises R denotes a ring with an identity element. It is not assumed that R is commutative unless there is an explicit statement to that effect.

1. E is a Noetherian R-module and γ a central element of R. Show that when m is sufficiently large the modules $\gamma^m E/\gamma^{m+1}E$ and $\gamma^{m+1}E/\gamma^{m+2}E$ are isomorphic.

2. Let R be a ring which satisfies the maximal condition for left ideals and let E be a finitely generated R-module. Show that every endomorphism of E *on to* itself is an isomorphism.

3. Let R be a *commutative* ring, E a finitely generated R-module, and $f : E \to E$ an endomorphism such that $f(E) = E$. Show that f is an isomorphism. (*Hint.* Replace R by a suitable Noetherian subring and apply the result of Exercise 2.)

4. Give an example of a ring R, an R-module E, and multiplicity system $\gamma_1, \gamma_2, ..., \gamma_s$ on E, such that $\gamma_1, \gamma_2, ..., \gamma_s$ fails to be a multiplicity system on at least one submodule of E.

5. Show that if E is a *Noetherian* R-module and γ is a central element, then $L_R(0 :_E \gamma) \leqslant L_R(E/\gamma E)$. Give an example of a ring R, an R-module M, and a central element γ, such that $L_R(M/\gamma M)$ and $L_R(0 :_M \gamma)$ are both finite and $L_R(0 :_M \gamma) > L_R(M/\gamma M)$.

6. Let E be an R-module and γ, γ' central elements of R. Show that there exists an exact sequence of the form

$$0 \to 0 :_E \gamma' \to 0 :_E \gamma\gamma' \to 0 :_E \gamma \to E/\gamma'E \to E/\gamma\gamma'E \to E/\gamma E \to 0.$$

Use this to give an alternative to the last part of the proof of Theorem 7.

7. Use Theorem 7 to show that if $\gamma_1, \gamma_2, ..., \gamma_s$ is a multiplicity system on a Noetherian R-module E and γ is an arbitrary central element, then

$$e_R(\gamma, \gamma_1, ..., \gamma_s | E) = 0.$$

8. Let $R = A[X_1, X_2, ..., X_s]$, where A is a (commutative) Artin ring and $X_1, X_2, ..., X_s$ are indeterminates. Show that

$$e_R(X_1^{n_1}, X_2^{n_2}, ..., X_s^{n_s} | R) = n_1 n_2 ... n_s L_A(A),$$

for arbitrary positive integers $n_1, n_2, ..., n_s$.

9. The polynomial $f(t)$ is of degree p, has complex coefficients, and is such that $f(n)$ is an integer whenever n is a large positive integer. Prove that

$$f(t) = \sum_{\nu=0}^{p} c_\nu \frac{(t+\nu)(t+\nu-1)\dots(t+1)}{\nu!},$$

where $c_0, c_1, \dots, c_p$ are integers which do not depend on t.

10. Let R be a commutative ring, E a Noetherian R-module, and $\gamma_1, \gamma_2, \dots, \gamma_s$ a multiplicity system on E. Assume that there exists a maximal ideal M, of R, such that $(\text{Ann}_R E, \gamma_1, \dots, \gamma_s) \subseteq M$ and $\text{rank}(M/\text{Ann}_R E) = s$. Now show that $e_R(\gamma_1, \dots, \gamma_s | E)$ is strictly positive.

11. Let R be a d-dimensional $(d \geqslant 0)$ local ring whose maximal ideal M can be generated by d elements. (A local ring with this property is called a *regular local ring*.) Prove that if $M = (u_1, u_2, \dots, u_d)$, then

$$e_R(u_1, u_2, \dots, u_d | R) = L_R(R/(u_1, \dots, u_d)) = 1.$$

Deduce that, for $0 \leqslant i \leqslant d$, the ideal $(u_1, u_2, \dots, u_i)$ is prime.

12. A proper left ideal P of a ring R will be called a *pseudo-prime ideal* if it satisfies the following condition: *whenever $r\gamma \in P$ and γ is a central element of R not contained in P, then $r \in P$.* Prove that if $E \neq 0$ is a Noetherian R-module, then there exists an element $e \in E$ and a pseudo-prime ideal P, such that Re and R/P are isomorphic R-modules.

13. $E \neq 0$ is a Noetherian R-module. Show that there exists an increasing sequence

$$0 = E_0 \subseteq E_1 \subseteq E_2 \subseteq \dots \subseteq E_m = E$$

of submodules of E and pseudo-prime ideals (see Exercise 12) $P_1, P_2, \dots, P_m$ such that for each $i\,(1 \leqslant i \leqslant m)$ we have an isomorphism $E_i/E_{i-1} \approx R/P_i$ of R-modules.

14. Use Exercise 13 to give an alternative proof of Lemma 4.

15. Let R be a (commutative) Artin ring and $\phi_1(X), \phi_2(X), \dots, \phi_p(X)$ homogeneous polynomials belonging to $R[X_1, X_2, \dots, X_s]$. Show that if the ideal $(\phi_1, \phi_2, \dots, \phi_p)$ has rank equal to p and the degree of $\phi_i(X)$ is $l_i\,(l_i > 0)$, then $H(n, R[X]/(\phi_1, \dots, \phi_p))$ is equal to the coefficient of t^n in the expansion of

$$L_R(R) \frac{(1-t^{l_1})(1-t^{l_2})\dots(1-t^{l_p})}{(1-t)^s}$$

as a power series in t. Show also that if $p < s$, then

$$\text{Ord}\,(R[X]/(\phi_1, \dots, \phi_p)) = l_1 l_2 \dots l_p L_R(R).$$

16. Let R be a commutative Noetherian ring and $I = (\gamma_1, \gamma_2, \dots, \gamma_s)$, where $s \geqslant 1$, a proper ideal satisfying $L_R(R/I) < \infty$. Prove that the following two statements are equivalent:

(a) $e_R(\gamma_1, \gamma_2, \dots, \gamma_s | R) = L_R(R/I)$;

(b) whenever $F(X_1, X_2, \dots, X_s)$ is a form, with coefficients in R, such that $F(\gamma_1, \gamma_2, \dots, \gamma_s) = 0$, then all the coefficients of F are in I.

(*Hint.* Use Theorems 12, 13 and 19.)

8

THE KOSZUL COMPLEX

General remarks. The aim of this chapter is to show how the properties of the Koszul complex throw light on certain aspects of the theory of multiplicities and the theory of grade. As usual R denotes a ring with an identity element. (It is not assumed that R is commutative unless there is a definite statement to the contrary.) The term *R-module* should always be interpreted as meaning a *left* R-module.

8.1 Complexes

By a *complex of R-modules* we shall understand a sequence

$$... \to X_{n+1} \xrightarrow{d_{n+1}} X_n \xrightarrow{d_n} X_{n-1} \xrightarrow{d_{n-1}} X_{n-2} \to ... \qquad (8.1.1)$$

of R-modules and R-homomorphisms extending indefinitely in both directions, with the property that

$$d_{n-1}d_n = 0 \qquad (8.1.2)$$

for all values of n. X_n will be called the *component module of degree n* or the *nth component* of the complex. We shall refer to d_n as the *nth boundary homomorphism* or the *nth differentiation homomorphism*.

In order to avoid writing out full details on every occasion, we shall refer to (8.1.1) as the complex (X, d) or, more simply, as the complex X. One way to put this on a more formal basis is to set

$$X = \bigoplus_{-\infty < n < \infty} X_n$$

and to define an R-endomorphism $d: X \to X$ so that $d(x_n) = d_n(x_n)$ for $x_n \in X_n$. In this way X becomes a graded module over R considered as a ring with a trivial grading, and d is such that it decreases the degree of a homogeneous element by unity. Furthermore the condition described in (8.1.2) now takes the form $d^2 = 0$.

Let (X, d) be a complex of R-modules and put

$$Z_n(X) = \text{Ker}\,(X_n \to X_{n-1}) = \text{Ker}\,d_n, \qquad (8.1.3)$$

$$B_n(X) = \text{Im}\,(X_{n+1} \to X_n) = \text{Im}\,d_{n+1}. \qquad (8.1.4)$$

The elements of $Z_n(X)$ are called the *n-cycles* of X and those of $B_n(X)$ the *n-boundaries*. It follows, from (8.1.2), that

$$B_n(X) \subseteq Z_n(X).$$

We now put $\qquad\qquad H_n(X) = Z_n(X)/B_n(X) \qquad\qquad$ (8.1.5)

and call $H_n(X)$ the *nth homology module* of the complex. Thus the sequence (8.1.1) is exact when and only when all the homology modules of X are zero.

Suppose that (X', d') and (X, d) are given complexes of R-modules. By *a mapping ϕ of the first complex into the second* we shall understand a family $\{\phi_n\}$ of R-homomorphisms

$$\phi_n : X'_n \to X_n \quad (-\infty < n < \infty)$$

such that the diagram

$$
\begin{array}{ccc}
X'_n & \xrightarrow{\;d'_n\;} & X'_{n-1} \\
\phi_n \downarrow & & \downarrow \phi_{n-1} \\
X_n & \xrightarrow{\;d_n\;} & X_{n-1}
\end{array}
\qquad (8.1.6)
$$

is commutative, i.e. such that

$$\phi_{n-1} d'_n = d_n \phi_n \qquad\qquad (8.1.7)$$

for every n. When this is the case we abbreviate the notation and write $\phi : X' \to X$. Such a ϕ is said to be an *isomorphism of complexes* if each ϕ_n is an isomorphism of R-modules.

Let $\phi : X' \to X$ be a mapping of the complex X' into the complex X. The fact that the diagram (8.1.6) is commutative shows that

$$\phi_n(\operatorname{Ker} d'_n) \subseteq \operatorname{Ker} d_n,$$

that is to say $\phi_n(Z_n(X')) \subseteq Z_n(X)$. On the other hand

$$
\begin{array}{ccc}
X'_{n+1} & \xrightarrow{\;d'_{n+1}\;} & X'_n \\
\phi_{n+1} \downarrow & & \downarrow \phi_n \\
X_{n+1} & \xrightarrow{\;d_{n+1}\;} & X_n
\end{array}
$$

is also commutative and therefore $\phi_n(\operatorname{Im} d'_{n+1}) \subseteq \operatorname{Im} d_{n+1}$. This may be rewritten as $\phi_n(B_n(X')) \subseteq B_n(X)$. Thus, in the language of section (1.5),

ϕ_n gives rise to a mapping of the pair $(B_n(X'), Z_n(X'))$ into the pair $(B_n(X), Z_n(X))$. It follows that there is induced a homomorphism

$$\phi_n^*: Z_n(X')/B_n(X') \to Z_n(X)/B_n(X).$$

Let us use (8.1.5) to rewrite this as

$$\phi_n^*: H_n(X') \to H_n(X). \tag{8.1.8}$$

Thus we see that a mapping $\phi: X' \to X$ of complexes induces homomorphisms $H_n(X') \to H_n(X)$ of the homology modules of X' into the homology modules of X.

Now let X', X, X'' be complexes of R-modules and let $\phi: X' \to X$ and $\psi: X \to X''$ be mappings of complexes We say that

$$X' \overset{\phi}{\to} X \overset{\psi}{\to} X''$$

is an *exact consequence of complexes* if

$$X_n' \overset{\phi_n}{\longrightarrow} X_n \overset{\psi_n}{\longrightarrow} X_n''$$

is an exact sequence of R-modules for every value of n. It will be convenient to use the hard-worked symbol 0 to denote a *null complex*, that is to say one which has the property that all its component modules are null modules. On this understanding we naturally say that

$$0 \to X' \overset{\phi}{\longrightarrow} X \overset{\psi}{\longrightarrow} X'' \to 0$$

is an exact sequence of complexes if

$$0 \to X_n' \overset{\phi_n}{\longrightarrow} X_n \overset{\psi_n}{\longrightarrow} X_n'' \to 0$$

is an exact sequence of R-modules for all n.

Lemma 1. *Suppose that (X', d'), (X, d) and (X'', d'') are complexes of R-modules and that the sequence*

$$0 \to X' \overset{\phi}{\to} X \overset{\psi}{\to} X'' \to 0$$

is exact. Then each of the induced sequences

$$H_n(X') \overset{\phi_n^*}{\longrightarrow} H_n(X) \overset{\psi_n^*}{\longrightarrow} H_n(X'')$$

is also exact.

Proof. The mappings

$$Z_n(X') \to Z_n(X) \to Z_n(X'')$$

arise by restriction from the homomorphisms

$$X_n' \overset{\phi_n}{\longrightarrow} X_n \overset{\psi_n}{\longrightarrow} X_n''.$$

Consequently the result of combining $Z_n(X') \to Z_n(X)$ with

$$Z_n(X) \to Z_n(X'')$$

is a null homomorphism. This in turn implies that $\psi_n^* \phi_n^* = 0$ or

$$\operatorname{Im} \phi_n^* \subseteq \operatorname{Ker} \psi_n^*.$$

Now let $\xi_n \in H_n(X)$ and suppose that ξ_n is represented by $u_n \in Z_n(X)$. Assume further that $\xi_n \in \operatorname{Ker} \psi_n^*$ so that $\psi_n^*(\xi_n) = 0$. Then

$$\psi_n(u_n) \in B_n(X'')$$

that is to say $\psi_n(u_n) = d_{n+1}''(x_{n+1}'')$, where x_{n+1}'' is some element of X_{n+1}''. Since $\psi_{n+1}: X_{n+1} \to X_{n+1}''$ is an epimorphism, $x_{n+1}'' = \psi_{n+1}(x_{n+1})$ for an element $x_{n+1} \in X_{n+1}$ and then

$$\psi_n(u_n) = d_{n+1}'' \psi_{n+1}(x_{n+1}) = \psi_n d_{n+1}(x_{n+1})$$

because $d_{n+1}'' \psi_{n+1} = \psi_n d_{n+1}$. Since $\operatorname{Im} \phi_n = \operatorname{Ker} \psi_n$, this shows that

$$u_n - d_{n+1}(x_{n+1}) = \phi_n(x_n') \tag{8.1.9}$$

for some $x_n' \in X_n'$. Thus

$$u_n \equiv \phi_n(x_n') \pmod{B_n(X)}. \tag{8.1.10}$$

Now, by (8.1.9) and the fact that $d_n \phi_n = \phi_{n-1} d_n'$, we have

$$\phi_{n-1} d_n'(x_n') = d_n \phi_n(x_n') = d_n(u_n - d_{n+1}(x_{n+1})) = 0$$

and therefore $d_n'(x_n') = 0$ because ϕ_{n-1} is a monomorphism. Accordingly $x_n' \in Z_n(X')$ and so it determines an element, η_n say, belonging to $H_n(X')$. Finally, by (8.1.10), $\phi_n^*(\eta_n) = \xi_n$. This shows that $\operatorname{Ker} \psi_n^* \subseteq \operatorname{Im} \phi_n^*$ and completes the proof of the lemma.

Once again let

$$0 \to X' \overset{\phi}{\to} X \overset{\psi}{\to} X'' \to 0$$

be an exact sequence of complexes. For each n we shall construct an R-homomorphism

$$\Delta_n : H_n(X'') \to H_{n-1}(X').$$

These homomorphisms are known as *connecting homomorphisms*. The details of the construction of Δ_n are important and are set out in the following paragraph for future reference.

Let $\zeta_n \in H_n(X'')$. We can then find $u_n'' \in Z_n(X'')$ so that u_n'' is a representative of ζ_n. Now ψ_n is an epimorphism; consequently there exists $x_n \in X_n$ such that $\psi_n(x_n) = u_n''$. Further

$$\psi_{n-1} d_n(x_n) = d_n'' \psi_n(x_n) = d_n''(u_n'') = 0$$

because $u_n'' \in Z_n(X'')$. However the sequence

$$0 \to X_{n-1}' \overset{\phi_{n-1}}{\longrightarrow} X_{n-1} \overset{\psi_{n-1}}{\longrightarrow} X_{n-1}'' \to 0$$

is exact and so $d_n(x_n) = \phi_{n-1}(x'_{n-1})$ for some $x'_{n-1} \in X'_{n-1}$. Next

$$\phi_{n-2}d'_{n-1}(x'_{n-1}) = d_{n-1}\phi_{n-1}(x'_{n-1}) = d_{n-1}d_n(x_n) = 0$$

because $d_{n-1}d_n = 0$. But ϕ_{n-2} is a monomorphism. It therefore follows that $d'_{n-1}(x'_{n-1}) = 0$ that is to say $x'_{n-1} \in Z_{n-1}(X')$. We now put

$$\Delta_n(\zeta_n) = \text{Image of } x'_{n-1} \quad \text{in} \quad H_{n-1}(X'). \tag{8.1.11}$$

This completes the construction of Δ_n. Of course there is a certain degree of freedom in the way the construction is carried out so it is not yet clear that the mapping Δ_n is well defined. This point will now be examined.

Once again let $\zeta_n \in H_n(X'')$ and suppose that we repeat the steps carried above and this time, instead of the elements u''_n, x_n, x'_{n-1}, obtain v''_n, y_n, y'_{n-1} in their places. Then u''_n and v''_n both represent ζ_n in $H_n(X'') = Z_n(X'')/B_n(X'')$ and therefore $u''_n - v''_n$ belongs to $B_n(X'')$. Let $u''_n - v''_n = d''_{n+1}(x''_{n+1})$, where $x''_{n+1} \in X''_{n+1}$. Then $x''_{n+1} = \psi_{n+1}(t_{n+1})$ for a suitable element $t_{n+1} \in X_{n+1}$ and so $u''_n - v''_n = \psi_n d_{n+1}(t_{n+1})$. Again $\psi_n(x_n) = u''_n$ and $\psi_n(y_n) = v''_n$. Hence $x_n - y_n - d_{n+1}(t_{n+1})$ belongs to $\text{Ker } \psi_n = \text{Im } \phi_n$. In view of this we may write

$$x_n - y_n - d_{n+1}(x_{n+1}) = \phi_n(t'_n),$$

where $t'_n \in X'_n$. Let us now operate on both sides with d_n and use the facts that $d_n(x_n) = \phi_{n-1}(x'_{n-1})$ and $d_n(y_n) = \phi_{n-1}(y'_{n-1})$. This shows that

$$\phi_{n-1}(x'_{n-1}) - \phi_{n-1}(y'_{n-1}) = \phi_{n-1}d'_n(t'_n).$$

However ϕ_{n-1} is a monomorphism. It follows that $x'_{n-1} - y'_{n-1} = d'_n(t'_n)$ whence $x'_{n-1} \equiv y'_{n-1} \pmod{B_{n-1}(X')}$. Accordingly x'_{n-1} and y'_{n-1} have the same image in $H_{n-1}(X')$ and with this the verification that Δ_n is a well defined mapping is complete. It is now a trivial matter to check that Δ_n is R-linear. The details are left to the reader.

Theorem 1. *Let $(X', d'), (X, d), (X'', d'')$ be complexes of R-modules and suppose that*

$$0 \to X' \overset{\phi}{\to} X \overset{\psi}{\to} X'' \to 0 \tag{8.1.12}$$

is an exact sequence. Then

$$\ldots \to H_n(X') \to H_n(X) \to H_n(X'') \to H_{n-1}(X') \to H_{n-1}(X) \to \ldots \tag{8.1.13}$$

is an exact sequence of R-modules, where

(i) $H_n(X') \to H_n(X)$ *and* $H_n(X) \to H_n(X'')$ *are the mappings* ϕ_n^* *and* ψ_n^* *respectively;*

(ii) $H_n(X'') \to H_{n-1}(X')$ *is the connecting homomorphism* Δ_n.

Remark. The sequence (8.1.13) will be referred to as the *exact homology sequence* arising from (8.1.12).

Proof. We first consider the triplet

$$H_n(X) \xrightarrow{\psi_n^*} H_n(X'') \xrightarrow{\Delta_n} H_{n-1}(X'). \tag{8.1.14}$$

Let $\xi_n \in H_n(X)$ and suppose that ξ_n is represented by $u_n \in Z_n(X)$. Then $\psi_n^*(\xi_n)$ is represented by the element $\psi_n(u_n)$ of $Z_n(X'')$. To find $\Delta_n \psi_n^*(\xi_n)$ we let $\psi_n(u_n)$ play the role of u_n'' in the above construction. The elements x_n and x_{n-1}' may then be taken as u_n and 0 respectively. It follows, from (8.1.11), that $\Delta_n \psi_n^*(\xi_n) = 0$. Thus $\operatorname{Im} \psi_n^* \subseteq \operatorname{Ker} \Delta_n$.

Now let $\zeta_n \in \operatorname{Ker} \Delta_n$ and choose u_n'', x_n, x_{n-1}' as in the construction of $\Delta_n(\zeta_n)$. Since $\Delta_n(\zeta_n) = 0$, (8.1.11) shows that $x_{n-1}' \in B_{n-1}(X')$. Hence $x_{n-1}' = d_n'(y_n')$, where $y_n' \in X_n'$. Next

$$d_n(x_n) = \phi_{n-1}(x_{n-1}') = \phi_{n-1}d_n'(y_n') = d_n \phi_n(y_n').$$

Thus $x_n - \phi_n(y_n')$ belongs to $\operatorname{Ker} d_n$ and represents an element ξ_n say in $H_n(X)$. Further $\psi_n^*(\xi_n)$ is represented by

$$\psi_n(x_n) - \psi_n \phi_n(y_n') = u_n'' - 0$$

and therefore $\psi_n^*(\xi_n) = \zeta_n$. This shows that $\operatorname{Ker} \Delta_n \subseteq \operatorname{Im} \psi_n^*$ and completes the verification that (8.1.14) is exact.

We now turn our attention to the triplet

$$H_n(X'') \xrightarrow{\Delta_n} H_{n-1}(X') \xrightarrow{\phi_{n-1}^*} H_{n-1}(X). \tag{8.1.15}$$

Let $\zeta_n \in H_n(X'')$ and choose u_n'', x_n, x_{n-1}' as in the construction of $\Delta_n(\zeta_n)$ described earlier. Then $\phi_{n-1}^* \Delta_n(\zeta_n)$ is the image of

$$\phi_{n-1}(x_{n-1}') \quad \text{in} \quad H_{n-1}(X).$$

But $\qquad\qquad \phi_{n-1}(x_{n-1}') = d_n(x_n) \in B_{n-1}(X).$

Consequently $\phi_{n-1}^* \Delta_n(\zeta_n) = 0$ and we have proved that

$$\operatorname{Im} \Delta_n \subseteq \operatorname{Ker} \phi_{n-1}^*.$$

Next suppose that $\eta_{n-1} \in H_{n-1}(X')$ and that $\phi_{n-1}^*(\eta_{n-1}) = 0$. Choose $y_{n-1}' \in Z_{n-1}(X')$ so that y_{n-1}' represents η_{n-1}. Then, because

$$\phi_{n-1}^*(\eta_{n-1}) = 0,$$

we have $\phi_{n-1}(y_{n-1}') \in B_{n-1}(X)$, say

$$\phi_{n-1}(y_{n-1}') = d_n(y_n),$$

where $y_n \in X_n$. Put $y_n'' = \psi_n(y_n)$. Then

$$d_n''(y_n'') = d_n'' \psi_n(y_n) = \psi_{n-1} d_n(y_n) = \psi_{n-1} \phi_{n-1}(y_{n-1}') = 0,$$

because $\psi_{n-1}\phi_{n-1} = 0$. Thus $y_n'' \in Z_n(X'')$ and so it represents an element $\zeta_n \in H_n(X'')$. Consider $\Delta_n(\zeta_n)$. If we carry through the construction described above we can take the elements u_n'', x_n, x_{n-1}' to be y_n'', y_n, y_{n-1}' respectively. This shows that $\Delta_n(\zeta_n)$ is the image of y_{n-1}' in $H_{n-1}(X')$ namely η_{n-1}. Accordingly $\operatorname{Ker}\phi_{n-1}^* \subseteq \operatorname{Im}\Delta_n$ and (8.1.15) has been shown to be exact.

Finally, the sequence

$$H_n(X') \xrightarrow{\phi_n^*} H_n(X) \xrightarrow{\psi_n^*} H_n(X'')$$

is exact by Lemma 1. Thus everything is proved and the theorem established.

We conclude this section by introducing some terminology which will be useful later. Let (X', d') and (X, d) be complexes of R-modules. We say that (X', d') is a *subcomplex* of (X, d) provided

(i) X_n' *is a submodule of* X_n *for each* n;
(ii) *the inclusion mappings* $X_n' \to X_n$ *constitute a mapping of the complex* (X', d') *into the complex* (X, d).

The latter condition means, of course, that, for each n, d_n coincides with d_n' on X_n'.

Suppose that (X', d') is a subcomplex of (X, d) and put $X_n'' = X_n/X_n'$. Then

$$d_n(X_n') = d_n'(X_n') \subseteq X_{n-1}'$$

and therefore d_n maps the pair (X_n', X_n) into the pair (X_{n-1}', X_{n-1}). Accordingly there is induced a homomorphism

$$d_n'': X_n'' \to X_{n-1}''.$$

Evidently $d_{n-1}'' d_n'' = 0$ so we have a new complex (X'', d''). This is called the *factor complex* of (X, d) with respect to (X', d'). Now the diagram

$$
\begin{array}{ccc}
X_n & \xrightarrow{d_n} & X_{n-1} \\
\downarrow & & \downarrow \\
X_n'' & \xrightarrow{d_n''} & X_{n-1}''
\end{array}
$$

is commutative, where the vertical mappings are natural mappings on to factor modules. Thus the natural mappings $X_n \to X_n''$ constitute a mapping of (X, d) into (X'', d''). Moreover, for each n, the sequence

$$0 \to X_n' \to X_n \to X_n'' \to 0$$

is exact. Thus, to sum up, *when (X', d') is a subcomplex of (X, d) we can form the factor complex (X'', d'') and then these will be related by the exact sequence*

$$0 \to X' \to X \to X'' \to 0$$

of complexes.

A complex (X, d) is called a *left* complex if $X_n = 0$ for all $n < 0$. Thus a typical left complex has the form

$$\ldots \to X_n \to X_{n-1} \to \ldots \to X_1 \to X_0 \to 0 \to 0 \to \ldots.$$

It will be seen presently that the Koszul complex is of this type. Suppose now that

$$0 \to X' \to X \to X'' \to 0$$

is an exact sequence of left complexes. By Theorem 1 this gives rise to an exact homology sequence. Now, in the situation being considered, $H_n(X')$, $H_n(X)$ and $H_n(X'')$ will be null modules if $n < 0$. Consequently the exact homology sequence takes the form

$$\ldots \to H_1(X) \to H_1(X'') \to H_0(X') \to H_0(X) \to H_0(X'') \to 0 \to 0 \to \ldots.$$

8.2 Construction of the Koszul complex

Let E be an R-module and $\gamma_1, \gamma_2, \ldots, \gamma_s$ ($s \geqslant 0$) central elements of R. Using these we propose to construct a complex of R-modules which will be called the *Koszul complex of E with respect to $\gamma_1, \gamma_2, \ldots, \gamma_s$*. The notation used for this complex will vary a little depending on just how explicit we wish to be. Thus if we wish to indicate all the ingredients in the construction we shall use $K(\gamma_1, \ldots, \gamma_s|R|E)$. This will be shortened to $K(\gamma_1, \ldots, \gamma_s|E)$ if there is no uncertainty about the ring of operators. Occasionally $K(\gamma|E)$ will be used if we are concerned with a *fixed* sequence $\gamma_1, \gamma_2, \ldots, \gamma_s$ of central elements and we need to avoid typographical complexities.

The Koszul complex $K(\gamma_1, \ldots, \gamma_s|E)$ has the form

$$\ldots \to 0 \to 0 \to K_s(\gamma|E) \to \ldots \to K_1(\gamma|E) \to K_0(\gamma|E) \to 0 \to 0 \to \ldots$$

so that, in particular, it is a left complex. For $0 \leqslant \mu \leqslant s$, the μth component $K_\mu(\gamma|E)$ is a direct sum of $\binom{s}{\mu}$ copies of E, where $\binom{s}{\mu}$ is the usual binomial coefficient. Thus when $s = 0$, $K(\cdot|E)$ is simply

$$\ldots \to 0 \to 0 \to E \to 0 \to 0 \to \ldots,$$

where E occurs as the component $K_0(\cdot|E)$.

For $s \geqslant 1$ it is necessary to describe the boundary homomorphisms and to assist in this we introduce some special notation. Let

$$T_1, T_2, \ldots, T_s$$

be new symbols and write

$$K_\mu(\gamma_1, \ldots, \gamma_s | E) = \bigoplus_{i_1 < i_2 < \ldots < i_\mu} E T_{i_1} T_{i_2} \ldots T_{i_\mu}, \qquad (8.2.1)$$

where $i_1, i_2, \ldots, i_\mu$ are integers which vary so that

$$1 \leqslant i_1 < i_2 < \ldots < i_\mu \leqslant s.$$

Thus, when $0 \leqslant \mu \leqslant s$, $K_\mu(\gamma_1, \ldots, \gamma_s | E)$ is effectively a direct sum of $\binom{s}{\mu}$ copies of E and a typical element of it has a *unique* representation in the form

$$\sum_{i_1 < i_2 < \ldots < i_\mu} e_{i_1 i_2 \ldots i_\mu} T_{i_1} T_{i_2} \ldots T_{i_\mu},$$

where $e_{i_1 i_2 \ldots i_\mu}$ belongs to E. Since it may not be quite clear how (8.2.1) is to be interpreted in the case $\mu = 0$, we state explicitly that

$$K_0(\gamma_1, \ldots, \gamma_s | E) = E. \qquad (8.2.2)$$

The next step is to define the boundary homomorphism

$$d_\mu \colon K_\mu(\gamma_1, \ldots, \gamma_s | E) \to K_{\mu-1}(\gamma_1, \ldots, \gamma_s | E)$$

and here it is only values of μ that satisfy $0 < \mu \leqslant s$ that need consideration. Suppose therefore that μ is in this range and let $i_1, i_2, \ldots, i_\mu$ be integers for which $1 \leqslant i_1 < \ldots < i_\mu \leqslant s$. Since each element of $K_\mu(\gamma_1, \ldots, \gamma_s | E)$ is uniquely expressible as a sum of elements of the form $e T_{i_1} T_{i_2} \ldots T_{i_\mu}$, where $e \in E$, we can specify d_μ by assigning a suitable meaning to $d_\mu(e T_{i_1} T_{i_2} \ldots T_{i_\mu})$. This we do by putting

$$d_\mu(e T_{i_1} T_{i_2} \ldots T_{i_\mu}) = \sum_{p=1}^{\mu} (-1)^{p-1} \gamma_{i_p} e T_{i_1} \ldots \hat{T}_{i_p} \ldots T_{i_\mu}, \qquad (8.2.3)$$

where the $\wedge$ over T_{i_p} means that this factor is to be omitted. (Note that the construction leads to an *R-linear* mapping because the γ_i belong to the *centre* of R.) Here it is understood that when $\mu = 1$ (8.2.3) asserts that

$$d_1(e T_{i_1}) = \gamma_{i_1} e. \qquad (8.2.4)$$

It is now necessary to show that

$$\ldots \to K_{\mu+1}(\gamma_1, \ldots, \gamma_s | E) \xrightarrow{d_{\mu+1}} K_\mu(\gamma_1, \ldots, \gamma_s | E)$$

$$\xrightarrow{d_\mu} K_{\mu-1}(\gamma_1, \ldots, \gamma_s | E) \to \ldots \qquad (8.2.5)$$

really is a complex which means that we must verify that $d_{\mu-1}d_\mu = 0$. For this we suppose that $0 \leqslant \mu-1 < \mu \leqslant s$ and check that

$$d_{\mu-1}d_\mu(eT_{i_1}T_{i_2}\dots T_{i_\mu}) = 0. \tag{8.2.6}$$

To this end note that if $1 \leqslant p < q \leqslant \mu$, then when we evaluate the left-hand side of (8.2.6) the term $\gamma_{i_p}\gamma_{i_q}eT_{i_1}\dots \hat{T}_{i_p}\dots \hat{T}_{i_q}\dots T_{i_\mu}$ occurs twice. On one occasion it is multiplied by $(-1)^{p-1}(-1)^{q-1}$ and on the other by $(-1)^{p-1}(-1)^{q-2}$. Thus the two terms cancel. Accordingly (8.2.6) has been established and (8.2.5) has been shown to be a complex.

Since $K(\gamma_1,\dots,\gamma_s|E)$ is a complex, we can form its homology modules. The μth homology module will be denoted by

$$H_\mu K(\gamma_1,\dots,\gamma_s|E)$$

which we shall sometimes abbreviate to $H_\mu K(\gamma|E)$. Evidently $H_\mu K(\gamma|E) = 0$ when $\mu > s$ or $\mu < 0$. Again a typical element of $K_1(\gamma_1,\dots,\gamma_s|E)$ has the form $\Sigma_i e_i T_i$, where $e_i \in E$, and by (8.2.4) we have

$$d_1\left(\sum_{i=1}^s e_i T_i\right) = \gamma_1 e_1 + \gamma_2 e_2 + \dots + \gamma_s e_s.$$

Thus $\operatorname{Im} d_1 = \gamma_1 E + \gamma_2 E + \dots + \gamma_s E$ whence, since

$$K_{-1}(\gamma_1,\dots,\gamma_s|E) = 0,$$

we have $\quad H_0 K(\gamma_1,\dots,\gamma_s|E) = E/(\gamma_1 E + \dots + \gamma_s E). \tag{8.2.7}$

Next each element of $K_s(\gamma_1,\dots,\gamma_s|E)$ has the form $eT_1T_2,\dots,T_s$. Also, by (8.2.3),

$$d_s(eT_1T_2\dots T_s) = \sum_{p=1}^s (-1)^{p-1}\gamma_p eT_1\dots \hat{T}_p\dots T_s.$$

Accordingly $d_s(eT_1T_2\dots T_s) = 0$ if and only if $\gamma_i e = 0$ for $i = 1,2,\dots,s$. In practice we identify $K_s(\gamma_1,\dots,\gamma_s|E)$ with E itself so that $eT_1T_2\dots T_s$ becomes identified with e. On this understanding

$$\operatorname{Ker} d_s = 0 :_E (\gamma_1 R + \dots + \gamma_s R).$$

Whence, since $K_{s+1}(\gamma_1,\dots,\gamma_s|E) = 0$,

$$H_s K(\gamma_1,\dots,\gamma_s|E) = 0 :_E (\gamma_1 R + \dots + \gamma_s R). \tag{8.2.8}$$

8.3 Properties of the Koszul complex

Let E be an R-module and $\gamma_1, \gamma_2, \dots, \gamma_s$ central elements. We shall show that the Koszul complex $K(\gamma_1, \gamma_2, \dots, \gamma_s|E)$ remains unchanged (to within isomorphism) if the sequence $\{\gamma_1, \gamma_2, \dots, \gamma_s\}$ is rearranged in any manner.

Proposition 1. *Let* $\{j_1, j_2, ..., j_s\}$ *be a permutation of* $\{1, 2, ..., s\}$. *Then the complexes* $K(\gamma_1, \gamma_2, ..., \gamma_s | E)$ *and* $K(\gamma_{j_1}, \gamma_{j_2}, ..., \gamma_{j_s} | E)$ *are isomorphic.*

Proof. Suppose that $1 \leqslant m < m+1 \leqslant s$. We shall show that

$$K(\gamma_1, ..., \gamma_m, \gamma_{m+1}, ..., \gamma_s | E)$$

and the complex $K(\gamma_1, ..., \gamma_{m+1}, \gamma_m, ..., \gamma_s | E)$, obtained from it by interchanging γ_m and γ_{m+1}, are isomorphic. From this the proposition will follow.

It is enough to establish that there exist isomorphisms

$$\phi_\mu : K_\mu(\gamma_1, ..., \gamma_m, \gamma_{m+1}, ..., \gamma_s | E) \to K_\mu(\gamma_1, ..., \gamma_{m+1}, \gamma_m, ..., \gamma_s | E)$$

which commute with the boundary homomorphisms. First we observe that the domain and range of the proposed ϕ_μ are the same module namely

$$\bigoplus_{1 \leqslant i_1 < ... < i_\mu \leqslant s} E T_{i_1} T_{i_2} ... T_{i_\mu}.$$

Next we describe ϕ_μ by saying what it does to an element

$$e T_{i_1} T_{i_2} ... T_{i_\mu}.$$

For this we must distinguish between the following four possibilities:

  (i) neither m or $m+1$ occurs among $\{i_1, i_2, ..., i_\mu\}$;

  (ii) m occurs among $\{i_1, i_2, ..., i_\mu\}$ but $m+1$ does not;

  (iii) $m+1$ occurs among $\{i_1, i_2, ..., i_\mu\}$ but m does not;

  (iv) both m and $m+1$ occur among $\{i_1, i_2, ..., i_\mu\}$.

The action of ϕ_μ now depends on which of these situation obtains. In case (i), the element $e T_{i_1} T_{i_2} ... T_{i_\mu}$ is left unchanged. In case (ii), $e T_{i_1} T_{i_2} ... T_{i_\mu} = e T_{i_1} ... T_m ... T_{i_\mu}$ and it is mapped into $e T_{i_1} ... T_{m+1} ... T_{i_\mu}$. In case (iii), $e T_{i_1} T_{i_2} ... T_{i_\mu} = e T_{i_1} ... T_{m+1} ... T_{i_\mu}$ and this time it is taken into $e T_{i_1} ... T_m ... T_{i_\mu}$. Finally, in case (iv), $e T_{i_1} T_{i_2} ... T_{i_\mu}$ is mapped into its negative. It is obvious that when ϕ_μ is defined in this way it produces an isomorphism of $K_\mu(\gamma_1, ..., \gamma_m, \gamma_{m+1}, ..., \gamma_s | E)$ on to

$$K_\mu(\gamma_1, ..., \gamma_{m+1}, \gamma_m, ..., \gamma_s | E)$$

and an easy verification shows that it commutes with the boundary homomorphisms. The proposition follows.

Corollary. *Let the assumptions be as in the proposition. Then for each μ we have an isomorphism*

$$H_\mu K(\gamma_1, \gamma_2, ..., \gamma_s | E) \approx H_\mu K(\gamma_{j_1}, \gamma_{j_2}, ..., \gamma_{j_s} | E)$$

of R-modules.

Once again let $\gamma_1, \gamma_2, ..., \gamma_s$ $(s \geqslant 0)$ be central elements of R but now suppose that both E' and E are R-modules. We can then form the

Koszul complexes $K(\gamma_1, ..., \gamma_s|E')$ and $K(\gamma_1, ..., \gamma_s|E)$. Suppose further that we are given a homomorphism $f: E' \to E$ of R-modules. Then from f we can derive homomorphisms

$$f_\mu: K_\mu(\gamma_1, ..., \gamma_s|E') \to K_\mu(\gamma_1, ..., \gamma_s|E) \qquad (8.3.1)$$

in which, with a self-explanatory notation,

$$f_\mu(e'T_{i_1}T_{i_2}... T_{i_\mu}) = f(e')T_{i_1}T_{i_2}... T_{i_\mu}. \qquad (8.3.2)$$

Evidently the f_μ will commute with the boundary homomorphisms of $K(\gamma_1, ..., \gamma_s|E')$ and $K(\gamma_1, ..., \gamma_s|E)$. Thus *each homomorphism $E' \to E$ of R-modules gives rise to a mapping $K(\gamma|E') \to K(\gamma|E)$ of the corresponding Koszul complexes.*

Finally assume that we have an exact sequence

$$0 \to E' \xrightarrow{f} E \xrightarrow{g} E'' \to 0 \qquad (8.3.3)$$

of R-modules. Then from f and g we obtain mappings

$$K(\gamma|E') \to K(\gamma|E)$$

and $K(\gamma|E) \to K(\gamma|E'')$ of complexes. Further, it is clear, from (8.3.2), that all sequences

$$0 \to K_\mu(\gamma|E') \xrightarrow{f_\mu} K_\mu(\gamma|E) \xrightarrow{g_\mu} K_\mu(\gamma|E'') \to 0$$

are exact. Accordingly the exact sequence (8.3.3) of R-modules gives rise to the exact sequence

$$0 \to K(\gamma|E') \to K(\gamma|E) \to K(\gamma|E'') \to 0 \qquad (8.3.4)$$

of complexes.

Theorem 2. *Let $\gamma_1, \gamma_2, ..., \gamma_s$ be central elements of R and let*

$$0 \to E' \to E \to E'' \to 0$$

be an exact sequence of R-modules. Then the homology modules of the Koszul complexes of E', E, E'' with respect to $\gamma_1, \gamma_2, ..., \gamma_s$ are connected by an exact sequence

$$0 \to H_s K(\gamma|E') \quad \to H_s K(\gamma|E) \quad \to H_s K(\gamma|E'') \quad \to ...$$

$$\to H_\mu K(\gamma|E') \quad \to H_\mu K(\gamma|E) \quad \to H_\mu K(\gamma|E'') \quad \to$$

$$H_{\mu-1} K(\gamma|E') \to H_{\mu-1} K(\gamma|E) \to H_{\mu-1} K(\gamma|E'') \to ...$$

$$\to H_0 K(\gamma|E') \quad \to H_0 K(\gamma|E) \quad \to H_0 K(\gamma|E'') \quad \to 0. \quad (8.3.5)$$

Proof. The exact sequence $0 \to E' \to E \to E'' \to 0$ yields the exact sequence (8.3.4) of complexes. If we now write down the exact homology sequence of (8.3.4) and remember that both $H_{s+1}K(\gamma|E'')$ and $H_{-1}K(\gamma|E')$ are null modules, then we obtain (8.3.5).

Let E be an R-module and $\gamma_1, \gamma_2, ..., \gamma_s$ central elements of R. Suppose further that j is an integer satisfying $1 \leqslant j \leqslant s$. For each integer μ we shall define a homomorphism

$$\sigma_\mu^{(j)}: K_\mu(\gamma|E) \to K_{\mu+1}(\gamma|E) \qquad (8.3.6)$$

of R-modules. Evidently the only values of μ for which details are required are those which satisfy $0 \leqslant \mu < s$. For such a μ it is necessary to define

$$\sigma_\mu^{(j)}(eT_{i_1}T_{i_2}...T_{i_\mu}), \qquad (8.3.7)$$

where $1 \leqslant i_1... < i_\mu \leqslant s$ and $e \in E$. Here there are two possibilities to consider namely (i) j does not occur among $i_1, i_2, ..., i_\mu$, and (ii) j is one of $i_1, i_2, ..., i_\mu$. In case (i) let ν denote the number of the integers in the sequence $i_1, i_2, ..., i_\mu$ which are *less* than j. Then

$$\sigma_\mu^{(j)}(eT_{i_1}T_{i_2}...T_{i_\mu}) = (-1)^\nu eT_{i_1}...T_{i_\nu}T_jT_{i_{\nu+1}}...T_{i_\mu}. \qquad (8.3.8)$$

In case (ii) we put $\sigma_\mu^{(j)}(eT_{i_1}T_{i_2}...T_{i_\mu}) = 0$. The homomorphisms (8.3.6) are now defined. Their importance arises from the following lemma.

Lemma 2. *Let x_μ belong to $K_\mu(\gamma_1, ..., \gamma_s|E)$, where μ is an arbitrary integer. Then*

$$d_{\mu+1}\sigma_\mu^{(j)}(x_\mu) = \gamma_j x_\mu - \sigma_{\mu-1}^{(j)}d_\mu(x_\mu). \qquad (8.3.9)$$

Proof. If $\mu > s$ or $\mu < 0$, then $x_\mu = 0$ and the assertion is trivial. Assume next that $0 \leqslant \mu < s$. Then in order to establish (8.3.9) we may suppose that $x_\mu = eT_{i_1}T_{i_2}...T_{i_\mu}$, where the notation is the same as in (8.3.7). First suppose that j does not occur among $i_1, i_2, ..., i_\mu$ and let ν be the number of these which are less than j. Then

$$d_\mu(x_\mu) = \sum_{p=1}^\nu (-1)^{p-1}\gamma_{i_p}eT_{i_1}...\hat{T}_{i_p}...T_{i_\mu}$$

$$+ \sum_{q=\nu+1}^\mu (-1)^{q-1}\gamma_{i_q}eT_{i_1}...\hat{T}_{i_q}...T_{i_\mu}$$

and therefore

$$\sigma_{\mu-1}^{(j)}d_\mu(x_\mu) = (-1)^{\nu-1}\sum_{p=1}^\nu (-1)^{p-1}\gamma_{i_p}eT_{i_1}...\hat{T}_{i_p}...T_j...T_{i_\mu}$$

$$+ (-1)^\nu \sum_{q=\nu+1}^\mu (-1)^{q-1}\gamma_{i_q}eT_{i_1}...T_j...\hat{T}_{i_q}...T_{i_\mu}.$$

Accordingly

$$\gamma_j x_\mu - \sigma^{(j)}_{\mu-1} d_\mu(x_\mu) = (-1)^\nu \sum_{p=1}^{\nu} (-1)^{p-1} \gamma_{i_p} eT_{i_1} \dots \hat{T}_{i_p} \dots T_j \dots T_{i_\mu}$$

$$+ (-1)^\nu (-1)^\nu \gamma_j eT_{i_1} \dots \hat{T}_j \dots T_{i_\mu}$$

$$+ (-1)^\nu \sum_{q=\nu+1}^{\mu} (-1)^q \gamma_{i_q} eT_{i_1} \dots T_j \dots \hat{T}_{i_q} \dots T_{i_\mu}$$

$$= (-1)^\nu d_{\mu+1}(eT_{i_1} \dots T_{i_\nu} T_j T_{i_{\nu+1}} \dots T_{i_\mu})$$

$$= d_{\mu+1} \sigma^{(j)}_\mu(x_\mu)$$

by (8.3.8).

Now assume that j is present among $i_1, i_2, \dots, i_\mu$, say $j = i_t$. Then

$$\sigma^{(j)}_{\mu-1} d_\mu(x_\mu) = \sigma^{(j)}_{\mu-1}((-1)^{t-1} \gamma_{i_t} eT_{i_1} \dots \hat{T}_{i_t} \dots T_{i\mu})$$

$$= \gamma_{i_t} eT_{i_1} \dots T_{i_t} \dots T_{i\mu}$$

$$= \gamma_j x_\mu.$$

Thus $\gamma_j x_\mu - \sigma^{(j)}_{\mu-1} d_\mu(x_\mu) = 0$ which is just what (8.3.9) reduces to in the present instance.

There remains only the case $\mu = s$ but for this situation the argument proceeds exactly as in the last paragraph.

Theorem 3. *Let E be an R-module and $\gamma_1, \gamma_2, \dots, \gamma_s$ central elements of R. Put $A = \gamma_1 R + \gamma_2 R + \dots + \gamma_s R$. Then A annihilates all the homology modules $H_\mu K(\gamma_1, \dots, \gamma_s | E)$ of the Koszul complex of E with respect to $\gamma_1, \gamma_2, \dots, \gamma_s$.*

Proof. Let $\xi_\mu \in H_\mu K(\gamma_1, \dots, \gamma_s | E)$ and suppose that $1 \leqslant j \leqslant s$. We shall show that $\gamma_j \xi_\mu = 0$ and this will complete the proof. Now a representative for ξ_μ will be an element x_μ of $K_\mu(\gamma_1, \dots, \gamma_s | E)$ which satisfies $d_\mu(x_\mu) = 0$. But then, by (8.3.9), $\gamma_j x_\mu = d_{\mu+1} \sigma^{(j)}_\mu(x_\mu)$ which shows that $\gamma_j x_\mu$ is a μ-boundary of $K(\gamma_1, \dots, \gamma_s | E)$. This means that the element of $H_\mu K(\gamma_1, \dots, \gamma_s | E)$ represented by $\gamma_j x_\mu$ is the zero element. Thus $\gamma_j \xi_\mu = 0$ and all is proved.

Corollary. *Let E be an R-module and $\gamma_1, \gamma_2, \dots, \gamma_s$ central elements of R. If now $\gamma_1 R + \gamma_2 R + \dots + \gamma_s R = R$, then all the homology modules $H_\mu K(\gamma_1, \dots, \gamma_s | E)$ of the Koszul complex of E with respect to $\gamma_1, \gamma_2, \dots, \gamma_s$ are zero.*

Let E be an R-module and $\gamma_1, \gamma_2, \dots, \gamma_s$ ($s \geqslant 1$) central elements of R. Evidently $K(\gamma_1, \dots, \gamma_{s-1} | E)$ is a subcomplex of $K(\gamma_1, \dots, \gamma_{s-1}, \gamma_s | E)$.

Furthermore we have, for each μ, an exact sequence

$$0 \to K_\mu(\gamma_1, ..., \gamma_{s-1}|E) \overset{f_\mu}{\to} K_\mu(\gamma_1, ..., \gamma_{s-1}, \gamma_s|E)$$

$$\overset{g_\mu}{\to} K_{\mu-1}(\gamma_1, ..., \gamma_{s-1}|E) \to 0$$

in which f_μ is the inclusion mapping and g_μ maps $eT_{i_1}T_{i_2}...T_{i_\mu}$ into zero if $i_\mu < s$ and into $eT_{i_1}T_{i_2}...T_{i_{\mu-1}}$ if $i_\mu = s$. Now the diagrams

$$
\begin{array}{ccc}
K_\mu(\gamma_1, ..., \gamma_{s-1}, \gamma_s|E) & \overset{g_\mu}{\longrightarrow} & K_{\mu-1}(\gamma_1, ..., \gamma_{s-1}|E) \\
\downarrow & & \downarrow \\
K_{\mu-1}(\gamma_1, ..., \gamma_{s-1}, \gamma_s|E) & \overset{g_{\mu-1}}{\longrightarrow} & K_{\mu-2}(\gamma_1, ..., \gamma_{s-1}|E)
\end{array}
\qquad (8.3.10)
$$

are all commutative it being understood that the vertical mappings are boundary homomorphisms. Put

$$X_\mu = K_{\mu-1}(\gamma_1, ..., \gamma_{s-1}|E)$$

and let $X_\mu \to X_{\mu-1}$ be the boundary homomorphism occurring in (8.3.10). Then X is a complex. Also the f_μ and g_μ provide us with mappings $K(\gamma_1, ..., \gamma_{s-1}|E) \to K(\gamma_1, ..., \gamma_{s-1}, \gamma_s|E)$ and

$$K(\gamma_1, ..., \gamma_{s-1}, \gamma_s|E) \to X$$

of complexes of such a sort that

$$0 \to K(\gamma_1, ..., \gamma_{s-1}|E) \to K(\gamma_1, ..., \gamma_{s-1}, \gamma_s|E) \to X \to 0 \qquad (8.3.11)$$

is an exact sequence. Further

$$H_\mu(X) = H_{\mu-1}K(\gamma_1, ..., \gamma_{s-1}|E). \qquad (8.3.12)$$

Proposition 2. *Let E be an R-module and $\gamma_1, \gamma_2, ..., \gamma_s$ $(s \geqslant 1)$ central elements of R. Then the homology modules of the Koszul complexes $K(\gamma_1, ..., \gamma_{s-1}|E)$ and $K(\gamma_1, ..., \gamma_{s-1}, \gamma_s|E)$ are connected through an exact sequence*

$$... \to H_{\mu+1}K(\gamma_1, ..., \gamma_{s-1}|E) \to H_{\mu+1}K(\gamma_1, ..., \gamma_{s-1}, \gamma_s|E)$$

$$\to H_\mu K(\gamma_1, ..., \gamma_{s-1}|E)$$

$$\to H_\mu K(\gamma_1, ..., \gamma_{s-1}|E) \to H_\mu K(\gamma_1, ..., \gamma_{s-1}, \gamma_s|E)$$

$$\to H_{\mu-1}K(\gamma_1, ..., \gamma_{s-1}|E) \to ....$$

Furthermore the typical connecting homomorphism

$$\Delta \colon H_\mu K(\gamma_1, ..., \gamma_{s-1}|E) \to H_\mu K(\gamma_1, ..., \gamma_{s-1}|E)$$

consists in multiplication by $(-1)^\mu \gamma_s$.

Proof. The exact sequence (8.3.11) yields the exact homology sequence

$$\ldots \to H_{\mu+1} K(\gamma_1, \ldots, \gamma_{s-1}|E) \to H_{\mu+1} K(\gamma_1, \ldots, \gamma_{s-1}, \gamma_s|E) \to H_{\mu+1}(X)$$

$$\to H_\mu K(\gamma_1, \ldots, \gamma_{s-1}|E) \to H_\mu K(\gamma_1, \ldots, \gamma_{s-1}, \gamma_s|E) \to H_\mu(X) \to \ldots.$$

If we now use (8.3.12) to substitute for the homology modules of X this at once gives the first assertion of the proposition.

For the consideration of the connecting homomorphism

$$\Delta: H_{\mu+1}(X) \to H_\mu K(\gamma_1. \ldots, \gamma_{s-1}|E)$$

we must refer back to the construction which culminated in (8.1.11). Let $\zeta_{\mu+1}$ belong to

$$H_{\mu+1}(X) = H_\mu K(\gamma_1, \ldots, \gamma_{s-1}|E).$$

Then $\zeta_{\mu+1}$ is represented by an element of $X_{\mu+1} = K_\mu(\gamma_1, \ldots, \gamma_{s-1}|E)$, say by

$$x_{\mu+1} = \sum_{1 \leqslant i_1 < \ldots < i_\mu \leqslant s-1} e_{i_1 i_2 \ldots i_\mu} T_{i_1} T_{i_2} \ldots T_{i_\mu},$$

and we have

$$\sum_{1 \leqslant i_1 < \ldots < i_\mu \leqslant s-1} \sum_{p=1}^{\mu} (-1)^{p-1} \gamma_{i_p} e_{i_1 i_2 \ldots i_\mu} T_{i_1} \ldots \hat{T}_{i_p} \ldots T_{i_\mu} = 0 \quad (8.3.13)$$

because $x_{\mu+1} \in Z_{\mu+1}(X)$. Now $x_{\mu+1} = g_{\mu+1}(y_{\mu+1})$, where

$$y_{\mu+1} = \sum_{1 \leqslant i_1 < \ldots < i_\mu \leqslant s-1} e_{i_1 i_2 \ldots i_\mu} T_{i_1} \ldots T_{i_\mu} T_s,$$

and, because of (8.3.13),

$$d_{\mu+1}(y_{\mu+1}) = (-1)^\mu \gamma_s \sum_{1 \leqslant i_1 < \ldots < i_\mu \leqslant s-1} e_{i_1 i_2 \ldots i_\mu} T_{i_1} T_{i_2} \ldots T_{i_\mu}$$

$$= (-1)^\mu \gamma_s x_{\mu+1}$$

$$= f_\mu((-1)^\mu \gamma_s x_{\mu+1}).$$

Thus $\Delta(\zeta_{\mu+1})$ is the image of $(-1)^\mu \gamma_s x_{\mu+1}$ in $H_\mu K(\gamma_1, \ldots, \gamma_{s-1}|E)$ that is to say $\Delta(\zeta_{\mu+1}) = (-1)^\mu \gamma_s \zeta_{\mu+1}$ as required. The proof is now complete.

We need one further result and then we shall be able to show the relevance of the Koszul complex to some of our earlier investigations. The extra property concerns the homology modules of

$$K(\gamma_1, \ldots, \gamma_{s-1}, \gamma_s|E) \quad \text{when} \quad 0:_E \gamma_s = 0,$$

that is to say when γ_s is not a zero-divisor on E. The information we need is obtained by considering, along with $K(\gamma_1, \ldots, \gamma_{s-1}, \gamma_s|E)$, the complex $K(\gamma_1, \ldots, \gamma_{s-1}, 1|E)$ obtained by replacing γ_s by the identity element of R.

First we note that the two complexes have the same component modules, the μth component being in each case,

$$\bigoplus_{1 \leqslant i_1 < \ldots < i_\mu \leqslant s} E T_{i_1} T_{i_2} \ldots T_{i_\mu}.$$

Next we define a homomorphism

$$f_\mu : K_\mu(\gamma_1, \ldots, \gamma_{s-1}, \gamma_s | E) \to K_\mu(\gamma_1, \ldots, \gamma_{s-1}, 1 | E)$$

by requiring that

$$f_\mu(e T_{i_1} T_{i_2} \ldots T_{i_\mu}) = \left. \begin{array}{ll} \gamma_s e T_{i_1} T_{i_2} \ldots T_{i_\mu} & (i_\mu = s) \\ e T_{i_1} T_{i_2} \ldots T_{i_\mu} & (i_\mu < s) \end{array} \right\}.$$

Evidently f_μ is a monomorphism in the case where γ_s is not a zero-divisor on E. Moreover, even without this assumption, the diagram

$$
\begin{array}{ccc}
K_\mu(\gamma_1, \ldots, \gamma_{s-1}, \gamma_s \mid E) & \xrightarrow{\;f_\mu\;} & K_\mu(\gamma_1, \ldots, \gamma_{s-1}, 1 \mid E) \\
\downarrow & & \downarrow \\
K_{\mu-1}(\gamma_1, \ldots, \gamma_{s-1}, \gamma_s \mid E) & \xrightarrow[f_{\mu-1}]{} & K_{\mu-1}(\gamma_1, \ldots, \gamma_{s-1}, 1 \mid E)
\end{array}
$$

is commutative. (Here it is understood that the vertical mappings are boundary homomorphisms.) Thus the f_μ constitute a mapping

$$f : K(\gamma_1, \ldots, \gamma_{s-1}, \gamma_s | E) \to K(\gamma_1, \ldots, \gamma_{s-1}, 1 | E)$$

of complexes.

Put $\bar{E} = E/\gamma_s E$ and for each $e \in E$ let $\bar{e}$ denote its natural image in $\bar{E}$. If now

$$X_\mu = \bigoplus_{1 \leqslant i_1 < \ldots < i_{\mu-1} \leqslant s-1} \bar{E} T_{i_1} T_{i_2} \ldots T_{i_{\mu-1}} = K_{\mu-1}(\gamma_1, \ldots, \gamma_{s-1} | \bar{E})$$

and

$$\delta_\mu : X_\mu \to X_{\mu-1}$$

is the homomorphism defined by

$$\delta_\mu(\bar{e} T_{i_1} T_{i_2} \ldots T_{i_{\mu-1}}) = \sum_{p=1}^{\mu-1} (-1)^{p-1} \gamma_{i_p} \bar{e} T_{i_1} \ldots \hat{T}_{i_p} \ldots T_{i_{\mu-1}},$$

then

$$\ldots \to X_{\mu+1} \xrightarrow{\delta_{\mu+1}} X_\mu \xrightarrow{\delta_\mu} X_{\mu-1} \to \ldots$$

is simply the complex $K(\gamma_1, \ldots, \gamma_{s-1} | \bar{E})$ displaced one place to the left. Accordingly

$$H_\mu(X) = H_{\mu-1} K(\gamma_1, \ldots, \gamma_{s-1} | \bar{E}) \qquad (8.3.14)$$

for all values of μ.

Consider the homomorphisms

$$g_\mu : K_\mu(\gamma_1, \dots, \gamma_{s-1}, 1 | E) \to X_\mu$$

defined by

$$g_\mu(eT_{i_1}T_{i_2} \dots T_{i_\mu}) = \begin{matrix} 0 & (i_\mu < s) \\ \bar{e}T_{i_1}T_{i_2} \dots T_{i_{\mu-1}} & (i_\mu = s) \end{matrix} \Big\} .$$

Evidently g_μ is an epimorphism, $\operatorname{Ker} g_\mu = \operatorname{Im} f_\mu$ and the diagram

$$
\begin{array}{ccc}
K_\mu(\gamma_1, \dots, \gamma_{s-1}, 1 \mid E) & \xrightarrow{\;g_\mu\;} & X_\mu \\
\downarrow & & \downarrow \\
K_{\mu-1}(\gamma_1, \dots, \gamma_{s-1}, 1 \mid E) & \xrightarrow[g_{\mu-1}]{} & X_{\mu-1}
\end{array}
$$

is commutative. (Once again the vertical mappings are understood to be boundary homomorphisms.) Thus the g_μ constitute a mapping

$$g : K(\gamma_1, \dots, \gamma_{s-1}, 1 | E) \to X$$

of complexes. Finally, *if γ_s is not a zero-divisor on E, then*

$$0 \to K(\gamma_1, \dots, \gamma_{s-1}, \gamma_s | E) \xrightarrow{f} K(\gamma_1, \dots, \gamma_{s-1}, 1 | E) \xrightarrow{g} X \to 0 \quad (8.3.15)$$

is an exact sequence of complexes.

Theorem 4. *Let E be an R-module and $\gamma_1, \gamma_2, \dots, \gamma_s$ $(s \geqslant 1)$ central elements of R. If now γ_s is not a zero-divisor on E, then we have an isomorphism*

$$H_\mu K(\gamma_1, \dots, \gamma_{s-1}, \gamma_s | E) \approx H_\mu K(\gamma_1, \dots, \gamma_{s-1} | E/\gamma_s E)$$

of R-modules for each value of μ.

Proof. If we apply Theorem 1 to the exact sequence (8.3.15), then we obtain an exact sequence

$$H_{\mu+1} K(\gamma_1, \dots, \gamma_{s-1}, 1 | E) \to H_{\mu+1}(X) \to H_\mu K(\gamma_1, \dots, \gamma_{s-1}, \gamma_s | E)$$
$$\to H_\mu K(\gamma_1, \dots, \gamma_{s-1}, 1 | E)$$

for each value of μ. However, by Theorem 3 Cor., both

$$H_{\mu+1} K(\gamma_1, \dots, \gamma_{s-1}, 1 | E) \quad \text{and} \quad H_\mu K(\gamma_1, \dots, \gamma_{s-1}, 1 | E)$$

are null modules and therefore we have an isomorphism

$$H_{\mu+1}(X) \approx H_\mu K(\gamma_1, \dots, \gamma_{s-1}, \gamma_s | E).$$

The desired result now follows from (8.3.14).

8.4 Connections with multiplicity theory

The Koszul complex can be used to give a new description of the multiplicity symbol $e_R(\gamma_1, ..., \gamma_s|E)$. However, before we can justify this statement we must establish some lemmas.

Lemma 3. *Let E be a Noetherian R-module and $\gamma_1, \gamma_2, ..., \gamma_s$ a multiplicity system† on E. Then each of the R-modules $H_\mu K(\gamma_1, ..., \gamma_s|E)$ has finite length.*

Proof. For each μ, $K_\mu(\gamma_1, ..., \gamma_s|E)$ is the direct sum of a finite number of copies of E. It is therefore Noetherian and admits $\gamma_1, \gamma_2, ..., \gamma_s$ as a multiplicity system. Next $H_\mu K(\gamma_1, ..., \gamma_s|E)$ is a factor module of a submodule of $K_\mu(\gamma_1, ..., \gamma_s|E)$. It therefore follows, by Proposition 3 of section (7.3), that $\gamma_1, \gamma_2, ..., \gamma_s$ is a multiplicity system on

$$H_\mu K(\gamma_1, ..., \gamma_s|E).$$

However, by Theorem 3, each γ_i annihilates $H_\mu K(\gamma_1, ..., \gamma_s|E)$. Accordingly $H_\mu K(\gamma_1, ..., \gamma_s|E)$ has finite length. This establishes the lemma.

Let E be a Noetherian R-module and $\gamma_1, \gamma_2, ..., \gamma_s$ ($s \geqslant 0$) a multiplicity system on E. By Lemma 3, each of the homology modules $H_\mu K(\gamma_1, ..., \gamma_s|E)$ has finite length. We may therefore put

$$\chi_R(\gamma_1, ..., \gamma_s|E) = \sum_\mu (-1)^\mu L_R\{H_\mu K(\gamma_1, ..., \gamma_s|E)\}. \quad (8.4.1)$$

(Observe that this is essentially a *finite* sum since the summand is zero if either $\mu > s$ or $\mu < 0$.) It will be shown that, in fact, $\chi_R(\gamma_1, ..., \gamma_s|E)$ is equal to $e_R(\gamma_1, ..., \gamma_s|E)$. Now for $s = 0$ (8.4.1) gives

$$\chi_R(\cdot|E) = L_R(E) \quad (8.4.2)$$

so there is no trouble in this particular case. Presently we shall prove the general result using induction on s, but first we make some remarks on terminology.

Suppose that (X, d) is a complex of R-modules such that all the component modules have finite length and at most a finite number of them are non-null. Then

$$\sum_\mu (-1)^\mu L_R(X_\mu)$$

is well defined. It is called the *Euler–Poincaré characteristic* of X. In the case of the Koszul complex, let us make a complex out of the homology modules $H_\mu K(\gamma_1, ..., \gamma_s|E)$ by taking each boundary homo-

† See section (7.3).

morphism to be a null map. The result is called the *homology complex* of
$K(\gamma_1, ..., \gamma_s|E)$. This enables us to describe the result we are going to
establish in the following terms: *the value of the multiplicity symbol*
$e_R(\gamma_1, ..., \gamma_s|E)$ *is equal to the Euler–Poincaré characteristic of the
homology complex of* $K(\gamma_1, ..., \gamma_s|E)$. The proof requires two lemmas.

Lemma 4. *Let $0 \to E' \to E \to E'' \to 0$ be an exact sequence of Noetherian
R-modules and suppose that each term admits $\gamma_1, \gamma_2, ..., \gamma_s$ as a multiplicity system. Then*

$$\chi_R(\gamma_1, ..., \gamma_s|E) = \chi_R(\gamma_1, ..., \gamma_s|E') + \chi_R(\gamma_1, ..., \gamma_s|E'').$$

Proof. By Lemma 3, all the modules in the exact sequence (8.3.5) have
finite length. The desired result now follows by applying Theorem 20 of
section (1.12) to this sequence.

Lemma 5. *Let E be a Noetherian R-module and $\gamma_1, \gamma_2, ..., \gamma_s$ a multiplicity system on E. If now $\gamma_s^m E = 0$, where $m > 0$, then*

$$\chi_R(\gamma_1, ..., \gamma_s|E) = 0.$$

Proof. Since $\gamma_1, ..., \gamma_{s-1}, \gamma_s$ is a multiplicity system on E the same is
true of $\gamma_1, ..., \gamma_{s-1}, \gamma_s^m$. But γ_s^m annihilates E. Consequently the
shortened sequence $\gamma_1, \gamma_2, ..., \gamma_{s-1}$ is also a multiplicity system
on E. It follows, by Lemma 3, that both $H_\mu K(\gamma_1, ..., \gamma_{s-1}|E)$ and
$H_\mu K(\gamma_1, ..., \gamma_{s-1}, \gamma_s|E)$ have finite length. We may therefore apply
Theorem 20 of section (1.12) to the exact sequence which occurs
in Proposition 2. This immediately yields the desired result.

Theorem 5. *Let E be a Noetherian R-module and $\gamma_1, \gamma_2, ..., \gamma_s$ a multiplicity system on E. Then*

$$e_R(\gamma_1, ..., \gamma_s|E) = \chi_R(\gamma_1, ..., \gamma_s|E).$$

Proof. We use induction on s. By (8.4.2), the theorem is true when
$s = 0$. It will therefore be supposed that $s > 0$ and that the theorem
has been established for multiplicity systems with only $s - 1$ elements.

Put $F = E/(0:_E \gamma_s^m)$, where m is chosen large enough to ensure that
γ_s is not a zero-divisor on F. This is possible by Lemma 5 of section
(7.4). Then, by applying Lemma 4 to the exact sequence

$$0 \to 0:_E \gamma_s^m \to E \to F \to 0,$$

we obtain

$$\chi_R(\gamma_1, ..., \gamma_s|E) = \chi_R(\gamma_1, ..., \gamma_s|F) + \chi_R(\gamma_1, ..., \gamma_s|0:_E \gamma_s^m)$$

which reduces to

$$\chi_R(\gamma_1, ..., \chi_s|E) = \chi_R(\gamma_1, ..., \gamma_s|F)$$

with the aid of Lemma 5. Since γ_s is not a zero-divisor on F, Theorem 4 shows that $H_\mu K(\gamma_1, ..., \gamma_{s-1}, \gamma_s | F)$ and $H_\mu K(\gamma_1, ..., \gamma_{s-1} | F/\gamma_s F)$ are isomorphic R-modules. It follows that

$$\chi_R(\gamma_1, ..., \gamma_{s-1}, \gamma_s | F) = \chi_R(\gamma_1, ..., \gamma_{s-1} | F/\gamma_s F)$$

and therefore

$$\chi_R(\gamma_1, ..., \gamma_{s-1}, \gamma_s | E) = \chi_R(\gamma_1, ..., \gamma_{s-1} | F/\gamma_s F).$$

On the other hand we have

$$e_R(\gamma_1, ..., \gamma_{s-1}, \gamma_s | E) = e_R(\gamma_1, ..., \gamma_{s-1}, \gamma_s | F) = e_R(\gamma_1, ..., \gamma_{s-1} | F/\gamma_s F)$$

by the basic properties of the multiplicity symbol. Finally

$$e_R(\gamma_1, ..., \gamma_{s-1} | F/\gamma_s F) = \chi_R(\gamma_1, ... \gamma_{s-1} | F/\gamma_s F)$$

by the inductive hypothesis. Thus the theorem is established.

8.5 Connections with the theory of grade

There are close connections between the theory of the Koszul complex and the theory of grade. This is shown by the next theorem.

Theorem 6. *Let R be a commutative ring, E a Noetherian R-module, and A a finitely generated ideal of R such that $AE \neq E$. Put*

$$q = \operatorname{gr}(A; E)$$

where, as in section (5.1), $\operatorname{gr}(A; E)$ denotes the grade of A on E. If now $A = (\gamma_1, \gamma_2, ..., \gamma_s)$, then $H_{s-q} K(\gamma_1, ..., \gamma_s | E) \neq 0$ and

$$H_\mu K(\gamma_1, ..., \gamma_s | E) = 0 \quad \text{for all} \quad \mu > s - q.$$

Remark. This theorem shows that the grade of A on E can be determined by looking for the last non-zero homology module of the Koszul complex of E with respect to any system of generators of A.

Proof. Let U be any R-module such that $AU \neq U$. By (8.2.7),

$$H_0 K(\gamma_1, ..., \gamma_s | U) = U/AU$$

which is not zero. Hence there is a largest integer $\lambda = \lambda(U)$ such that $H_\lambda K(\gamma_1, ..., \gamma_s | U) \neq 0$. Obviously $0 \leqslant \lambda(U) \leqslant s$. We have to show that $q + \lambda(E) = s$.

First suppose that $\operatorname{gr}(A; E) = 0$. This means that every element of A is a zero-divisor on E. Accordingly, by Theorem 14 Cor. 2 of section (4.4), each element of A is contained in some prime ideal belonging to the zero submodule of E. It follows, by Proposition 5 of section (2.3), that A itself is contained by one of these prime ideals and therefore, $0:_E A \neq 0$. However, by (8.2.8),

$$H_s K(\gamma_1, \ldots, \gamma_s | E) = 0:_E A$$

and so we see that $\lambda(E) = s$. Conversely if $\lambda(E) = s$, then $0:_E A$ is a non-null module and therefore every element of A is a zero-divisor on E. Thus to sum up : gr $(A;E) = 0$ *when and only when* $\lambda(E) = s$.

Now assume that gr $(A;E) > 0$ or equivalently that $\lambda(E) < s$. Then there exists an element $\beta_1 \in A$ such that β_1 is not a zero-divisor on E. It follows that

$$\text{gr}\,(A;E/\beta_1 E) = \text{gr}\,(A;E) - 1. \tag{8.5.1}$$

Next we have an exact sequence

$$0 \to E \overset{f}{\to} E \to E/\beta_1 E \to 0,$$

where f consists in multiplication by β_1. Accordingly, by Theorem 2, this gives rise to exact sequences of the form

$$H_\mu K(\gamma_1, \ldots, \gamma_s | E) \to H_\mu K(\gamma_1, \ldots, \gamma_s | E) \to H_\mu K(\gamma_1, \ldots, \gamma_s | E/\beta_1 E)$$
$$\to H_{\mu-1} K(\gamma_1, \ldots, \gamma_s | E) \to H_{\mu-1} K(\gamma_1, \ldots, \gamma_s | E).$$

Again, since

$$H_\mu K(\gamma_1, \ldots, \gamma_s | E) \to H_\mu K(\gamma_1, \ldots, \gamma_s | E) \tag{8.5.2}$$

is induced by f, it too consists in multiplication by β_1. But, by Theorem 3, A annihilates $H_\mu K(\gamma_1, \ldots, \gamma_s | E)$ and, by hypothesis, $\beta_1 \in A$. It follows that (8.5.2) is a null map and therefore our exact sequence yields the simplified exact sequence

$$0 \to H_\mu K(\gamma_1, \ldots, \gamma_s | E) \to H_\mu K(\gamma_1, \ldots, \gamma_s | E/\beta_1 E)$$
$$\to H_{\mu-1} K(\gamma_1, \ldots, \gamma_s | E) \to 0.$$

Put $\lambda = \lambda(E)$. Then for $\mu > \lambda + 1$ both $H_\mu K(\gamma_1, \ldots, \gamma_s | E)$ and $H_{\mu-1} K(\gamma_1, \ldots, \gamma_s | E)$ are null modules. Hence $H_\mu K(\gamma_1, \ldots, \gamma_s | E/\beta_1 E) = 0$. On the other hand, when $\mu = \lambda + 1$ we obtain an isomorphism

$$H_{\lambda+1} K(\gamma_1, \ldots, \gamma_s | E/\beta_1 E) \approx H_\lambda K(\gamma_1, \ldots, \gamma_s | E).$$

This shows, in particular, that $H_{\lambda+1} K(\gamma_1, \ldots, \gamma_s | E/\beta_1 E) \neq 0$. Accordingly

$$\lambda(E/\beta_1 E) = \lambda(E) + 1. \tag{8.5.3}$$

Finally suppose that $\beta_1, \beta_2, \ldots, \beta_p$ are elements of A such that, for each i $(i = 0, 1, \ldots p - 1)$, β_{i+1} is not a zero divisor on

$$E/(\beta_1, \ldots, \beta_i)E.$$

Then, by repeated applications of (8.5.1) and (8.5.3),

$$\mathrm{gr}\,(A\,;E/(\beta_1,...,\beta_p)E) = \mathrm{gr}\,(A\,;E)-p$$

and
$$\lambda(E/(\beta_1,...,\beta_p)E) = \lambda(E)+p.$$

Also, the sequence $\beta_1, \beta_2, ..., \beta_p$ is incapable of being extended when $\mathrm{gr}\,(A\,;E/(\beta_1,...,\beta_p)E) = 0$, which, by our earlier remarks, is equivalent to the condition that $\lambda(E/(\beta_1,...,\beta_p)E) = s$. It follows that $\lambda(E)+\mathrm{gr}\,(A\,;E) = s$ and with this the proof is complete.

For the final results of this chapter we suppose, once again, that R is a general ring, that is we no longer require that R be commutative.

Theorem 7. *Let E be an R-module and $\gamma_1, \gamma_2, ..., \gamma_s$ $(s \geqslant 0)$ central elements of R such that, for each i $(1 \leqslant i \leqslant s)$, γ_i is not a zero-divisor on $E/(\gamma_1 E+\gamma_2 E+...+\gamma_{i-1}E)$. Then $H_\mu K(\gamma_1, ..., \gamma_s|E) = 0$ for all $\mu \neq 0$.*

Proof. We use induction on s and begin by observing that the assertion is trivial when $s = 0$. Now suppose that $s \geqslant 1$ and make the obvious inductive hypothesis. By Proposition 1 Cor. and Theorem 4, we have isomorphisms

$$H_\mu K(\gamma_1, \gamma_2, ..., \gamma_s|E) \approx H_\mu K(\gamma_2, ..., \gamma_s, \gamma_1|E)$$
$$\approx H_\mu K(\gamma_2, ..., \gamma_s|E/\gamma_1 E)$$

for every value of μ. However if $\bar{E} = E/\gamma_1 E$, then, for $2 \leqslant i \leqslant s$, γ_i is not a zero-divisor on $\bar{E}/(\gamma_2\bar{E}+...+\gamma_{i-1}\bar{E})$. Hence, by the inductive hypothesis, $H_\mu K(\gamma_2, ..., \gamma_s|E/\gamma_1 E) = 0$ for all $\mu \neq 0$. The desired result follows.

Our final theorem is striking in that it tells us that, under appropriate conditions, it is the vanishing of $H_1 K(\gamma_1, ..., \gamma_s|E)$ that is all important. First, however, we must prove a lemma.

Lemma 6. *Let E be a Noetherian R-module and $\gamma_1, \gamma_2, ..., \gamma_s$ central elements of R. If now $H_p K(\gamma_1, ..., \gamma_{s-1}, \gamma_s|E) = 0$ and γ_s is in the Jacobson radical of R, then $H_p K(\gamma_1, ..., \gamma_{s-1}|E) = 0$ as well.*

Proof. By Proposition 2, we have an exact sequence

$$H_p K(\gamma_1, ..., \gamma_{s-1}|E) \to H_p K(\gamma_1, ..., \gamma_{s-1}|E) \to H_p K(\gamma_1, ..., \gamma_{s-1}, \gamma_s|E)$$

in which the first mapping consists in multiplication by $(-1)^p \gamma_s$. Since $H_p K(\gamma_1, ..., \gamma_{s-1}, \gamma_s|E) = 0$, this shows that

$$H_p K(\gamma_1, ..., \gamma_{s-1}|E) = (\gamma_s) H_p K(\gamma_1, ..., \gamma_{s-1}|E).$$

Consequently

$$H_p K(\gamma_1, ..., \gamma_{s-1}|E) = \bigcap_{n=1}^{\infty} (\gamma_s)^n H_p K(\gamma_1, ..., \gamma_{s-1}|E).$$

Now $H_p K(\gamma_1, ..., \gamma_{s-1}|E)$ is a factor module of a submodule of a finite direct sum of copies of E. It follows that $H_p K(\gamma_1, ..., \gamma_{s-1}|E)$ is a Noetherian R-module. That $H_p K(\gamma_1, ..., \gamma_{s-1}|E) = 0$ now follows from Theorem 4 of section (7.2).

Theorem 8. *Let E be a Noetherian R-module and $\gamma_1, \gamma_2, ..., \gamma_s$ $(s \geqslant 1)$ central elements contained in the Jacobson radical of R. Then the following statements are equivalent:*

(a) *for each i $(1 \leqslant i \leqslant s)$, γ_i is not a zero-divisor on*
$$E/(\gamma_1 E + ... + \gamma_{i-1} E);$$

(b) $H_\mu K(\gamma_1, ..., \gamma_s|E) = 0$ *for all $\mu \neq 0$;*

(c) $H_1 K(\gamma_1, ..., \gamma_s|E) = 0.$

Proof. Theorem 7 shows that (a) implies (b) and it is evident that (b) implies (c). It remains for us to show that (c) implies (a). This will be accomplished using induction on s.

Suppose first that $s = 1$. Then, by (8.2.8), the hypothesis
$$H_1 K(\gamma_1|E) = 0$$
means that $0 :_E \gamma_1 = 0$, i.e. that γ_1 is not a zero-divisor on E. Thus the result in question is certainly true in this case.

From now on it will be assumed that $s > 1$ and that the assertion '(c) implies (a)' has been established when the number of central elements involved is only $s - 1$. By repeated applications of Lemma 6, we see that all the modules $H_1 K(\gamma_1, ..., \gamma_i|E)$ $(i = 1, 2, ..., s)$ are null modules. In particular we have $H_1 K(\gamma_1|E) = 0$ from which, as we saw above, it follows that γ_1 is not a zero-divisor on E. Put $\bar{E} = E/\gamma_1 E$. Then, by Proposition 1 Cor. and Theorem 4, we have isomorphisms

$$H_1 K(\gamma_1, \gamma_2, ..., \gamma_s|E) \approx H_1 K(\gamma_2, ..., \gamma_s, \gamma_1|E) \approx H_1 K(\gamma_2, ..., \gamma_s|\bar{E}).$$

Thus $H_1 K(\gamma_2, ..., \gamma_s|\bar{E}) = 0$ and therefore, by the inductive hypothesis, γ_i is not a zero-divisor on $\bar{E}/(\gamma_2 \bar{E} + ... + \gamma_{i-1} \bar{E})$ for any i satisfying $2 \leqslant i \leqslant s$. Since $\bar{E}/(\gamma_2 \bar{E} + ... + \gamma_{i-1} \bar{E})$ is isomorphic to

$$E/(\gamma_1 E + \gamma_2 E + ... + \gamma_{i-1} E),$$

this completes the proof.

Exercises on Chapter 8

As usual R denotes a ring with an identity element. It is not assumed that R is commutative unless the formulation of an exercise contains an explicit statement that this condition is imposed. If E is an R-module and $\gamma_1, \gamma_2, ..., \gamma_s$ are central elements of R, then $K(\gamma_1, \dots, \gamma_s | E)$ denotes the Koszul complex of E with respect to $\gamma_1, \gamma_2, \dots, \gamma_s$.

1. X is a complex of R-modules such that $X_\mu = 0$ for almost all μ and $L_R(X_\mu) < \infty$ for every μ. Prove that

$$\Sigma(-1)^\mu L_R(X_\mu) = \Sigma(-1)^\mu L_R(H_\mu(X)).$$

2. Let R be a commutative ring and S a non-empty multiplicatively closed subset of R. If X is the complex

$$\dots \to X_{n+1} \to X_n \to X_{n-1} \to X_{n-2} \to \dots$$

of R-modules, denote by X_S the complex

$$\dots \to (X_{n+1})_S \to (X_n)_S \to (X_{n-1})_S \to (X_{n-2})_S \to \dots$$

of R_S-modules obtained by employing the fraction functor associated with S. Prove that, for each integer μ, $H_\mu(X_S)$ and $(H_\mu(X))_S$ are isomorphic R_S-modules.

3. Let R be a commutative ring, E an R-module, $\gamma_1, \gamma_2, \dots, \gamma_s$ elements of R, and S a non-empty multiplicatively closed subset of R. If now

$$X = K(\gamma_1, \dots, \gamma_s | E) \quad \text{and} \quad \phi : R \to R_S$$

is the canonical ring-homomorphism, show that X_S (see Exercise 2) and $K(\phi\gamma_1, \dots, \phi\gamma_s | E_S)$ are isomorphic complexes of R_S-modules.

4. Show that the exact sequence which occurs in Lemma 3 of section (7.4) is a special case of the exact sequence which occurs in Theorem 2.

5. E is an R-module and $\gamma_1, \gamma_2, \dots, \gamma_s \, (s \geqslant 1)$ central elements of R. Show that if $\gamma_s E = E$, then for each value of μ there exists an isomorphism

$$H_\mu K(\gamma_1, \dots, \gamma_{s-1}, \gamma_s | E) \approx H_{\mu-1} K(\gamma_1, \dots, \gamma_{s-1} | 0 :_E \gamma_s)$$

of R-modules.

6. Let E be a Noetherian R-module and $\gamma_1, \gamma_2, \dots, \gamma_s \, (s \geqslant 1)$ a multiplicity system on E. Prove directly from the properties of the Koszul complex that

$$\chi_R(\gamma_1, \dots, \gamma_{s-1}, \gamma_s | E) = \chi_R(\gamma_1, \dots, \gamma_{s-1} | E/\gamma_s E) - \chi_R(\gamma_1, \dots, \gamma_{s-1} | 0 :_E \gamma_s),$$

where the notation is the same as that used in section (8.4). (*Hint.* Use induction on m, where $m \geqslant 0$ is the smallest integer such that $0 :_E \gamma_s^m = 0 :_E \gamma_s^{m+1}$.)

7. Let E be a Noetherian R-module and K an R-submodule. Show that if E and K are regarded as modules over the centre of R, then K has a primary decomposition in E.

8. Let $E \neq 0$ be a Noetherian R-module and A a finitely generated central ideal satisfying $0 :_E A = 0$. Show that there exists a central element belonging to A which is not a zero-divisor on E.

9. E is an R-module and A an ideal generated by s central elements. Further $\beta_1, \beta_2, \dots, \beta_q$ are central elements contained in A which are such that, for $1 \leqslant i \leqslant q$, β_i is not a zero-divisor on $E/(\beta_1 E + \dots + \beta_{i-1} E)$. Prove that if $AE \neq E$, then q does not exceed s.

10. Let E be an R-module and $\gamma_1, \gamma_2, \dots, \gamma_s$ central elements of R such that $AE \neq E$, where $A = \gamma_1 R + \gamma_2 R + \dots + \gamma_s R$. Further let $\beta_1, \beta_2, \dots, \beta_q$ be central elements contained in A such that, for each $i \, (1 \leqslant i \leqslant q)$, β_i is not a zero-divisor on $E/(\beta_1 E + \beta_2 E + \dots + \beta_{i-1} E)$. Prove that there exists an isomorphism

$$H_{s-q} K(\gamma_1, \dots, \gamma_s | E) \approx ((\beta_1 E + \dots + \beta_q E) :_E A)/(\beta_1 E + \dots + \beta_q E)$$

of R-modules.

11. Let R be a commutative ring, E a Noetherian R-module, and

$$\gamma_1, \gamma_2, \dots, \gamma_s$$

elements of R. Show that if $H_1 K(\gamma_1, \dots, \gamma_s | E) = 0$, then

$$H_\mu K(\gamma_1, \dots, \gamma_s | E) = 0 \quad \text{for all} \quad \mu \neq 0.$$

12. Let R be a (commutative) local ring and $u_1, u_2, \dots, u_s$ elements which generate its maximal ideal. Show that $\operatorname{Dim} R = s$ if and only if

$$H_1 K(u_1, u_2, \dots, u_s | R) = 0.$$

13. Let R be a commutative ring, A a finitely generated ideal, and

$$0 \to E' \to E \to E'' \to 0$$

an exact sequence of Noetherian R-modules such that $AE' \neq E'$, $AE \neq E$, and $AE'' \neq E''$. If $\operatorname{gr}(A;E) > \operatorname{gr}(A;E'')$, show that $\operatorname{gr}(A;E') = \operatorname{gr}(A;E'') + 1$.

9

FILTERED RINGS AND MODULES

General remarks. As usual R denotes a ring with an identity element. The only section throughout which R is assumed to be commutative is (9.11). Wherever this assumption is not made, the term *R-module* should be interpreted as meaning *left R-module*.

A substantial part of the theory of commutative rings came into being through the study of local and semi-local rings, these being regarded as rings with a natural topology. In this context, the natural topologies arise from ideals. These investigations are now incorporated in a more general study which it is convenient to describe as *the theory of rings and modules with filtrations*. What follows is an introduction to this subject.

9.1 Topological prerequisites

It is assumed that the reader has some familiarity with the basic concepts of general topology. However, for his convenience, we recall briefly those that are relevant here.

A set U becomes endowed with a *topology* as soon as there is given a collection Ω of subsets of U having the following properties:

(a) *both U and its empty subset belong to Ω;*

(b) *if $\{O_i\}_{i \in I}$ is an arbitrary family of subsets of U such that each O_i belongs to Ω, then $\bigcup_{i \in I} O_i$ also belongs to Ω;*

(c) *if $O_1, O_2, ..., O_n$ (n is an integer) belong to Ω, then $O_1 \cap O_2 \cap ... \cap O_n$ belongs to Ω as well.*

The members of Ω are then referred to as the *open sets* of the topology and (a), (b), (c) are known as the *open sets axioms*. They can be summarized by saying that the class of open sets is closed under arbitrary unions and finite intersections.

Let U be a topological space. (This means that U is a set and we are given a collection of subsets which obeys the axioms for open sets.) A subset F, of U, is called a *closed subset* if its complement with respect to U is open. The union of a finite number of closed sets is necessarily a closed set and so too is the intersection of a completely arbitrary family of closed sets. Of course it can happen that a set is both open and closed.

Suppose that K is a subset of U. The intersection of all the closed sets containing K is itself a closed set which will be denoted by $\bar{K}$. This is called the *closure* or *adherence* of K in U. If $u \in U$, then it is easy to show that u belongs to $\bar{K}$ if and only if every open set containing u contains at least one element of K. If it happens that $\bar{K} = U$, then one says that K *is everywhere dense in* U or that K *is an everywhere dense subset of* U.

Now assume that V is a subset of a topological space U. It is then possible to define a topology on V whose open sets are precisely the intersections of V with the open subsets of U. This is known as the *induced topology*. When V is endowed with the induced topology, we say that V is a *subspace* of U.

The elements of a topological space are often referred to as *points*. By a *neighbourhood* of a point u, of U, is meant any set which contains an open set to which u belongs. Suppose that U has the following property: *whenever u and u' are distinct points there exists a neighbourhood N of u and a neighbourhood N' of u' such that $N \cap N'$ is empty*. We then say that U is a *Hausdorff space* or a *separated space*. For the more general parts of our theory this condition is not needed, but it comes into prominence in connection with the study of completions.

Once again, let U be a general topological space. A collection $\mathfrak{B}$ of subsets of U is called a *base for the topology of* U or a *base for the open sets of* U if (i) every set in $\mathfrak{B}$ is an open set of U, and (ii) every open set of U is the union of a (possibly empty) family of sets belonging to $\mathfrak{B}$. Of course, one can obtain a base by letting $\mathfrak{B}$ consist of *all* the open sets, but, in practice, interesting topologies tend to be those which have a base whose members are (in some sense) of a rather special kind.

The notion of a base is useful in defining the *product* of two topological spaces. To this end let U and U' be topological spaces and suppose that $\mathfrak{B}$ and $\mathfrak{B}'$ are bases for their respective topologies. The Cartesian product $U \times U'$ consists of all ordered pairs (u, u'), where $u \in U$ and $u' \in U'$. If now $B \in \mathfrak{B}$ and $B' \in \mathfrak{B}'$, then $B \times B'$ is a subset of $U \times U'$. It is easy to see that there is a unique topology on $U \times U'$ which has the subsets $B \times B'$ as a base. Furthermore, this topology is unaltered if we replace $\mathfrak{B}$ and $\mathfrak{B}'$ by other bases for the open sets of U and U' respectively. The topological space $U \times U'$, so obtained, is called the *product* of U and U'. The topology on the product space is referred to as the *product topology*.

We conclude by recalling certain terms and definitions connected with the idea of a *continuous mapping*. Let U and U^* be topological spaces and $f: U \to U^*$ a mapping. Then f is said to be *continuous at the point u*, of U, if for each neighbourhood N^* of $f(u)$ there exists a neighbourhood N of u such that $f(N) \subseteq N^*$. Should it happen that we are given bases $\mathfrak{B}$ and $\mathfrak{B}^*$ for the topologies of U and U^* respectively, then the above definition is equivalent to the following: *given B^* in $\mathfrak{B}^*$ such that $f(u) \in B^*$, it is always possible to find B in $\mathfrak{B}$ such that $u \in B$ and $f(B) \subseteq B^*$*. When f is continuous at every point of U we say simply that *f is continuous*. It is well known that the following three statements are equivalent:

(a) *the mapping $f: U \to U^*$ is continuous*;

(b) *for every open set O^* in U^*, $f^{-1}(O^*)$ is open in U*;

(c) *for every closed set F^* in U^*, $f^{-1}(F^*)$ is closed in U*.

Particularly important is the situation in which f is a one-one mapping of U on to U^* and both f and f^{-1} are continuous. Such a mapping f is called a *homeomorphism* of U on to U^* and f^{-1} is known as the *inverse homeomorphism*. Naturally we then say that the spaces U and U^* are *homeomorphic*. If a one-one correspondence between U and U^* is a homeomorphism then *a subset of one space is open if and only if the corresponding subset of the other space is open*.

9.2 Filtered modules

Let R be a ring with an identity element and E a left R-module. By a *filtration on E* will be meant a family $\{E_n\}_{n \geqslant 0}$ of submodules, indexed by the non-negative integers, and such that $E_n \supseteq E_{n+1}$ for all n. In other words a filtration on E is just a decreasing sequence of submodules of E.

It will be shown presently that every filtration on E determines a topology on E. First, however, we make some preliminary remarks so as to facilitate the description of a base of this topology. Let $x \in E$ and let K be a submodule of E. The set of all elements of E which can be expressed in the form $x + k$, with k in K, is just that coset of K in E which contains x. As in section (1.5), we denote it by $x + K$. Such a subset of E is sometimes known as an *affine subset*. Note that *if y belongs to $x + K$, then $x + K = y + K$*. Also

$$(x + K_1) \cap (x + K_2) \cap \ldots \cap (x + K_p) = x + (K_1 \cap K_2 \cap \ldots \cap K_p) \quad (9.2.1)$$

if $K_1, K_2, \ldots, K_p$ are submodules of E.

Now suppose that $\{E_n\}_{n \geqslant 0}$ is a given filtration on E and let us form a collection Ω of subsets of E in the following way: *a subset V of E is to belong to Ω provided that whenever $v \in V$ there exists an integer n such that $v + E_n$ is wholly contained in V.* Obviously E and its empty subset belong to Ω, and Ω is closed under arbitrary unions. Now assume that $V_1, V_2, \ldots, V_q$ belong to Ω and let v be an element of their intersection. For each i ($1 \leqslant i \leqslant q$), $v + E_{n_i}$ is contained in V_i, where n_i is a suitably chosen non-negative integer. Put $n = \max (n_1, n_2, \ldots, n_q)$. Then $v + E_n$ is contained in $V_1 \cap V_2 \cap \ldots \cap V_q$. This proves that Ω is also closed under finite intersections. Accordingly there is a topology on E whose open sets are precisely those subsets of E which belong to Ω. We shall refer to it either as the *topology derived from the filtration* or, more briefly, as the *filtration topology*.

Proposition 1. *Let E be an R-module with filtration $\{E_n\}_{n \geqslant 0}$. Then the sets of the form $e + E_n$, where $e \in E$ and n is a non-negative integer, form a base for the filtration topology. The sets of this base which contain a given element x all have the form $x + E_n$.*

Proof. It is clear that every open set is a union of sets of the form $e + E_n$. Next if e' belongs to $e + E_n$, then $e' + E_n = e + E_n$. This proves that $e + E_n$ is open. The first assertion is therefore proved and the second assertion is clear from our preceding remarks.

Proposition 2. *Let E be an R-module with filtration $\{E_n\}_{n \geqslant 0}$ and let K be a submodule of E. Then K is open in E (with respect to the filtration topology) if and only if $E_m \subseteq K$ for some non-negative integer m. If a submodule is open, then it is also closed.*

Proof. First suppose that K is open. Since $0 \in K$, we see that

$$0 + E_m = E_m$$

is contained in K for some non-negative integer m. Moreover the complement of K in E is the union of all sets $e + E_m$, where $e \in E$, $e \notin K$. Since each of these sets is open, the complement of K is open and therefore K itself is closed.

Next suppose that $E_m \subseteq K$. Then K is the union of the open sets $k + E_m$, where $k \in K$. Hence K is open.

Corollary. *Each of the submodules E_n, of the filtration, is both open and closed in E.*

We now consider the closure of a submodule of a filtered module.

Proposition 3. *Let E be an R-module with filtration $\{E_n\}_{n \geqslant 0}$ and let K be a submodule of E. Then the closure $\bar{K}$ of K (with respect to the filtration topology) is given by*

$$\bar{K} = \bigcap_{n=0}^{\infty} (K + E_n).$$

In particular, the closure of K is also a submodule of E.

Proof. By Proposition 2, $K + E_n$ is a closed submodule of E. Since $\bar{K}$ is the intersection of all the closed subsets of E that contain K, we have

$$\bar{K} \subseteq \bigcap_{n=0}^{\infty} (K + E_n).$$

Now suppose that e belongs to $K + E_n$ for all n. Then $e + E_n$ contains an element of K. As this holds for every n, it follows that each neighbourhood of e meets K. Accordingly $e \in \bar{K}$ and the proof is complete.

Corollary. *The closure of the zero submodule of E is $\bigcap\limits_{n=0}^{\infty} E_n$.*

We next give a criterion for a filtered module to be a Hausdorff space.

Proposition 4. *Let E be an R-module with filtration $\{E_n\}_{n \geqslant 0}$. Then in order that E, when endowed with the filtration topology, should be a Hausdorff space it is necessary and sufficient that $\bigcap\limits_{n=0}^{\infty} E_n = 0$.*

Proof. First assume that the intersection of the E_n is the zero submodule and let e, e' be distinct elements of E. Then there exists an integer m ($m \geqslant 0$) such that $e - e'$ is not in E_m. Thus $e + E_m$ and $e' + E_m$ are neighbourhoods of e and e' respectively and they have no point in common. It follows that E is a Hausdorff space.

We now suppose that E is a Hausdorff space. Let $e^* \neq 0$ belong to E. Then there exists an open set containing 0 but not containing e^*, and therefore there is an integer $s \geqslant 0$ such that $e^* \notin E_s$. Accordingly the intersection of all the E_n does not contain e^*. Since e^* was an arbitrary non-zero element of E, this means that $\bigcap\limits_{n=0}^{\infty} E_n = 0$.

Corollary. *In order that E should be a Hausdorff space it is necessary and sufficient that the set which consists solely of the zero element be closed.*

We shall now examine, from the point of view of continuity, some of the basic operations that one performs with the elements of an

R-module. First, however, let us agree that, unless the contrary is stated explicitly, *a reference to a topology on a filtered module always refers to the filtration topology*. This convention will help us to avoid excessive verbosity and will be used continually from here on.

Proposition 5. *Let E be an R-module with filtration. Denote by $\nu : E \to E$ the mapping in which $\nu(e) = -e$ and, for each $x \in E$, let $\tau_x : E \to E$ be defined by $\tau_x(e) = x + e$. Then ν and τ_x are homeomorphisms of E on to itself.*

Proof. Let $\{E_n\}_{n \geqslant 0}$ be the filtration on E. Then

$$\nu(e + E_n) = -e + E_n = \nu(e) + E_n$$

and $$\tau_x(e + E_n) = (x + e) + E_n = \tau_x(e) + E_n.$$

These show that ν and τ_x are continuous and it is obvious that each of them is a one-one mapping of E on to itself. Since ν is its own inverse, the inverse is continuous. Next, the inverse of τ_x is τ_y, where $y = -x$. Thus τ_x also has a continuous inverse. The proof is now complete.

Proposition 6. *Let E be a filtered R-module and let $E \times E$ be endowed with the product topology. Then the mapping*

$$\sigma : E \times E \to E$$

in which $\sigma(e, e') = e + e'$ is continuous.

Remark. It is customary to describe this result by saying that *addition on E is a continuous operation*.

Proof. If x belongs to $e + E_m$ and y to $e' + E_m$, then $x + y$ is in $(e + e') + E_m$. Hence the image, under σ, of $(e + E_m) \times (e' + E_m)$ is contained in $\sigma(e, e') + E_m$. This establishes the proposition.

Proposition 7. *Let E be a filtered R-module and let r belong to R. Further let the mapping $\mu_r : E \to E$ be defined by $\mu_r(e) = re$. Then μ_r is continuous.*

Proof. If x belongs to $e + E_n$, then rx belongs to $re + E_n$. Consequently $\mu_r(e + E_n) \subseteq \mu_r(e) + E_n$ and the proposition follows.

9.3 Continuous, compatible and strict homomorphisms

We must now turn our attention to homomorphisms of one filtered module into another. It is clear that for such a mapping to be significant it must have some relation to the filtrations. However there are various forms which such a relation can take.

Proposition 8. *Let E and E' be filtered R-modules and let $f:E \to E'$ be an R-homomorphism. Then in order that f should be continuous it is necessary and sufficient that it be continuous at the zero element of E.*

Proof. The condition is necessary for trivial reasons. We shall therefore assume that f is continuous at the zero element of E and deduce that it is continuous everywhere else.

Let $\{E_n\}_{n \geqslant 0}$ and $\{E'_n\}_{n \geqslant 0}$ be the filtrations on E and E' respectively. Suppose that $e \in E$ and that $k \geqslant 0$ is a given integer. Then, since f is continuous at the zero element of E, there exists an integer $m \geqslant 0$ such that $f(E_m) \subseteq E'_k$. But now

$$f(e + E_m) = f(e) + f(E_m) \subseteq f(e) + E'_k.$$

The required result follows.

Once again, let E and E' be filtered R-modules whose filtrations are $\{E_n\}_{n \geqslant 0}$ and $\{E'_n\}_{n \geqslant 0}$ respectively.

Definition. *An R-homomorphism $f:E \to E'$ is said to be 'compatible with the filtrations' if*

$$f(E_n) \subseteq E'_n \quad (all\ n \geqslant 0). \tag{9.3.1}$$

It is said to be 'strict' if

$$f(E_n) = f(E) \cap E'_n \tag{9.3.2}$$

for every $n \geqslant 0$.

It is clear that if f is strict, then it is also compatible with the filtrations. Again, by (9.3.1), it is obvious that a compatible homomorphism is continuous at the zero element of its domain. It therefore follows, by Proposition 8, that such a mapping is continuous everywhere.

Let $\{E_n\}_{n \geqslant 0}$ be a filtration on an R-module E and let K be a submodule of E. Then the family $\{K \cap E_n\}_{n \geqslant 0}$ of submodules of K constitutes a filtration on K. This is known as the *induced filtration*. Observe that if $k \in K$, then $(k + E_n) \cap K = k + (E_n \cap K)$. It follows that a subset of K is open in the topology derived from the filtration $\{K \cap E_n\}_{n \geqslant 0}$ if and only if it is the intersection of an open subset of E with K. In other words, *the topology derived from the induced filtration coincides with the topology that K acquires as a subspace of E.*

Again $\{(E_n + K)/K\}_{n \geqslant 0}$ is a family of submodules of E/K which constitutes a filtration on that module. This is called the *factor filtration* on E/K. Note that in the canonical exact sequence

$$K \to E \to E/K$$

both homomorphisms are strict if K and E/K are endowed with the induced and factor filtrations respectively.

Let E and E' be filtered R-modules with $\{E_n\}_{n \geqslant 0}$ and $\{E'_n\}_{n \geqslant 0}$ as their respective filtrations. Further let $f \colon E \to E'$ be an R-homomorphism which is compatible with these filtrations. Then, since

$$f(E_n) \subseteq E'_n,$$

f induces a homomorphism $f_n \colon E/E_n \to E'/E'_n$ which is such that the diagram

$$
\begin{array}{ccc}
E & \xrightarrow{\ f\ } & E' \\
\phi_n \downarrow & & \downarrow \phi'_n \\
E/E_n & \xrightarrow[f_n]{} & E'/E'_n
\end{array}
$$

is commutative. Here, of course, ϕ_n is the natural mapping of E on to E/E_n and ϕ'_n has a similar significance. It follows that if e belongs to $\operatorname{Ker} f$, then $\phi_n(e)$ belongs to $\operatorname{Ker} f_n$. Hence for each $n \geqslant 0$, ϕ_n gives rise to a homomorphism $\operatorname{Ker} f \to \operatorname{Ker} f_n$. We now use these mappings to obtain a useful criterion for a compatible homomorphism to be strict.

Lemma 1. *Let E and E' be filtered R-modules and $f \colon E \to E'$ an R-homomorphism that is compatible with the filtrations. Then in order that f should be strict it is necessary and sufficient that, for each $n \geqslant 0$, the associated mapping $\operatorname{Ker} f \to \operatorname{Ker} f_n$ be an epimorphism.*

Proof. First suppose that f is strict. With the previous notation, a typical element of $\operatorname{Ker} f_n$ has the form $\phi_n(e)$, where

$$e \in E \quad \text{and} \quad f_n \phi_n(e) = 0.$$

It follows that $\phi'_n f(e) = 0$ and therefore $f(e)$ belongs to

$$f(E) \cap E'_n = f(E_n).$$

Hence there exists $e_n \in E_n$ such that $f(e) = f(e_n)$. Accordingly $e - e_n$ is in $\operatorname{Ker} f$ and $\phi_n(e - e_n) = \phi_n(e)$. Thus $\phi_n(e)$ is the image of the element $e - e_n$ of $\operatorname{Ker} f$ and therefore $\operatorname{Ker} f \to \operatorname{Ker} f_n$ is an epimorphism.

Now assume that $\operatorname{Ker} f \to \operatorname{Ker} f_n$ is an epimorphism for all n and let x' belong to $f(E) \cap E'_n$. In order to complete the proof we have only to show that x' is in $f(E_n)$. Since $x' \in f(E)$, we have $x' = f(x)$ for a suitable element $x \in E$ and now

$$f_n \phi_n(x) = \phi'_n f(x) = \phi'_n(x') = 0$$

because $x' \in E_n'$. Thus $\phi_n(x) \in \mathrm{Ker} f_n$. But $\mathrm{Ker} f \to \mathrm{Ker} f_n$ is an epimorphism. Consequently $\phi_n(x) = \phi_n(y)$, where $y \in \mathrm{Ker} f$. It follows that $x - y$ belongs to E_n and $f(x) = f(x - y)$. Thus $x' = f(x)$ belongs to $f(E_n)$ and now the proof is complete.

9.4 Complete filtered modules

As before we suppose that E is an R-module and $\{E_n\}_{n \geqslant 0}$ a filtration on E. Now let x be an element of E and $\{x_m\}$, where $m = 1, 2, 3, \ldots$, a sequence of elements of the module.

Definition. *The sequence $\{x_m\}$ is said to 'converge to x' or to 'have x as a limit' if the following condition is satisfied: given any integer $k \geqslant 0$ there always exists a positive integer N such that $(x - x_m) \in E_k$ whenever $m > N$. In these circumstances we often write $x_m \to x$.*

It is easy to verify that if $x_m \to x$ and $y_m \to y$, then

$$(x_m + y_m) \to (x + y), \tag{9.4.1}$$

and

$$(x_m - y_m) \to (x - y), \tag{9.4.2}$$

and

$$r x_m \to r x \tag{9.4.3}$$

for every r in R. Suppose now that $f : E \to E'$ is a continuous mapping of E into a second filtered module E'. (It is not necessary that f should be a homomorphism.) A simple application of the definitions shows that the sequence $\{f(x_m)\}$ converges to $f(x)$. Hence

$$x_m \to x \text{ implies that } f(x_m) \to f(x) \tag{9.4.4}$$

whenever f is continuous.

Lemma 2. *Let K be a subset of a filtered R-module E and let x belong to E. Then x belongs to the closure of K in E if and only if there exists a sequence $\{x_m\}$ such that $x_m \in K$ for all m and $x_m \to x$.*

Proof. If such a sequence exists, then every neighbourhood of x contains at least one x_m and therefore it contains a point of K. Accordingly x belongs to the closure of K.

Suppose now that x belongs to the closure of K. Then $x + E_m$ meets K and therefore we can find $x_m \in K$ such that $(x - x_m) \in E_m$. Evidently $x_m \to x$, so the proof is complete.

Once again, let $\{x_m\}$ be a sequence of elements of E and let us suppose that both $x_m \to x$ and $x_m \to x'$. If now $k \geqslant 0$ is an integer, then $x - x' = (x - x_m) - (x' - x_m)$ will belong to E_k provided that m is sufficiently large. It follows that $x - x'$ is in $\bigcap_k E_k$. However should it happen that E is a Hausdorff space then this intersection is zero by Proposition 4. Hence *when E is a Hausdorff space, limits are unique.* In these circumstances we shall use $\lim x_m = x$ as an alternative to $x_m \to x$.

There is another important concept, relating to sequences, that we shall need. This is embodied in the following

Definition. *A sequence $\{x_m\}$ of elements (of the filtered module E) is called a 'Cauchy sequence' if given any integer k ($k \geqslant 0$) there exists an integer N such that $(x_n - x_m) \in E_k$ whenever both m and n are greater than N.*

If $\{x_m\}$ is a convergent sequence, say $x_m \to x$, then it is also a Cauchy sequence. For given $k \geqslant 0$ there exists an integer N such that

$$(x - x_m) \in E_k \quad \text{for all} \quad m > N.$$

It follows that if $m > N$ and $n > N$, then $x_m - x_n = (x - x_n) - (x - x_m)$ belongs to E_k. This proves the assertion.

Definition. *An R-module E with filtration $\{E_n\}_{n \geqslant 0}$ is said to be 'complete' if the following two conditions hold:*

(a) $\bigcap\limits_{n=0}^{\infty} E_n = 0$, *that is to say E is a Hausdorff space;*

(b) *every Cauchy sequence converges to some element of E.*

Let $\{x_m\}$ be a sequence of elements of E and suppose that it converges to x. Then $x_2, x_3, x_4, \ldots$ also converges to x and therefore, by (9.4.2), $(x_{m+1} - x_m) \to 0$. It will now be shown that in the case of a complete module the converse holds.

Proposition 9. *Let E be a filtered module which is complete with respect to its filtration and let $\{x_m\}$ be a sequence of elements of E. Then $\{x_m\}$ converges if and only if $\lim (x_{m+1} - x_m) = 0$.*

Proof. We shall assume that $\lim(x_{m+1} - x_m) = 0$ and deduce that $\{x_m\}$ is a Cauchy sequence and hence a convergent sequence. This will establish the proposition because the reverse implication has already been established.

Let an integer $k \geqslant 0$ be given. There exists an integer N such that $x_{m+1} - x_m$ belongs to E_k whenever $m \geqslant N$. Suppose now that $s > N$ and $t > N$. Then

$$x_s - x_N = (x_{N+1} - x_N) + (x_{N+2} - x_{N+1}) + \ldots + (x_s - x_{s-1})$$

which belongs to E_k. Similarly $x_t - x_N$ is in E_k. Accordingly $(x_s - x_t) \in E_k$. This shows that $\{x_m\}$ is a Cauchy sequence and completes the proof.

It is sometimes convenient to make use of the notion of a convergent *series*. To explain what this means let $\{u_m\}$ be a sequence of elements of the filtered module E and put $s_m = u_1 + u_2 + \ldots + u_m$. We then say that *the series*

$$u_1 + u_2 + u_3 + \ldots + u_m + \ldots$$

converges to s if $s_m \to s$. If this is the case and if also E is a Hausdorff space, then we write

$$s = u_1 + u_2 + u_3 + \ldots$$

or

$$s = \sum_{i=1}^{\infty} u_i.$$

Proposition 10. *Let E be a filtered R-module which is complete with respect to its filtration. Then a series*

$$u_1 + u_2 + u_3 + \ldots + u_m + \ldots,$$

of elements of E, converges if and only if $\lim u_m = 0$.

This follows at once by applying Proposition 9 to the partial sums $u_1 + u_2 + \ldots + u_m$ of the series.

We recall that by an *affine* subset of an R-module E is meant a subset of the form $e + K$, where $e \in E$ and K is a submodule of E.

Theorem 1. *Let E be an R-module and $\{E_n\}_{n \geqslant 0}$ a filtration on E. Suppose that E is complete with respect to its filtration and that, for each $n \geqslant 0$, E/E_n satisfies the minimal condition for submodules. If now*

$$M_1 \supseteq M_2 \supseteq M_3 \supseteq \ldots \supseteq M_p \supseteq \ldots$$

is an infinite decreasing sequence of closed, affine subsets of E, then the intersection of these affine subsets is not empty.

Proof. Let $M_p = u_p + K_p$, where $u_p \in E$ and K_p is a submodule of E. Since each element of K_{p+1} is the difference of two elements of M_{p+1} and since $M_{p+1} \subseteq M_p$, it follows that $K_{p+1} \subseteq K_p$. Thus

$$K_1 \supseteq K_2 \supseteq K_3 \supseteq \ldots$$

is a decreasing sequence of submodules of E.

By hypothesis, E/E_1 satisfies the minimal condition for submodules. It follows that there exists an integer ν_1 such that

$$K_t + E_1 = K_{\nu_1} + E_1 \quad \text{for all} \quad t \geqslant \nu_1.$$

Next, since E/E_2 satisfies the minimal condition for submodules, there exists an integer ν_2 such that $\nu_2 > \nu_1$ and $K_t + E_2 = K_{\nu_2} + E_2$ for all $t \geqslant \nu_2$. It is now possible to find $\nu_3 > \nu_2$ with the property that

$K_t + E_3 = K_{\nu_3} + E_3$ whenever $t \geqslant \nu_3$. In this way we obtain an infinite strictly increasing sequence $\nu_1, \nu_2, \nu_3, \ldots$ of positive integers such that $K_t + E_s = K_{\nu_s} + E_s$ whenever $t \geqslant \nu_s$.

We shall now construct elements $x_1, x_2, x_3, \ldots$ *in succession* so that $x_i \in M_{\nu_i}$ and $x_{i+1} \equiv x_i \pmod{E_i}$. To start the sequence we take x_1 to be any element of M_{ν_1}. Suppose now that $x_1, x_2, x_3, \ldots, x_r$ have all been found and satisfy the prescribed conditions. Then

$$M_{\nu_r} = u_{\nu_r} + K_{\nu_r} = x_r + K_{\nu_r} \subseteq x_r + K_{\nu_r} + E_r = x_r + K_{\nu_{r+1}} + E_r,$$

and therefore $\qquad M_{\nu_{r+1}} \subseteq M_{\nu_r} \subseteq x_r + K_{\nu_{r+1}} + E_r.$

It follows that $u_{\nu_{r+1}} = x_r + k_{\nu_{r+1}} + e_r$, where the notation is self-explanatory. If therefore we put $x_{r+1} = u_{\nu_{r+1}} - k_{\nu_{r+1}}$, then $x_{r+1} \in M_{\nu_{r+1}}$ and $x_{r+1} \equiv x_r \pmod{E_r}$. This establishes the existence of a sequence $x_1, x_2, x_3, \ldots$ having the properties enumerated above.

By construction, $\lim(x_{r+1} - x_r) = 0$. Hence, by Proposition 9, the sequence $\{x_r\}$ converges. Put $x = \lim x_r$. For a fixed integer p, x_r belongs to M_p once r is sufficiently large. Hence, by Lemma 2, x belongs to the closure of M_p. But M_p is closed. Consequently $x \in M_p$. As this is true for all values of p, the proof is complete.

Theorem 2. *Let E be an R-module and $\{E_n\}_{n \geqslant 0}$ a filtration on E. Suppose that E is complete with respect to its filtration and that, for each $n \geqslant 0$, E/E_n satisfies the minimal condition for submodules. If now*

$$K_1 \supseteq K_2 \supseteq K_3 \supseteq \ldots \supseteq K_p \supseteq \ldots$$

is an infinite decreasing sequence of closed submodules of E such that $\bigcap_{p=1}^{\infty} K_p = 0$, then given any integer $s > 0$ there exists an integer ν_s such that $K_{\nu_s} \subseteq E_s$.

Proof. As in the proof of Theorem 1, there exists an increasing sequence $\nu_1, \nu_2, \nu_3, \ldots$ of positive integers such that $K_t + E_s = K_{\nu_s} + E_s$ provided that $t \geqslant \nu_s$. Put $L_p = K_{\nu_p}$. Then the L_p form a decreasing sequence of closed submodules of E such that $\bigcap_p L_p = 0$ and $L_{p+1} + E_p = L_p + E_p$ for all p.

Let $s > 0$ be given. It will be shown that $L_s \subseteq E_s$ and this will complete the proof. To this end suppose that $x \in L_s$. Then x belongs to $L_{s+1} + E_s$, say $x = e_s + y_{s+1}$, where $e_s \in E_s$ and $y_{s+1} \in L_{s+1} \subseteq L_{s+2} + E_{s+1}$. Accordingly $x = e_s + e_{s+1} + y_{s+2}$, where now $e_{s+1} \in E_{s+1}$ and

$$y_{s+2} \in L_{s+2} \subseteq L_{s+3} + E_{s+2}.$$

Proceeding in this way we obtain sequences

$$e_s, e_{s+1}, e_{s+2}, \ldots$$

and

$$y_{s+1}, y_{s+2}, y_{s+3}, \ldots,$$

where $\quad e_{s+i} \in E_{s+i} \quad (i \geqslant 0), \qquad y_{s+j} \in L_{s+j} \quad (j \geqslant 1),$

and for all $n \geqslant 1$

$$x = (e_s + e_{s+1} + \ldots + e_{s+n-1}) + y_{s+n}.$$

By Proposition 10, the series

$$e_s + e_{s+1} + e_{s+2} + \ldots$$

converges. Let e be its sum. Then, since all the partial sums belong to E_s and E_s is a closed subset of E (Proposition 2 Cor.), we have $e \in E_s$. Furthermore, the sequence $y_{s+1}, y_{s+2}, y_{s+3}, \ldots$ converges to $x - e$. Now for a fixed value of p all the terms of this sequence belong to L_p with at most a finite number of exceptions. Hence, by Lemma 2 and because L_p is closed, $x - e$ is in L_p. This is true for every value of p. Accordingly $x - e$ belongs to $\bigcap_p L_p = 0$ and therefore $x = e \in E_s$. The proof is now complete.

9.5 The completion of a filtered module

Let E be an R-module and $\{E_n\}_{n \geqslant 0}$ a filtration on E. The notion of *a completion of E with respect to its filtration* is central to the ideas being developed in this chapter. The precise definition is as follows.

Definition. *A 'completion of E' with respect to the filtration $\{E_n\}_{n \geqslant 0}$ is a composite object consisting of an R-module $\hat{E}$, a filtration $\{\hat{E}_n\}_{n \geqslant 0}$ on $\hat{E}$, and an R-homomorphism $\psi : E \to \hat{E}$. These are required to satisfy the following conditions*:

(i) *$\hat{E}$ is complete with respect to $\{\hat{E}_n\}_{n \geqslant 0}$;*

(ii) *the homomorphism $\psi : E \to \hat{E}$ is strict;*

(iii) *$\psi(E)$ is everywhere dense in $\hat{E}$;*

(iv) *$\operatorname{Ker} \psi = \bigcap_{n=0}^{\infty} E_n.$*

In practice one tends to use a rather casual form of expression when referring to completions because it is so very tedious to have to say everything in full. Thus one might say '*let $\hat{E}$ with filtration $\{\hat{E}_n\}_{n \geqslant 0}$ be a* *completion of the filtered module E'*. A more extreme example would be the statement '*let $\hat{E}$ be a completion of E*'. In neither case is there a reference to the all important homomorphism ψ. However, if at some later stage it becomes necessary to mention this mapping, then it will be introduced by some such phrase as '*where $\psi: E \to \hat{E}$ is the canonical homomorphism*'. Again, assume that subsequent to making the informal remark '*suppose that $\hat{E}$ is a completion of E*' we need to mention the special filtration $\{\hat{E}_n\}_{n \geqslant 0}$ that is implicit in the notion of a completion. We accomplish this by calling $\{\hat{E}_n\}_{n \geqslant 0}$ *the canonical filtration on $\hat{E}$*.

So far we do not know whether every filtered module possesses a completion. It will be shown that this is so in the next section. We also require to know to what extent completions are unique. This question will be dealt with shortly but it is convenient to deal with some other matters first.

Theorem 3. *Let E be an R-module with filtration $\{E_n\}_{n \geqslant 0}$ and let $\hat{E}$ with filtration $\{\hat{E}_n\}_{n \geqslant 0}$ be one of its completions. If now $\psi: E \to \hat{E}$ is the canonical homomorphism, then $\hat{E}_n$ is the closure of $\psi(E_n)$ in $\hat{E}$ and $E_n = \psi^{-1}(\hat{E}_n)$.*

Proof. Since $\psi: E \to \hat{E}$ is strict, we have $\psi(E_n) = \psi(E) \cap \hat{E}_n$ and therefore $\psi(E_n) \subseteq \hat{E}_n$. Next, by Proposition 2 Cor., $\hat{E}_n$ is closed in $\hat{E}$. Consequently the closure of $\psi(E_n)$ in $\hat{E}$ is contained in $\hat{E}_n$.

Let ξ belong to $\hat{E}_n$. By the definition of a completion, $\psi(E)$ is everywhere dense in $\hat{E}$. Hence for every integer $k \geqslant 0$, the intersection of $\xi + \hat{E}_{n+k}$ and $\psi(E)$ is not empty. Choose $\eta_k \in \hat{E}_{n+k}$ so that

$$\xi + \eta_k \in \psi(E).$$

Then $\xi + \eta_k$ belongs to $\psi(E) \cap \hat{E}_n = \psi(E_n)$. Now $\lim \eta_k = 0$ and therefore $\lim(\xi + \eta_k) = \xi$. Thus ξ is the limit of a sequence of elements each of which belongs to $\psi(E_n)$. Accordingly, by Lemma 2, ξ belongs to the closure of $\psi(E_n)$ in $\hat{E}$. It follows that the closure of $\psi(E_n)$ is $\hat{E}_n$ as was asserted. Finally

$$\psi^{-1}(\hat{E}_n) = \psi^{-1}(\psi(E) \cap \hat{E}_n) = \psi^{-1}(\psi(E_n)) = E_n + \operatorname{Ker} \psi = E_n$$

because $\operatorname{Ker} \psi$ is contained in E_n. This completes the proof.

Theorem 4. *Let E be a filtered R-module with completion $\hat{E}$. Then there is a one-one correspondence between the open R-submodules U of E and the open R-submodules V of $\hat{E}$. This is such that if U and V correspond and*

$\psi: E \to \hat{E}$ is the canonical mapping, then $U = \psi^{-1}(V)$ and V is the closure of $\psi(U)$ in $\hat{E}$. In this situation ψ induces an isomorphism of the R-module E/U on to the R-module $\hat{E}/V$.

Proof. We employ the usual notation for the filtrations on E and $\hat{E}$. First suppose that V is an open submodule of $\hat{E}$ and put $U = \psi^{-1}(V)$. Then, since ψ is continuous, U is an open submodule of E. Choose s large enough to ensure that $\hat{E}_s \subseteq V$. By Proposition 2, $\psi(E) + \hat{E}_s$ is an open and therefore also a closed submodule of $\hat{E}$. But $\psi(E)$ is everywhere dense in $\hat{E}$. Consequently $\psi(E) + \hat{E}_s = \hat{E}$. Now $\psi(U) = \psi(E) \cap V$ and therefore

$$\psi(U) + \hat{E}_s = (\psi(E) \cap V) + \hat{E}_s$$
$$= (\psi(E) + \hat{E}_s) \cap V$$
$$= \hat{E} \cap V$$
$$= V.$$

Since this holds for all large values of s, it follows, from Proposition 3, that the closure of $\psi(U)$ in $\hat{E}$ is V. Again ψ induces a homomorphism $E/U \to \hat{E}/V$. Also, if s is large, $\psi(E) + V$ contains $\psi(E) + \hat{E}_s = \hat{E}$. Thus $\psi(E) + V = \hat{E}$ which shows that the homomorphism in question is an epimorphism. On the other hand, if $\psi(e) \in V$, then $e \in U$. This shows that $E/U \to \hat{E}/V$ is also a monomorphism and hence an isomorphism.

To complete the proof we must now show that if U' is a given open submodule of E, then there exists an open submodule V' of $\hat{E}$ such that $U' = \psi^{-1}(V')$. By Proposition 2, we can find an integer n so that $E_n \subseteq U'$. Put $V' = \psi(U') + \hat{E}_n$. Then V' is certainly an open submodule of $\hat{E}$ and $U' \subseteq \psi^{-1}(V')$. Moreover, if $e \in \psi^{-1}(V')$, then there exists $u' \in U'$ such that $\psi(e - u') \in \hat{E}_n$. Accordingly, by Theorem 3, $(e - u') \in E_n$ and hence $e \in U'$. It follows that $U' = \psi^{-1}(V')$ and now the proof is complete.

Corollary. *With the usual notation, ψ induces an isomorphism*

$$E/E_n \approx \hat{E}/\hat{E}_n,$$

of R-modules, for each $n \geqslant 0$.

Proof. By Theorem 3, we may take $U = E_n$ and $V = \hat{E}_n$ in Theorem 4.

We now consider a continuous homomorphism between two filtered modules in relation to completions of these modules.

Theorem 5. *Let E and E' be filtered R-modules and $\hat{E}$, $\hat{E}'$ completions of E and E' respectively. If now $f: E \to E'$ is a continuous R-homomorphism, then there exists a unique continuous mapping $\hat{f}: \hat{E} \to \hat{E}'$ such that the diagram*

$$
\begin{array}{ccc}
E & \xrightarrow{\ f\ } & E' \\
\psi \downarrow & & \downarrow \psi' \\
\hat{E} & \xrightarrow[\hat{f}]{} & \hat{E}'
\end{array}
$$

is commutative. (Here ψ and ψ' are the canonical mappings of E and E' into their completions.) The mapping $\hat{f}$, so obtained, is necessarily a homomorphism of R-modules.

Proof. The relevant filtrations on E, E', $\hat{E}$ and $\hat{E}'$ will be denoted by $\{E_n\}$, $\{E'_n\}$, $\{\hat{E}_n\}$ and $\{\hat{E}'_n\}$ respectively. Since f is continuous, we can find a strictly increasing sequence $\nu_0, \nu_1, \nu_2, \ldots$ of non-negative integers such that

$$f(E_{\nu_k}) \subseteq E'_k \tag{9.5.1}$$

for all values of k. It follows that

$$f\Big(\bigcap_{k=0}^{\infty} E_k\Big) \subseteq \bigcap_{k=0}^{\infty} E'_k$$

that is to say $f(\mathrm{Ker}\,\psi) \subseteq \mathrm{Ker}\,\psi'$. Accordingly f induces an R-homomorphism

$$f^*: \psi(E) \to \psi'(E') \tag{9.5.2}$$

such that the diagram

$$
\begin{array}{ccc}
E & \xrightarrow{\ f\ } & E' \\
\downarrow & & \downarrow \\
\psi(E) & \xrightarrow[f^*]{} & \psi'(E')
\end{array}
\tag{9.5.3}
$$

is commutative. Further, by (9.5.1), we have

$$f^*(\psi(E) \cap \hat{E}_{\nu_k}) \subseteq \psi'(E') \cap \hat{E}'_k. \tag{9.5.4}$$

Let $\xi \in \hat{E}$. Since $\psi(E)$ is everywhere dense in $\hat{E}$, there exists a sequence $\{u_m\}$ of elements of $\psi(E)$ such that $\lim u_m = \xi$. Accordingly $\{u_m\}$ is a Cauchy sequence and now we see, from (9.5.4), that $\{f^*(u_m)\}$ is a Cauchy sequence in $\hat{E}'$. But $\hat{E}'$ is complete. Accordingly $\{f^*(u_m)\}$ has a limit in $\hat{E}'$ and, since $\hat{E}'$ is a Hausdorff space, this limit is uniquely determined by $\{u_m\}$.

Suppose that we have a second sequence, $\{v_m\}$ say, such that $v_m \in \psi(E)$ for all m and $\lim v_m = \xi$. Then, of course, $\lim f^*(v_m)$ also exists. Now $\lim(u_m - v_m) = 0$. Hence by (9.5.4), $\lim f^*(u_m - v_m) = 0$ and therefore $\lim f^*(u_m) = \lim f^*(v_m)$. This conclusion may be described by saying that $\lim f^*(u_m)$ *depends only on ξ and is otherwise independent of the choice of the sequence $\{u_m\}$.* We can therefore define a mapping $\hat{f}: \hat{E} \to \hat{E}'$ by means of the equation

$$\hat{f}(\xi) = \lim f^*(u_m). \tag{9.5.5}$$

If $r \in R$, then, by (9.4.3), both $\lim(ru_m) = r\xi$ and

$$\lim f^*(ru_m) = r \lim f^*(u_m).$$

Thus $\hat{f}(r\xi) = r\hat{f}(\xi)$. Again, if $\xi^{(1)}$ and $\xi^{(2)}$ belong to $\hat{E}$, then we can find sequences $\{u_m^{(1)}\}$ and $\{u_m^{(2)}\}$, of elements of $\psi(E)$, which converge to $\xi^{(1)}$ and $\xi^{(2)}$ respectively. But then, by (9.4.1), $\lim(u_m^{(1)} + u_m^{(2)}) = \xi^{(1)} + \xi^{(2)}$ and so

$$\hat{f}(\xi^{(1)} + \xi^{(2)}) = \lim f^*(u_m^{(1)} + u_m^{(2)}) = \lim f^*(u_m^{(1)}) + \lim f^*(u_m^{(2)}).$$

Accordingly $\qquad \hat{f}(\xi^{(1)} + \xi^{(2)}) = \hat{f}(\xi^{(1)}) + \hat{f}(\xi^{(2)}). \tag{9.5.6}$

Thus we see that $\hat{f}: \hat{E} \to \hat{E}'$ is a homomorphism of R-modules.

Next assume that $\xi \in \hat{E}_{\nu_k}$. By Theorem 3, $\hat{E}_{\nu_k}$ is the closure, in $\hat{E}$, of $\psi(E_{\nu_k})$ and therefore we can arrange that the sequence $\{u_m\}$ is composed of elements of $\psi(E_{\nu_k}) = \psi(E) \cap \hat{E}_{\nu_k}$. But in that case $f^*(u_m) \in \hat{E}'_k$ for all m by virtue of (9.5.4). Now, by Proposition 2 Cor., $\hat{E}'_k$ is a closed submodule of $\hat{E}'$ and therefore $\hat{f}(\xi) = \lim f^*(u_m)$ belongs to $\hat{E}'_k$ as well. It follows that

$$\hat{f}(\hat{E}_{\nu_k}) \subseteq \hat{E}'_k \tag{9.5.7}$$

for all values of k. This shows that the homomorphism $\hat{f}$ is continuous at the zero element. Hence, by Proposition 8, it is continuous everywhere.

Again, if $\xi \in \psi(E)$, then we may take for $\{u_m\}$ the sequence whose terms are all equal to ξ and from this we see that $\hat{f}(\xi) = f^*(\xi)$. Thus $\hat{f}$ and f^* agree on $\psi(E)$ and therefore the diagram

$$\begin{array}{ccc} E & \xrightarrow{f} & E' \\ \psi \downarrow & & \downarrow \psi' \\ \hat{E} & \xrightarrow{\hat{f}} & \hat{E}' \end{array}$$

is commutative.

For the final step we observe that if two mappings, $\hat{f}_1$ and $\hat{f}_2$ say, have the properties described in the statement of the theorem, then they must agree with f^* on $\psi(E)$. Thus $\hat{f}_1$ and $\hat{f}_2$ are continuous and agree on an everywhere dense subset of $\hat{E}$. They must therefore agree everywhere. This completes the proof.

Corollary 1. *Let the situation be as in the statement of the theorem and suppose that f is compatible with the filtrations on E and E'. Then $\hat{f}: \hat{E} \to \hat{E}'$ is compatible with the filtrations on $\hat{E}$ and $\hat{E}'$.*

Proof. Using the same notation as was employed in the proof of Theorem 5, we note that the sequence $\nu_0, \nu_1, \nu_2, \dots$ may be taken to consist of the integers $0, 1, 2, \dots$ in their natural order. In that case (9.5.7) becomes $\hat{f}(\hat{E}_k) \subseteq \hat{E}'_k$ which is what we have to prove.

Corollary 2. *Let the situation be as described in the statement of Theorem 5 and suppose that the homomorphism $f: E \to E'$ is strict. Then $\hat{f}: \hat{E} \to \hat{E}'$ is also strict.*

Proof. By Corollary 1, $\hat{f}$ is compatible with the filtrations on $\hat{E}$ and $\hat{E}'$. Consequently $\hat{f}(\hat{E}_k) \subseteq \hat{f}(\hat{E}) \cap \hat{E}'_k$. Now assume that ξ' belongs to

$$\hat{f}(\hat{E}) \cap \hat{E}'_k.$$

Then there exists $\xi \in \hat{E}$ such that $\hat{f}(\xi) = \xi'$. Next, since $\psi(E)$ is everywhere dense in $\hat{E}$, there exists $e \in E$ such that $(\xi - \psi(e)) \in \hat{E}_k$. It follows that $\hat{f}(\psi(e)) \in \hat{E}'_k$. But $\hat{f}(\psi(e)) = \psi'(f(e))$ and, by Theorem 3,

$$\psi'^{-1}(\hat{E}'_k) = E'_k.$$

Thus $f(e) \in E'_k$ and therefore $f(e)$ is in $f(E) \cap E'_k = f(E_k)$ because f is strict. We now see that there is an element $x \in E_k$ such that $f(x) = f(e)$ and therefore $\hat{f}(\psi(x)) = \hat{f}(\psi(e))$. Thus $\psi(e) - \psi(x)$ belongs to $\mathrm{Ker}\,\hat{f}$ and so

$$\xi = (\xi - \psi(e)) + \psi(x) + (\psi(e) - \psi(x))$$

is in $\hat{E}_k + \mathrm{Ker}\,\hat{f}$. Finally

$$\xi' = \hat{f}(\xi) \in \hat{f}(\hat{E}_k + \mathrm{Ker}\,\hat{f}) = \hat{f}(\hat{E}_k).$$

The corollary follows.

    Theorem 5 shows that each continuous homomorphism $E \to E'$ gives rise to a well defined continuous homomorphism between a completion of E and a completion of E'. From here on the latter mapping will automatically be designated by putting a $\wedge$ over the symbol for the original mapping. This will be done without further explanatory comment.

Theorem 6. *Let E, E' and E'' be filtered R-modules having $\hat{E}$, $\hat{E}'$ and $\hat{E}''$ respectively as completions. Suppose further that $f: E \to E'$ and*

$$g: E' \to E''$$

are continuous R-homomorphisms. Then $gf: E \to E''$ is a continuous R-homomorphism and $\widehat{gf} = \hat{g}\hat{f}$.

This is obvious.

Corollary. *Let E and E' be filtered R-modules with completions $\hat{E}$ and $\hat{E}'$ respectively. If now $f: E \to E'$ is a continuous isomorphism of E on to E' and if the inverse isomorphism $g: E' \to E$ is also continuous, then $\hat{f}: \hat{E} \to \hat{E}'$ is a continuous isomorphism of $\hat{E}$ on to $\hat{E}'$ and its inverse is $\hat{g}$.*

Proof. We have $gf = i$, where i denotes the identity mapping of E. Hence, by Theorem 6, $\hat{g}\hat{f} = \hat{i}$. However it is obvious from the definition that $\hat{i}$ is the identity mapping of $\hat{E}$. Similarly $\hat{f}\hat{g}$ is the identity mapping of $\hat{E}'$. This completes the proof.

The next theorem tells us that any two completions of a filtered module are virtually identical. In view of this it is customary to speak of *the* completion of a filtered module rather than of *a* completion.

Theorem 7. *Let E be a filtered R-module and suppose that $\hat{E}$ with filtration $\{\hat{E}_n\}_{n \geqslant 0}$ and $\hat{E}'$ with filtration $\{\hat{E}'_n\}_{n \geqslant 0}$ are two completions of E. Then there exists one and only one continuous mapping $\omega: \hat{E} \to \hat{E}'$ such that the diagram*

is commutative, where ψ and ψ' are the canonical homomorphisms associated with the completions. The mapping ω has the further properties that it is an isomorphism of the R-module $\hat{E}$ on to the R-module $\hat{E}'$ and $\omega(\hat{E}_n) = \hat{E}'_n$ for every n.

Proof. The existence and uniqueness of ω both follow from Theorem 5. Next, taking $E = E'$ and f to be the identity mapping of E, we can apply the corollary to Theorem 6. This shows us that ω is an isomorphism of R-modules. Finally, by Theorem 5 Cor. 2, ω is strict which in the present instance means that $\omega(\hat{E}_n) = \hat{E}'_n$. The proof is now complete.

Let $f: E \to E'$ be a homomorphism of an R-module E into an R-module E' and let r belong to R. In general, the mapping $\phi: E \to E'$ defined by $\phi(e) = rf(e)$ is not a homomorphism because we are not assuming that R is commutative. However, in the special case where r belongs to the *centre* of R, ϕ will be R-linear. This observation has been made in preparation for

Theorem 8. *Let E and E' be filtered R-modules having $\hat{E}$ and $\hat{E}'$ as their respective completions. Further, let f, f_1, f_2 be continuous homomorphisms of E into E' and let γ belong to the centre of R. Then $f_1 + f_2$, $f_1 - f_2$ and γf are continuous homomorphisms of E into E' and $\widehat{f_1 + f_2} = \hat{f}_1 + \hat{f}_2$, $\widehat{f_1 - f_2} = \hat{f}_1 - \hat{f}_2$, and $\widehat{\gamma f} = \gamma \hat{f}$.*

Proof. Let s be a given non-negative integer. We can then find $t \geqslant 0$ so that $f_i(E_t) \subseteq E'_s$ for $i = 1, 2$, where we have employed the usual notation for the filtrations on E and E'. It follows that $(f_1 + f_2)(E_t) \subseteq E'_s$ and therefore $f_1 + f_2$ is continuous. Similarly $\hat{f}_1 + \hat{f}_2$ is continuous. Since the diagram

$$
\begin{array}{ccc}
E & \xrightarrow{f_1 + f_2} & E' \\
\psi \downarrow & & \downarrow \psi' \\
\hat{E} & \xrightarrow{\hat{f}_1 + \hat{f}_2} & \hat{E}'
\end{array}
$$

is commutative (ψ and ψ' are canonical mappings), this proves that $\widehat{f_1 + f_2} = \hat{f}_1 + \hat{f}_2$. The other two relations follow by means of similar considerations.

Corollary. *Suppose that $f : E \to E'$ is a null homomorphism. Then so is $\hat{f} : \hat{E} \to \hat{E}'$.*

Suppose now that E, E', E'' are filtered R-modules and that f, g are homomorphisms

$$
E'' \xrightarrow{g} E \xrightarrow{f} E'
$$

which are compatible with the appropriate filtrations. This gives rise to the situation

$$
\hat{E}'' \xrightarrow{\hat{g}} \hat{E} \xrightarrow{\hat{f}} \hat{E}'
$$

in which, by Theorem 5 Cor. 1, $\hat{f}$ and $\hat{g}$ satisfy similar compatibility conditions. (We are assuming that each of the given modules possesses a completion.) Next, for each $n \geqslant 0$, we have a commutative diagram

$$
\begin{array}{ccc}
E''/E''_n \longrightarrow E/E_n \longrightarrow E'/E'_n \\
\downarrow \qquad\qquad \downarrow \qquad\qquad \downarrow \\
\hat{E}''/\hat{E}''_n \longrightarrow \hat{E}/\hat{E}_n \longrightarrow \hat{E}'/\hat{E}'_n
\end{array}
$$

$$(9.5.8)$$

Here the mappings in the upper row are induced by f and g whereas those in the lower row arise similarly from $\hat{f}$ and $\hat{g}$. The vertical mappings, on the other hand, come from the canonical mappings of E'', E and E' into their respective completions. Accordingly, by Theorem 4 Cor., *all the vertical mappings are isomorphisms.*

Lemma 3. *Let the situation be as described above and suppose, in addition, that g is strict and*

$$E''/E''_n \to E/E_n \to E'/E'_n$$

is exact for every $n \geqslant 0$. Then the sequence

$$\hat{E}'' \xrightarrow{\hat{g}} \hat{E} \xrightarrow{\hat{f}} \hat{E}'$$

is also exact.

Proof. The properties of the diagram (9.5.8) show that

$$\hat{E}''/\hat{E}''_n \to \hat{E}/\hat{E}_n \to \hat{E}'/\hat{E}'_n$$

is exact for every $n \geqslant 0$. Also, by Theorem 5 Cor. 2, $\hat{g}$ is strict. This means that for the purposes of the proof we may assume that E, E', E'' are themselves complete and by deducing that

$$E'' \xrightarrow{g} E \xrightarrow{f} E'$$

is exact, establish the lemma.

Let e'' belong to E''. Since the result of composing $E''/E''_n \to E/E_n$ and $E/E_n \to E'/E'_n$ is a null map, it follows that $fg(e'') \in E'_n$ for all n. But $\bigcap_0^\infty E'_n = 0$ because E' is complete. Hence $fg(e'') = 0$ and therefore fg is a null homomorphism.

Let $\xi \in \operatorname{Ker} f$. We shall show that $\xi = g(\eta)$ for some $\eta \in E''$ and then the lemma will be established. To this end we construct a sequence $\eta_0, \eta_1, \eta_2, \ldots$ of elements of E'' so as to satisfy

(a) $\xi \equiv g(\eta_n) \pmod{E_n}$,

(b) $\eta_{n+1} \equiv \eta_n \pmod{E''_n}$,

for every value of n. Assume, for the moment, that we have such a sequence. By Proposition 9 and (b), it follows that $\lim \eta_n = \eta$ (say) exists. Next from (a) and the fact that g is continuous, we obtain

$$\xi = \lim g(\eta_n) = g(\lim \eta_n) = g(\eta)$$

as required. Thus it only remains to be shown that a sequence $\eta_0, \eta_1, \eta_2, \ldots$ with the necessary properties can be found. This will be done by constructing the terms in succession.

We start by taking η_0 to be an arbitrary element of E''. Now suppose that we have obtained $\eta_0, \eta_1, \ldots, \eta_k$ and that these satisfy (a) and (b) so far as these conditions relate to these elements. Since $\xi \in \mathrm{Ker} f$, the natural image of ξ in E/E_{k+1} belongs to

$$\mathrm{Ker}\,(E/E_{k+1} \to E'/E'_{k+1}) = \mathrm{Im}\,(E''/E''_{k+1} \to E/E_{k+1})$$

and so there exists $\eta^* \in E''$ such that $g(\eta^*) \equiv \xi(\mathrm{mod}\,E_{k+1})$. Accordingly $g(\eta^*) \equiv g(\eta_k)\,(\mathrm{mod}\,E_k)$ and so $g(\eta^* - \eta_k)$ belongs to $g(E'') \cap E_k = g(E''_k)$ because g is strict. Let $g(\eta^* - \eta_k) = g(e''_k)$, where $e''_k \in E''_k$, and put $\eta_{k+1} = \eta_k + e''_k$. Then $\eta_{k+1} \equiv \eta_k\,(\mathrm{mod}\,E''_k)$ and, because $g(\eta_{k+1}) = g(\eta^*)$, we also have $g(\eta_{k+1}) \equiv \xi(\mathrm{mod}\,E_{k+1})$. This shows that the sequence $\eta_0, \eta_1, \ldots, \eta_k$ can be continued and completes the proof.

Theorem 9. *Let E, E' and E'' be R-modules with filtrations $\{E_n\}_{n \geqslant 0}$, $\{E'_n\}_{n \geqslant 0}$ and $\{E''_n\}_{n \geqslant 0}$ and completions $\hat{E}$, $\hat{E}'$ and $\hat{E}''$. Furthermore, let*

$$E'' \overset{g}{\to} E \overset{f}{\to} E'$$

be an exact sequence in which f and g are strict homomorphisms. Then, for each $n \geqslant 0$, the induced sequence

$$E''/E''_n \to E/E_n \to E'/E'_n \tag{9.5.9}$$

is exact. In addition, $\hat{f}$ and $\hat{g}$ are strict homomorphisms and

$$\hat{E}'' \overset{\hat{g}}{\longrightarrow} \hat{E} \overset{\hat{f}}{\longrightarrow} \hat{E}'$$

is also an exact sequence.

Proof. It is only necessary to show that (9.5.9) is an exact sequence for then the theorem will follow by virtue of Theorem 5 Cor. 2 and Lemma 3. Obviously $\mathrm{Im}\,(E''/E''_n \to E/E_n)$ is contained in

$$\mathrm{Ker}\,(E/E_n \to E'/E'_n).$$

In what follows $\phi_n : E \to E/E_n$ and $\phi''_n : E'' \to E''/E''_n$ denote natural mappings.

Let $\omega \in \mathrm{Ker}\,(E/E_n \to E'/E'_n)$. Since f is strict, it follows from Lemma 1 that $\omega = \phi_n(e)$ for some $e \in \mathrm{Ker} f$. But $\mathrm{Ker} f = \mathrm{Im} g$. Consequently $e = g(e'')$ for a suitable element $e'' \in E''$. Thus $\omega = \phi_n g(e'')$ which shows that ω is the image of $\phi''_n(e'')$ under the mapping $E''/E''_n \to E/E_n$. Accordingly $\omega \in \mathrm{Im}\,(E''/E''_n \to E/E_n)$ and the exactness of (9.5.9) is demonstrated. This completes the proof.

9.6 The existence of completions

In the last section we established a number of properties of the completions of filtered modules, but we still have to show that a filtered module always possesses a completion. There is more than one way in which this can be done. The method we shall use employs the notion of a *projective limit*.

Let $\{E_n\}_{n \geqslant 0}$ be a filtration on an R-module E. Then the identity mapping of E induces an R-homomorphism

$$\sigma_n : E/E_n \to E/E_{n-1}$$

for each $n \geqslant 1$. Accordingly we have mappings

$$\ldots \to E/E_{n+1} \xrightarrow{\sigma_{n+1}} E/E_n \xrightarrow{\sigma_n} E/E_{n-1} \to \ldots \to E/E_1 \to E/E_0. \quad (9.6.1)$$

Suppose that $\xi = \{x_n\}_{n \geqslant 0}$ is a sequence in which $x_n \in E/E_n$ for all $n \geqslant 0$ and $\sigma_n(x_n) = x_{n-1}$ for all $n \geqslant 1$. The totality of all such sequences will be denoted by $\hat{E}$. It is called the *projective limit* of the system (9.6.1). Obviously $\hat{E}$ is not empty. Further we can give it the structure of an R-module in such a way that if $\{x_n\}$ and $\{x'_n\}$ belong to $\hat{E}$ and $r \in R$, then

$$\{x_n\} + \{x'_n\} = \{x_n + x'_n\} \quad (9.6.2)$$

and

$$r\{x_n\} = \{rx_n\}. \quad (9.6.3)$$

For $k \geqslant 0$, let $\hat{E}_k$ be the R-submodule of $\hat{E}$ consisting of all elements $\{x_n\}$ of $\hat{E}$ for which $x_k = 0$. Then

$$\hat{E}_0 \supseteq \hat{E}_1 \supseteq \hat{E}_2 \supseteq \ldots \supseteq \hat{E}_k \supseteq \ldots.$$

Thus $\{\hat{E}_k\}_{k \geqslant 0}$ is a filtration on E and we have

$$\bigcap_{k=0}^{\infty} \hat{E}_k = 0.$$

It will now be shown that $\hat{E}$ is complete with respect to its filtration.

Let $\{\xi^{(s)}\}$, where $\xi^{(s)} = \{x_n^{(s)}\}_{n \geqslant 0}$ and $s = 1, 2, 3, \ldots$, be a Cauchy sequence of elements of $\hat{E}$ and suppose that $k \geqslant 0$ is a given integer. Then $\xi^{(s+1)} - \xi^{(s)}$ belongs to $\hat{E}_k$ when s is large and therefore

$$x_k^{(s+1)} = x_k^{(s)}.$$

Thus the sequence $x_k^{(1)}, x_k^{(2)}, x_k^{(3)}, \ldots$ eventually becomes constant. Let the terminal value be x_k and put $\xi = \{x_k\}_{k \geqslant 0}$. Obviously $\xi \in \hat{E}$ and $\xi - \xi^{(s)}$ belongs to $\hat{E}_k$ when s is sufficiently large. Accordingly $\xi^{(s)} \to \xi$ and $\hat{E}$ has been shown to be complete.

Let $\phi_n: E \to E/E_n$ be the natural mapping. Then for each $e \in E$ the sequence $\{\phi_n(e)\}_{n \geqslant 0}$ is an element of $\hat{E}$. We can therefore define a mapping $\psi: E \to \hat{E}$ so that

$$\psi(e) = \{\phi_n(e)\}_{n \geqslant 0}.$$

This is clearly an R-homomorphism and $\mathrm{Ker}\,\psi = \bigcap_{n=0}^{\infty} E_n$. *We claim that ψ is strict.* Indeed suppose that $\xi = \{x_n\}_{n \geqslant 0}$ belongs to $\psi(E) \cap \hat{E}_k$. Then there exists $e \in E$ such that $\psi(e) = \xi$ and therefore $x_n = \phi_n(e)$ for all $n \geqslant 0$. But $x_k = 0$. Consequently $e \in E_k$ and $\xi = \psi(e) \in \psi(E_k)$. It follows that $\psi(E) \cap \hat{E}_k \subseteq \psi(E_k)$ and as the opposite inclusion is obvious, our claim that ψ is strict has been established.

There is one final property to be checked before we can be sure that we have a completion of E, namely we must show that the closure $\overline{\psi(E)}$, of $\psi(E)$, coincides with $\hat{E}$. To this end let $\eta = \{y_n\}_{n \geqslant 0}$ be a given element of $\hat{E}$ and let $k \geqslant 0$ be a given integer. Then $y_k \in E/E_k$ and therefore there exists $e \in E$ such that $\phi_k(e) = y_k$. It now follows that $\eta - \psi(e)$ belongs to $\hat{E}_k$ and therefore $\eta + \hat{E}_k$ meets $\psi(E)$. Moreover this is true for all k. This shows that $\eta \in \overline{\psi(E)}$ and thus we see that $\overline{\psi(E)} = \hat{E}$ as required. We summarize and record our conclusions in

Theorem 10. *Let $\{E_n\}_{n \geqslant 0}$ be a filtration on an R-module E. Then E possesses a completion with respect to this filtration. Moreover a completion can be obtained as the projective limit of the system*

$$\ldots \to E/E_{n+1} \xrightarrow{\sigma_{n+1}} E/E_n \xrightarrow{\sigma_n} E/E_{n-1} \to \ldots \to E/E_1 \to E/E_0$$

in the manner described above.

9.7 Filtered rings

The notion of a *filtered ring* is similar in many respects to that of a filtered module, but it has to be adjusted to take account of the multiplicative structure. In fact, if R is a ring, then by a *filtration on R* is meant a family $\{A_n\}_{n \geqslant 0}$ of two-sided ideals satisfying $A_n \supseteq A_{n+1}$ for all $n \geqslant 0$.

Suppose that we have this situation. We can, of course, regard R as a (left) module with respect to itself and then R is a filtered module, in the sense used previously, with $\{A_n\}_{n \geqslant 0}$ as filtration. This at once enables us to carry over a number of results from the preceding theory. For example, there is a filtration topology on R which has the sets of the form $\alpha + A_n$, where $\alpha \in R$ and $n \geqslant 0$, as a base. Again, by Proposition 5, the mapping $R \to R$ in which each element is mapped into its negative is a homeomorphism and so too is the mapping $\tau_\rho : R \to R$, where $\tau_\rho(r) = r + \rho$. Here, of course, ρ denotes a fixed element of the ring.

Let $R \times R$ be endowed with the product topology. We already know, from Proposition 6, that the mapping $\sigma : R \times R \to R$ defined by $\sigma(r, r') = r + r'$ is continuous. On the present occasion we also have a multiplication mapping $\mu : R \times R \to R$ in which $\mu(r, r') = rr'$. This too is continuous. For let α, β belong to R and let $k \geqslant 0$ be an integer. Then when x belongs to $\alpha + A_k$ and y belongs to $\beta + A_k$, xy will belong to $\alpha\beta + A_k$. The continuity of μ follows from this observation.

Let $\{\alpha_m\}$ and $\{\beta_m\}$ be sequences of elements of the filtered ring R and suppose that $\alpha_m \to \alpha$ and $\beta_m \to \beta$. We already know that

$$\alpha_m + \beta_m \to \alpha + \beta \tag{9.7.1}$$

and

$$\alpha_m - \beta_m \to \alpha - \beta. \tag{9.7.2}$$

To these we can add

$$\alpha_m \beta_m \to \alpha\beta \tag{9.7.3}$$

because, as we have just observed, multiplication is a continuous operation.

It is clear what is to be understood by a Cauchy sequence of elements of a filtered ring and likewise the motion of a complete filtered ring presents no problem. However the concept of a completion of a filtered ring requires a few comments.

Let $\{A_n\}_{n \geqslant 0}$ be a filtration on a ring R. By a *completion of R with respect to its filtration* one understands a composite object consisting of a ring $\hat{R}$, a filtration $\{\hat{A}_n\}_{n \geqslant 0}$ of two-sided ideals of $\hat{R}$, and a *ring-homomorphism* $\psi : R \to \hat{R}$. These are required to satisfy the following conditions:

(i) $\hat{R}$ is complete with respect to $\{\hat{A}_n\}_{n \geqslant 0}$;
(ii) $\psi(A_n) = \psi(R) \cap \hat{A}_n$ for all $n \geqslant 0$;
(iii) $\psi(R)$ is everywhere dense in $\hat{R}$;
(iv) $\operatorname{Ker} \psi = \bigcap_{n=0}^{\infty} A_n$.

As in the case of modules, we refer to $\{\hat{A}\}_{n \geqslant 0}$ as the canonical filtration on $\hat{R}$ and to $\psi : R \to \hat{R}$ as the canonical homomorphism. (Remember that it is part of our definition of a ring-homomorphism that it maps identity element into identity element.) Let us note, in passing, that *if R is a commutative ring, then so too is $\hat{R}$.* For it is clear that if ξ, η belong to $\psi(R)$, then $\xi\eta = \eta\xi$. But every element of $\hat{R}$ is the limit of a sequence of elements of $\psi(R)$. Consequently the assertion follows from (9.7.3) and the uniqueness of limits.

Suppose that $\hat{R}$ is a completion of the ring R. If $r \in R$ and $\xi \in \hat{R}$, then we can turn $\hat{R}$ into a left R-module by putting $r\xi = \psi(r)\,\xi$, where ψ is the canonical homomorphism. The $\hat{R}$-ideals $\hat{A}_n$ then become R-submodules of $\hat{R}$. We may therefore regard both R and $\hat{R}$ as *filtered R-modules* having $\{A_n\}_{n \geqslant 0}$ and $\{\hat{A}_n\}_{n \geqslant 0}$ as their respective filtrations. On this understanding $\hat{R}$ is a completion of R in the sense that this was understood in section (9.5). Because of this we may make use of a number of earlier results without the trouble of examining their proofs to see whether the change from modules to rings makes any essential difference. Thus from Theorem 3 it follows that $\hat{A}_n$ is the closure of $\psi(A_n)$ in $\hat{R}$ and $A_n = \psi^{-1}(\hat{A}_n)$. Next, by the corollary to Theorem 4, ψ induces a one-one mapping

$$R/A_n \to \hat{R}/\hat{A}_n$$

of R/A_n on to $\hat{R}/\hat{A}_n$. Since ψ is a ring-homomorphism and A_n, $\hat{A}_n$ are two-sided ideals, the mapping is in fact a *ring-isomorphism*.

The question of the *uniqueness* of the completion of a filtered ring can also be settled with the aid of earlier results. For suppose that $\hat{R}$ with filtration $\{\hat{A}_n\}_{n \geqslant 0}$ and $\hat{R}'$ with filtration $\{\hat{A}'_n\}_{n \geqslant 0}$ are two completions of the filtered ring R and let $\psi : R \to \hat{R}$ and $\psi' : R \to \hat{R}'$ be the canonical mappings. The theory of filtered modules tells us that there exists one and only one continuous mapping $\omega : \hat{R} \to \hat{R}'$ such that the diagram

$$(9.7.4)$$

is commutative. We contend that $\omega(\hat{A}_n) = \hat{A}'_n$ *for all n and ω itself is a ring-isomorphism*. Indeed, by regarding R, $\hat{R}$ and $\hat{R}'$ as filtered modules and applying Theorem 7, it follows that (i) $\omega(\hat{A}_n) = \hat{A}'_n$, (ii) ω is a one-one mapping of $\hat{R}$ on to $\hat{R}'$, and (iii) if ξ, $\eta \in \hat{R}$, then

$$\omega(\xi + \eta) = \omega(\xi) + \omega(\eta).$$

Now, because (9.7.4) is commutative and ψ, ψ' are ring-homomorphisms, we see that $\omega(1_{\hat{R}}) = 1_{\hat{R}'}$. Accordingly it is only necessary to show that $\omega(\xi\eta) = \omega(\xi)\,\omega(\eta)$ for all ξ, η in $\hat{R}$. However, when ξ, η belong to $\psi(R)$ this is a consequence of the commutative properties of the diagram (9.7.4). In the general case, each of ξ and η is the limit of a sequence of elements belonging to $\psi(R)$. Since ω is continuous, the desired result follows by making use of (9.7.3) and (9.4.4).

The existence of completions of filtered rings can be established by modifying the method used in the case of modules. To see this, let R be a filtered ring with filtration $\{A_n\}_{n \geqslant 0}$. Then the identity map induces a surjective ring-homomorphism

$$\sigma_n \colon R/A_n \to R/A_{n-1}$$

for each $n \geqslant 1$. Now let $\hat{R}$ be the projective limit of the system

$$\ldots \to R/A_{n+1} \xrightarrow{\sigma_{n+1}} R/A_n \xrightarrow{\sigma_n} R/A_{n-1} \to \ldots \to R/A_1 \xrightarrow{\sigma_1} R/A_0$$

and let $\{x_n\}_{n \geqslant 0}$, $\{y_n\}_{n \geqslant 0}$ be typical elements of $\hat{R}$. Then $\hat{R}$ has a natural structure as a ring with identity element in which

$$\{x_n\}_{n \geqslant 0} + \{y_n\}_{n \geqslant 0} = \{x_n + y_n\}_{n \geqslant 0}$$

and
$$\{x_n\}_{n \geqslant 0}\{y_n\}_{n \geqslant 0} = \{x_n y_n\}_{n \geqslant 0}.$$

The identity element is the sequence $\{x_n\}_{n \geqslant 0}$ in which, for each $n \geqslant 0$, x_n is the identity element of the ring R/A_n.

For $k \geqslant 0$ let $\hat{A}_k$ be the two-sided $\hat{R}$-ideal consisting of those sequences $\{y_n\}_{n \geqslant 0}$ for which $y_k = 0$. Then $\{\hat{A}_k\}_{k \geqslant 0}$ is a filtration on $\hat{R}$ and $\hat{R}$ is complete with respect to this filtration.

Finally, let $\phi_n \colon R \to R/A_n$ be the natural mapping and define $\psi \colon R \to \hat{R}$ by $\psi(r) = \{\phi_n(r)\}_{n \geqslant 0}$. Then ψ is a ring-homomorphism. Indeed $\hat{R}$, the filtration $\{\hat{A}_n\}_{n \geqslant 0}$, and the homomorphism ψ constitute a completion of the filtered ring R. The reasons why they have the necessary properties are essentially the same as those encountered as the same stage in the theory of filtered modules.

We next record certain other facts about filtered rings which have significance for our subsequent investigations.

Lemma 4. *Let R be a filtered ring, $\hat{R}$ its completion, and $\psi \colon R \to \hat{R}$ the canonical ring-homomorphism. If now γ belongs to the centre of R, then $\psi(\gamma)$ belongs to the centre of $\hat{R}$.*

Proof. If η belongs to $\psi(R)$, then $\eta\psi(\gamma) = \psi(\gamma)\eta$. However, $\psi(R)$ is everywhere dense in $\hat{R}$ and multiplication is a continuous operation. It follows that $\xi\psi(\gamma) = \psi(\gamma)\xi$ for all $\xi \in \hat{R}$. This establishes the lemma.

Lemma 5. *Let R be a filtered ring. Denote by Ω the subset of R consisting of all elements x such that x^n converges to zero as n tends to infinity. Then Ω is closed with respect to the filtration topology.*

Proof. Let $\{A_n\}_{n \geqslant 0}$ be the filtration on R and suppose that y belongs to the closure $\bar{\Omega}$ of Ω. If now an integer $k \geqslant 0$ is given, then $y + A_k$ meets Ω and therefore $y \equiv x \pmod{A_k}$ for some $x \in \Omega$. But A_k is a two-sided ideal. Consequently $y^n \equiv x^n \pmod{A_k}$ for *all* n. However $x^n \in A_k$ for all large n and therefore $y^n \in A_k$ when n is large. Thus $y^n \to 0$, that is to say $y \in \Omega$. The lemma follows.

Proposition 11. *Let R be a complete filtered ring and B a left ideal of R. Suppose that for every $x \in B$, x^n converges to zero as n tends to infinity. Then B is contained in the Jacobson radical of R.*

Proof. Let $x \in B$ and let L be a maximal left ideal. It will suffice to show that x belongs to L. Assume the contrary. Then $Rx + L = R$ and so $rx + \lambda = 1$ for suitable elements $r \in R$ and $\lambda \in L$. Put $rx = y$. Then $\lambda = 1 - y$ and $y \in B$. It follows that $y^n \to 0$ and therefore, by Proposition 10, the series
$$1 + y + y^2 + y^3 + \cdots$$
converges. Let z be its sum. Then, since
$$(1 + y + y^2 + \cdots + y^n)(1 - y) = 1 - y^{n+1}$$
for all $n \geqslant 0$, it follows that $z(1 - y) = 1$. Thus $z\lambda = 1$ and now we have a contradiction because L is a proper left ideal. This completes the proof.

9.8 Filtered modules over filtered rings

Let R be a filtered ring with filtration $\{A_n\}_{n \geqslant 0}$ and let E be an R-module with filtration $\{E_n\}_{n \geqslant 0}$. We can then complete R to obtain a ring $\hat{R}$ and we can also complete E thereby obtaining an R-module $\hat{E}$. Naturally we expect that $\hat{E}$ will take on the structure of an $\hat{R}$-module whenever the two filtrations are suitably connected. It is this idea which will now be explored.

Since E is an R-module, we have a multiplication mapping $R \times E \to E$ in which (r, e) is carried into re. Further each of R and E is endowed with a filtration topology, so we can endow $R \times E$ with the product topology.

Definition. *We shall say that the filtration on E is 'compatible' with the filtration on the ring R if the multiplication mapping $R \times E \to E$ is continuous.*

Suppose for the moment that this is the case and let $\alpha_m \to \alpha$ and $x_m \to x$, where $\{\alpha_m\}$ is a sequence of elements of R and $\{x_m\}$ a sequence of elements of E. Then, of course,

$$\alpha_m x_m \to \alpha x. \tag{9.8.1}$$

Lemma 6. *The following two statements are equivalent*:

(a) *the multiplication mapping $R \times E \to E$ is continuous*;

(b) *given $e \in E$ and $k \geqslant 0$ there exists an integer $p \geqslant 0$ such that $A_p e \subseteq E_k$.*

Proof. Suppose that (a) is true and let $e \in E$ and $k \geqslant 0$ be given. Since multiplication is continuous at $(0, e)$, there exist integers $p \geqslant 0$, $q \geqslant 0$ such that $A_p(e + E_q) \subseteq E_k$. In particular, $A_p e \subseteq E_k$. Thus (a) implies (b).

Now assume that (b) is true. We shall show that multiplication is continuous at (r, e). To this end suppose that an integer $k \geqslant 0$ is given. Then $A_p e \subseteq E_k$ for a suitable integer p. If now α belongs to $r + A_p$ and x belongs to $e + E_k$, then αx is in $re + E_k$. Accordingly we have continuity at (r, e) and with this the proof is complete.

Consider a situation in which the filtration on E is compatible with that on R, and let $\hat{E}$ and $\hat{R}$ be the respective completions. We shall use $\psi_R: R \to \hat{R}$ and $\psi_E: E \to \hat{E}$ to denote the canonical mappings. (Thus ψ_R is a ring-homomorphism and ψ_E a homomorphism of R-modules.) The notation for the filtrations on R, E, $\hat{R}$ and $\hat{E}$ will be as usual.

Let $\hat{e} \in \hat{E}$ and let $k \geqslant 0$ be a given integer. We can then find $e \in E$ so that $\hat{e} \equiv \psi_E(e) \,(\mathrm{mod}\,\hat{E}_k)$. Next, by Lemma 6, there exists $p \geqslant 0$ such that $A_p e \subseteq E_k$ and therefore $A_p \psi_E(e) \subseteq \hat{E}_k$. It follows that

$$A_p \hat{e} \subseteq \hat{E}_k. \tag{9.8.2}$$

This shows that the filtration on $\hat{E}$ is compatible with the filtration on R.

Next let $\alpha \in \hat{R}$ and $\hat{e} \in \hat{E}$. There exists a sequence $\{a_m\}$ of elements of R such that $\alpha = \lim \psi_R(a_m)$. Suppose now that $k \geqslant 0$ is given and choose $p \geqslant 0$ so that (9.8.2) holds. There is then a positive integer N such that, when $m, q \geqslant N$, the difference $\psi_R(a_m) - \psi_R(a_q)$ belongs to $\hat{A}_p$. In this case $a_m - a_q$ belongs to $\psi_R^{-1}(\hat{A}_p) = A_p$ and therefore $a_m \hat{e} - a_q \hat{e}$ is in $\hat{E}_k$. This shows that $\{a_m \hat{e}\}$ is a Cauchy sequence and hence a convergent sequence. Also, if $\{a_m'\}$ is another sequence of elements of R for which $\alpha = \lim \psi_R(a_m')$, then we can easily check that $\{a_m \hat{e} - a_m' \hat{e}\}$ converges to zero. It is therefore in order to put

$$\alpha \hat{e} = \lim (a_m \hat{e}). \tag{9.8.3}$$

Observe that if $a \in R$, then

$$\psi_R(a)\hat{e} = a\hat{e}. \tag{9.8.4}$$

We next suppose that $\alpha \in \hat{A}_p$. Then, since $\hat{A}_p$ is the closure of $\psi_R(A_p)$, we can arrange that all the elements a_m are in A_p. But in that case $a_m \hat{e} \in \hat{E}_k$ by (9.8.2) and so $\alpha \hat{e} \in \hat{E}_k$ because $\hat{E}_k$ is closed. This shows that

$$\hat{A}_p \hat{e} \subseteq \hat{E}_k. \tag{9.8.5}$$

Again if $\hat{e} \in \hat{E}_k$ and α is now arbitrary, then $a_m \hat{e} \in \hat{E}_k$ because $\hat{E}_k$ is an R-module. Thus we obtain $\alpha \hat{e} \in \hat{E}_k$ and we have shown that

$$\alpha \hat{E}_k \subseteq \hat{E}_k. \tag{9.8.6}$$

We claim that $\hat{E}$ has become an $\hat{R}$-module. To this end suppose that α, $\beta \in \hat{R}$ and $\hat{e} \in \hat{E}$. We shall show that $\alpha(\beta \hat{e})$ coincides with $(\alpha \beta) \hat{e}$. (The verification of the other module axioms is quite straightforward and will be left to the reader.) Choose sequences $\{a_m\}$ and $\{b_m\}$, of elements of R, so that $\alpha = \lim \psi_R(a_m)$ and $\beta = \lim \psi_R(b_m)$ and let $k \geqslant 0$ be a given integer. When m is sufficiently large, we have $\beta \hat{e} \equiv b_m \hat{e} \pmod{\hat{E}_k}$ and therefore $a_m(\beta \hat{e}) - (a_m b_m) \hat{e}$ belongs to $\hat{E}_k$. By (9.8.3), $a_m(\beta \hat{e}) \to \alpha(\beta \hat{e})$. Also $\psi_R(a_m b_m) = \psi_R(a_m) \psi_R(b_m)$ tends to $\alpha \beta$. Hence $(a_m b_m) \hat{e} \to (\alpha \beta) \hat{e}$. It follows that $\alpha(\beta \hat{e}) - (\alpha \beta) \hat{e}$ belongs to $\hat{E}_k$. But this is true for all k. Consequently $\alpha(\beta \hat{e}) = (\alpha \beta) \hat{e}$ and our claim is justified.

By (9.8.6), $\hat{E}_k$ is an $\hat{R}$-module. Thus $\{\hat{E}_n\}_{n \geqslant 0}$ is a filtration on $\hat{E}$ when $\hat{E}$ is regarded as an $\hat{R}$-module. By (9.8.5), $\{\hat{E}_k\}_{k \geqslant 0}$ is compatible with $\{\hat{A}_k\}_{k \geqslant 0}$. We sum up and slightly extend our conclusions in

Theorem 11. *Let R be a filtered ring, E a filtered R-module, and suppose that the filtration on E is compatible with that on R. If now $\hat{R}$ and $\hat{E}$ are completions of R and E respectively and $\psi_R : R \to \hat{R}$ is the canonical ring-homomorphism, then $\hat{E}$ can be given the structure of an $\hat{R}$-module so that* (i) $\psi_R(r) \hat{e} = r \hat{e}$ *for all $r \in R$ and $\hat{e} \in \hat{E}$, and* (ii) *the filtration on $\hat{E}$ consists of $\hat{R}$-submodules and is compatible with that on $\hat{R}$. Moreover, these two requirements completely determine the structure of $\hat{E}$ as an $\hat{R}$-module.*

Only the final assertion needs comment. However when all the conditions are satisfied it is clear that we must have $\alpha \hat{e} = \lim (a_m \hat{e})$ provided that $\{a_m\}$ is a sequence of elements of R such that $\{\psi_R(a_m)\}$ converges to α. This shows that there is only one possible value for the product of α and $\hat{e}$.

Corollary 1. *Let the situation be as described in the statement of the theorem. Then for $r \in R$ and $e \in E$ we have $\psi_E(re) = \psi_R(r) \psi_E(e)$, where $\psi_E : E \to \hat{E}$ is the canonical homomorphism of R-modules.*

This follows at once because, by the theorem, $\psi_R(r)\psi_E(e) = r\psi_E(e)$, and $r\psi_E(e) = \psi_E(re)$ because ψ_E is an R-homomorphism.

Corollary 2. *With the same assumptions as in Theorem 11, the open $\hat{R}$-submodules of $\hat{E}$ coincide with the open R-submodules of $\hat{E}$.*

Proof. Let V be an open R-submodule of $\hat{E}$. We shall show that it is also an open $\hat{R}$-submodule. The converse is obvious because *any* $\hat{R}$-submodule of $\hat{E}$ is an R-submodule.

Let $\alpha \in \hat{R}$ and $v \in V$. Since V is open in $\hat{E}$, there exists an integer k such that $\hat{E}_k \subseteq V$. Now the filtration on $\hat{E}$ is compatible with that on $\hat{R}$. Consequently, by Lemma 6, there exists an integer p such that $\hat{A}_p v \subseteq \hat{E}_k$. Choose $r \in R$ so that $\alpha \equiv \psi_R(r) \pmod{\hat{A}_p}$. Then

$$(\alpha - \psi_R(r))\, v \in \hat{E}_k \subseteq V.$$

But $\psi_R(r)\, v = rv \in V$ because V is an R-module. It follows that $\alpha v \in V$. Accordingly V is an $\hat{R}$-submodule of $\hat{E}$ and the proof is complete.

In future, if R is a filtered ring, E an R-module with a compatible filtration and we speak of $\hat{E}$ as an $\hat{R}$-module, then it is always to be understood that we refer to the structure, as an $\hat{R}$-module, provided by Theorem 11. This applies, in particular, to the theorem which follows.

Theorem 12. *Let R be a filtered ring and E, E' filtered R-modules whose filtrations are compatible with that on R. If now $f : E \to E'$ is a continuous R-homomorphism, then the associated mapping $\hat{f} : \hat{E} \to \hat{E}'$ (see Theorem 5) is a homomorphism of $\hat{R}$-modules.*

Proof. Let $\alpha \in \hat{R}$ and $\hat{e} \in \hat{E}$. There exists a sequence $\{a_m\}$ of elements of R such that $\alpha = \lim \psi_R(a_m)$ and then $a_m \hat{e} \to \alpha \hat{e}$. Accordingly

$$\lim \hat{f}(a_m \hat{e}) = \hat{f}(\alpha \hat{e})$$

because $\hat{f}$ is continuous. However, we know that $\hat{f}$ is an R-homomorphism. Hence $\hat{f}(a_m \hat{e}) = a_m \hat{f}(\hat{e})$ and therefore $\lim \hat{f}(a_m \hat{e}) = \alpha \hat{f}(\hat{e})$. Thus $\hat{f}(\alpha \hat{e}) = \alpha \hat{f}(\hat{e})$ and the theorem is proved.

9.9 Multiplicative filtrations

So far the filtrations we have considered have been extremely general in character. We shall now investigate filtrations of a rather special kind. These are such that the theory of graded rings and modules provides us with a powerful tool for studying the properties of completions.

Let $\{A_n\}_{n \geqslant 0}$ be a filtration of two-sided ideals on a ring R. We shall say that $\{A_n\}_{n \geqslant 0}$ is a *multiplicative filtration* if

$$A_0 = R \tag{9.9.1}$$

and $$A_m A_n \subseteq A_{m+n} \tag{9.9.2}$$

for all $m \geqslant 0$, $n \geqslant 0$. Thus if I is a two-sided ideal of R, then the powers I^n $(n = 0, 1, 2, \ldots)$ of I constitute† a multiplicative filtration on R. Indeed this is the most important example of such a filtration.

Suppose now that we are given a multiplicative filtration $\{A_n\}_{n \geqslant 0}$ on R. Put

$$G(R) = (A_0/A_1) \oplus (A_1/A_2) \oplus (A_2/A_3) \oplus \ldots \tag{9.9.3}$$

In the first instance $G(R)$ is just an abelian group but we can turn it into a ring without difficulty. For let $\rho \in (A_m/A_{m+1})$ and $\rho' \in (A_n/A_{n+1})$ have representatives r and r' respectively, where $r \in A_m$ and $r' \in A_n$. By (9.9.2), rr' belongs to A_{m+n} and one can verify without difficulty that the natural image of rr' in A_{m+n}/A_{m+n+1} depends only on ρ and ρ' and is independent of the choice of representatives r and r'. After this it is a simple matter to check that $G(R)$ has a natural structure as a graded ring in which

$$\rho\rho' = \text{image of } rr' \text{ in } A_{m+n}/A_{m+n+1} \tag{9.9.4}$$

and the homogeneous elements of degree m $(m \geqslant 0)$ are just those in A_m/A_{m+1}. (Note that $G(R)$ possesses as identity element the image of 1_R in $A_0/A_1 = R/A_1$.) We shall refer to $G(R)$ as *the graded ring associated with the multiplicatively filtered ring R*.

Now let E be an R-module and $\{E_n\}_{n \geqslant 0}$ a filtration on E. This filtration will be said to be *strongly compatible* with the multiplicative filtration on R if

$$E_0 = E \tag{9.9.5}$$

and $$A_m E_n \subseteq E_{m+n} \tag{9.9.6}$$

for all $m, n \geqslant 0$. Since this implies that $A_m E \subseteq E_m$ it follows that such a filtration is compatible with the filtration on R in the weaker sense of section (9.8).

Assume that $\{E_n\}_{n \geqslant 0}$ is strongly compatible with $\{A_n\}_{n \geqslant 0}$. We put

$$G(E) = (E_0/E_1) \oplus (E_1/E_2) \oplus (E_2/E_3) \oplus \ldots \tag{9.9.7}$$

To start with $G(E)$ has the structure of an additive abelian group. Now suppose that $\rho \in (A_m/A_{m+1})$ and $\eta \in (E_n/E_{n+1})$. Select a representative

---

† By I^0 we mean, of course, R itself. This convention will be used in the sequel without further comment.

$r \in A_m$ for ρ and a representative $y \in E_n$ for η. Then, by (9.9.6), $ry \in E_{m+n}$. Also the image of ry in E_{m+n}/E_{m+n+1} depends only on ρ and η and is independent of the choice of their representatives r and y respectively. We can now, in a unique manner, endow $G(E)$ with the structure of a graded (left) $G(R)$-module in which

$$\rho\eta = \text{image of } ry \text{ in } E_{m+n}/E_{m+n+1}, \tag{9.9.8}$$

and the elements of $G(E)$ which are homogeneous of degree n ($n \geqslant 0$) are those which belong to E_n/E_{n+1}.

Theorem 13. *Let R be complete with respect to a multiplicative filtration $\{A_n\}_{n \geqslant 0}$ and let E be an R-module with a filtration $\{E_n\}_{n \geqslant 0}$ which is strongly compatible with the filtration on R and which satisfies*

$$\bigcap_{n=0}^{\infty} E_n = 0.$$

Further let $e_1, e_2, \ldots, e_s$ be elements of E, where $e_i \in E_{p_i}$, and denote by $\bar{e}_i$ the natural image of e_i in E_{p_i}/E_{p_i+1} so that $\bar{e}_1, \bar{e}_2, \ldots, \bar{e}_s$ are homogeneous elements of $G(E)$. If now $\bar{e}_1, \bar{e}_2, \ldots, \bar{e}_s$ generate $G(E)$ as a $G(R)$-module, then

(a) $E = Re_1 + Re_2 + \ldots + Re_s$,
(b) $E_k = A_{k-p_1}e_1 + A_{k-p_2}e_2 + \ldots + A_{k-p_s}e_s$ *and*
(c) E *is complete with respect to* $\{E_n\}_{n \geqslant 0}$.

Remark. We put $A_n = R$ when $n < 0$. This convention is needed to give meaning to (b) for values of k that are smaller than

$$\max(p_1, p_2, \ldots, p_s).$$

It will also be useful in the course of the proof.

Proof. Let $k \geqslant 0$ be a given integer and let $e \in E_k$. We shall begin by showing that there exist infinite sequences

$$a^{(i)}_{k-p_i}, a^{(i)}_{k-p_i+1}, a^{(i)}_{k-p_i+2}, \ldots, a^{(i)}_{k-p_i+m}, \ldots,$$

one for each value of i ($1 \leqslant i \leqslant s$), such that

$$a^{(i)}_{k-p_i+m} \equiv 0 \pmod{A_{k-p_i+m}}$$

and

$$e \equiv \sum_{i=1}^{s} \left(\sum_{\mu=0}^{m} a^{(i)}_{k-p_i+\mu} \right) e_i \pmod{E_{k+m+1}} \tag{9.9.9}$$

for all $m \geqslant 0$. To this end suppose that, for $1 \leqslant i \leqslant s$,

$$a^{(i)}_{k-p_i}, a^{(i)}_{k-p_i+1}, \ldots, a^{(i)}_{k-p_i+m-1}$$

have already been constructed in conformity with the above requirements. It will be shown that the sequences can be carried one stage further. Since the method by which this is done can readily be adapted to show that the sequences can be started, this will establish the existence of *infinite sequences* having the necessary properties.

Consider the element

$$z = e - \sum_{i=1}^{s} \left(\sum_{\mu=0}^{m-1} a_{k-p_i+\mu}^{(i)} \right) e_i.$$

Because of our hypotheses, we have $z \in E_{k+m}$. Let $\bar{z}$ denote the natural image of z in E_{k+m}/E_{k+m+1}. Then $\bar{z}$ is a homogeneous element of $G(E)$ whose degree is equal to $k+m$. It can therefore be expressed in the form

$$\bar{z} = \omega_1 \bar{e}_1 + \omega_2 \bar{e}_2 + \ldots + \omega_s \bar{e}_s,$$

where ω_i belongs to $A_{k-p_i+m}/A_{k-p_i+m+1}$ if $k-p_i+m \geq 0$ and is zero otherwise. In the former case let $a_{k-p_i+m}^{(i)}$ be a representative of ω_i in A_{k-p_i+m} and in the latter case let it be zero. Then, in any event, we have
$$a_{k-p_i+m}^{(i)} \equiv 0 \pmod{A_{k-p_i+m}}.$$
We also have

$$z \equiv a_{k-p_1+m}^{(1)} e_1 + a_{k-p_2+m}^{(2)} e_2 + \ldots + a_{k-p_s+m}^{(s)} e_s \pmod{E_{k+m+1}}.$$

Thus the elements $a_{k-p_i+m}^{(i)}$ provide the desired continuations of our finite sequences.

Since R is complete with respect to its filtration, it follows, from Proposition 10, that the series

$$\sum_{\mu=0}^{\infty} a_{k-p_i+\mu}^{(i)}$$

converges. Let $a^{(i)}$ be its sum. Then all the terms and partial sums of the series belong to A_{k-p_i} and this is a closed subset of R. Thus $a^{(i)} \in A_{k-p_i}$. Also the sequence whose mth term is

$$\sum_{i=1}^{s} \left(\sum_{\mu=0}^{m} a_{k-p_i+\mu}^{(i)} \right) e_i$$

converges to $\sum_{i=1}^{s} a^{(i)} e_i$ and, by (9.9.9), the same sequence converges to e. But, by hypothesis, E is a Hausdorff space. Consequently the two limits are the same and therefore $e = a^{(1)} e_1 + a^{(2)} e_2 + \ldots + a^{(s)} e_s$. The relation (b) follows and we derive (a) from (b) by taking $k = 0$.

Let $\{\xi_m\}$ be a Cauchy sequence of elements of E. Then

$$\lim (\xi_{m+1} - \xi_m) = 0.$$

It follows that there exists an infinite sequence $\nu_1, \nu_2, \nu_3, \ldots$ of non-negative integers such that $\nu_m \to \infty$ and $(\xi_{m+1} - \xi_m) \in E_{\nu_m}$ for all m. By (b), we have

$$\xi_{m+1} - \xi_m = c_{m1} e_1 + c_{m2} e_2 + \ldots + c_{ms} e_s,$$

where $c_{mi} \in A_{\nu_m - p_i}$. Accordingly each sequence $c_{1i}, c_{2i}, c_{3i}, \ldots$ converges to zero and therefore, since R is complete, the series

$$c_{1i} + c_{2i} + c_{3i} + \ldots + c_{mi} + \ldots$$

converges (Proposition 10). Let its sum be c_i. Then, as $m \to \infty$,

$$\xi_{m+1} - \xi_1 = \sum_{i=1}^{s} \left(\sum_{\mu=1}^{m} c_{\mu i} \right) e_i$$

tends to $\sum_{i=1}^{s} c_i e_i$. It follows that $\{\xi_m\}$ is a convergent sequence and hence that the filtered module E is complete. The theorem is therefore established.

Theorem 14. *Let R be complete with respect to a multiplicative filtration $\{A_n\}_{n \geqslant 0}$ and let E be an R-module with a filtration $\{E_n\}_{n \geqslant 0}$ which is strongly compatible with the filtration on R and which satisfies $\bigcap_{n=0}^{\infty} E_n = 0$. Further let $G(E)$ be a Noetherian $G(R)$-module. Then E is complete with respect to its filtration, it is a Noetherian R-module, and every R-submodule of E is closed in E.*

Proof. $G(E)$ is a finitely generated $G(R)$-module. It can therefore be generated by a finite number of homogeneous elements. Theorem 13 now shows that E is complete.

Let K be a submodule of E and put $K_n = K \cap E_n$. This yields a filtration $\{K_n\}_{n \geqslant 0}$ on K which is strongly compatible with the filtration on R. Now

$$(K_0 + E_1)/E_1 \oplus (K_1 + E_2)/E_2 \oplus (K_2 + E_3)/E_3 \oplus \ldots \qquad (9.9.10)$$

is a homogeneous $G(R)$-submodule of $G(E)$. Also $K_n \cap E_{n+1} = K_{n+1}$. Consequently the inclusion mapping $K_n \to (K_n + E_{n+1})$ induces an isomorphism

$$K_n / K_{n+1} \approx (K_n + E_{n+1})/E_{n+1}$$

of R-modules for each $n \geqslant 0$. Using these mappings we derive an isomorphism between

$$G(K) = (K_0/K_1) \oplus (K_1/K_2) \oplus (K_2/K_3) \oplus \ldots$$

considered as an R-module and the R-module in (9.9.10). Indeed examination of the mapping shows that this is a degree-preserving isomorphism of graded $G(R)$-modules. Accordingly we may conclude that $G(K)$ is generated, as a $G(R)$-module, by a finite number of homogeneous elements and, since $\bigcap_{n=0}^{\infty} K_n = 0$, we can apply Theorem 13. This shows, among other things, that K is a finitely generated R-module and that it is complete with respect to the filtration

$$\{K \cap E_n\}_{n \geqslant 0}.$$

It remains for us to show that K is closed in E. Let ξ belong to the closure of K in E. Then $\xi = \lim \xi_m$, where $\xi_1, \xi_2, \xi_3, \dots$ is a sequence of elements of K and the limit is taken with respect to the filtration on E. Further $\{\xi_m\}$ is a Cauchy sequence relative to the filtration on E and therefore also relative to the filtration on K. But K is complete with respect to its filtration. Accordingly there exists $\xi' \in K$ such that $\{\xi_m\}$ converges to ξ' with respect to the filtration on K. However this implies that $\{\xi_m\}$ converges to ξ' with respect to the filtration on E. Thus $\xi' = \xi$ because E is a Hausdorff space. It follows that $\xi \in K$ and the proof is complete.

These results will now be applied to the theory of completions. Let $\{A_n\}_{n \geqslant 0}$ be a multiplicative filtration on a ring R and let $\hat{R}$ be the completion of R. Then $\hat{R}$ possesses a canonical filtration by virtue of being a completion. This filtration will be denoted by $\{\hat{A}_n\}_{n \geqslant 0}$. Now, for $m, n \geqslant 0$, $A_m A_n \subseteq A_{m+n}$ and so, with the usual notation,

$$\psi_R(A_m)\, \psi_R(A_n) \subseteq \psi_R(A_{m+n}) \subseteq \hat{A}_{m+n}.$$

But the closures of $\psi_R(A_m)$ and $\psi_R(A_n)$ are $\hat{A}_m$ and $\hat{A}_n$ respectively and $\hat{A}_{m+n}$ is closed in $\hat{R}$. It follows that $\hat{A}_m \hat{A}_n \subseteq \hat{A}_{m+n}$. Further $\hat{A}_0$, since it is the closure of $\psi_R(A_0) = \psi_R(R)$, coincides with $\hat{R}$. Thus *the canonical filtration on $\hat{R}$ is also a multiplicative filtration.*

The ring-homomorphism $\psi_R : R \to \hat{R}$ induces a homomorphism $R \to \hat{R}/\hat{A}_{n+1}$. Indeed the fact that ψ_R gives rise to an isomorphism $R/A_{n+1} \approx \hat{R}/\hat{A}_{n+1}$ shows that R is mapped *on to* $\hat{R}/\hat{A}_{n+1}$ and that its kernel is A_{n+1}. Now the inverse image of $\hat{A}_n/\hat{A}_{n+1}$ is $\psi_R^{-1}(\hat{A}_n) = A_n$ (see Theorem 3). Consequently, for each $n \geqslant 0$, ψ_R induces an isomorphism

$$A_n/A_{n+1} \approx \hat{A}_n/\hat{A}_{n+1}$$

of abelian groups. Let us use these isomorphisms to set up a one-one mapping of

$$G(R) = (A_0/A_1) \oplus (A_1/A_2) \oplus (A_2/A_3) \oplus \cdots$$

on to

$$G(\hat{R}) = (\hat{A}_0/\hat{A}_1) \oplus (\hat{A}_1/\hat{A}_2) \oplus (\hat{A}_3/\hat{A}_3) \oplus \cdots.$$

It is then found that this mapping is a degree-preserving ring-isomorphism of the graded ring $G(R)$ on to the graded ring $G(\hat{R})$. *Thus $G(R)$ and $G(\hat{R})$ may be identified (as graded rings) in a natural manner.*

Next let E be an R-module and $\{E_n\}_{n \geqslant 0}$ a filtration on E which is strongly compatible with the multiplicative filtration on R. Since $A_m E_n \subseteq E_{m+n}$, it follows, from Theorem 11 Cor. 1, that

$$\psi_R(A_m)\psi_E(E_n) \subseteq \hat{E}_{m+n},$$

where $\psi_E : E \to \hat{E}$ is the canonical homomorphism of E into its completion. But $\psi_R(A_m)$ has closure $\hat{A}_m$ in $\hat{R}$, $\psi_E(E_n)$ has closure $\hat{E}_n$ in $\hat{E}$, and the filtration on $\hat{E}$ is compatible with that on $\hat{R}$. From this we conclude that $\hat{A}_m \hat{E}_n \subseteq \hat{E}_{m+n}$. Further $\hat{E}_0$, since it is the closure of $\psi_E(E_0) = \psi_E(E)$ in $\hat{E}$, coincides with $\hat{E}$. *Thus when the filtration on E is strongly compatible with that on R, the canonical filtration on $\hat{E}$ is strongly compatible with the canonical filtration on $\hat{R}$.*

It has already been observed that ψ_R gives rise to isomorphisms $A_n/A_{n+1} \approx \hat{A}_n/\hat{A}_{n+1}$. In a similar manner, using ψ_E in place of ψ_R, we obtain isomorphisms $E_n/E_{n+1} \approx \hat{E}_n/\hat{E}_{n+1}$ of R-modules. These can be used, in an obvious way, to yield an isomorphism between

$$G(E) = (E_0/E_1) \oplus (E_1/E_2) \oplus (E_2/E_3) \oplus \cdots$$

and $\qquad G(\hat{E}) = (\hat{E}_0/\hat{E}_1) \oplus (\hat{E}_1/\hat{E}_2) \oplus (\hat{E}_2/\hat{E}_3) \oplus \cdots$

where, in the first instance, $G(E)$ and $G(\hat{E})$ are regarded as R-modules. Now $G(E)$ is a $G(R)$-module and $G(\hat{E})$ a $G(\hat{R})$-module, and we have a natural ring-isomorphism of $G(R)$ on to $G(\hat{R})$. Further it is easily verified that the product of an element of $G(R)$ with an element of $G(E)$ is mapped into the product of the corresponding element of $G(\hat{R})$ with the corresponding element of $G(\hat{E})$. Accordingly the isomorphism $G(E) \approx G(\hat{E})$ matches the $G(R)$-submodules of $G(E)$ with the $G(\hat{R})$-submodules of $G(\hat{E})$. In particular, $G(\hat{E})$ *is a Noetherian $G(\hat{R})$-module when, and only when, $G(E)$ is a Noetherian $G(R)$-module.* By combining this observation with Theorem 14 we obtain

Theorem 15. *Let R be a ring with a multiplicative filtration and E a filtered R-module whose filtration is strongly compatible with the filtration on R. Further, let $G(E)$ be a Noetherian $G(R)$-module. Then $\hat{E}$ is a Noetherian $\hat{R}$-module and every $\hat{R}$-submodule of $\hat{E}$ is closed in $\hat{E}$.*

Corollary 1. *Let the assumptions be as in the statement of the theorem and suppose, in addition, that K is an R-submodule of E. Then, with the usual notation, the closure of $\psi_E(K)$ in $\hat{E}$ is the $\hat{R}$-submodule, of $\hat{E}$, generated by $\psi_E(K)$.*

Proof. Denote the closure of $\psi_E(K)$ in $\hat{E}$ by $\overline{\psi_E(K)}$ and let the $\hat{R}$-submodule of $\hat{E}$ generated by $\psi_E(K)$ be designated by $\hat{R}\psi_E(K)$. If

$$r_1, r_2, \ldots, r_p$$

belong to R and $k_1, k_2, \ldots, k_p$ to K, then, by Theorem 11 Cor. 1,

$$\psi_R(r_1)\,\psi_E(k_1) + \psi_R(r_2)\,\psi_E(k_2) + \ldots + \psi_R(r_p)\,\psi_E(k_p)$$

belongs to $\psi_E(K)$. But $\psi_R(R)$ is everywhere dense in $\hat{R}$. Hence, by continuity, $\hat{R}\psi_E(K) \subseteq \overline{\psi_E(K)}$. On the other hand, $\hat{R}\psi_E(K)$ is closed in $\hat{E}$ because it is an $\hat{R}$-submodule. We also have $\psi_E(K) \subseteq \hat{R}\psi_E(K)$. Accordingly $\overline{\psi_E(K)} \subseteq \hat{R}\psi_E(K)$ and now the proof is complete.

The special case in which $E = R$ is important enough to deserve a separate mention.

Corollary 2. *Let R be a ring with a multiplicative filtration and suppose that the associated graded ring $G(R)$ satisfies the maximal condition for left ideals. Then $\hat{R}$ satisfies the maximal condition for left ideals and every left ideal of $\hat{R}$ is closed in $\hat{R}$. Further, if B is a left R-ideal and $\psi: R \to \hat{R}$ is the canonical ring-homomorphism, then the left $\hat{R}$-ideal $\hat{R}\psi(B)$ is also the closure of $\psi(B)$ in $\hat{R}$. In particular, if $\{\hat{A}_n\}_{n \geqslant 0}$ is the canonical filtration on $\hat{R}$, then $\hat{A}_n = \hat{R}\psi(A_n)$ for all $n \geqslant 0$.*

Let I be a two-sided ideal of R. Then, as we have already noted, $\{I^n\}_{n \geqslant 0}$ is a multiplicative filtration on R. This is known as the *I-adic filtration* on R. If now E is an R-module, then the submodules $\{I^n E\}_{n \geqslant 0}$, of E, provide a filtration which is strongly compatible with the I-adic filtration on R. We call $\{I^n E\}_{n \geqslant 0}$ the *I-adic filtration* on E. The next section will be devoted to a study of the properties of I-adic filtrations when I is a central ideal. There are, however, some questions concerning these filtrations which it is more convenient to treat in the present context.

Theorem 16. *Let I be a two-sided ideal of R and let $\hat{R}$ be the completion of R with respect to its I-adic filtration. If now $\psi: R \to \hat{R}$ is the canonical ring-homomorphism, then the closure of $\psi(I)$ in $\hat{R}$ is contained in the Jacobson radical of $\hat{R}$. Further there is a one-one correspondence between the maximal left ideals of L of R containing I and the maximal left ideals Λ of $\hat{R}$. This is such that if L corresponds to Λ, then $L = \psi^{-1}(\Lambda)$ and Λ is the closure of $\psi(L)$ in $\hat{R}$.*

Remark. Naturally, in this theorem, the expression 'maximal left ideal' can be replaced by 'maximal right ideal' wherever it occurs and this will not affect the truth of the assertions.

Proof. Let $\overline{\psi(I)}$ denote the closure of $\psi(I)$ in $\hat{R}$ and let $\{\hat{A}_n\}_{n \geqslant 0}$ be the canonical filtration on $\hat{R}$. Then $\overline{\psi(I)} = \hat{A}_1$. Now if $y \in \psi(I)$, then $y^n \in \psi(I^n) \subseteq \hat{A}_n$ and therefore the sequence $\{y^n\}$ converges to zero. It therefore follows, from Lemma 5, that if x belongs to $\overline{\psi(I)} = \hat{A}_1$, then $\{x^n\}$ converges to zero. Accordingly, by Proposition 11, $\hat{A}_1$ is contained in the Jacobson radical of $\hat{R}$.

We recall that we can regard both R with filtration $\{I^n\}_{n \geqslant 0}$ and $\hat{R}$ with filtration $\{\hat{A}_n\}_{n \geqslant 0}$ as filtered (left) R-modules. On this understanding $\hat{R}$ will be the completion of R in the sense of the theory of filtered modules. Now let L be a maximal left ideal of R and suppose that $I \subseteq L$. Then L is open in R and therefore, by Theorem 4, the closure $\overline{\psi(L)}$, of $\psi(L)$, will be the corresponding open R-submodule of $\hat{R}$. However, by Theorem 11 Cor. 2, the open R-submodules of $\hat{R}$ are the same as the open left $\hat{R}$-ideals. Thus $\overline{\psi(L)}$ is a left ideal of $\hat{R}$. Again, by Theorem 4, R/L and $\hat{R}/\overline{\psi(L)}$ are isomorphic R-modules and so there can be no R-submodule of $\hat{R}$ strictly between $\overline{\psi(L)}$ and $\hat{R}$. It follows that there is no $\hat{R}$-ideal strictly between $\overline{\psi(L)}$ and $\hat{R}$. Hence $\overline{\psi(L)}$ is a maximal left ideal of $\hat{R}$.

Finally suppose that Λ' is a maximal left ideal of $\hat{R}$ and put

$$L' = \psi^{-1}(\Lambda').$$

Then
$$\overline{\psi(I)} = \hat{A}_1 \subseteq \Lambda'$$

because $\hat{A}_1$ is contained in the Jacobson radical of $\hat{R}$. It follows that Λ' is an open R-submodule of $\hat{R}$, L' is the corresponding open left ideal of R, and $\overline{\psi(L')} = \Lambda'$. Further, there can be no R-submodule of $\hat{R}$ which is strictly between $\hat{R}$ and Λ' for such a submodule would be open and hence a left $\hat{R}$-ideal. Thus $\hat{R}/\Lambda'$ is a simple R-module. Evidently $I \subseteq L'$ and, since (Theorem 4) R/L' and $\hat{R}/\Lambda'$ are isomorphic R-modules, L' must be a maximal left ideal of R. This completes the proof.

Once again let I be a two-sided ideal of R. It will have been noticed that, so far, we have said very little about the nature of the canonical filtration on the I-adic completion of R. This will now be investigated *under the assumption that I can be generated by a finite set $\gamma_1, \gamma_2, ..., \gamma_s$ of central elements*. This, of course, imposes a severe restriction in the case of general rings. For commutative rings the limitation caused by this assumption is less noticeable.

The graded ring associated with R through the I-adic filtration is

$$G(R) = R/I \oplus I/I^2 \oplus I/I^3 \oplus ... \qquad (9.9.11)$$

and R/I is the subring of $G(R)$ formed by the elements of degree zero. Denote by $\bar{\gamma}_i$ the natural image of γ_i in I/I^2. Then $\bar{\gamma}_1, \bar{\gamma}_2, ..., \bar{\gamma}_s$ belong to the centre of $G(R)$ and they are homogeneous of degree one. Further each element of $G(R)$ can, in an obvious sense, be expressed as a polynomial in $\bar{\gamma}_1, \bar{\gamma}_2, ..., \bar{\gamma}_s$ with coefficients in R/I. Let us indicate this by writing

$$G(R) = (R/I)[\bar{\gamma}_1, \bar{\gamma}_2, ..., \bar{\gamma}_s]. \tag{9.9.12}$$

Suppose now that k is a positive integer. The ideal of $G(R)$ generated by all the power-products of $\bar{\gamma}_1, \bar{\gamma}_2, ..., \bar{\gamma}_s$ of degree k is

$$I^k/I^{k+1} \oplus I^{k+1}/I^{k+2} \oplus I^{k+2}/I^{k+3} \oplus .... \tag{9.9.13}$$

Let $\hat{R}$ be the I-adic completion of R, $\{\hat{A}_n\}_{n \geqslant 0}$ the canonical filtration on $\hat{R}$, and $\psi: R \to \hat{R}$ the canonical ring-homomorphism. We have already seen that there exists a natural ring-isomorphism $G(R) \approx G(\hat{R})$. Put $\psi(\gamma_i) = \omega_i \ (1 \leqslant i \leqslant s)$. Then $\omega_i \in \hat{A}_1$ and, by Lemma 4, ω_i belongs to the centre of $\hat{R}$. We shall use $\bar{\omega}_i$ to denote the natural image of ω_i in $\hat{A}_1/\hat{A}_2$. With this notation, the isomorphism $G(R) \approx G(\hat{R})$ makes $\bar{\gamma}_i$ correspond to $\bar{\omega}_i$ and it matches the ideal (9.9.13) with

$$\hat{A}_k/\hat{A}_{k+1} \oplus \hat{A}_{k+1}/\hat{A}_{k+2} \oplus \hat{A}_{k+2}/\hat{A}_{k+3} \oplus ....$$

Hence if $\nu_1, \nu_2, ..., \nu_s$ denote non-negative integers satisfying

$$\nu_1 + \nu_2 + ... + \nu_s = k,$$

then

$$\sum_{(\nu)} G(\hat{R}) \bar{\omega}_1^{\nu_1} \bar{\omega}_2^{\nu_2} ... \bar{\omega}_s^{\nu_s} = \hat{A}_k/\hat{A}_{k+1} \oplus \hat{A}_{k+1}/\hat{A}_{k+2} \oplus .... \tag{9.9.14}$$

Put $E = \hat{A}_k$ and let us regard E as an $\hat{R}$-module. We obtain a filtration $\{E_n\}_{n \geqslant 0}$ on E by setting $E_n = \hat{A}_k$ for $0 \leqslant n < k$ and $E_n = \hat{A}_n$ for $n \geqslant k$. This filtration is strongly compatible with $\{\hat{A}_n\}_{n \geqslant 0}$ and $\bigcap_{n=0}^{\infty} E_n = 0$. Further

$$G(E) = 0 \oplus ... \oplus 0 \oplus \hat{A}_k/\hat{A}_{k+1} \oplus \hat{A}_{k+1}/\hat{A}_{k+2} \oplus ....$$

Now if $\nu_1 + \nu_2 + ... + \nu_s = k$, then

$$\omega_1^{\nu_1} \omega_2^{\nu_2} ... \omega_s^{\nu_s} = \psi(\gamma_1^{\nu_1} \gamma_2^{\nu_2} ... \gamma_s^{\nu_s}) \in \hat{A}_k = E_k.$$

Further, the natural image of $\omega_1^{\nu_1} \omega_2^{\nu_2} ... \omega_s^{\nu_s}$ in $\hat{A}_k/\hat{A}_{k+1}$ is $\bar{\omega}_1^{\nu_1} \bar{\omega}_2^{\nu_2} ... \bar{\omega}_s^{\nu_s}$ and, by (9.9.14), such elements generate $G(E)$ as a $G(\hat{R})$-module. Accordingly, by Theorem 13,

$$\hat{A}_k = E = \sum_{(\nu)} \hat{R} \omega_1^{\nu_1} \omega_2^{\nu_2} ... \omega_s^{\nu_s}.$$

Now $I\hat{R} = \omega_1 \hat{R} + \omega_2 \hat{R} + ... + \omega_s \hat{R}$. The above relation may now be rewritten as $\hat{A}_k = (I\hat{R})^k$. We have therefore established

Theorem 17. *Let $\hat{R}$ be the I-adic completion of R, where I is a finitely generated central ideal of R. Then $I\hat{R}$ is a finitely generated central ideal of $\hat{R}$. Furthermore, if $\{\hat{A}_n\}_{n \geqslant 0}$ is the canonical filtration which $\hat{R}$ possesses by virtue of being a completion, then $\hat{A}_n = (I\hat{R})^n$ for all $n \geqslant 0$.*

Proposition 12. *Let I be a finitely generated central ideal of R and E an R-module. If now E/IE is a Noetherian R-module, then $G(E)$ is a Noetherian $G(R)$-module.*

Remark. In this proposition it is understood that $G(R)$ and $G(E)$ are derived from the I-adic filtrations on R and E respectively.

Proof. Let I be generated by the central elements $\gamma_1, \gamma_2, \ldots, \gamma_s$ and let $X_1, X_2, \ldots, X_s$ be indeterminates. (In what follows we employ the same notation as was used in the discussion leading up to Theorem 17.) By (9.9.12),
$$G(R) = (R/I)[\overline{\gamma}_1, \overline{\gamma}_2, \ldots, \overline{\gamma}_s].$$
Hence, there exists a ring-homomorphism of the polynomial ring $(R/I)[X_1, X_2, \ldots, X_s]$ on to $G(R)$ in which the elements of R/I are left fixed and X_i is mapped into $\overline{\gamma}_i$. Using this homomorphism, we may regard $G(E)$ as a module with respect to $(R/I)[X_1, X_2, \ldots, X_s]$ and it is then sufficient to establish that
$$G(E) = (E/IE) \oplus (IE/I^2E) \oplus (I^2E/I^3E) \oplus \ldots$$
is Noetherian as an $(R/I)[X_1, X_2, \ldots, X_s]$-module. Now E/IE can be regarded as an R/I-module and as such it is Noetherian. It follows, by Theorem 1 of section (4.1), that $(E/I)[X_1, X_2, \ldots, X_s]$ is a Noetherian $(R/I)[X_1, X_2, \ldots, X_s]$-module. It is therefore enough to show that there exists a mapping
$$(E/IE)[X_1, X_2, \ldots, X_s] \to G(E)$$
which is an epimorphism of $R[X_1, X_2, \ldots, X_s]$-modules. But if
$$\Sigma \eta_{\nu_1 \nu_2 \ldots \nu_s} X_1^{\nu_1} X_2^{\nu_2} \ldots X_s^{\nu_s},$$
where $\eta_{\nu_1 \nu_2 \ldots \nu_s}$ belongs to E/IE, is a typical element of
$$(E/IE)[X_1, X_2, \ldots, X_s],$$
then a mapping with the requisite property is obtained by transporting this element into
$$\Sigma \overline{\gamma}_1^{\nu_1} \overline{\gamma}_2^{\nu_2} \ldots \overline{\gamma}_s^{\nu_s} \eta_{\nu_1 \nu_2 \ldots \nu_s}.$$
The proof is now complete.

9.10 I-adic filtrations

Throughout this section I will denote a *central ideal* of R and we shall consider I-adic filtrations on R-modules and on the ring R itself. Generally speaking, we shall be interested in *Noetherian* R-modules. Now if $E_1, E_2, ..., E_p$ are Noetherian R-modules, then it is possible to choose central elements $\gamma_1, \gamma_2, ..., \gamma_s$, in I, so that

$$IE_j = \gamma_1 E_j + \gamma_2 E_j + ... + \gamma_s E_j$$

for $j = 1, 2, ..., p$. Put $I_0 = \gamma_1 R + \gamma_2 R + ... + \gamma_s R$. This secures that I_0 is a *finitely generated* central ideal and $IE_j = I_0 E_j$ for each j. It follows that

$$I^2 E_j = II_0 E_j = I_0 I E_j = I_0^2 E_j$$

and generally $I^n E_j = I_0^n E_j$ for all $n \geqslant 0$. Thus the I-adic and I_0-adic filtrations on E_j are the same †. For this reason, *we shall assume throughout section* (9.10) *that I is a finitely generated central ideal*. The advantage that we gain from this is that our main theorems can now be stated with comparatively little variation in the basic conditions. It may be objected that, by doing this, we sacrifice generality. However this is not serious for in those situations where we could have derived a more general result we can easily recover what we have lost by means of the above remark.

Let $\hat{R}$ be the I-adic completion of R. Since we are assuming that I is finitely generated, Theorem 17 shows that $I\hat{R}$ is a finitely generated central ideal of $\hat{R}$ and that its powers provide the filtration which $\hat{R}$ possesses by virtue of being a completion. This can be viewed in two ways. On the one hand it says that the canonical filtration on $\hat{R}$ is the $I\hat{R}$-adic filtration; but, equally well, it says that the canonical filtration on $\hat{R}$ is its I-adic filtration when $\hat{R}$ is regarded as an R-module.

In the results which follow, all references to completions of R and of R-modules refer to completions with respect to the I-adic filtrations which they carry; unless, of course, there is an explicit statement concerning some other filtration.

Theorem 18. *Let I be a finitely generated central ideal, of R, and E an R-module. Suppose that E/IE is a Noetherian R-module. If now $\hat{R}$ and $\hat{E}$ are the I-adic completions of R and E respectively, then $\hat{E}$ is a Noetherian $\hat{R}$-module and all the $\hat{R}$-submodules of $\hat{E}$ are closed in $\hat{E}$. Further, the canonical filtration which $\hat{E}$ possesses by virtue of being*

---

† Of course, the I-adic and I_0-adic filtrations on R will normally be different.

a completion is $\{I^n \hat{E}\}_{n \geqslant 0}$ or, equivalently, $\{(I\hat{R})^n \hat{E}\}_{n \geqslant 0}$. Finally, $\hat{E}$ is generated, as an $\hat{R}$-module, by the image of E under the canonical homomorphism $\psi_E : E \to \hat{E}$.

Proof. By Proposition 12, $G(E)$ is a Noetherian $G(R)$-module, where $G(E)$ and $G(R)$ have their usual meanings. Hence, by Theorem 15, $\hat{E}$ is a Noetherian $\hat{R}$-module and all its $\hat{R}$-submodules are closed. Again, by Corollary 1 of the same theorem, the closure $\overline{\psi_E(I^n E)}$, of $\psi_E(I^n E)$ in $\hat{E}$, is the $\hat{R}$-submodule of $\hat{E}$ generated by $\psi_E(I^n E)$. Thus we may write

$$\overline{\psi_E(I^n E)} = \hat{R}\psi_E(I^n E) = \hat{R}(I^n \psi_E(E)) = I^n(\hat{R}\psi_E(E))$$

because ψ_E is a homomorphism of R-modules. From the value $n = 0$ we obtain $\hat{E} = \hat{R}\psi_E(E)$ which is the final assertion of the theorem. Moreover, it has now been shown that

$$\overline{\psi_E(I^n E)} = I^n \hat{E}$$

and, as it is evident that $I^n \hat{E} = (I\hat{R})^n \hat{E}$, the rest follows by virtue of Theorem 3.

It is convenient to state, as a separate result, the special case of Theorem 18 in which $E = R$.

Theorem 19. *Let I be a finitely generated central ideal of the ring R and suppose that the ring R/I satisfies the maximal condition for left ideals. If now $\hat{R}$ denotes the I-adic completion of R, then $\hat{R}$ satisfies the maximal condition for left ideals and every left ideal of $\hat{R}$ is closed in $\hat{R}$.*

Let $f : E' \to E$ be a homomorphism of R-modules. Then

$$f(I^n E') = I^n f(E') \subseteq I^n E.$$

Hence f is compatible with the I-adic filtrations and if f is an epimorphism, then it is strict.

The next result is of the greatest importance for our theory.

Theorem 20. *Let I be a finitely generated central ideal of R and let*

$$0 \to E' \xrightarrow{f} E \xrightarrow{g} E'' \to 0$$

be an exact sequence of Noetherian R-modules. If now completions are taken with respect to the I-adic filtrations, then

$$0 \to \hat{E}' \xrightarrow{\hat{f}} \hat{E} \xrightarrow{\hat{g}} \hat{E}'' \to 0$$

is an exact sequence of Noetherian $\hat{R}$-modules.

Proof. Theorem 18 shows that $\hat{E}'$, $\hat{E}$ and $\hat{E}''$ are Noetherian $\hat{R}$-modules. Also $\hat{f}$ and $\hat{g}$ are well defined because f and g are compatible with the appropriate I-adic filtrations. Indeed $\hat{f}, \hat{g}$ are $\hat{R}$-homomorphisms by virtue of Theorem 12.

Put $K = \operatorname{Im} f$. Then we have an exact sequence

$$0 \to K \overset{\mu}{\to} E \overset{g}{\to} E'' \to 0, \qquad (9.10.1)$$

where μ is the inclusion mapping. In this sequence, let E and E'' be endowed with the I-adic filtrations and let the filtration on K be taken as $\{K_n\}_{n \geqslant 0}$, where $K_n = K \cap I^n E$. During the remainder of the proof we shall ignore the filtration on R and work within the framework of theory of section (9.5).

Bearing in mind that a null module has only one filtration, we see that all the homomorphisms in (9.10.1) are strict. It follows, from Theorem 9, that

$$0 \to \hat{K} \overset{\hat{\mu}}{\longrightarrow} \hat{E} \overset{\hat{g}}{\longrightarrow} \hat{E}'' \to 0 \qquad (9.10.2)$$

is an exact sequence of R-modules. Here $\hat{E}$ and $\hat{E}''$ denote completions taken with respect to the I-adic filtrations on E and E'', whereas $\hat{K}$ denotes the completion of K with respect to $\{K_n\}_{n \geqslant 0}$.

The mapping f induces an isomorphism $\phi : E' \to K$, where $f = \mu\phi$. Now $\phi(I^n E') = I^n K$ which is contained in K_n. This shows that if we regard E' as being filtered by the I-adic filtration, then ϕ *is continuous*.

By hypothesis, E is a Noetherian R-module. We may therefore apply Theorem 1 of section (7.2). This shows that there exists an integer $q \geqslant 0$ such that

$$K_{m+q} = I^{m+q} E \cap K = I^m (I^q E \cap K) \subseteq I^m K$$

for all $m \geqslant 0$. It follows that $\phi^{-1}(K_{m+q}) \subseteq I^m E'$ and therefore $\phi^{-1} : K \to E'$ *is also continuous*. Hence, by Theorem 6 Cor., $\hat{\phi} : \hat{E}' \to \hat{K}$ is an isomorphism of R-modules. Accordingly, from (9.10.2), the sequence

$$0 \to \hat{E}' \overset{\hat{\mu}\hat{\phi}}{\longrightarrow} \hat{E} \overset{\hat{g}}{\longrightarrow} \hat{E}'' \to 0$$

is exact. But $\hat{f} = \hat{\mu}\hat{\phi}$ (see Theorem 6) and with this the proof is complete.

Corollary. *Suppose that*

$$E' \overset{f}{\to} E \overset{g}{\to} E''$$

is an exact sequence of Noetherian R-modules and that completions are *formed with respect to their I-adic filtrations, where I is a finitely generated central ideal. Then the resulting sequence*

$$\hat{E}' \to \hat{E} \to \hat{E}''$$

is also exact.

Proof. The corollary follows at once if we apply the theorem to the exact sequences

$$0 \to \mathrm{Ker}\, f \to E' \to \mathrm{Im}\, f \to 0,$$
$$0 \to \mathrm{Im}\, f \to E \to \mathrm{Im}\, g \to 0$$

and

$$0 \to \mathrm{Im}\, g \to E'' \to E''/\mathrm{Im}\, g \to 0,$$

of Noetherian *R*-modules, and afterwards combine the results.

Let *E* be a Noetherian *R*-module and *K* a submodule of *E*. The inclusion mapping $K \to E$ induces an $\hat{R}$-homomorphism $\hat{K} \to \hat{E}$. Furthermore, since $0 \to K \to E$ is exact, the same is true of $0 \to \hat{K} \to \hat{E}$. This shows that $\hat{K}$ *can be identified with an* $\hat{R}$-*submodule of* $\hat{E}$. The next result gives an alternative description of the submodule in question.

Theorem 21. *Let I be a finitely generated central ideal of R and U a subset of a Noetherian R-module E. Put $K = RU$ (i.e. K is the R-submodule of E generated by U) and denote by $\hat{R}$, $\hat{K}$ and $\hat{E}$ the I-adic completions of R, K and E respectively. If now $\psi_E : E \to \hat{E}$ is the canonical homomorphism and $\hat{K}$ is regarded as an $\hat{R}$-submodule of $\hat{E}$, then*

$$\hat{K} = \hat{R}\psi_E(U).$$

Proof. By Theorem 18, $\hat{K} = \hat{R}\psi_K(K)$, where $\psi_K : K \to \hat{K}$ is the canonical homomorphism. But $K = RU$. Hence, by Theorem 11 Cor. 1, $\hat{K} = \hat{R}\psi_K(U)$. The theorem now follows from the fact that the diagram

$$\begin{array}{ccc} K & \longrightarrow & E \\ \psi_K \downarrow & & \downarrow \psi_E \\ \hat{K} & \longrightarrow & \hat{E} \end{array}$$

is commutative.

Let $\{K_\lambda\}_{\lambda \in \Lambda}$ be a family of *R*-submodules of a Noetherian *R*-module *E*. Then $\sum_\lambda K_\lambda$ is also a submodule. It follows that we can regard the *I*-adic completion of $\sum_\lambda K_\lambda$, as well as the completions of the individual K_λ, as an $\hat{R}$-submodule of $\hat{E}$. On this understanding we have

Theorem 22. *Let the situation be as described above. Then*

$$\widehat{\sum_{\lambda \in \Lambda} K_\lambda} = \sum_{\lambda \in \Lambda} \hat{K}_\lambda, \tag{9.10.3}$$

where it is understood that I is a finitely generated central ideal and the completions are I-adic completions.

Proof. We have $\sum\limits_{\lambda} K_\lambda = R(\bigcup\limits_{\lambda} K_\lambda)$ and therefore, by Theorem 21,

$$\widehat{\sum\limits_{\lambda} K_\lambda} = \hat{R}\psi_E(\bigcup\limits_{\lambda} K_\lambda)$$

$$= \hat{R}\,(\bigcup\limits_{\lambda} \psi_E(K_\lambda))$$

$$= \sum\limits_{\lambda} \hat{R}\psi_E(K_\lambda)$$

$$= \sum\limits_{\lambda} \hat{K}_\lambda.$$

Theorem 23. *Let I be a finitely generated central ideal of R, E a Noetherian R-module, and A a left ideal of R. If now the I-adic completion of AE is regarded as an $\hat{R}$-submodule of the I-adic completion of E, then, with an obvious notation,*

$$\widehat{AE} = (\hat{R}A)\,\hat{E}. \qquad (9.10.4)$$

Remark. By $\hat{R}A$ is meant the left $\hat{R}$-ideal generated by the image of A under the canonical ring-homomorphism $\psi_R : R \to \hat{R}$.

Proof. The R-module AE is generated by all products ae, where $a \in A$ and $e \in E$. Hence, by Theorem 21, $\widehat{AE}$ is the $\hat{R}$-submodule of $\hat{E}$ generated by the elements $\psi_E(ae) = \psi_R(a)\,\psi_E(e)$ where $\psi_E : E \to \hat{E}$ is the usual canonical mapping. It follows that $\widehat{AE} = (\hat{R}A)\,\psi_E(E)$, that is the set of all elements which can be expressed in the form

$$\alpha_1 x_1 + \alpha_2 x_2 + \ldots + \alpha_p x_p$$

with $\alpha_i \in \hat{R}A$ and $x_i \in \psi_E(E)$. In particular, we see that $\widehat{AE} \subseteq (\hat{R}A)\,\hat{E}$. On the other hand, $\psi_E(E)$ is everywhere dense in $\hat{E}$ and $\hat{E}$ is a filtered $\hat{R}$-module. Hence $(\hat{R}A)\,\hat{E}$ is contained in the closure of

$$(\hat{R}A)\,\psi_E(E) = \widehat{AE}.$$

But $\widehat{AE}$ is an $\hat{R}$-submodule of $\hat{E}$ and therefore it is already closed in $\hat{E}$ by virtue of Theorem 18. Accordingly $(\hat{R}A)\,\hat{E} \subseteq \widehat{AE}$ and now the proof is complete.

It has probably occurred to the reader that I-adic completions provide an excellent example of a functor. However, before we can make this quite precise, a few explanatory remarks are needed.

If E is an R-module, then its I-adic completion is unique to the extent indicated by Theorem 7. Let us take, for each E, one definite completion. For example, we could agree to use the particular completion provided by the theory of projective limits as explained in section (9.6). We also take a particular completion of R and keep it fixed in what follows.

As a temporary device put $T(E) = \hat{E}$, where $\hat{E}$ denotes the chosen completion of E. Then with each homomorphism $f: E \to E'$ of R-modules, there is associated a homomorphism $\hat{f}: \hat{E} \to \hat{E}'$ of $\hat{R}$-modules. Put $T(f) = \hat{f}$ so that $T(f): T(E) \to T(E')$ and observe that T has the following properties:

(i)   *if i is the identity mapping of E, then $T(i)$ is the identity mapping of $T(E)$;*

(ii) *if $f: E \to E'$ and $g: E' \to E''$ are homomorphisms of R-modules, then $T(gf) = T(g)\,T(f)$;*

(iii) *if $f_1, f_2: E \to E'$ are homomorphisms of R-modules, then*

$$T(f_1 + f_2) = T(f_1) + T(f_2);$$

(iv) *if γ belongs to the centre of R and $f: E \to E'$ is a homomorphism of R-modules, then $T(\gamma f) = \gamma T(f)$.*

Indeed (i) is an immediate consequence of the definition of $T(i)$ and the fact that we have circumvented the difficulty which arises because, strictly speaking, a filtered module has more than one completion. So far as the remaining statements are concerned, (ii) follows from Theorem 6 whereas (iii) and (iv) are obtained from Theorem 8.

These observations show that we have here the kind of situation that was discussed in section (3.4). Indeed, using the language employed there, we can summarize (i), (ii) and (iii) by saying that T is an additive (covariant) functor from the category of R-modules to the category of $\hat{R}$-modules. We shall refer to T as the *I-adic functor*.

Now suppose that

$$0 \to E' \overset{f}{\to} E \overset{g}{\to} E'' \to 0$$

is an exact sequence of *Noetherian* R-modules. Then, by Theorem 20, the sequence

$$0 \to T(E') \xrightarrow{\; T(f) \;} T(E) \xrightarrow{\; T(g) \;} T(E'') \to 0$$

is also exact. Hence, in the terminology of section (3.4), *the I-adic functor is exact on the category of Noetherian R-modules.*

In Theorem 24, $K_1, K_2, \ldots, K_s$ denote R-submodules of a Noetherian R-module E. Since $K_1 \cap K_2 \cap \ldots \cap K_s$ is a submodule of E, its completion, as well as $\hat{K}_1, \hat{K}_2, \ldots, \hat{K}_s$, may be regarded as an $\hat{R}$-submodule of $\hat{E}$. On this understanding, we have

Theorem 24. *Let the situation be as described above. Then*

$$\widehat{\bigcap_{i=1}^{s} K_i} = \bigcap_{i=1}^{s} \hat{K}_i. \tag{9.10.5}$$

Here completions are taken with respect to the I-adic filtrations, where I is a finitely generated central ideal.

This follows from Theorem 8 of section (3.4) when we take account of the remarks made above.

Let K be a submodule of a Noetherian R-module E. From the exact sequence

$$0 \to K \to E \to E/K \to 0$$

we obtain the exact sequence

$$0 \to \hat{K} \to \hat{E} \to \widehat{E/K} \to 0 \tag{9.10.6}$$

by passing to the appropriate I-adic completions. In particular we see that $\hat{E} \to \widehat{E/K}$ is an epimorphism whose kernel is $\hat{K}$ regarded as an $\hat{R}$-submodule of $\hat{E}$. Thus we have an $\hat{R}$-isomorphism $\hat{E}/\hat{K} \approx \widehat{E/K}$. Accordingly, with a suitable identification,

$$\hat{E}/\hat{K} = \widehat{E/K} \tag{9.10.7}$$

provided always that E is a *Noetherian* R-module.

Lemma 7. *Let K be a submodule of a Noetherian R-module E and let α belong to the centre of R. If now the I-adic completion $\widehat{K:_E \alpha}$ of $K:_E \alpha$ is regarded as an $\hat{R}$-submodule of $\hat{E}$, then*

$$\widehat{K:_E \alpha} = \hat{K}:_{\hat{E}} \alpha. \tag{9.10.8}$$

Proof. Let $i: E \to E$ be the identity mapping of E and $\phi: E \to E/K$ the homomorphism obtained by combining $\alpha i: E \to E$ with the natural mapping $E \to E/K$. Then

$$K:_E \alpha \to E \xrightarrow{\phi} E/K$$

is an exact sequence of Noetherian R-modules and therefore, by Theorem 20 Cor., it gives rise to the exact sequence

$$\widehat{K:_E \alpha} \to \hat{E} \xrightarrow{\hat{\phi}} \widehat{E/K}.$$

Thus $\text{Ker}\,\hat{\phi} = \widehat{K:_E \alpha}$, where the latter is regarded as a submodule of $\hat{E}$. But $\hat{\phi}$ may be obtained by combining $\widehat{\alpha i}$ with the mapping $\hat{E} \to \widehat{E/K}$ corresponding to the natural homomorphism $E \to E/K$. By (9.10.6), $\hat{E} \to \widehat{E/K}$ has kernel $\hat{K} \subseteq \hat{E}$. Since $\hat{i}$ is the identity mapping of $\hat{E}$, it follows that $\text{Ker}\,\hat{\phi} = \hat{K}:_{\hat{E}} \alpha$ and now the proof is complete.

Proposition 13. *Let I and A be finitely generated central ideals of R and let K be a submodule of a Noetherian R-module E. If now completions are formed with respect to the appropriate I-adic filtrations, and $\hat{K}$ and $\widehat{K:_E A}$ are regarded as $\hat{R}$-submodules of $\hat{E}$, then*

$$\widehat{K:_E A} = \hat{K}:_{\hat{E}}(\hat{R}A). \qquad (9.10.9)$$

Remark. This is a partial result which it is useful to have when dealing with a commutative ring. Note that if A is generated by central elements $\alpha_1, \alpha_2, \ldots, \alpha_p$ and $\psi_R : R \to \hat{R}$ is the canonical ring-homomorphism, then $\psi_R(\alpha_1), \psi_R(\alpha_2), \ldots, \psi_R(\alpha_p)$ are central elements of $\hat{R}$ (see Lemma 4) and

$$\hat{R}A = \hat{R}\psi_R(\alpha_1) + \hat{R}\psi_R(\alpha_2) + \ldots + \hat{R}\psi_R(\alpha_p) = A\hat{R}.$$

Proof. We have

$$K:_E A = (K:_E \alpha_1) \cap (K:_E \alpha_2) \cap \ldots \cap (K:_E \alpha_p)$$

and so, by Theorem 24 and (9.10.8),

$$\widehat{K:_E A} = (\hat{K}:_{\hat{E}} \alpha_1) \cap (\hat{K}:_{\hat{E}} \alpha_2) \cap \ldots \cap (\hat{K}:_{\hat{E}} \alpha_p)$$
$$= (\hat{K}:_{\hat{E}} \psi_R(\alpha_1)) \cap (\hat{K}:_{\hat{E}} \psi_R(\alpha_2)) \cap \ldots \cap (\hat{K}:_{\hat{E}} \psi_R(\alpha_p)).$$

Since the expression on the right-hand side is equal to $\hat{K}:_{\hat{E}}(\hat{R}A)$, the proof is complete.

Corollary. *Let I be a finitely generated central ideal of R, E a Noetherian R-module, and α a central element of R which is not a zero-divisor on E. Then, with the usual notation, $\psi_R(\alpha)$ is not a zero-divisor on $\hat{E}$.*

In fact, this is immediately clear from (9.10.8) if we take K to be the zero submodule of E.

It is convenient, at this stage, to include two results concerning I-adic filtrations on *commutative* rings. This is in preparation for the next section where such matters will be the main object of consideration.

Propositon 14. *Let I be an ideal of a commutative Noetherian ring R and let E be a finitely generated R-module. Then*

$$(\mathrm{Ann}_R E)\,\hat{R} = \mathrm{Ann}_{\hat{R}}\hat{E}, \qquad (9.10.10)$$

where $\hat{R}$ and $\hat{E}$ denote the I-adic completions of R and E respectively.

Proof. Let $e \in E$ and consider the exact sequence

$$0 \to \mathrm{Ann}_R(Re) \to R \overset{\phi}{\to} E,$$

where $\phi: R \to E$ is the R-homomorphism in which $\phi(1) = e$. Since R and E are both Noetherian, there results an exact sequence

$$0 \to \widehat{\mathrm{Ann}_R(Re)} \to \hat{R} \overset{\hat{\phi}}{\longrightarrow} \hat{E}$$

of $\hat{R}$-modules. Furthermore, since we have a commutative diagram

$$
\begin{array}{ccc}
R & \overset{\phi}{\longrightarrow} & E \\
\psi_R \downarrow & & \downarrow \psi_E \\
\hat{R} & \underset{\hat{\phi}}{\longrightarrow} & \hat{E}
\end{array}
$$

it follows that $\hat{\phi}$ is the $\hat{R}$-homomorphism in which the identity element of $\hat{R}$ is mapped into $\psi_E(e)$. Now $\hat{R}$ is a commutative ring. Consequently $\mathrm{Ker}\,\hat{\phi} = \mathrm{Ann}_{\hat{R}}(\hat{R}\psi_E(e))$ and therefore

$$\widehat{\mathrm{Ann}_R(Re)} = \mathrm{Ann}_{\hat{R}}(\hat{R}\psi_E(e)). \qquad (9.10.11)$$

After these preliminaries, let $E = Re_1 + Re_2 + \ldots + Re_p$ and put $A = \mathrm{Ann}_R E$. Then

$$A = \mathrm{Ann}_R(Re_1) \cap \mathrm{Ann}_R(Re_2) \cap \ldots \cap \mathrm{Ann}_R(Re_p)$$

and so, by Theorem 24 and (9.10.11),

$$\hat{A} = \mathrm{Ann}_{\hat{R}}(\hat{R}\psi_E(e_1)) \cap \mathrm{Ann}_{\hat{R}}(\hat{R}\psi_E(e_2)) \cap \ldots \cap \mathrm{Ann}_{\hat{R}}(\hat{R}\psi_E(e_p))$$

$$= \mathrm{Ann}_{\hat{R}}\hat{E}$$

because, by Theorem 21, $\psi_E(e_1), \psi_E(e_2), \ldots, \psi_E(e_p)$ generate $\hat{E}$ as an $\hat{R}$-module. In this equation, $\hat{A}$ is the I-adic completion of A and is considered as an $\hat{R}$-submodule of $\hat{R}$. Hence, by Theorem 21,

$$\hat{A} = \hat{R}\psi_R(A) = \hat{R}A.$$

This completes the proof.

Corollary. *Let the assumptions be as in the statement of the theorem and, in addition, let K be an R-submodule of E. Then*

$$(K:E)\,\hat{R} = \hat{K}:\hat{E},$$

where both sides are regarded as ideals of $\hat{R}$.

Proof. We have only to apply the theorem with E/K in place of E. The desired result then follows by virtue of (9.10.7).

Theorem 25. *Let R be a commutative ring, I a finitely generated ideal, $\hat{R}$ the I-adic completion of R, and $\psi : R \to \hat{R}$ the canonical ring-homomorphism. Then there is a one-one correspondence between those maximal ideals M, of R, which contain I and the maximal ideals Π of $\hat{R}$. This is such that if M and Π correspond, then $\Pi = M\hat{R}$ and $M = \psi^{-1}(\Pi)$. In particular, if I is contained in the Jacobson radical of R, then this provides a one-one correspondence between the maximal ideals of R and those of its completion.*

Remark. It is worth while noting that, for this theorem, it is not necessary to assume that R is Noetherian.

Proof. By Theorem 16, there is a one-one correspondence between the M's and the Π's which is such that, when M and Π correspond, $M = \psi^{-1}(\Pi)$ and Π is the closure, $\overline{\psi(M)}$, of $\psi(M)$ in $\hat{R}$. It is therefore enough to prove that $\overline{\psi(M)}$ coincides with $\hat{R}\psi(M)$. Now it is obvious that

$$\hat{R}I = \hat{R}\psi(I) \subseteq \hat{R}\psi(M) \subseteq \overline{\psi(M)}.$$

On the other hand, Theorem 17 shows that $\{(\hat{R}I)^n\}_{n \geqslant 0}$ is the canonical filtration on $\hat{R}$. Accordingly $\hat{R}\psi(M)$ is an open $\hat{R}$-ideal and therefore a closed $\hat{R}$-ideal (see Proposition 2). But $\psi(M) \subseteq \hat{R}\psi(M)$. Consequently $\overline{\psi(M)} \subseteq \hat{R}\psi(M)$ and now the proof is complete.

We return to the consideration of rings which are not necessarily commutative.

Theorem 26. *Let I be a finitely generated central ideal contained in the Jacobson radical of R, and let E be a Noetherian R-module. If now E is endowed with the topology arising from its I-adic filtration, then E is a Hausdorff space and all its R-submodules are closed in E. Furthermore, if $\hat{E}$ is the I-adic completion of E, then the canonical mapping $E \to \hat{E}$ is a monomorphism. It follows that E and all its submodules may be regarded, in a natural way, as R-submodules of $\hat{E}$.*

Proof. Let K be an R-submodule of E. Then, by Theorem 4 of section (7.2),

$$\bigcap_{n=0}^{\infty} I^n(E/K) = 0$$

and therefore

$$\bigcap_{n=0}^{\infty} (K + I^n E) = K.$$

It follows, from Proposition 3, that K is closed in E. In particular we see that

$$\bigcap_{n=0}^{\infty} I^n E = 0.$$

Accordingly E is a Hausdorff space and $\mathrm{Ker}\,(E \to \hat{E}) = 0$. This completes the proof.

Theorem 27. *Let I be a finitely generated central ideal contained in the Jacobson radical of R, and let K be an R-submodule of a Noetherian R-module E. If now K and E are regarded as R-submodules of $\hat{E}$ (see Theorem 26), then*

$$\hat{K} = \hat{R}K \tag{9.10.12}$$

and

$$\hat{K} \cap E = \hat{R}K \cap E = K. \tag{9.10.13}$$

Proof. The first assertion is a special case of Theorem 21. Suppose now that $x \in (\hat{R}K \cap E)$ and let an integer $n \geqslant 0$ be given. Then x is the limit, in $\hat{E}$, of a sequence of elements of K and so, by Theorem 18, x belongs to $K + I^n\hat{E}$. Accordingly x is in

$$(K + I^n\hat{E}) \cap E = K + (I^n\hat{E} \cap E).$$

However $I^n\hat{E} \cap E = I^n E$ by virtue of Theorem 3. It follows that

$$x \in \bigcap_{n=0}^{\infty} (K + I^n E)$$

which yields $x \in K$ because, by Theorem 26, K is closed in E. The theorem follows.

9.11 I-adic completions of commutative Noetherian rings

In this section R will always denote a *commutative Noetherian* ring. Let I be an ideal of R, $\hat{R}$ the I-adic completion of R and $\psi : R \to \hat{R}$ the canonical ring-homomorphism. Since I is necessarily finitely generated, the results of section (9.10) provide information about $\hat{R}$ and ψ. However the main facts may not be immediately obvious because they have to be regarded as special cases of more general results. We shall therefore spend a little time summarizing the principal conclusions which carry over in this way.

The ring $\hat{R}$ is commutative and Noetherian (Theorem 19), all its ideals are closed (Theorem 19), and the canonical filtration on $\hat{R}$ consists of the powers of $I\hat{R}$ (Theorem 17). If M ranges over all the maximal ideals of R containing I, then $M\hat{R}$ ranges over all the maximal ideals of $\hat{R}$ and, for each such M, we have $M = \psi^{-1}(M\hat{R})$ (Theorem 25). If A is an ideal of R, then, because $A = AR$, the I-adic completion $\hat{A}$, of A, is given by

$$\hat{A} = A\hat{R}. \qquad (9.11.1)$$

This is a special case of (9.10.4).

Suppose now that $A, A_1, A_2, \ldots, A_m$ and B are R-ideals. Then, by (9.11.1) and Theorem 24,

$$(A_1 \cap A_2 \cap \ldots \cap A_m)\,\hat{R} = A_1\hat{R} \cap A_2\hat{R} \cap \ldots \cap A_m\hat{R}. \qquad (9.11.2)$$

Next, $A + B$ is a finitely generated R-module which admits B as a submodule. Hence, by Proposition 14 Cor.,

$$(B{:}(A+B))\,\hat{R} = \hat{B}{:}\widehat{A+B}.$$

But $B{:}(A+B) = B{:}A$ and, by (9.11.1), $\hat{B} = B\hat{R}$ and

$$\widehat{A+B} = (A+B)\,\hat{R} = A\hat{R} + B\hat{R}.$$

It follows that

$$(B{:}A)\,\hat{R} = B\hat{R}{:}A\hat{R} \qquad (9.11.3)$$

this being a relation between ideals of $\hat{R}$.

Again, the ring-homomorphism $R \to \hat{R}$ induces the further ring-homomorphism $R/A \to \hat{R}/A\hat{R}$. On the other hand, we have a commutative diagram

$$
\begin{array}{ccccccccc}
0 & \longrightarrow & A & \longrightarrow & R & \longrightarrow & R/A & \longrightarrow & 0 \\
& & \downarrow & & \downarrow & & \downarrow & & \\
0 & \longrightarrow & \hat{A} & \longrightarrow & \hat{R} & \longrightarrow & \widehat{R/A} & \longrightarrow & 0
\end{array}
$$

with exact rows. From this and (9.11.1) we obtain the diagram

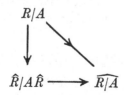

which is also commutative. Here the vertical mapping is the ring-homomorphism to which reference has already been made and the horizontal mapping is an isomorphism of $\hat{R}$-modules. The oblique mapping is the canonical homomorphism of the R-module R/A into its I-adic completion $\widehat{R/A}$. It follows that $\hat{R}/A\hat{R}$ can also be regarded as an I-adic completion of R/A and in such a way that the ring-homomorphism $R/A \to \hat{R}/A\hat{R}$ becomes the canonical homomorphism which is implicit in the notion of a completion. Now R/A is a *ring*, $(I+A)/A$ is an ideal of this ring, and $I^n(R/A)$ is the nth power of $(I+A)/A$. Thus the I-adic filtration on R/A is the same as its $(I+A)/A$-adic filtration. Next $\hat{R}/A\hat{R}$, regarded as the I-adic completion of R/A, has $\{I^n(\hat{R}/A\hat{R})\}_{n \geqslant 0}$ as its canonical filtration (see Theorem 18). But $I^n(\hat{R}/A\hat{R})$ is the nth power of the ideal $(I\hat{R}+A\hat{R})/A\hat{R}$ of the ring $\hat{R}/A\hat{R}$. This shows that the ring $\hat{R}/A\hat{R}$ with the $(I\hat{R}+A\hat{R})/A\hat{R}$-adic filtration is the completion of the ring R/A with its $(I+A)/A$-adic filtration. Thus we have proved

Proposition 15. *Let R be a commutative Noetherian ring and I, A ideals of R. Further let $\hat{R}$ be the I-adic completion of R. Then the $(I+A)/A$-adic completion of the ring R/A is the ring $\hat{R}/A\hat{R}$. Furthermore the canonical ring-homomorphism $R/A \to \hat{R}/A\hat{R}$, associated with this completion, is the one induced by the canonical homomorphism $R \to \hat{R}$.*

The next result follows from Theorems 26 and 27.

Proposition 16. *Let R be a commutative Noetherian ring and I an ideal contained in its Jacobson radical. Further let $\hat{R}$ be the I-adic completion of R. Then the canonical ring-homomorphism $R \to \hat{R}$ is a monomorphism and therefore R may be regarded as a subring of $\hat{R}$. On this understanding, if A is an ideal of R, then $A\hat{R} \cap R = A$. Furthermore*

$$\bigcap_{n=0}^{\infty} (A+I^n) = A. \tag{9.11.4}$$

Of course, (9.11.4) asserts that A is a closed subset of R relative to the I-adic topology.

We now turn our attention to semi-local rings. Let R be a semi-local ring. Then it is a non-null, commutative, Noetherian ring with only a finite number of maximal ideals. Let the different maximal ideals be $M_1, M_2, ..., M_s$ and put

$$J = M_1 \cap M_2 \cap ... \cap M_s = M_1 M_2 ... M_s$$

so that J is the (Jacobson) radical of R.

Definition. *By the 'natural filtration' on a semi-local ring R is meant the J-adic filtration, where J is the radical of R. The completion of R with respect to its natural filtration will be called the 'natural completion' of R.*

Thus the natural filtration on a local ring consists of the powers of its maximal ideal.

Theorem 28. *Let R be a semi-local ring with maximal ideals*

$$M_1, M_2, \ldots, M_s$$

and radical J. Further, let $\hat{R}$ be the natural completion of R. Then $\hat{R}$ is a semi-local ring having $J\hat{R}$ as its radical; $M_1\hat{R}, M_2\hat{R}, \ldots, M_s\hat{R}$ are its maximal ideals; and the canonical filtration which $\hat{R}$ possesses by virtue of being a completion is the same as its natural filtration.

Proof. We already know that $\hat{R}$ is a commutative Noetherian ring and the statement identifying its maximal ideals is an immediate consequence of Theorem 25. Again, the filtration which $\hat{R}$ possesses by virtue of being a completion is the $J\hat{R}$-adic filtration. But

$$J\hat{R} = (M_1\hat{R})(M_2\hat{R}) \ldots (M_s\hat{R})$$

which is the radical of $\hat{R}$. This completes the proof.

Corollary. *The completion of a local ring with respect to the powers of its maximal ideal is also a local ring.*

Let R be a semi-local ring with radical J and let $\hat{R}$ be its J-adic completion. We then have a special case of the situation envisaged in Proposition 16. Hence R is a subring of $\hat{R}$ and, for every R-ideal A, we have

$$A\hat{R} \cap R = A \qquad (9.11.5)$$

and

$$\bigcap_{n=0}^{\infty} (A + J^n) = A. \qquad (9.11.6)$$

Theorem 29. *Let R be a semi-local ring and $\hat{R}$ its natural completion. If now A is a proper ideal of R, then R/A is a semi-local ring which can be regarded as a subring of $\hat{R}/A\hat{R}$. Furthermore $\hat{R}/A\hat{R}$ is the natural completion of R/A.*

Proof. By (9.11.5), the inclusion mapping $R \to \hat{R}$ induces a mono-morphism $R/A \to \hat{R}/A\hat{R}$ and this allows us to regard R/A as a subring of $\hat{R}/A\hat{R}$. Next it is clear that R/A is a semi-local ring with radical $(J+A)/A$, where J is the radical of R. The final assertion of the theorem now follows from Proposition 15.

Proposition 17. *Let R be a semi-local ring and A an ideal of R containing a power of the radical J. If now $\hat{R}$ is the completion of R with respect to the natural filtration, then the inclusion mapping $R \to \hat{R}$ induces a ring-isomorphism $R/A \approx \hat{R}/A\hat{R}$.*

Proof. The induced mapping $R/A \to \hat{R}/A\hat{R}$ is certainly a ring-homomorphism and by (9.11.5) it is a monomorphism. Now let $\hat{r}$ belong to $\hat{R}$ and choose $n \geqslant 0$ so that $J^n \subseteq A$. Then there exists $r \in R$ so that

$$\hat{r} \equiv r(\mathrm{mod}\,J^n\hat{R}) \quad \text{and therefore} \quad \hat{r} \equiv r(\mathrm{mod}\,A\hat{R}).$$

Accordingly each element of $\hat{R}/A\hat{R}$ is the image of some element of R/A and with this the proof is complete.

Corollary. *If $M_1, M_2, \ldots, M_s$ are the maximal ideals of the semi-local ring R, then the residue fields R/M_i and $\hat{R}/M_i\hat{R}$ are isomorphic for $i = 1, 2, \ldots, s$.*

Proposition 18. *Let R be a semi-local ring and $\hat{R}$ its natural completion. Let X be an $\hat{R}$-ideal which contains a power of the radical of $\hat{R}$. If now $A = X \cap R$, then $X = A\hat{R}$ and A contains a power of the radical of R. Furthermore, if X can be generated by p elements, then so can the R-ideal A.*

Proof. Let J be the radical of R. Then $J\hat{R}$ is the radical of $\hat{R}$ and $J^n\hat{R} \subseteq X$ if n is sufficiently large. Suppose that $X = \hat{R}\omega_1 + \hat{R}\omega_2 \ldots, \hat{R}\omega_p$. Then, since $J^{n+1}\hat{R} \subseteq X(J\hat{R})$ and R is everywhere dense in $\hat{R}$, we can choose elements $\alpha_1, \alpha_2, \ldots, \alpha_p$ in R so that $\omega_i - \alpha_i$ belongs to $X(J\hat{R})$ for $i = 1, 2, \ldots, p$. This secures that

$$X \subseteq (\hat{R}\alpha_1 + \hat{R}\alpha_2 + \ldots + \hat{R}\alpha_p) + X(J\hat{R}).$$

It now follows, by induction on m, that

$$X \subseteq (\hat{R}\alpha_1 + \hat{R}\alpha_2 + \ldots + \hat{R}\alpha_p) + X(J^m\hat{R})$$

for $m \geqslant 1$. Accordingly

$$X \subseteq \bigcap_{m=0}^{\infty} (\hat{R}\alpha_1 + \ldots + \hat{R}\alpha_p + J^m\hat{R})$$

whence $X \subseteq \hat{R}\alpha_1 + \hat{R}\alpha_2 + \ldots + \hat{R}\alpha_p$ because $J\hat{R}$ is the radical of $\hat{R}$ (see Proposition 16). However the opposite inclusion is obvious and so we have proved that $X = \hat{R}\alpha_1 + \hat{R}\alpha_2 + \ldots + \hat{R}\alpha_p$.

Put $A = R\alpha_1 + R\alpha_2 + \ldots + R\alpha_p$. Then $A\hat{R} = X$ and therefore $A = X \cap R$ by (9.11.5). Finally

$$J^n = J^n\hat{R} \cap R \subseteq X \cap R = A.$$

The proof is now complete.

In section (4.9) we introduced terminology whereby an ideal B of a semi-local ring R was called an *ideal of definition* if $J^n \subseteq B \subseteq J$ for

COMPLETIONS OF COMMUTATIVE RINGS   433

some positive integer n. It is clear that if B is an ideal of definition of R and $\hat{R}$ is the natural completion of R, then $B\hat{R}$ is an ideal of definition of $\hat{R}$. Indeed Proposition 18 shows that every ideal of definition of $\hat{R}$ can be obtained in this way. Now suppose that Dim $R = d$. It will be remembered that by a *system of parameters* of R is meant a set of d elements which generates an ideal of definition.

Theorem 30. *Let R be a semi-local ring and $\hat{R}$ its natural completion. Then* Dim $R =$ Dim $\hat{R}$.

Proof. Let $a_1, a_2, ..., a_d$ be a system of parameters of R. Then $d = $ Dim R and $Ra_1 + Ra_2 + ... + Ra_d$ is an ideal of definition of R. Consequently $\hat{R}a_1 + \hat{R}a_2 + ... + \hat{R}a_d$ is an ideal of definition of $\hat{R}$ and therefore, by Theorem 23 of section (4.9), Dim $\hat{R} \leqslant d = $ Dim R.

Now suppose that $\omega_1, \omega_2, ..., \omega_p$ is a system of parameters of $\hat{R}$. Then $p = $ Dim $\hat{R}$ and $X = \hat{R}\omega_1 + \hat{R}\omega_2 + ... + \hat{R}\omega_p$ is an ideal of definition of $\hat{R}$. Accordingly $A = X \cap R$ is an ideal of definition of R. Moreover, by Proposition 18, A can be generated by p elements. It follows, again by Theorem 23 of section (4.9), that Dim $\hat{R} = p \geqslant$ Dim R. This completes the proof.

Corollary. *Let $b_1, b_2, ..., b_q$ be elements of the semi-local ring R. Then the b_i form a system of parameters of R if and only if they form a system of parameters of its natural completion.*

Let R be a semi-local ring, A an ideal of R, and b an element of R. By (9.11.3), we have $(A :_R b) \hat{R} = A\hat{R} :_{\hat{R}} b$. Also, by (9.11.5), $A\hat{R} \cap R = A$ and $(A :_R b) \hat{R} \cap R = A :_R b$. It follows that $A :_R b = A$ when and only when $A\hat{R} :_{\hat{R}} b = A\hat{R}$.

By Theorem 9 of section (5.2), if one system of parameters of R forms an R-sequence, then the same is true of every system of parameters. The next result shows that every semi-local ring which has this property shares it with its completion.

Theorem 31. *Let R be a semi-local ring and $\hat{R}$ its natural completion. Then the following two statements are equivalent:*

(a) *every system of parameters of R forms an R-sequence;*
(b) *every system of parameters of $\hat{R}$ forms an $\hat{R}$-sequence.*

Proof. Let $a_1, a_2, ..., a_d$ be a system of parameters of R. Then, by Theorem 30 Cor., it is also a system of parameters of $\hat{R}$. Further, by the above remarks,

$$(Ra_1 + ... + Ra_{i-1}) :_R a_i = Ra_1 + ... + Ra_{i-1}$$

if and only if $(\hat{R}a_1 + \ldots + \hat{R}a_{i-1}) :_{\hat{R}} a_i = \hat{R}a_1 + \ldots + \hat{R}a_{i-1}$.

Accordingly $a_1, a_2, \ldots, a_d$ is an R-sequence if and only if it is an $\hat{R}$-sequence. The theorem follows.

Theorem 32. *Let R be a commutative Noetherian ring and M a maximal ideal of R. Let $\widehat{R_M}$ be the natural completion of the local ring R_M obtained by localizing R at M. Then $\widehat{R_M}$ can be regarded as the M-adic completion of R and in such a way that the canonical ring-homomorphism of R into its completion is obtained by combining the canonical ring-homomorphisms $R \to R_M$ and $R_M \to \widehat{R_M}$.*

Proof. By section (9.7), an M-adic completion $\hat{R}$, of R, can be obtained as the projective limit of the system

$$\ldots \to R/M^{n+1} \to R/M^n \to R/M^{n-1} \to \ldots.$$

On the other hand $\widehat{R_M}$ can be identified with the projective limit of

$$\ldots \to R_M/M^{n+1}R_M \to R_M/M^nR_M \to R_M/M^{n-1}R_M \to \ldots.$$

Now the mapping $R/M^n \to R_M/M^nR_M$ induced by $R \to R_M$ is a ring-isomorphism.† Since each diagram

$$
\begin{array}{ccc}
R/M^{n+1} & \longrightarrow & R/M^n \\
\downarrow & & \downarrow \\
R_M/M^{n+1}R_M & \longrightarrow & R_M/M^nR_M
\end{array}
$$

is commutative, this leads to a ring-isomorphism of $\hat{R}$ on to $\widehat{R_M}$. Finally the diagram

$$
\begin{array}{ccc}
 & R & \\
\swarrow & & \searrow \\
\hat{R} & & R_M \\
\searrow & & \swarrow \\
 & \widehat{R_M} &
\end{array}
$$

is commutative and so the theorem follows.

The next result is somewhat more general in character.

Proposition 19. *Let R be a commutative Noetherian ring and $I_1, I_2, \ldots, I_p$ ideals of R. Put*

$$I = I_1 \cap I_2 \cap \ldots \cap I_p$$

and suppose that $I_\mu + I_\nu = R$ whenever $1 \leqslant \mu < \nu \leqslant p$. Then the I-adic completion of R is ring-isomorphic to the direct sum $\hat{R}_1 \oplus \hat{R}_2 \oplus \ldots \oplus \hat{R}_p$, where $\hat{R}_t$ denotes the I_t-adic completion of R.

† See the proof of Proposition 10 of section (4.5).

Remark. The ring $\hat{R}_1 \oplus \hat{R}_2 \oplus \ldots \oplus \hat{R}_p$ admits the family

$$I_1^n \hat{R}_1 \oplus I_2^n \hat{R}_2 \oplus \ldots \oplus I_p^n \hat{R}_p \quad (n \geqslant 0) \tag{9.11.7}$$

of ideals as a filtration and, with respect to this filtration, it is complete. Again, for each $i \, (1 \leqslant i \leqslant p)$, we have a canonical ring-homomorphism $\psi_i : R \to \hat{R}_i$. Define the ring-homomorphism

$$\psi : R \to \hat{R}_1 \oplus \ldots \oplus \hat{R}_p \quad \text{by} \quad \psi(r) = \{\psi_1(r), \psi_2(r), \ldots, \psi_p(r)\}.$$

The proof will show that when $\hat{R}_1 \oplus \hat{R}_2 \oplus \ldots \oplus \hat{R}_p$ is regarded as the I-adic completion of R, then (i) ψ may be taken as the canonical homomorphism of R into its completion, and (ii) in that case the canonical filtration on the completion is the one described in (9.11.7).

Proof. Suppose that $1 \leqslant \mu < \nu \leqslant p$ and let $n \geqslant 0$ be an integer. Then $I_\mu^n + I_\nu^n = R$. (For otherwise, we should have $n > 0$ and there would exist a maximal ideal containing both I_μ^n and I_ν^n. But this means that the maximal ideal would contain $I_\mu + I_\nu$ which, however, is impossible.) Hence, by Proposition 4 of section (4.2),

$$I_1^n \cap I_2^n \cap \ldots \cap I_p^n = (I_1 I_2 \ldots I_p)^n.$$

In particular we see that $I = I_1 I_2 \ldots I_p$ and therefore

$$I_1^n \cap I_2^n \cap \ldots \cap I_p^n = I^n.$$

Next $\text{Ker}\,\psi = \text{Ker}\,\psi_1 \cap \text{Ker}\,\psi_2 \cap \ldots \cap \text{Ker}\,\psi_p$

and $\text{Ker}\,\psi_\mu = \bigcap\limits_{n=0}^{\infty} I_\mu^n.$

Accordingly

$$\text{Ker}\,\psi = \bigcap_{n=0}^{\infty} (I_1^n \cap I_2^n \cap \ldots \cap I_p^n) = \bigcap_{n=0}^{\infty} I^n.$$

Again $\psi_\mu(r)$ is in $I_\mu^n \hat{R}_\mu$ if and only if $r \in I_\mu^n$. It follows that $\psi(r)$ is in $I_1^n \hat{R}_1 \oplus I_2^n \hat{R}_2 \oplus \ldots \oplus I_p^n \hat{R}_p$ if and only if r is in $I_1^n \cap I_2^n \cap \ldots \cap I_p^n = I^n$. Hence

$$\psi(R) \cap (I_1^n \hat{R}_1 \oplus \ldots \oplus I_p^n \hat{R}_p) = \psi(\psi^{-1}(I_1^n \hat{R}_1 \oplus \ldots \oplus I_p^n \hat{R}_p)) = \psi(I^n).$$

To complete the proof it suffices to check that $\psi(R)$ is everywhere dense in $\hat{R}_1 \oplus \hat{R}_2 \oplus \ldots \oplus \hat{R}_p$. To this end let $x = \{x_1, x_2, \ldots, x_p\}$ belong to the direct sum and let $k \geqslant 0$ be an integer. For each $\mu \, (1 \leqslant \mu \leqslant p)$ there exists $r_\mu \in R$ so that $x_\mu \equiv \psi_\mu(r_\mu) \pmod{I_\mu^k \hat{R}_\mu}$. Again†

$$I_\mu^k + (I_1^k \cap \ldots \cap I_{\mu-1}^k \cap I_{\mu+1}^k \cap \ldots \cap I_\rho^k) = R.$$

† See (4.2.2).

Hence we can write $1 = a_\mu + b_\mu$, where (i) $a_\mu \equiv 0 \pmod{I_\mu^k}$, and (ii) $b_\mu \equiv 0 \pmod{I_\nu^k}$ for $\nu \ne \mu$. Put $r = r_1 b_1 + r_2 b_2 + \ldots + r_p b_p$. Then $r - r_\mu$ belongs to I_μ^k and therefore

$$\psi_\mu(r) \equiv \psi_\mu(r_\mu) \equiv x_\mu (\text{mod } I_\mu^k \hat{R}_\mu).$$

Accordingly $x - \psi(r)$ is in $I_1^k \hat{R}_1 \oplus \ldots \oplus I_p^k \hat{R}_p$. This shows that $\psi(R)$ is everywhere dense in $\hat{R}_1 \oplus \hat{R}_2 \oplus \ldots \oplus \hat{R}_p$ relative to the filtration (9.11.7) and, with this established, the proof is complete.

Theorem 33. *Let R be a semi-local ring with maximal ideals $M_1, M_2, \ldots, M_s$. If now $\hat{R}$ is the natural completion of R and $\widehat{R_{M_i}}$ is the natural completion of the local ring R_{M_i}, then $\hat{R}$ is isomorphic to the direct sum*

$$\widehat{R_{M_1}} \oplus \widehat{R_{M_2}} \oplus \ldots \oplus \widehat{R_{M_p}}.$$

Proof. Let $\hat{R}_i$ denote the M_i-adic completion of R. By Proposition 19, $\hat{R}$ is ring-isomorphic to $\hat{R}_1 \oplus \hat{R}_2 \oplus \ldots \oplus \hat{R}_p$. The theorem now follows from Theorem 32.

It will next be shown that a semi-local ring is semi-regular if and only if its completion is semi-regular. For this a few preparatory remarks are necessary.

Let $R^{(1)}, R^{(2)}, \ldots, R^{(p)}$ be commutative rings and put

$$R = R^{(1)} \oplus R^{(2)} \oplus \ldots \oplus R^{(p)}.$$

For $1 \le i \le p$ denote by e_i that element of R whose projection on to $R^{(j)}$ is zero if $j \ne i$ and whose projection on to $R^{(i)}$ is the identity element of that ring. Then $e_1 + e_2 + \ldots + e_p = 1_R$ and $e_i e_j = 0$ if $1 \le i < j \le p$. Now suppose that M is a maximal ideal of R. We can then find ν so that $1 \le \nu \le p$ and $e_\nu \notin M$. But in that case we must have $e_i \in M$ for all $i \ne \nu$. It is now clear that

$$M = R^{(1)} \oplus \ldots \oplus M^{(\nu)} \oplus \ldots \oplus R^{(p)}, \tag{9.11.8}$$

where $M^{(\nu)}$ is a maximal ideal of $R^{(\nu)}$. Conversely if $M^{(\nu)}$ is a maximal ideal of $R^{(\nu)}$, then (9.11.8) defines a maximal ideal of R. Next the projection mapping $R \to R^{(\nu)}$ is a ring-homomorphism from which it is easy to derive a *ring-isomorphism* $R_M \approx R^{(\nu)}{}_{M^{(\nu)}}$. Hence *the set of rings obtained by localizing R at its maximal ideals is (up to isomorphism) the same as the set obtained by localizing the various $R^{(i)}$ at their maximal ideals.*

Now suppose that $R^{(1)}, R^{(2)}, \ldots, R^{(p)}$ are non-null, commutative, Noetherian rings. By Theorem 10 of section (4.2),

$$R = R^{(1)} \oplus R^{(2)} \oplus \ldots \oplus R^{(p)}$$

is also a non-null commutative, Noetherian ring. It therefore follows, from Theorem 11 of section (5.3) and the remarks made in the last paragraph, that R *is semi-regular if and only if each $R^{(i)}$ is semi-regular.*

Theorem 34. *Let R be a semi-local ring and $\hat{R}$ its natural completion. Then R is semi-regular if and only if $\hat{R}$ is semi-regular.*

Proof. Let $M_1, M_2, \ldots, M_p$ be the maximal ideals of R and put $R_i = R_{M_i}$. Then R_i is a local ring and so it has a natural completion $\hat{R}_i$ say. By Theorem 11 of section (5.3), R is semi-regular if and only if each R_i is semi-regular. Also, by Theorem 33, $\hat{R}$ is isomorphic to

$$\hat{R}_1 \oplus \hat{R}_2 \oplus \ldots \oplus \hat{R}_p$$

and so, by the above remarks, $\hat{R}$ is semi-regular if and only if each $\hat{R}_i$ is semi-regular. But, by Theorem 10 of section (5.3), a local ring R' is semi-regular if and only if every system of parameters forms an R'-sequence. Accordingly, by Theorem 31, R_i is semi-regular if and only if $\hat{R}_i$ is semi-regular. The theorem follows.

Theorem 35. *Let R be a semi-local ring with radical J which is complete with respect to its natural filtration, and let R' be a commutative extension ring of R with the properties that $\bigcap_{n=0}^{\infty} J^n R' = 0$ and R'/JR' is a finitely generated R-module. Then R' is a finitely generated R-module and therefore, by Proposition 18 of section (4·9), a semi-local ring. Moreover if each of R and R' is endowed with its natural filtration, then R' is complete and R is a subspace of R'.*

Proof. We put $E = R'$ and regard E as a module with respect to R. Then, since E/JE is a Noetherian R-module, Theorem 18 shows that $\hat{E}$ is a Noetherian $\hat{R}$-module, where $\hat{E}$ and $\hat{R}$ denote the J-adic completions of E and R respectively. Next $\bigcap_{n=0}^{\infty} J^n E = 0$ by hypothesis. Accordingly the canonical homomorphism $E \to \hat{E}$ is a monomorphism. Thus E may be regarded as being embedded in $\hat{E}$ and then, again by Theorem 18, we have $\hat{E} = \hat{R}E$. However we are given that R is complete. This means that we can take $\hat{R}$ to be R itself and now it follows that

$$\hat{E} = \hat{R}E = RE = E.$$

But we already know that $\hat{E}$ is a Noetherian $\hat{R}$-module. Accordingly E is a finitely generated R-module.

The discussion up to this point shows that R' is a finitely generated R-module which is complete with respect to the J-adic filtration. From

Proposition 18 of section (4.9), we conclude that R' is a semi-local ring. Let us denote its radical by J'.

By Theorem 5 of section (2.5), R' is an integral extension of R. It follows that every maximal ideal of R' contracts to a maximal ideal of R and therefore $JR' \subseteq J'$. On the other hand, if a prime ideal of R' contains JR', then its contraction in R is a prime ideal containing J. The contraction is therefore a maximal ideal of R. Accordingly, by Theorem 11 of section (2.6), every prime ideal containing JR' is a maximal ideal of R'. Thus the prime ideals containing JR' are precisely the maximal ideals of the larger ring. This shows that

$$\mathrm{Rad}\,(JR') = J'$$

and therefore

$$J'^k \subseteq JR' \subseteq J'$$

for a suitable integer k. But we have already established that R' is complete with respect to the filtration $\{(JR')^n\}_{n \geqslant 0}$. It follows that it is also complete with respect to its natural filtration. The fact that R is a subspace of R' can now be derived as a special case of the following more general result.

Proposition 20. *Let R and R' be semi-local rings whose radicals are J and J' respectively. Further let R be a subring of R' and suppose that R is complete with respect to its natural filtration. If now $JR' \subseteq J'$, then R is a subspace of R'.*

Remark. It is, of course, understood that each of R and R' is to be endowed with the topology arising from its natural filtration.

Proof. By (9.11.6), $\bigcap\limits_{n=0}^{\infty} J'^n = 0$. Consequently we have $\bigcap\limits_{n=0}^{\infty} (J'^n \cap R) = 0$. Next, by Proposition 7 Cor. of section (4.4), for each $n \geqslant 0$ the R-module R/J^n has finite length. Also $J'^n \cap R$ is an R-ideal which, since it contains J^n, is both open and closed in R. We may therefore apply Theorem 2. This shows that for each integer $n \geqslant 0$ there is an integer $\nu_n \geqslant 0$ such that

$$J'^{\nu_n} \cap R \subseteq J^n \subseteq J'^n \cap R.$$

Let U be a subset of R. Then U is open with respect to the natural topology on R if whenever $r \in U$ we have $r + J^n \subseteq U$ for some $n \geqslant 0$. On the other hand, in order that U be open with respect to the topology induced from R' it is necessary and sufficient that whenever $r \in U$ we have $r + (J'^n \cap R) \subseteq U$ for some $n \geqslant 0$. However the above considerations show that these two conditions are equivalent. The proof is therefore complete.

Exercises on Chapter 9

As usual, R denotes a ring with an identity element (R need not be commutative). For Exercises 6 to 10 inclusive, $X_1, X_2, ..., X_n$ represents indeterminates and, as in section (4.1), $R[X_1, X_2, ..., X_n]$ denotes the ring of polynomials in $X_1, X_2, ..., X_n$ with coefficients in R. The symbol $R[[X_1, X_2, ..., X_n]]$ is used for the ring of *power series* in $X_1, X_2, ..., X_n$ with coefficients in R. (The difference between a polynomial and a power series is that for the former we require that the coefficients of almost all power-products

$$X_1^{\nu_1} X_2^{\nu_2} ... X_n^{\nu_n} \quad (\nu_1 \geqslant 0, \nu_2 \geqslant 0, ..., \nu_n \geqslant 0)$$

be zero; whereas for a power series no condition is imposed on the coefficients save that they must belong to R.) For power series, addition and multiplication are defined in the obvious way so that $R[X_1, X_2, ..., X_n]$ becomes a subring of $R[[X_1, X_2, ..., X_n]]$.

In the case of a non-commutative ring, the term *module* should always be interpreted as meaning a *left* module. The terms 'local ring' and 'semi-local ring' are used only in connection with commutative Noetherian rings.

1. R is a complete filtered ring and A a two-sided R-ideal with the property that if $x \in A$, then x^n converges to zero as n tends to infinity. Show that an element $u \in R$ has a two-sided inverse if and only if its natural image in R/A has a two-sided inverse in R/A.

2. The ring R satisfies the minimal condition for left ideals and J is its Jacobson radical. Show that $J^m = 0$ for some positive integer m.

3. Let E be a Noetherian R-module and $\gamma_1, \gamma_2, ..., \gamma_s$ a multiplicity system on E. Further let I be a finitely generated central ideal such that some power of I is contained in $\gamma_1 R + \gamma_2 R + ... + \gamma_s R$. Show that $\psi\gamma_1, \psi\gamma_2, ..., \psi\gamma_s$ is a multiplicity system on the Noetherian $\hat{R}$-module $\hat{E}$ and that

$$e_{\hat{R}}(\psi\gamma_1, \psi\gamma_2, ..., \psi\gamma_s | \hat{E}) = e_R(\gamma_1, \gamma_2, ..., \gamma_s | E).$$

Here $\hat{R}$ and $\hat{E}$ denote the I-adic completions of R and E respectively and $\psi : R \to \hat{R}$ the canonical ring-homomorphism.

4. Let I be a central ideal of R which is not necessarily finitely generated. Show that every exact sequence

$$0 \to E' \to E \to E'' \to 0$$

of Noetherian R-modules gives rise to an exact sequence

$$0 \to \hat{E}' \to \hat{E} \to \hat{E}'' \to 0$$

of $\hat{R}$-modules, where $\hat{E}', \hat{E}$ and $\hat{E}''$ denote the I-adic completions of E', E and E'' respectively and $\hat{R}$ denotes the I-adic completion of R.

5. The central ideal I is contained in the Jacobson radical of the ring R which satisfies the maximal condition for left ideals. If now E is a finitely generated R-module and $\hat{E}$ is its I-adic completion, show that E is a free R-module if and only if $\hat{E}$ is a free $\hat{R}$-module.

6. Show that if I is the ideal of $R[X_1, X_2, ..., X_n]$ that is generated by the X_i, then the I-adic completion of $R[X_1, X_2, ..., X_n]$ is the power series ring

$$R[[X_1, X_2, ..., X_n]].$$

7. Show that an element of the power series ring $R[[X_1, X_2, ..., X_n]]$ has a two-sided inverse if and only if its 'constant term' has a two-sided inverse in R.

8. Show that if L is a maximal left ideal of R, then the left ideal of

$$R[[X_1, X_2, \ldots, X_n]]$$

which is generated by L and $X_1, X_2, \ldots, X_n$ is a maximal left ideal of the power series ring. Show that all the maximal left ideals of $R[[X_1, X_2, \ldots, X_n]]$ arise in this way.

9. By using Theorem 19 and Exercise 6, show that if R satisfies the maximal condition for left ideals, then the power series ring $R[[X_1, X_2, \ldots, X_n]]$ also satisfies the maximal condition for left ideals and, moreover, all its left ideals are closed.

10. Let F be a field and ϕ an element of the power series ring

$$F[[X_1, X_2, \ldots, X_n]]$$

which is not a unit. Suppose that ϕ contains terms of the form cX_n^k, where $c \in F$, $c \neq 0$, and let s ($s \geqslant 1$) be the smallest value of k for which such a term can be found. If now χ belongs to $F[[X_1, X_2, \ldots, X_n]]$ show that there exists θ in $F[[X_1, X_2, \ldots, X_n]]$ and $\psi_0, \psi_1, \ldots, \psi_{s-1}$ in $F[[X_1, X_2, \ldots, X_{n-1}]]$ such that

$$\chi = \theta \phi + \sum_{i=0}^{s-1} \psi_i X_n^i.$$

11. A commutative ring R has a single maximal ideal M and it is complete with respect to its M-adic filtration. Show that R is a Noetherian ring if and only if M is finitely generated.

12. Let R be a commutative ring, I an ideal of R, E a Noetherian R-module, and K a submodule of E. Further, let $K = N_1 \cap N_2 \cap \ldots \cap N_s$, where N_ν is a P_ν-primary submodule of E. If E is endowed with the I-adic filtration, show that the closure of K in E is $\bigcap_j N_j$, where j varies so that $1 \leqslant j \leqslant s$ and $I + P_j \neq R$.

13. Let R be a commutative ring and A, I finitely generated ideals of R. Suppose that I is contained in the Jacobson radical of R and that E is a Noetherian R-module such that $AE \neq E$. If now $\hat{R}$ and $\hat{E}$ are the I-adic completions of R and E respectively, show that $(A\hat{R})\hat{E} \neq \hat{E}$ and $\mathrm{gr}\,(A; E) = \mathrm{gr}\,(A\hat{R}; \hat{E})$.

14. Let Z denote the ring of integers and let p be a prime number. Show that the (p)-adic completion of Z is a one-dimensional local ring which is also an integral domain.

15. R is a commutative Noetherian ring and I is an ideal contained in its Jacobson radical. If now $\hat{R}$ is the I-adic completion of R show that the full ring of fractions of R may be regarded as a subring of the full ring of fractions of $\hat{R}$. Show also that R is the intersection of its full ring of fractions and the ring $\hat{R}$.

16. Let I and A be ideals of a commutative Noetherian ring R and let $\hat{R}$ be the I-adic completion of R. Further let P be a prime R-ideal which belongs to A and Π a prime $\hat{R}$-ideal which belongs to $P\hat{R}$. Prove that Π belongs to $A\hat{R}$.

17. I is an ideal of a commutative Noetherian ring R whose I-adic completion is $\hat{R}$. Further Q is a P-primary ideal of the ring R. Show that if the prime $\hat{R}$-ideal Π belongs to $Q\hat{R}$, then it also belongs to $P\hat{R}$.

18. Let I be an ideal of a commutative Noetherian ring R and let $\hat{R}$ be the I-adic completion of R. Suppose that A is an R-ideal and Π a prime ideal of $\hat{R}$. By using Exercises 16 and 17 show that Π belongs to $A\hat{R}$ if and only if there is a prime R-ideal P such that P belongs to A and Π belongs to $P\hat{R}$.

19. R is a commutative Noetherian ring and A, B are ideals of R such that $B \subseteq A$. Show that if R is complete with respect to the A-adic filtration, then it is also complete with respect to the B-adic filtration.

INDEX

444INDEX

strongly compatible filtration, 408
subcomplex, 357
submodule, 7
submodule generated by a set, 9
subring, 85
subring generated by a set of elements, 87
subspace, 378
sum of a family of submodules, 9
supplement of a direct summand, 142
surjection, 5
system of generators of a module, 9
system of parameters, 226

torsionless grading monoid, 118
totally ordered set, 71
total order compatible with a monoid structure, 119

trivial grading on a ring, 113
two-sided ideal, 26

unit of a ring, 222
unmixed ideals, 257
upper bound of a subset of a partially ordered set, 71

vector space, 167

Wedderburn's Theorem on simple rings, 50
well ordered set, 128
Wright's inequality, 296

zero-divisor, 74
zero-divisor on a module, 235
zero module, 8
Zeros Theorem, 285
Zorn's Lemma, 71